HANS KREBS

MONOGRAPHS ON THE HISTORY AND PHILOSOPHY OF BIOLOGY

Editors

RICHARD BURIAN, RICHARD BURKHARDT, JR.
RICHARD LEWONTIN, JOHN MAYNARD SMITH

HANS KREBS

Architect of Intermediary Metabolism 1933–1937

Volume II

FREDERIC LAWRENCE HOLMES

New York   Oxford
OXFORD UNIVERSITY PRESS
1993

Oxford University Press

Oxford   New York   Toronto
Delhi   Bombay   Calcutta   Madras   Karachi
Kuala Lumpur   Singapore   Hong Kong   Tokyo
Nairobi   Dar es Salaam   Cape Town
Melbourne   Auckland   Madrid

and associated companies in
Berlin   Ibadan

Published by Oxford University Press, Inc.
200 Madison Avenue, New York, New York 10016

Oxford is a registered trademark of Oxford University Press

Library of Congress Cataloging-in-Publication Data
(Revised for vol. 2)
Holmes, Frederic Lawrence.
Hans Krebs.
(Monographs on the history and philosophy of biology)
Includes bibliographical references and indexes.
Contents: v. 1. The formation of a scientific life, 1900–1933—
v.   2. Architect of intermediary metabolism, 1933–1937.
1. Krebs, Hans Adolf, Sir.
2. Biochemists—Germany—Biography.   I. Title. II. Series.
QP5l1.8.K73H65   1991   574.19'2'092 [B] 91-2201
ISBN 0-19-507072-0 (v. 1)
ISBN 0-19-507657-5 (v. 2)

1 3 5 7 9 8 6 4 2

Printed in the United States of America
on acid-free paper

For Catherine, Susan, and Rebecca

NOTES ON SOURCES

Documents designated "KC" (Krebs Collection), were, at the time I used them, in the personal possession of Sir Hans Krebs. They were grouped by categories in boxes, but were not formally organized. They have since then been catalogued and deposited in the Sheffield University Library. I have not had opportunity to visit this archive in order to identify the catalogue entries under which the individual documents can now be found in that collection. Readers who wish to examine these materials should be able to locate them with facility by means of the comprehensive *Catalogue of the Papers and Correspondence of Sir Hans Adolf Krebs FRS (1900–1981),* compiled by Jeannine Alton and Peter Harper. Inquiries concerning the catalogue and the collection should be addressed to The Librarian, Sheffield University, Sheffield S10 2TN, UK. For documents that I consulted after the collection had been catalogued in Oxford, I have included the entry numbers.

I have referred to the laboratory notebooks of Hans Krebs by numbers that he and I affixed to their covers during the course of my visits to Oxford so that he could easily retrieve those that I wished to examine further. These numbers do not coincide with the numbering system adopted in the *Catalogue and Correspondence.* They can most easily be identified with the *Catalogue* entries through the dates covered by each notebook.

References to transcribed conversations are given according to the participants, in the form "person interviewed-FLH," the date of the conversation, the number of the tape, and the page number of the taped transcription. Because there are some irregularities in the latter numbers some ambiguities result, and these references are given only as an informal guide to those who may wish to consult the transcripts. Microfilms of most, but not all, of the transcripts are on deposit in the Sheffield archive. The complete set of transcripts is presently in my possession.

All translations from published and unpublished sources originally written in German, including Hans Krebs's laboratory notebooks, are my own.

During the period covered by this work, the notations for organic compounds were less standardized than they have since become. I have, in general reproduced the representations of structural formulas originally used by the authors whose writings I cite, rather than to modernize them.

In the footnotes the following abbreviations are used for titles of journals repeatedly cited:

Ann. Chem.	Annalen der Chemie und Pharmacie
Arch. exp. Path.	Archiv für experimentelle Pathologie und Pharmakologie
Arch. ges. Physiol.	Pflüger's Archiv für die gesammte Physiologie des Menschen und der Thiere
Arch. mikros. Anat.	Archiv für mikroskopische Anatomie
Arch. Pharm.	Naunyn-Schmiedeberg's Archiv für Pharmakologie und experimentelle Pathologie
Beitr. chem. Physiol.	Beiträge zur chemischen Physiologie und Pathologie
Ber. chem. Ges.	Berichte der deutschen chemischen Gesellschaft
Biochem. J.	Biochemical Journal
Biochem. Z.	Biochemische Zeitschrift
C. r. Acad. Sci.	Comptes rendus hebdomadaires des séances de l'Académie des Sciences
C. r. Soc. Biol.	Comptes rendus de la Société de Biologie
Erg. Physiol.	Ergebnisse der Physiologie
J. Biol. Chem.	Journal of Biological Chemistry
Journ. de physiol.	Journal de physiologie et de pathologie générale
J. Gen. Physiol.	The Journal of General Physiology
J. Physiol.	The Journal of Physiology
J. prakt. Chem.	Journal für praktische Chemie
Klin. Woch.	Berliner Klinische Wochenschrift
Physiol. Rev.	Physiological Reviews
Proc. Roy. Soc. Lond.	Proceedings of the Royal Society of London
Skand. Arch. Physiol.	Skandinavisches Archiv für Physiologie
Z. Biol.	Zeitschrift für Biologie
Z. exp. Med.	Zeitschrift für die gesamte experimentelle Medizin
Z. physiol. Chem.	Hoppe-Seyler's Zeitschrift für physiologischen Chemie

ACKNOWLEDGMENTS

Much of the content of the acknowledgments in volume I applies also to this volume. Here, therefore, I add only my special thanks to those whose help was particularly associated with volume II.

Among the friends and colleagues of Hans Krebs whom I interviewed, those whose recollections have been incorporated into volume II include, Hermann Blaschko, Vernon Booth, Lady Margaret Krebs, E. Wolfgang Krebs, David Nachmansohn, Norman W. Pirie, Antoinette Pirie, Juda H. Quastel, Sir Edward and Lady Nan Wayne, and Hans Weil-Malherbe.

Mrs. David Nachmansohn kindly gave me a group of letters written by Hans Krebs to her late husband. Mr. W. J. Hitchins, director of the Hans Krebs archive in Sheffield, was graciously hospitable during my visit there during the final preparation of volume II.

It has been a special pleasure to work with two successive editors at Oxford University Press: William F. Curtis, who encouraged me to submit a very long manuscript, and who astonished me by agreeing with enthusiasm to my tentative suggestion that it might be better to publish it in two volumes; and Kirk Jensen, who has effectively and graciously followed both volumes to completion. Careful management by Susan Hannan, and meticulous copyediting by Randi Laisi have contributed greatly to the effective production of volume II.

Robert Berger, owner of the *Write Way*, capably produced the camera-ready copy for both volumes. He and I shared the frustrations of encountering many unanticipated problems, and the satisfaction of resolving most of them.

Selda Lippa, who typed all of volume I and most of volume II, retired before the latter was finished. Her successor, Joanna Gorman, completed the work with great care and efficiency. In the last stages of preparation of the manuscript, when I was preoccupied with other matters and could barely look back at what I already viewed as finished, she and Pat Johnson took great care to ensure that all of the final details were accurately completed.

In January 1992, my wife Harriet and I celebrated the appearance of volume I by traveling to England to visit again Hans Krebs's wife Margaret, and his brother Wolf. The joy of these reunions was tinged only by my regret that I could not share the occasion with Hans Krebs himself. I have wondered often how he would have regarded the outcome of a project to which he had contributed so much. I feel certain that he would have questioned the length of the volumes: in part because he valued succinctness in all forms of writing, in part because he would have doubted that the fine details of his scientific and personal life would be of interest to others. But I like to believe that he would otherwise have approved the way in which I have portrayed him, even if he would not have agreed with every interpretation. He understood that the historian's viewpoint was, and should be, distinct from his own, and he trusted that I would treat his work and his life more objectively than he himself could. In spite of the biases that might arise from the warm regard for him that I acquired, the value I placed on his friendship, and the impossibility of true historical objectivity, I have done everything I could to fulfill that trust.

New Haven, Connecticut   F. L. H.
August 1992

CONTENTS

INTRODUCTION

This volume reconstructs the investigative pathway and the professional and personal life of Hans Krebs from the time of his arrival in England in June 1933 until the time in 1937 at which he made the discovery for which he is best known. Driven from his native Germany by the racist policies of the National Socialist regime, Krebs came to England as a young biochemist already recognized for his pathfinding discovery of the ornithine cycle of urea synthesis. The next four years were as dramatic for him personally as they were productive scientifically. During that interval he consolidated the experimental style that attained for him a long-lasting leadership in the field of metabolic biochemistry. He identified specific metabolic problems that determined the direction of a career in which he remained active for over 40 more years. It is therefore particularly rewarding to follow in fine detail the day-by-day development of Krebs's investigative enterprise during these critical years. This was also a time in which Krebs's future professional base and his personal attachments remained uncertain from year to year. The year 1937 brought him not only a climactic scientific success, but a resolution of his hitherto unsettled professional and personal circumstances that provided relative stability for the remainder of his career. That year marks, therefore, a natural conclusion for a study of the formation of a scientific life.

This volume forms the sequel to *Hans Krebs: The Formation of a Scientific Life, 1900–1933*, which follows him from childhood until his forced departure from Germany. The introduction with which volume I begins is intended to frame this one as well. Some of the thoughtful suggestions made by the readers who reviewed this volume for Oxford University Press prompt me, however, to add here several supplementary remarks.

Chapter 1 of the first volume provides a concise overview of developments in intermediary metabolism up to 1930, the time at which Krebs entered the field. I have not extended this overview into the decade in which I follow Krebs's investigative activity in the field, although I have summarized work of many other biochemists whose results were relevant to his own. One of the readers argues cogently that, because the approaches, methods, and results of a number of his colleagues were important in shaping Krebs's intellectual approach, a framework at a higher level of generality than Krebs's personal laboratory activities might be more critical to understanding what he did than is the exhaustive tracking of his day-to-day experimental pathway.

The organizing framework chosen for any book precludes alternative perspectives that might have been equally or more illuminating. The fortunate survival of the complete set of laboratory notebooks covering Hans Krebs's experimental work from the beginning of his scientific apprenticeship through his most significant discoveries set for me the goal of reconstructing an unbroken research trail that I believed would reveal much about what I have sometimes called the "fine structure of scientific creativity."[1] I did not know at the beginning what patterns would emerge from this reconstruction, nor to what extent the whole pathway preceding any given stage of Krebs's research would turn out to be essential to understand that stage. Some readers may feel that it would have been possible to eliminate or drastically foreshorten some of the intervening segments without seriously detracting, for example, from our view of the developments leading to the discovery of the citric acid cycle. To account for the origins of the citric acid cycle was, however, only one of the objectives of these volumes. The period between the two peak achievements of Krebs's formative years—the ornithine cycle of 1932 and the citric acid cycle of 1937—was not an interlude of lesser creative effort. During those years he made at least two other discoveries that opened important subareas of intermediary metabolism. Moreover, his responses during those phases of his research in which he made no conclusive advances, or his aims were frustrated, illuminate the character of his creative activity as deeply as do the periods of auspicious success, and I have therefore given as much attention to such episodes in his scientific life as to the more dramatic events.

I agree, on the other hand, that a fuller portrayal of the approaches of figures central to contemporary developments surrounding Krebs's work, such as that of Albert Szent-Györgyi, would enhance our understanding of events such as the formulation of the citric acid cycle, for which Szent-Györgyi's results and views provided a crucial precondition. I had at one time hoped to treat the development of Szent-Györgyi's investigative pathway and those of a number of other contemporaries in considerable detail. My only good reason for not doing so is that such a plan would have made a narrative that is already intricate nearly unmanageable.

To write simultaneously on the scale of events that penetrates to the day-by-day activities of the individual scientist and on the scale at which the collective activities of a group of such individuals are fitted into the broader movement of a research frontier is a daunting task. I believe that these two objectives can best be realized in separate ventures. A history of intermediary metabolism in which a condensed version of the account of Hans Krebs's investigative pathway presented here would be integrated with those of many other individuals into an account of the formation of a scientific field, would be a vast project. I have elsewhere used the example of intermediary metabolism to argue for the importance and the difficulty of describing the growth of modern scientific fields and subfields as collective investigative streams.[2] I hope that my effort to portray Hans Krebs pursuing his personal scientific trajectory within such a network of contemporary activity may encourage some young historian of science to take up the larger task.

A similar admission of incompleteness surrounds my decision to end this two-volume biography midway through Hans Krebs's life. That is a natural transition point between the formative and the mature years, both of his career and his life. More compelling to me personally was that I did not believe I could sustain over a much longer time span the finely detailed treatment of his experimental enterprise that I have followed across the length of the first decade that he pursued it.

If these volumes are viewed as the first half of a potential full scientific biography, the volumes covering Krebs's later life would necessarily differ in approach and scale from the first two. During the 1930s Krebs worked mainly as an individual bench scientist, even when he had a few students or assistants. His personal investigative trail therefore provides the central organizing theme. After 1940 he ceased to perform regularly his own experiments, becoming instead the head of a unit in which a growing number of associates carried out investigations under varying degrees of supervision by him. He remained the guiding intellectual force for this group, and kept in close daily contact with the work he directed. Treatment of his post-war career would nevertheless have to focus on the identity and functions of a scientific research school; on the style of Krebs's leadership; on the students and postdoctoral fellows who came to his laboratory and their subsequent careers; on Krebs's expanding international network of scientific contacts and his influence on work in other centers; and on his increasingly public role as a famous, sometimes "lionized" scientist. The archival and other documentary material that such a study must assimilate is enormous. The number of people who can still give testimony to their interactions with Krebs during the later stages of his life is legion. It is unlikely that I shall soon be in a position to undertake such a project. Until someone does, those readers who wish to learn in more succinct terms what Hans Krebs did during the 44 years not included in these two volumes can turn to his autobiography, *Reminiscences and Reflections,* to the excellent short biography prepared for *Biographical Memoirs of Fellows of the Royal Society* by Sir Hans Kornberg and D.H. Williamson, or to my entry on Krebs in the *Dictionary of Scientific Biography.*[3]

In the final chapter I have reflected on some aspects of Hans Krebs's scientific style. These comments do not add up, however, to the analysis that one of the readers called for concerning what this case reveals about "the logic of discovery." My hope is that this study, together with others like it, can reveal much, not only about the logic of discovery, but about the nature of experimentation in science, the nature of individual creativity, and a range of related issues. I have not introduced such a discussion here, however, because I do not feel that I am yet well enough prepared to generalize broadly. I have now completed historical reconstructions of extended investigative pathways of three outstanding scientists active respectively in three different centuries. I am presently engaged in three further studies of detailed experimental activity. I hope eventually to stand back from these specific cases far enough to examine systematically how they can illuminate some of the questions about experimental science to which historians of science, philosophers, sociologists, and cognitive scientists are now addressing themselves. The urge to generalize quickly is, however, a danger into which some of the current writings on these subjects are falling. Some recent discussions about the nature of experimental science have mounted confident interpretative generalizations on the analysis of a handful of experiments picked out of the investigative pathway of a single scientist, a short time spent as an observer in one laboratory, or other similarly limited experiences. We should be more reticent than that. To discern the general in the specific, to separate the representative from the idiosyncratic, to situate the micro-event within its macroscopic dimensions requires many more bases of comparison than we now possess. One contribution that I hope this study and my previous studies of individual scientists can make to the ongoing discussions is to show that a particular experiment is almost always a link in an extended investigative chain. It can only be fully understood within that context, as well as within the social contexts that presently receive greater attention from "students of science."

* * *

Biochemists who have read this text have found the technical aspects of the investigations described in it straightforward and rather elementary. Readers without such background have found them formidable. Because I hope that there will be readers of both types, the question of whether to provide aids for nonspecialists is not easy to resolve. There is no way to incorporate an adequate substitute for some basic knowledge of areas of biochemistry relevant to intermediary metabolism. For a reader not well acquainted with the subject, the most difficult problem is to keep in mind the structural formulas of the numerous organic compounds mentioned in the text. To ease this problem there is included at the back of this volume an alphabetical list of compounds frequently cited in the text, together with their formulas.

A lesser problem, of which biochemists will be particularly aware, is that some of the notations common during the 1930s are no longer used. Where no significant conceptual difference is involved, I have occasionally modernized, for example, the ways in which Krebs and his contemporaries represented concentrations of solutions. Otherwise I have retained the usages of the time. Many of the structural formulas are reproduced directly from the original texts. Some of the configurations may appear unfamiliar, but will cause little confusion.

One notation that does require explanation, because it has since disappeared from the literature, is the Q value introduced by Otto Warburg as the measure of the rate of gaseous exchanges in tissue slices, and later generalized to other metabolic reactions.

Warburg defined Q as the quotient:

$$\frac{\text{cubic millimeters of gaseous substance}}{\text{milligrams of tissue} \times \text{hours}}$$

He converted weight measurements, even of nongaseous metabolites or products, to volumes, for easy comparison to rates of respiration, and he measured nongaseous reactions whenever possible, by forming a secondary gaseous product, such as CO_2, that could be measured manometrically.

HANS KREBS

1

A New Home for a Career

Hans Krebs arrived at Victoria Station in London at 7:45 a.m. on June 20, 1933. He had his books and his personal possessions, but very little money. Under German regulations he was permitted to bring less than 10 marks with him, and he had smuggled an additional 200 marks between the pages of his books.[1]

Hermann Blaschko met Krebs at the train and took him to the home of his aunt, Helene Nauheim, where he could stay for several days while he began to pursue the contacts he had made in advance. Krebs brought Blaschko up to date on the last stages of his departure from Freiburg and discussed the possibilities he had in Cambridge and Oxford. There was also time to talk shop.[2]

Moving quickly, Krebs first followed up on the letter he had written Rudolph Peters from Freiburg, inquiring if there was a possibility for him to work in Oxford. On June 22 he received an answer from Peters, inviting him to come to discuss the situation.[3] Before leaving, Krebs wrote to Frederick Gowland Hopkins:

London, 22 June, 1933

Dear Sir Frederick Hopkins

This is to tell you that I have arrived in England, since it became almost dangerous for me to stay in Freiburg any longer.

Today, I received an answer from Dr. Peters which I enclose.

I shall be very grateful if you could make it possible for me to see you at Cambridge this weekend.[4]

Krebs drafted this handwritten note, the earliest surviving document by him in English, with the assistance of Helene Nauheim.[5]

The next day, in Oxford, Peters explained that he had no position to offer, but that he could make a place for Krebs in his laboratory, if the Master of Balliol College were to give him a lectureship. From his interview at Balliol, however, Krebs concluded that his chances were not good. On his return to

3

London he received a reply from Hopkins inviting him to call at his home on Sunday, June 24. Krebs traveled to Cambridge on Saturday. After booking a hotel room, he started on foot for the Biochemical Laboratory. On the way, he met Allan Elliott, a South African who had been working in Hopkins's laboratory for six years, and whom Krebs had met the year before at a symposium in Heidelberg. Elliott introduced him to David Keilin, the discoverer of the cytochromes, who worked in the Molteno Institute, in a building adjacent to the Biochemistry Department. With his customary warmth, Keilin welcomed Krebs and insisted that he move into his own home until he could find permanent lodgings.[6]

Hopkins's daughter Barbara Holmes, with whom Krebs had earlier corresponded about the formation of ammonia in the kidney, invited him to come to the Hopkins home the next day, for their customary Sunday tea. There she met him particularly warmly, and expressed great interest in his coming. Hopkins told Krebs that he would be very glad to have him with him, but because of the small salary that the Rockefeller grant could afford, he hesitated to urge Krebs to come for so much less a position than his scientific standing should merit. To Krebs, the amount—£300 annual salary—seemed adequate, a little better than he had had in Freiburg. Strongly attracted as he was to Cambridge from the beginning, he accepted "without a moment's thought."[7] The next day he went back to London to prepare to move, and to arrange for the further shipment of his laboratory equipment. On the 28th he left London, sending Helene Nauheim a bouquet of roses for her hospitality, and returned to settle in at Cambridge.[8]

His new salary provided Krebs the means for only a simple way of life. He found a one-room bedroom apartment at 46 Hills Road, in a home at which he could also eat breakfast and some other meals. For transportation he purchased a secondhand bicycle, with a basket, for about £3. In the flat surroundings of Cambridge a bicycle was an easy and important way to get around.[9]

Soon after deciding for Cambridge, Krebs heard from Peters that there had been, after all, a very good prospect that a place could be made available for him at Oxford.[10] He probably did not regret the lost opportunity, for it was the warmth of Hopkins's exchanges with him and the nature of the Biochemistry Department at Cambridge that most appealed to him. He received also in Cambridge a letter from Joseph Aub, at the Harvard Medical School, dated June 30. Aub asked, "What your plans are for the coming winter, and whether the possibility of coming here to work would interest you." He made it clear that he was not making an actual offer, but that it was "within the realm of possibility" that he could find about $2000 for Krebs for the next year. On July 15 Krebs replied that "for the coming year, until the 1st of July, 1934, I shall work in Sir Frederick Hopkins's Institute. But I do not know yet anything about the further future, so I would be grateful if you keep me in mind and let me know if you have any possibility for me. I shall be very glad going to Boston again and working with you."[11] Growing accustomed to short-run opportunities without long-term prospects, Krebs was also growing astute about making his choices without closing off alternative possibilities.

Of more direct utility to Krebs was that through Aub he also received an Ella Sachs Plotz Foundation award of $350 for "continuation of studies on the breakdown of protein in animals and on the metabolism involved in the action of insulin."[12] The immediate benefit was that Krebs would be able to hire a technician to assist him with his experiments. The topics specified in the grant seem unusual, since neither of them were major themes in his most recent work. Perhaps they were agreed upon as those aspects of his various research activities that were of most interest to the foundation.

The Biochemical Laboratory that Krebs now entered was not only one of the most prominent, but one of the best-funded, best-equipped institutions of its type in the world. With an ample endowment bestowed from the estate of Sir William Dunn, Hopkins had been able, in 1924, to establish the Dunn Institute of Biochemistry in a large, new building, with operating funds to support extensive experimental activities. There were two central laboratory rooms, one on the ground floor, one on the first floor, providing between them enough bench space for over 30 workers. In addition there were several classrooms and lecture rooms, a library, workshops, and animal quarters.[13]

I

At the time Krebs came to Cambridge, Frederick Gowland Hopkins was 72 years old, but he showed few signs of slowing down. His scientific reputation had been established early in the century. Among his most prominent achievements had been the first decisive demonstration, in 1907, of the association between muscular contraction and the formation of lactic acid. His discovery in 1912 of "accessory food factors," necessary in trace amounts, led to the "vitamine hypothesis" and to a Nobel Prize for him in 1929. During the 1920s his isolation of glutathione, a substance thought to play a catalytic role in cellular oxidations, attracted great interest. Nevertheless, Hopkins had seldom been in a position to carry on long-sustained personal experimental investigations. For many years he had inadequate facilities and support. By the time that the Dunn endowment rectified that problem, he had acquired demanding administrative and other duties. It was as head of the Cambridge school more than through his own investigations that Hopkins maintained his leadership in biochemistry into the 1930s. In addition he had a special ability to formulate the general problems, possibilities, and domains of biochemistry, in lucid, elegant form, in lectures for which he became famous in his field.[14]

The manner in which Hopkins ran his laboratory was nearly the opposite from what Krebs had seen in Otto Warburg's laboratory. Where Warburg had deliberately kept his groups of assistants small, Hopkins brought in as many investigators as his resources could absorb. Where Warburg had subordinated everyone to his own interests, Hopkins encouraged those working in his laboratory to pursue a wide diversity of problems. Where Warburg had been authoritarian, Hopkins was gentle and tolerant. Whereas Warburg was singleminded and everpresent in his laboratory, Hopkins had so many responsibilities that he was often away or inaccessible. He spent much of his time dur-

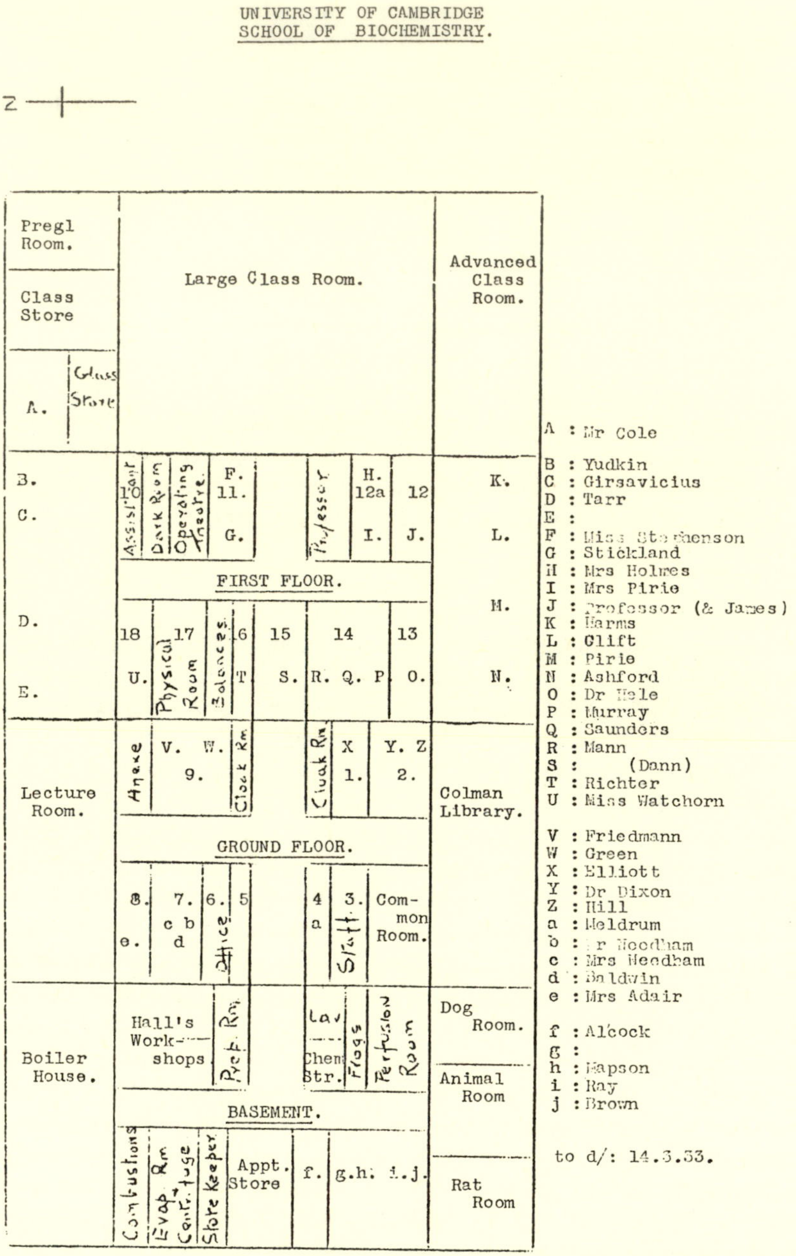
UNIVERSITY OF CAMBRIDGE
SCHOOL OF BIOCHEMISTRY.
N
Pregl Room.
Class Store
Large Class Room.
Advanced Class Room.
Glass Store
A.
B.
C.
D.
E.
Assistant 10
Dark Room
Operating Theatre
F. 11.
G.
Professor
H. 12a
I.
12
J.
K.
L.
M.
N.
FIRST FLOOR.
18 U.
17
Physical Room
Balances
16 T
15 S.
14 R. Q. P
13 O.
Lecture Room.
Annexe
V. W. 9.
Cloak Rm
Cloak Rm
X 1.
Y. Z 2.
Colman Library.
GROUND FLOOR.
8. e.
7. c b d
6.
Office
5
4 a
3.
Staff
Com-mon Room.
Boiler House.
Hall's Work-shops
Prep. Rm.
Lav
Chem Str.
Frogs
Perfusion Room
Dog Room.
Animal Room
BASEMENT.
Combustions
Evap. Rm
Centrifuge
Store Keeper
Appt. Store
f.
g.h.
i.j.
Rat Room
A : Mr Cole
B : Yudkin
C : Girsavicius
D : Tarr
E :
F : Miss Stephenson
G : Stickland
H : Mrs Holmes
I : Mrs Pirie
J : Professor (& James)
K : Harms
L : Clift
M : Pirie
N : Ashford
O : Dr Hele
P : Murray
Q : Saunders
R : Mann
S : (Dann)
T : Richter
U : Miss Watchorn
V : Friedmann
W : Green
X : Elliott
Y : Dr Dixon
Z : Hill
a : Meldrum
b : Dr Needham
c : Mrs Needham
d : Baldwin
e : Mrs Adair
f : Alcock
g :
h : Mapson
i : Ray
j : Brown
to d/: 14.3.33.

Figure 1-1 Floor plan of Biochemistry Laboratory, University of Cambridge, with places of experimental workers in 1933, just prior to arrival of Krebs. From University Archives, Cambridge.

Figure 1-2 Exterior of Biochemistry Laboratory, Cambridge.

ing the 1930s in London, carrying on his duties as president of the Royal Society. Those who came to work in Hopkins's laboratory were, therefore, often thrown on their own resources, or had to turn to others in the laboratory for advice. Nevertheless, when he was around, Hopkins had the capacity to ask probing questions, or provide insights, that could bring clarity and focus to a problem that the investigators had not previously been able to reach on their own. During the weekly teas at which individual investigators presented their results informally to the assembled group, Hopkins presided. Through his comments, his questions, and his encouragement, he was able to sustain an intellectual leadership over the efforts of investigators who otherwise maintained a degree of independence that would not have been imaginable in Warburg's laboratory. Sometimes Hopkins was overly optimistic about the talents of those in his laboratory. He was said to regard "all of his geese as swans," but his inspiration led many of those who came under his influence to do the best work of which they were capable.[15]

Figure 1-3 Frederick Gowland Hopkins in 1933. From *Hopkins and Biochemistry 1861-1947* ed. J. Needham, and E. Baldwin (Cambridge: W. Heffer, 1949), opp. p. 244.

Into this community of investigators Hans Krebs was received with great warmth. After the weeks of rejection and isolation that he had endured in Germany, the friendly, relaxed atmosphere of Cambridge struck him with special force.[16] Many individuals went out of their way to make him feel welcome. The eminent physiologist Joseph Barcroft was among those who came over from neighboring buildings especially to greet him.[17] Within the laboratory he was quickly comfortable in a stimulating environment of scientific and personal exchange. On July 16, he wrote to his former chief, Siegfried Thannhauser,

> Well, I have landed in Cambridge. I feel very much at home here already, and am deeply immersed in work.
> At first I was in London for several days, until the decision for Cambridge was made. Shortly after the decision for Cambridge had taken place, another offer came from Oxford, which offered, to be sure, 100 pounds more per year, and the possibility to teach besides; but I am not sorry that I am here. Everyone takes care to make life easier for me, and I have in the Institute everything that I need for my work.[18]

The reception that Krebs found in Cambridge was in part a manifestation of a very special generosity of spirit with which scientists there and elsewhere in Britain responded to the plight of their German colleagues who had been made victims by the Nazis. To the Cambridge biochemists, however, Krebs was far more than another young German refugee scientist. It was not only Hopkins who had admired his work; it was well known to the entire laboratory. In biochemistry courses at Cambridge the discovery of the way in which urea is synthesized was taught as a major event in the field. The volume of *Hoppe-Seyler's Zeitschrift* in the library containing the paper by Krebs and Kurt Henseleit was conspicuously soiled from frequent use. Krebs came not as a stranger to the Biochemical Laboratory, but as one who was taken to honor it by his presence. This feeling was recalled in 1977 by Vernon Booth, who had been a student there when Krebs arrived:

> I joined the Part II class in biochemistry, in Cambridge, in . . . about 1932 or '33. . . . We had, of course, heard various theories about how urea was produced, and then *suddenly* the problem was solved while I was an undergraduate. And then, to our astonishment, Hans Krebs came to work in Cambridge.[19]

II

It is not certain just when Krebs resumed his experimental research in Cambridge. In his autobiography he wrote, "I started experimental work in the laboratory on 7 July—two and a half weeks after leaving Freiburg and twelve weeks after being expelled from my laboratory—a very brief interruption considering the circumstances."[20] There is, however, no entry in his laboratory notebook for July 7. The most likely explanation for his statement is that he looked briefly back at his old notebook, saw on page 260 the heading "Cambridge," with the date "4.7.33," and misread this as July 7, rather than July 4. All that is entered on the page, however, are a few calibrations, and on the facing page there is an index to certain previous experiments.[21] New experiments do not begin until July 23, so the resumption of his research in Cambridge was not nearly so rapid as he remembered. Since he had no other duties, it is most likely that Krebs was occupied in the laboratory during the extra two weeks, but that it took him most of that time to receive, unpack, and set up his apparatus and to make other preparations to begin.

The index of earlier experiments that Krebs drew up on July 4 can be viewed also as a preparation for a research program that he was setting forth for himself as he contemplated where to reenter the investigative activity from which he had been so rudely separated three months before. He did not choose to follow up the unexpected findings with respect to the consumption of ammonia by glutamic and aspartic acid in kidney tissue that had engaged his attention in his last days in his Freiburg laboratory; rather, he chose to continue instead the several overlapping lines of investigation within the area of oxidative metabolism. The topics that he listed included the "consumption of acetic acid," the "formation of acetoacetic acid," the "acetate-succinate series," the "determination of ketone bodies," "fatty acids," and "butyric acid."[22]

On July 23 Krebs initiated his experimental activity in Cambridge by comparing the formation of ketone bodies in liver slices from a starved guinea pig, in the presence of butyric acid and acetic acid. He had found in Freiburg that butyric acid can form large quantities of ketone body in liver slices and that acetic acid "apparently" forms acetoacetic acid. He was now, therefore, mainly extending those preliminary results. In the new experiment, butyric acid raised the rate of formation substantially over the control, but acetate had only a small effect. Repeating the experiments on the 25th and 26th with liver tissue, respectively, from a nourished guinea pig and a nourished rat, he obtained similar results. On the 28th, with tissue from a nourished rat, on the other hand, both butyric and acetic acid doubled the control rate.[23] Thus far Krebs had achieved little more than to support earlier evidence, dating back to Gustav Embden's perfusion experiments three decades before, that ketone bodies can arise from even-numbered fatty acids through β-oxidation or from acetic acid through some synthetic reaction. Krebs probably carried out these experiments, however, as a base for comparing the actions of other substances and conditions on the formation of acetoacetic acid.

On July 29 Krebs tested the effects on this process of two substances that one would expect to exert significant actions. Carbohydrates had been known since late in the nineteenth century to be important "antiketogenic" agents, because their removal from the diet caused even normal subjects to excrete ketone bodies. In diabetics, glucose together with insulin was an effective antidote to ketosis. Krebs tested insulin alone, and insulin plus glucose, on liver slices from starved rats in the presence of butyric acid and acetic acid. Neither substance significantly affected the respiration. Glucose decreased the consumption of butyric acid, but not that of acetic acid. Insulin was "without definite effect" on the consumption of either acid.[24] Finding no immediate opening here, Krebs turned to other questions, but his interest in the actions of glucose and insulin remained strong.

As he pursued his study of the formation of ketone bodies, Krebs found himself in need of better analytical procedures for determining them. The method of Donald Van Slyke that he had adapted for microquantities was clumsy and difficult to carry out on the scale he required. Moreover, as he noted after the experiment of July 23, it gave only "minimal values, since β-hydroxybutyric acid does not react quantitatively." On July 31 he began examining ways in which he might further modify the method. The central difficulty lay in the titration of the acetone-mercury compound separated out in the Van Slyke method. He tested an older method of Erwin Rupp for titrating the mercury with iodine, and found it more satisfactory. During the analysis he precipitated the mercury compound of acetoacetic and β-hydroxybutyric acid separately so that he could then determine the quantities of both acids by means of the Rupp titration method. With this procedure worked out, he returned to the main line of investigation. On August 2 he added to the usual acetic and butyric acid the unsaturated C_4 fatty acid crotonic acid (CH_3-CH=CH-COOH), with liver slices from a starved rat. The crotonic acid increased the formation of acetoacetate and β-hydroxybutyrate more than either of the other substances. On August 4 he repeated the experiment anaerobically, and obtained mostly higher values than for the corresponding set of aerobic results. This must have been somewhat puzzling, since the reactions involved oxidations, but Krebs recorded no response to the outcome.[25]

Still seeking more convenient analytical methods, Krebs tested the reaction of aniline with acetoacetic acid the next day. In Freiburg, Pawel Ostern had developed for him a manometric method for determining oxaloacetic acid, using aniline to decarboxylate the acid (see Vol. I, p. 393). The only other β-ketonic acid on which aniline had acted was acetoacetic acid, and that reaction had been too slow to interfere with the oxaloacetic acid determination. Krebs nevertheless now tried to apply the aniline method to acetoacetic acid. He noted that there was a "rapid decomposition with aniline" and almost no decomposition without the addition. The idea therefore appeared feasible. He planned to examine further the conditions for the reaction on the following day, but did not carry out the experiments. Instead he again refined the procedures for his microscale version of the Van Slyke mercury method.[26]

Krebs next turned to a set of experiments on the "acetic acid series," with rat kidney slices. Four of the five substances he tested—acetic, butyric, crotonic, and β-hydroxybutyric acid—were obvious choices from the perspective of his investigation of the formation of acetoacetic acid up to that point. The fifth, citric acid, was conspicuously anomalous from that standpoint. As before, he measured the effects of each substance on the respiration and the formation of acetoacetic and β-hydroxybutyric acid. In this experiment only crotonic and β-hydroxybutyric acid produced measurable quantities of acetoacetic acid. Citric acid produced none. That substance did appear to form β-hydroxybutyric acid, but Krebs concluded that the result was probably an artifact due to a reaction of the citrate with the potassium dichromate used as a reagent in the analytical procedure.[27]

As we have seen (in Vol. I, pp. 23, 34–35), citric acid was one of the few substances long known to increase strongly the respiration of isolated tissues, but no place for it had been found in the various metabolic reaction schemes that had been proposed over the preceding two decades. Krebs too had mentioned the substance only in passing in his review essay for Carl Oppenheimer's *Handbuch*, and had rarely included it in his experiments. It appears that here, for the first time, he tried to incorporate citric acid into a broader framework. The rationale that he had in mind is not recorded, but can be reconstructed from the logic of the situation and his own recollections when he and I reexamined the record of this experiment. Citric acid was known to yield, through various chemical decomposition reactions, acetone dicarboxylic acid:

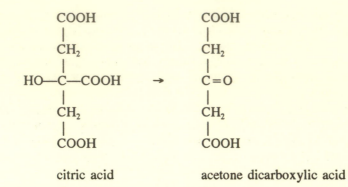

citric acid   acetone dicarboxylic acid

Moreover, Thorsten Thunberg, who had discovered citricodehydrogenase in plant seeds in 1929, had postulated that this enzymatic dehydrogenation gives rise to acetone dicarboxylic acid (see Vol. I, pp. 34-35). Acetone dicarboxylic acid was in turn known to decompose spontaneously, in a double decarboxylation reaction, to give acetone:

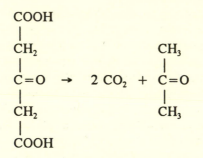

This reaction was not likely to occur physiologically, since acetone was not regarded as a natural metabolic intermediate. Krebs apparently reasoned, however, that if only one carboxyl group were removed in the tissues, the product might be acetoacetic acid:

COOH
|
CH$_2$
|
C=O → CO$_2$ +
|
CH$_2$
|
COOH

CH$_3$
|
C=O
|
CH$_2$
|
COOH

If this were so, it would be possible to lead citric acid "into normal pathways of metabolism."[28] It is easy to see how attractive such a possibility would be. If the reaction could be demonstrated, it would be possible to move citric acid out of the scientific byway it had occupied for so many years and into the main line of metabolic reaction schemes. Unlike many of his experiments on oxidative metabolism, he was not merely testing theories proposed in the literature, but visualizing a potential connection that others had missed. If the connection actually existed, it would be a major step toward integrating the known intermediates into a coherent scheme. The first result was negative, but Krebs had clearly become interested in the role of citric acid, and he did not drop the question, as he often did when confronted with such an initial outcome.

On August 11 Krebs pursued the subject of citric acid with another experiment on rat kidney tissue. This time he included aconitic acid, a substance closely related to citric acid and formed readily by heating citric acid in a retort. Undoubtedly he had in mind the possibility that aconitic acid might constitute an early step in the metabolic decomposition of citric acid as well. In the experiment, however, aconitic acid depressed the respiration, whereas citric and acetic acid raised it slightly.[29]

* * *

When we discussed this experiment in 1977, Krebs commented, "I tried this [aconitic acid] here, but, as I now know, I tried the wrong isomer. This is *trans*-aconitic acid, one can buy in a shop."[30] The *trans*-aconitic acid only *became* the wrong isomer as a result of later events; nevertheless, the course of events might have been different, if it had happened to be easier to purchase *cis*-aconitic acid than *trans*-aconitic acid.

* * *

In the same set of experiments, Krebs compared the rate of respiration in the presence of combinations of acetic acid with pyruvic and lactic acid, respectively, to the rate in the presence of each of the latter two substances alone. According to his recollection in 1977, these tests were "based on the idea that there may be some kind of interaction, just as in the urea cycle, that ornithine is essential because it combines with one of the other reactants."[31]

We cannot infer from this statement that Krebs actually modeled the experiment, at the time, upon his previous success with urea synthesis; that may merely have been the example that came to mind as he discussed the situation with me. As we have seen, it had long been assumed that acetic acid undergoes some kind of synthetic reaction. The possibilities that had attracted the most attention were that two molecules of acetic acid combine, to yield either succinic acid, or acetoacetic acid. Krebs may have been seeking to extend the range of possibilities by considering that one molecule of acetic acid may instead combine with some other small metabolite. Acetic plus pyruvic acid did, in fact, nearly double the rate of the respiration with either substance alone. Krebs noted, however, that he had used "poor pyruvic acid," so he probably regarded the experiments as defective.[32]

On the page following these experiments, Krebs wrote out in his laboratory

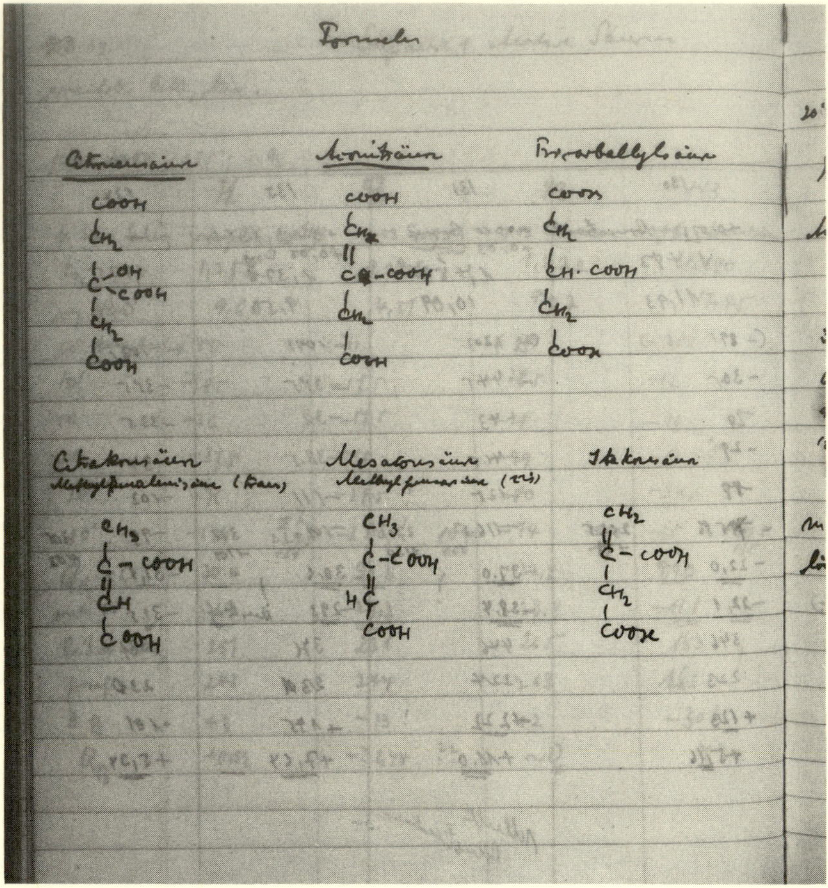

Figure 1-4 Krebs laboratory notebook No "10," p. 294.

notebook the formulas for six organic acids. In the top row were three C_6 tricarboxylic acids—citric, aconitic, and tricarballylic acid. Below them were three C_5 dicarboxylic acids—citaconic, mesaconic, and itaconic acid:[33]

There are no statements accompanying these formulas to indicate his purpose in putting them down. Their placement, however, and their temporal relationship to the preceding experiments are suggestive. The top row of C_6 acids are related in a manner analogous to the relationship between the C_4 dicarboxylic acids, malic, fumaric, and succinic acid; it seems plausible to infer that Krebs wondered whether citric acid might be incorporated into some analogous sequence of metabolic reactions. The lower row includes a compound—itaconic acid—to which citric acid gave rise chemically when heated with water in a closed vessel,[34] so he may have been looking for possible steps in the metabolic decomposition of citric acid. There is, however, an intriguing further possible connection with the experiment just carried out. If acetic acid were to combine with either pyruvic or lactic acid, the product could be a C_5 compound, perhaps a dicarboxylic acid. It is, therefore, possible that Krebs was looking for another way in which he could link the decomposition of citric acid with the "normal reaction pathways" of these central metabolic intermediates, while at the same time finding an alternative solution to the long-standing problem of the synthetic reaction that acetic acid was presumed to undergo. If this were what he had in mind, it would have been a bold conjecture. There is no indication that he was able to envision a specific hypothetical reaction scheme, and it would have been characteristic of him merely to have had a very general idea that he could hope to bring into sharper focus through experiments, rather than through prolonged contemplation.

It is probably indicative of his intention to focus on the metabolism of citric acid at this time that Krebs tried to utilize the reactions converting citric acid into acetonedicarboxylic acid as the basis for a manometric method to determine citric acid. He wrote down the following equations for known reactions (see page 16). The first reaction took place in concentrated H_2SO_4. He hoped that the process could be organized in such a way that the carbon monoxide arising from the formic acid shown at the right could be measured manometrically. Using a measured quantity of citric acid, he tried out the reaction in the manometer, using NaOH to absorb the CO_2 that might form. The pressure change recorded, however, was too large for the quantity of CO that the reaction could yield, and he concluded that some of the CO_2 must not have been absorbed.[35]

Next Krebs thought of a possible means to simplify the procedures for measuring the disappearance of the acids he had been testing with kidney slices. "In small V_F" [that is, small fluid volumes], he asked, "can one measure the CO_2 absorption directly, (as in aerobic fermentation, when the fermentation is large in comparison to the respiration)?" That is, he thought that perhaps the CO_2 absorbed, as the consumption of the organic acids neutralized the solutions, might be so large in comparison to the oxygen absorbed in respiration under these conditions that he could neglect the latter and calculate the former directly from the pressure change in the manometer. He tried out the idea, using acetic, pyruvic, citric, butyric, crotonic, pyruvic + acetic, aconitic, β-hydroxybutyric, and acetoacetic acid—nearly the whole range of acids whose metabolic relationships he had been studying. He found, however, the "disappearance of the acids not measurable" by this method, "because it is too small in comparison to the respiration!" There were "large errors in the calculation!!"[36] With none of his efforts to devise new analytical procedures being successful, Krebs reverted to his usual methods on August 14, when he again tested the same set of acids, this time with guinea pig kidney slices. He measured the respiration directly and the disappearance of the acids by the changes in the quantities of bicarbonate in the vessels during the course of the experiment. The largest changes in the latter occurred with acetic and citric acid. The changes in the rest were small, and for acetoacetic acid there was very little change. This result was unfavorable for two of the reaction pathways Krebs had been entertaining. He noted that "acetate and citrate react very rapidly. Acetoacetic acid more

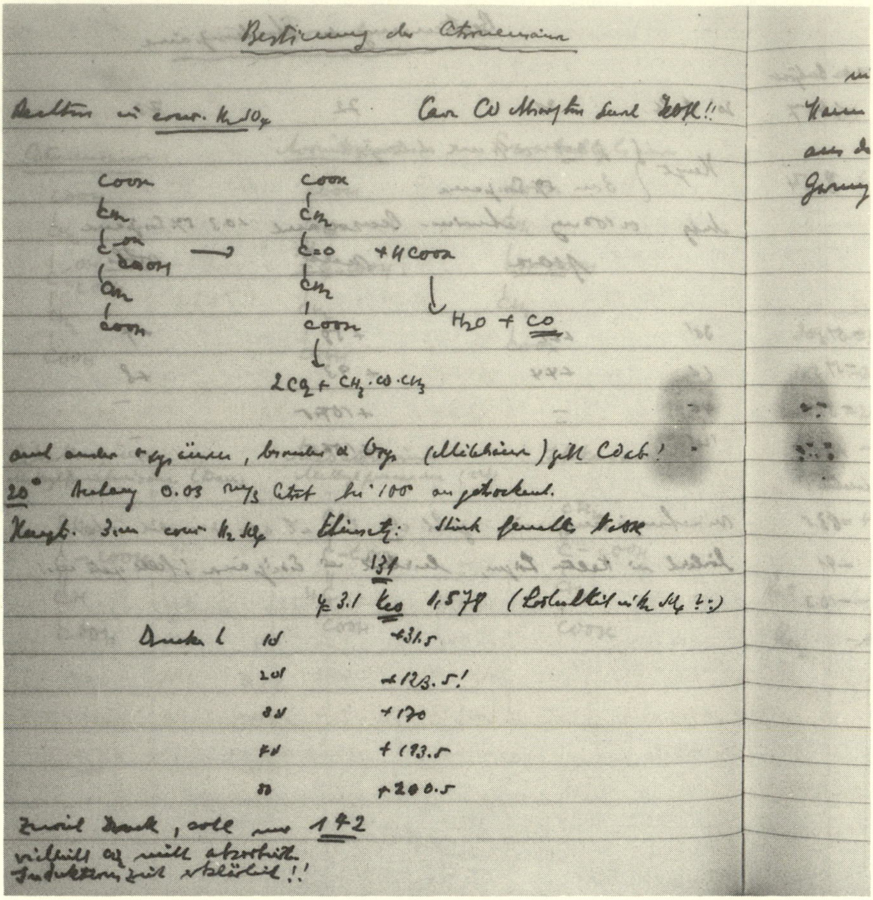

Figure 1-5 Krebs laboratory notebook No "10," p. 296.

slowly, can therefore not be an intermediate product."[37] Thus it appeared unlikely either that two molecules of acetic acid combine to form a molecule of acetoacetate, or that citric acid is decomposed to acetoacetic acid.

Regarding the same experiment, Krebs also noted, "Butyric, β-hydroxy-butyric, and acetoacetic acid [disappearance] about the same, all slow." Although this result indicated, on the one hand, that guinea pig tissue was not favorable to study the "butyric acid" question, it was compatible with the view that butyric, β-hydroxybutyric, and acetoacetic acid are intermediates in the same reaction sequence. Accordingly, Krebs narrowed his attention the next day to these three acids, utilizing rat kidney tissue in place of the guinea pig tissue. All three substances increased the respiration, but the bicarbonate determinations were "spoiled," because the solutions became acidified.[38] This experiment therefore did not advance the question.

On the same day Krebs again attempted to utilize the acetonedicarboxylic acid reaction of citric acid to determine the latter by the "splitting off of CO." He compared the reaction of citric acid, in concentrated sulfuric acid, to that of lactic acid, which could give rise to an analogous reaction. This time he was apparently successful in absorbing all of the CO_2 to which the acetonic dicarboxylic acid might give rise, because the pressure change of 201 mm^3 was very close to the theoretical 199 mm^3 for the CO that the reaction should yield. Krebs wrote, "Citric acid reacts more rapidly than lactic acid, in 120 minutes quantitatively." This would appear to have been an outcome favorable to his objective, but Krebs did not pursue it to a workable method.[39]

On August 16 Krebs followed up the question of the relative rates of disappearance of butyric, β-hydroxybutyric, and acetoacetic acid in rat kidney slices. He dispensed with measurements of the respiration, concentrating on the change in bicarbonate, and did duplicate experiments for each acid. The averages for the duplicate results were butyric acid, $Q_{bicarbonate} = -8.44$; β-hydroxybutyric acid, $Q_{bicarbonate} = -1.04$; acetoacetic acid, $Q_{bicarbonate} = +3.92$. This unexpected result seemed to contradict long-held assumptions about the metabolic connections involved. Krebs noted that "butyric acid disappears much more rapidly as an acid than do the β-oxidation products. From acetoacetic acid arises acid!!! That must be checked. It follows that butyric acid is decomposed in a different way. Succinic acid?" Thus, by the basic criterion that an intermediate must be consumed at at least the same rate as the overall reaction, this experiment appeared to rule out the application of Franz Knoop's long-established β-oxidation theory of fatty acid decomposition to the crucial C_4 stage in that process. Krebs responded by entertaining a possible alternative route, incorporating a mechanism that had attracted his interest a few months earlier (see Vol. I, pp. 399–400). A Pieter Verkade type ω-oxidation might convert the butyric acid to succinic acid:

$$
\begin{array}{ccc}
CH_3 & & COOH \\
| & & | \\
CH_2 & \rightarrow & CH_2 \\
| & & | \\
CH_2 & & CH_2 \\
| & & | \\
COOH & & COOH
\end{array}
$$

This hypothesis would break the connection between butyric acid and the ketone bodies, linking it instead with the well-known "succinate series."[40] In preparation for an exploration of such a connection, Krebs returned on the next day to explore that reaction sequence itself.

It is indicative of the intricate interplay between the temporal sequence of Krebs's experimental pathway and the conceptual relationships of the subproblems into which it was divided that this new development in his study of the metabolism of butyric, β-hydroxybutyric, and acetoacetic acid induced him to pick up his study of the succinate series at almost the exact point at which he had broken it off in March, when the behavior of acetic acid had led him toward the metabolism of acetoacetic acid (see Vol. I, pp. 408–409, 413). Just prior to that time he had begun to employ inhibitors such as malonate and arsenite in an effort to block the further decomposition of the expected products of the succinate series of reactions—that is, of oxaloacetic and pyruvic acid. We may recall that, in one of these experiments in March, he had tested only for oxaloacetic acid and regretted afterward that he had not tested for pyruvic acid. Now, in August, he set out to do what he had then omitted to do.

On August 17, Krebs examined the reaction "succinic acid → pyruvic acid" in kidney slices from a nourished rat. He tested succinate by itself and in the presence of arsenite, as well as with another possible inhibitor, hydroxylamine. After measuring the respiration under each of these conditions, he determined the quantities of pyruvate present at the end by means of the manometric fermentation method. It was "questionable whether hydroxylamine inhibits" the disappearance of pyruvate. Arsenite, on the other hand, clearly did, for the measurable rate of pyruvic acid formed in that case gave a Q value of 2.25, compared to 0.67 with no inhibitor. There was, Krebs wrote, "more O_2 consumption with succinate in the presence of As_2O_3. Q_{O_2} = 15, pyruvic acid formation = + 2.25, therefore somewhat more than 10% of the quantity possible according to the O_2 measurement (1 O_2 → 1 pyruvic acid from succinic acid)."[41]

Building what he could on this result, Krebs tested succinic acid the next day in a similar experiment, with several concentrations of arsenite, and found that "very small quantities of arsenite (0.0003 M) do not inhibit the respiration, but cause the formation of pyruvic acid." It appeared that the pyruvate formed only at the beginning of the reaction. He was very far from a demonstration that all or most of the succinic acid consumed was converted to pyruvic acid. Nevertheless, he felt ready to compare the effects of butyric acid on the formation of pyruvic acid with what he had observed with succinic acid—that is, to test his working hypothesis that butyric acid may be decomposed by way of succinic acid. Using rat kidney slices, he tested both compounds in the presence of two concentrations of arsenite. Only with succinic acid did he obtain measurable quantities of pyruvic acid. The result, overall, was "a small amount of arsenite does not completely inhibit the oxidation of butyric acid!! But no pyruvic acid formation. From succinic acid little pyruvic acid (about 10%)."[42] The latter result, that even with succinate he could detect only 10 percent of the pyruvic acid expected to form, made the experiment less than decisive. It must, however, have been enough to discourage Krebs's hypothesis that butyric acid enters into the succinic acid pathway.

After this experiment Krebs interrupted his work for a week. During that time the Third International Congress for Experimental Cytology met in Cambridge. Unaware that the congress was to take place until after it was under way, he attended only a few sessions. He missed the opportunity to see and hear such eminent scientists as Leonor Michaelis, Heinrich Wieland, and Karl Landsteiner. Krebs did, however, meet Albert Szent-Györgyi there. He must have expressed his appreciation to Szent-Györgyi for what the latter had done to help initiate the process that had brought him to Cambridge, and he presumably caught up on Szent-Györgyi's most recent work on cellular metabolism.[43]

If Krebs reflected, during this pause, on his first month of research in Cambridge, he would not have been able to point to any clear marks of progress. His efforts to connect acetic acid and citric acid to known metabolic reactions had so far led nowhere. Even experiments expected to confirm the connections long believed to exist between butyric acid and the ketone bodies had led to contradictions, and the alternative hypothesis he had thought of in response appeared unpromising after the initial test to which he subjected it. Moreover, his attempts to devise new analytical methods, especially for citric acid, and to simplify the procedures for measuring the disappearance of acids in his experiments had fallen short of their mark. Nowhere in what he had done during those weeks would there appear to be an opening to pursue. As in similar situations in the past, he probably carried on after the international congress in the same manner as before, in the optimistic belief that if he kept trying things, something new would turn up.

III

Perhaps to make up some of the time he had missed, Krebs came to the laboratory on Sunday, August 27, to prepare three substances—glycollic acid, glyoxylic acid, and acetonedicarboxylic acid—that he intended to employ for experiments the next day. On Monday he measured the effects of each of these compounds on the respiration of guinea pig kidney slices, and estimated, as usual, the disappearance of the acids from the change in the quantity of bicarbonate in the fluid medium. One of his objectives was to test alternative possibilities for the metabolic pathways for acetic acid that his recent experiments had seemed to eliminate from consideration. Glyoxylic and glycollic acid were the two theoretically plausible intermediates that could arise if acetic acid were oxidized without an intervening synthetic reaction. They had repeatedly been rejected in the past as metabolites and he had already tested them in February with negative results. That he should nevertheless turn to them once more is perhaps a measure of how little conceptual room for maneuver remained for him at this point. Acetonedicarboxylic acid was, as we have seen, the product of the action of citricodehydrogenase on citric acid according to Thunberg. A little earlier Krebs had thought that this sequence might lead further, through acetoacetic acid, to "known pathways." That idea appeared to have been refuted in his first experimental trial, but Thunberg's reaction itself remained to be tested in animal tissue. The reaction probably seemed metabolically less promising to him now, because even if verified it would leave the two compounds isolated from the main oxidative sequences. Krebs did not have to face that problem, however, for all three of the tests turned out negative. "Glycollic and glyoxylic acid," he recorded, "scarcely react (10–100 times slower than acetic acid). Acetone dicarboxylic acid does not increase respiration. Disappearance of the acid not provable."[44] These results therefore only closed down two more possible avenues, leaving him with fewer leads than ever.

After doing these experiments, on a Wednesday, Krebs carried out no more for the rest of the week. Perhaps in the wake of so many negative results he needed a few days to decide what to do next. On Monday, September 4, he performed an experiment on "pyruvic acid and kidney." Despite the fact that he had recently measured the formation of pyruvic acid in kidney, from succinic and butyric acid, the new experiment was probably not a continuation of previous lines of investigation. He appeared rather to be examining some of the new views emanating from the work of Embden and others on glycolysis in the previous spring. Krebs had become particularly interested in the "many recent works on pyruvic acid" during his period of enforced leave in Freiburg. A paper by Otto Meyerhof and Donald McEachern, written just prior to the publication of Embden's glycolytic pathway, had particularly attracted Krebs's attention. In it they had provided evidence that in frog muscle tissue pyruvic acid arises as an intermediate in the formation of lactic acid. In an appendix Meyerhof and McEachern remarked that their experiments confirmed Embden's new scheme. Krebs had also taken note of papers by Szent-Györgyi and his associates demonstrating in heart muscle tissue an enzyme system catalyzing the reversible interconversion of pyruvic and lactic acid.[45]

In the experiment, Krebs measured the consumption of pyruvic acid in kidney tissue on a relatively large scale, utilizing 47 mg of tissue in about 50 cm^3. of medium. At intervals of one, two, and three hours he measured the change in the bicarbonate and the quantity of pyruvic acid in 1 cm^3. portions of the fluid. Comparing the two changes, the one a measure of the total change in acid content, the other in the content of pyruvic acid, he observed that "somewhat more pyruvic acid consumed than acid disappeared. There must arise, therefore, another non-fermentable acid from pyruvic acid. Lactic acid?" His surmise that lactic acid may have been produced obviously derived from the conclusions of Embden, and of Meyerhof, that pyruvic acid is an intermediate in the glycolytic formation of lactic acid. If he could demonstrate the appearance of lactic acid in his experiment, Krebs might provide a useful confirmation and a demonstration that the reaction can take place aerobically. When he tested for lactic acid with the "CO-method" (that is, by means of a manometric measurement of the carbon monoxide assumed to be yielded in the decomposition of lactic acid in concentrated sulfuric acid), however, he obtained "after one hour = 0." Although he regarded his analytical procedure as "not entirely certain," he concluded "but probably no lactic acid."[46] His initial venture into this area therefore added nothing to what others had already found.

For the moment, Krebs pursued the pyruvic acid question no further, but returned to try again to straighten out the tangled relationship between acetic and acetoacetic acid. On September 5, he measured the effects on the respiration of rat kidney slices of acetate, acetoacetate, acetate + acetoacetate, and acetate in the presence of glucose. For the first and last of these he determined also the change in bicarbonate. It is not evident whether he was testing a particular idea or just looking for effects that might give new direction to his thought. Testing acetate and acetoacetate together may have been a means to assess whether they exert independent or interconnected effects upon the respiration. Each of the four additions in fact raised the respiration by about the same amount, whereas glucose partially inhibited the consumption of acetic acid. Krebs regarded it as noteworthy that there was "also an increase in respiration with acetoacetic acid," but recorded no further conclusions. A summary on the next page of the Q values for the consumption of acetic acid in all of his experiments up to that point suggests that he intended to concentrate his attention now on the metabolic role of that compound.[47] When we reexamined the section of the laboratory notebook following these pages, in 1977, Krebs interpreted his activity in pragmatic terms: "I spent a lot of time, obviously, in finding conditions which affect the rate of utilization of acetate."[48]

Pursuing the relation of acetate to acetoacetate, Krebs again compared their effects on rat kidney tissue on September 7. In a medium containing a high concentration of bicarbonate, acetate and acetoacetate raised the respiration and were, according to the change in bicarbonate, themselves consumed, at the same rate; whereas in a low–bicarbonate medium acetoacetate disappeared at only half the rate of acetate. In both mediums glucose had no effect. Probably because he felt the need to measure specifically the quantity of acetoacetic acid consumed in such experiments, Krebs paused on the next day to make an improvement in the aniline method for determining that compound. By substituting aniline citrate for the aniline sulfate he had previously used, he was able to get the reagent into solution more readily, therefore making the procedure easier to carry out. The next day he employed the modified method to measure the consumption of acetoacetic acid in rat kidney tissue. Using the compound in several concentrations, and carrying out double experiments for each condition, he measured the respiration and the change in bicarbonate, as well as the acetoacetic acid present at the end. Good agreement in the duplicate measurements for acetoacetic acid gave him confidence in his aniline citrate method. Overall, he found, the acetoacetic acid was consumed "very slowly," only "half as fast as acetic or butyric acid, if one looks at the Q_B [the rate of change in bicarbonate]." Because the rate of disappearance of the acetoacetic acid as measured by the aniline citrate method, however, was greater than the rate of disappearance as measured by the change in bicarbonate ($Q_{acetoacetic} = 6.45$ and 6.51; $Q_B = 2.97$), he inferred that a "part of the acetoacetate [consumed] remained as an acid, probably β-hydroxybutyric acid."[49] That is, he concluded that a part had undergone a reduction reaction, whereas the remainder had been oxidized. This reasoning was exactly analogous to that through which he had inferred a few days earlier that some of the pyruvic acid consumed in kidney tissue might be reduced to lactic acid. For the present case, there is no indication that he had defined the relationship between the oxidation and reduction processes more specifically in terms of paired oxidative and reductive reactions. In view of the prominence of such dismutation reactions in past and current theories of metabolism, however, such a connection was probably at least in the back of his mind.

Changing course again, on September 12 and 13 Krebs measured the respiratory quotients of rat kidney slices in the presence of acetic, succinic, pyruvic, and malic acid. According to the R.Q.s, he found, these substances were incompletely oxidized, but he decided that the "correct measure" of their oxidation was not the R.Q., but the ratio of oxygen consumed to the increase in bicarbonate (X_{O_2}/X_B). That is, by comparing the quantity of oxygen theoretically required to oxidize one molecule of the compound in question to the ratio between the measured oxygen consumption and the increase in bicarbonate, regarded as equivalent to the acid consumed, he could determine what proportion of the acid that disappeared had been oxidized. The remainder he regarded as "resynthesized," in the manner that Meyerhof treated the lactic acid not oxidized in muscle as resynthesized to carbohydrate. It is not fully apparent what Krebs had in mind; perhaps that, if acetic, succinic, malic, and pyruvic acid were intermediates in pathways leading to lactic acid, then they too could be treated as participants in the Meyerhof lactic acid cycle. At any rate, he did not pursue this direction of thought, but turned on September 14 to the more recent theories of glycolysis of Embden and Meyerhof. In the appendix to the paper of Meyerhof and McEachern that had impressed Krebs three months earlier, they wrote that they had "fully confirmed" Embden's "hypothesis that the formation of lactic acid arises from a reaction between glycerolphosphoric acid and pyruvic acid." When they added equivalent quantities of the latter two compounds to isolated muscle tissue, the pyruvic acid disappeared and lactic acid appeared. Krebs now tried an experiment along the same lines, on rat kidney slices, but under aerobic conditions. Measuring the respiration and the change in the quantity of pyruvic acid in the medium, in the presence of pyruvic acid alone and pyruvic acid together with glycerol phosphoric acid, he found that with the combination there was "increased respiration, increased disappearance of pyruvic acid!!" So far his results fitted the new views of Embden and Meyerhof, but since the reaction was supposed to be included in glycolysis, the crucial demonstration would be that under anaerobic conditions there is increased formation of lactic acid. When he tried this experiment the next day (testing glycerol plus pyruvic acid as well as glycerol phosphoric acid plus pyruvic acid), he observed "no effect from glycerol or glycerol phosphoric acid!!" There is no indication that he thought his result a potential disproof of Embden's theory that glycerol phosphoric acid reacts with pyruvic acid to form lactic acid. Most likely he merely regarded the experiment as inconclusive. He did not pursue the question further.[50]

In place of the expected effect that did not appear, Krebs observed a special effect of pyruvic acid itself. By comparison with the control, there was in the vessel to which that compound was added a "sensible increase in the CO_2 driven off." To this statement he added "Mendel effect?" His query was an allusion to a paper his friend Bruno Mendel had published in 1931, showing that under anaerobic conditions the addition of pyruvic acid to isolated animal tissues can raise the rate of "fermentation" by 100 percent or more.[51]

It was probably the observation of this "Mendel effect" that gave Krebs another idea for an experiment concerning glycolysis. When he had written the review essay on "The Rate of Respiration and Glycolysis in Living Cells" for Oppenheimer's *Handbuch* during the previous winter, he had included a section on the "activation of fermentation" in which he had discussed Mendel's results with pyruvic acid among the substances that increase the rate of alcoholic fermentation in yeast or of glycolysis in animal tissues. The result that reminded him of this action of pyruvic acid may also have reminded him that in this same section he had devoted a paragraph to an unknown "activator" of glycolysis postulated by Otto Rosenthal.[52] In 1932 Rosenthal had followed up Mendel's result, studying in great detail the conditions under which pyruvic acid can produce an "extra fermentation" in liver tissue slices. He concluded that there must be an activator involved that is produced in the tissue itself while it respires aerobically. The activator must be different from the pyruvic acid itself, but Rosenthal left its chemical identity in doubt.[53] Now Krebs asked himself, "Is Rosenthal's activator acetoacetic acid? It is unlikely, since 'extra fermentation' goes on only in nourished animals." He did not state his reasons for considering acetoacetic acid, but they were probably that (1) it is, like pyruvic acid, a keto acid and (2) it was assumed to be produced by an oxidative metabolic process. In spite of his skepticism about his own hypothesis, Krebs tested it by measuring the anaerobic formation of lactic acid in rat liver slices in the presence of and the absence of acetoacetic acid. The result was as he had foreseen: "No influence of acetoacetic acid, unfortunately [*leider*]."[54]

* * *

This rare interjection of an expression of feeling into his laboratory notebook led to a comment by Krebs, when we examined this experiment in 1977, that is more interesting than the result itself:

FLH: *Leider*, does that mean "unfortunately"?
HAK: Yes, because it would be nice—*leider* means "unfortunately"—it would be nice to explain the nature of what he called the activator.
FLH: I see.
HAK: Yes, occasionally I express some sentiments in there, as you notice.
FLH: Well, rarely. . . .
HAK: Well, I accepted the facts as they are, without lamenting, without also getting too exhilarated.[55]

Not only did Krebs rarely express "sentiments" in his notebook record, he rarely admitted to them in conversations, when I asked him about his responses to particular experimental results that I would have expected to be exciting, or disappointing. It was in keeping with his character not to "lament," or to be "exhilarated." He was unusually even tempered, and not given to "shows of emotion." Nevertheless, those who knew him well could tell, from his animation when an event in the laboratory caught his interest, that, in subtler ways than some people, he did become excited about promising developments; he must have responded inwardly with a *leider* more often than he wrote that word down in his notebook. Krebs had the capacity to accept large and small disappointments stoically, but his retrospective view that he generally "accepted the facts" without emotional involvement was, I believe, influenced by his conviction that that is how a scientist should behave.

* * *

With his experiment on Rosenthal's activator, Krebs ended another brief foray into the area of glycolysis. It had yielded him no foothold on current developments. Again he returned to his persistent problems concerning the oxidative pathways of acetic and acetoacetic acid.

The results of an experiment are seldom simply the data that can be read off directly or calculated mechanically. Even the immediate conclusions that can be drawn are inferences in which the meaning of the data is mediated by a background of prior knowledge and opinions. The position in which Krebs found himself as he continued his study of acetic and acetoacetic acid illustrates the complexity of what appear on the surface to be straightforward judgments. We may recall that, when he tested the consumption of citric, acetic, and acetoacetic acid in kidney slices, on August 14, the "result" he had written down was "acetate and citrate react very rapidly. Acetoacetic acid more slowly, can therefore not be an intermediate product." From this particular experiment alone, the conclusion that acetoacetic acid cannot be an intermediate applied equally to a reaction sequence beginning with citric acid and one beginning with acetic acid. Krebs's subsequent actions show, however, that he treated the outcome differently for the two cases. With respect to citric acid the result was decisive; he discarded the hypothesis that would have linked the decomposition of that compound to other metabolic pathways. As he took up the question of acetic and acetoacetic acid again in September, however, it was clear that for this case he had not accepted the same experimental verdict. He continued to devise experiments based on the assumption that acetic acid may give rise to acetoacetic acid in animal tissues. The difference in his responses rested, of course, on differences in the weight of prior reasons for supporting those respective supposed pathways. The conversion of citric to acetoacetic acid was only an idea that had recently occurred to Krebs himself as he sought to link the first of these compounds to known metabolic reactions. There was not enough momentum behind the idea to withstand a negative test. The conversion of acetic to acetoacetic acid was, however, a view that had been entertained repeatedly by biochemists for nearly three decades. It could not be so readily dismissed in the face of an unfavorable experiment.

Similarly, despite the fact that in his experiments of the summer he had most often found glucose to have "no effect" on the processes yielding acetoacetic acid, the long-established antiketogenic properties of carbohydrates prevented Krebs from accepting such results as definitive. On September 18, the same day that he performed the Rosenthal experiment, he undertook a further test of the effects of glucose—and of glycogen—on the formation of acetoacetic acid in the presence of acetic acid. He measured the respiration, the disappearance of acid (increase in bicarbonate), and the formation of acetoacetic acid, probably by his aniline citrate method. The results appeared indecisive—glucose decreased the quantity of acetoacetic acid but not the disappearance of acid, whereas glycogen had the inverse effects. Moreover, he had neglected to provide controls employing glucose and glycogen without acetic acid. He therefore drew no conclusions, but repeated the experiment on the same day, rectifying this omission. The inclusion of these control experiments proved to be significant, in that they permitted him to distinguish two modes of formation of acetoacetic acid, one affected by the carbohydrates, the other unaffected. Among the four "results" he listed, the second was that "the spontaneous formation of [acetoacetic] acid is reduced by glucose and glycogen." The third result was that "the increase in bicarbonate through acetate is, by glucose and glycogen, *not influenced* [he then crossed out the two words here placed in italics and added instead] diminished. It appears, therefore, that glucose and glycogen inhibit the consumption of acetic acid. Still, the investigation is not unambiguous."[56]

Here we have a glimpse of an investigator starting to interpret his data in one way, but in the process of writing down his conclusion changing his mind about the meaning of the numbers. The data in question were Q_B ("corrected") for acetate alone, $+2.64$; for acetate + glucose, $+1.82$; for acetate + glycogen, $+0.91$. He must have at first judged that the difference between 2.64 and 1.82 was not significant and discounted the result for glycogen as less reliable than that for glucose. As he put down this evaluation he realized that these differences probably were significant after all. His qualification that the investigation was "not unambiguous" probably reflected the fact that the change in bicarbonate measured the change in total acids, not that of acetic acid alone, and that in previous experiments when he had added an acid whose disappearance he had measured directly, the quantity consumed did not always coincide with the overall decrease in the quantity of acid.

Turning to the measurements of the acetoacetic acid present at the end of the experiment, Krebs reached his fourth, and most interesting immediate conclusion:

Acetoacetic acid formation is inhibited by glucose and glycogen, insofar as it is <u>oxidation</u>, but <u>not</u> the formation of acetoacetic acid from acetic acid, for

without addition (that is, sugar)	3.16 - 1.37	= 1.74!
with glucose	[2.35 - 0.70]	= 1.65!
with glycogen	[2.40 - 0.70]	= 1.70!

To arrive at this conclusion Krebs subtracted the $Q_{acetoacetate}$ values for the controls—that is, for no addition, for glucose, and for glycogen—respectively from the $Q_{acetoacetate}$ values for acetic acid, acetic acid + glucose, and acetic acid + glycogen. This showed that the acetic acid had *raised* the rate of formation of acetoacetic acid by the same amount, whether or not glucose or glycogen was present. The $Q_{acetoacetate}$ values for glucose (0.70) and for glycogen (0.20) alone, on the other hand, were substantially lower than for the control with no additions. The two carbohydrates therefore did inhibit the formation of that part of the acetoacetic acid whose appearance was independent of the addition of acetic acid.[57]

After enumerating the specific "results" of the experiment, Krebs summarized what he thought its overall consequence to be: "Glucose and glycogen appear to inhibit the formation of acetoacetic acid from substances other than acetic acid."[58] Here, for the first time in his numerous attacks on the problem, Krebs appeared to have attained a result that advanced his understanding of the metabolism of acetoacetic acid and that provided a promising direction for further inquiry. Underlying his efforts all along had been a general assumption that the formation of acetoacetic acid from acetic acid, if it took place, would be a special, synthetic process, different in nature from the oxidative process leading from the longer fatty acids to acetoacetic acid. Now he had at hand a means to differentiate these processes experimentally, an agent that inhibited the latter but not the former. At the same time, this result might lead toward a more specific identification of the antiketogenic action of the two carbohydrates.

Since it was the addition of the "controls," including glucose and glycogen without acetic acid, to the second of the two otherwise identical experiments performed on September 18 that made possible his new insight, we naturally wonder whether Krebs foresaw that potential outcome and added these controls *in order* to permit the calculation contained in the conclusion reproduced above, or whether he added the controls as a routine procedure that he had neglected, and noticed the regularity in the results afterward. The record does not permit a decision between these two alternatives.

In the light of this experiment, one can look back and see that in an earlier experiment, that of July 29, a hint of the differential action of glucose on these two processes had already appeared. Then Krebs had found that glucose decreased the consumption of butyric acid in liver tissue, but not the consumption of acetic acid. If one assumes that butyric acid is a source of acetoacetic acid through an oxidative process, and acetic acid is a source of acetoacetic acid through a special process, then this result fits into the same pattern as the later experiment through which Krebs drew the above conclusion. Did he himself perceive that clue in July, and only follow it up in September, or is it a clue whose significance is only apparent in the light of further developments? There is no way of knowing—no way to tell even whether Krebs himself ever looked back at the earlier experiment.

From the course of his experiments over the next several weeks, it appears evident that his conclusion about the action of glucose and glycogen on the oxidative formation of acetoacetic acid induced Krebs to pursue intensively the question of the nature of that action. He did not immediately incorporate glucose or glycogen into the ensuing experiments, but began with a more general reexamination of the process itself. Nor did he fix his attention, as he had done during the summer, on the assumed immediate precursors of acetoacetic acid in the oxidative process—butyric, β-hydroxybutyric, or crotonic acid—but began to test the effects of longer-chain fatty acids on the rate of formation of acetoacetic acid. There are at least two plausible rationales for shifting his focus at this point to the longer metabolic pathway. One reason could have been to obviate for the time being the experimental difficulties he had encountered in his recent efforts to establish the relationships between the C_4 acids supposed to comprise the last steps. By backing off from these problems, he could utilize his new aniline citrate method for acetoacetate instead to verify the overall conception that ω-oxidation, and perhaps also β-oxidation, leads to its formation. A second reason could be that, since Krebs had observed only that glucose and glycogen inhibit the "spontaneous" formation of acetoacetic acid—that is, from fatty acids presumed to be present in the tissue—he did not know at what stage in the oxidative process these substances act. It would, therefore, be best to study that action first in experiments that bore on long segments of the overall process.

On September 21, Krebs measured the effects of caproic acid [$CH_3(CH_2)_4C-OOH$] and adipic acid [$HOOC(CH_2)_4COOH$] upon the respiration, the change in bicarbonate, and the formation of acetoacetic acid. He employed each of the two acids in three concentrations. This comparison between the normal C_6 carboxylic acid and the C_6 dicarboxylic acid suggests that he may have been testing the relative importance of ordinary β-oxidation and the ω-oxidation postulated by Verkade. If caproic acid underwent the latter type of oxidation, it should give rise to adipic acid, which would be further oxidized by "double-ended" β-oxidations. The result was clear: "Caproic acid causes a large increase in respiration, the formation of acetoacetic acid, and the formation of acid!! Adipic acid scarcely reacts!!" This outcome must simultaneously have discouraged Krebs's interest in ω-oxidation and encouraged him that his system was suitable to demonstrate the more conventional conception of β-oxidation. In two other experiments carried out at the same time, he measured the decomposition of acetoacetic acid added to rat liver slices. There was a substantial consumption of the compound, in quantities proportional to the two concentrations he employed, but since the respiration did not increase, he suspected that "the largest part was decomposed not in the cells, but spontaneously."[59]

The next day Krebs extended the same type of experiment to three fatty acids—caproic (C_6), enanthic (C_7), and caprylic (C_8). Each of the acids was "attacked" and formed "much acid." The two even-numbered fatty acids also formed acetoacetic acid, but the odd-numbered one did not, a result that conformed nicely with the β-oxidation theory. (Krebs also repeated the comparison between caproic and adipic acid, but with kidney tissue, and observed "no sensible decomposition.")[60] Expanding the scope of the experiment still further, on September 25 he tested three longer-chain fatty acids: pelargonic (C_9), myristic (C_{14}), and stearic (C_{16}), each employed in three concentrations. Now the results became more complicated. At the higher concentrations each of the acids inhibited the respiration. Moreover, the measurements of the acetoacetic acid produced were this time not in accord with expectations:

Acetoacetic acid also from pelargonic acid [C_9]!! (impure substance?). From stearic acid in high concentrations there is inhibition of ketone formation, and in intermediate concentration little ketone. From myristic acid no more ketone than calculated from the last $\underline{4}$ carbon atoms!![61]

Thus, in contrast to the preceding experiments, the results were in contradiction to the β-oxidation theory. The even-numbered fatty acid yielded either no acetoacetic acid or less than expected; the odd-numbered acid, which should not give rise to it, yielded more than the even-numbered ones did. As his remark suggests, Krebs suspected that the pelargonic acid contained an impurity that was the actual source of the acetoacetic acid.

Continuing along the same line, Krebs tested on September 26 enanthic (C_7), pelargonic (C_9), myristic (C_{14}), and stearic (C_{16}) acid. None of them caused acetoacetic acid to form at a rate larger than the control. Moreover, with pelargonic acid, no acetoacetic acid formed at all, and the respiration fell to one third its normal rate. Krebs wrote, "Result: no substantial extra-formation of acetoacetic acid from fatty acids: toxicity of nonylic [pelargonic] acid." Thus the new result introduced further anomalies into the situation; now none of the fatty acids tested appeared to lead to the formation of the supposed product of their oxidation. Perhaps attempting to obviate this negative result by modifying the conditions, he tested "the influence of bicarbonate (pH) on the formation of ketones from stearic acid." Nevertheless, he obtained no "extra formation" of acetoacetic acid at any of the three pHs that he tried.[62]

Despite these unsettled difficulties, Krebs moved on, on September 28, to the main objective of these experiments, the action of glucose on the formation of acetoacetic acid from fatty acids. He also tested the action of a narcotic, phenylurea. Again he used rat liver slices. Whereas the narcotic strongly inhibited the production of acetoacetic acid from caprylic acid, there was "no noticeable glucose effect!!" Clearly unwilling to accept this as a definitive outcome, he tried again on the 29th, again using caprylic acid. Along with the glucose he tested a different narcotic, phenylurethane. In other respects the conditions were the same as in the preceding experiment. The results, however, were significantly different. "One % glucose," he wrote, "increases the respiration by about 45%. It inhibits the spontaneous formation of ketone almost completely (1/3). The formation of ketone from caprylic acid is relatively little influenced, but clearly!!" ($Q_{acetoacetic\ acid}$ in the control was 2.18; $Q_{acetoacetate}$ with glucose alone, 0.68, or one third the control rate. $Q_{acetoacetic\ acid}$ with caprylic acid was 5.46, with caprylic acid plus glucose 4.32. Thus the reduction caused by glucose with caprylic acid was proportionately much less than the reduction without caprylic acid, but nevertheless there was a "clear" drop.) Regarding the narcotic, Krebs noted, "Narcotics + fatty acids, stronger inhibition [of respiration] than fatty acids and narcotics alone: difficult to explain." He did not pursue this puzzle, however. The tests of narcotics were clearly subordinate to the investigation of the effect of glucose. Having moved from "no effect" to a small effect of glucose, he pushed on, seeking conditions that would make the effect more definite.[63]   On September 30 he fixed his attention entirely upon the action of glucose on the formation of acetoacetic acid, spontaneously and from caprylic acid. He employed two concentrations of caprylic acid and three concentrations of glucose for each condition.   The results he obtained for acetoacetic acid were:

			$Q_{acetoacetic\ acid}$ =	1.00
Control				
+ glucose	0.15 mm^3	20%		0.42
	0.45 mm^3	20%		0.30
+ caprylic acid	0.15 mm^3	0.1 M		5.36
+ caprylic acid	+ glucose	0.15		3.37
+ caprylic acid	+ glucose	0.45		2.05
+ caprylic acid	0.05 mm^3	0.1 M		3.01
+ caprylic acid	+ glucose	0.15		2.23
+ caprylic acid	+ glucose	0.45		1.44

Krebs summarized the result simply: "Overall, distinct effect of glucose on the formation of ketone."[64]   Scanning the above data we can see that, in terms of his objective, this must have been an extremely satisfying outcome. Under each of the three primary conditions tested—the control, or "spontaneous" formation of acetoacetic acid, and two concentrations of caprylic acid—glucose substantially lowered the rate, the larger concentration of glucose having the greater effect. The experiment must have seemed to him to provide strong support for the conclusion he had drawn 12 days earlier, that glucose inhibits the formation of acetoacetic acid from substances other than acetate.

Looking back over the experiments of these 12 days, we can see that Krebs had reached this favorable vantage point by skirting, rather than overcoming, the obstacles he had encountered on the way. Confronted with experiments in which the fatty acids he tested failed to behave as they should have according to the β-oxidation theory, or as he needed them to do for his purposes, he had singled out for his tests of the effects of glucose on the process, one of the fatty acids that caused the least difficulty. The question of the anomalous behaviors of the other fatty acids he simply postponed. In this case he did not allow himself to be diverted from the experimental question he had already taken up.

Experimentalists encounter unanticipated "effects" so frequently that they must make almost daily judgments about which ones to treat as side effects and which ones as new problems to explore. This series had begun with a judgment that a difference observed between the effect of glucose or glycogen on the formation of acetoacetic acid when acetate was, and was not, present, offered a problem worth following up. As will be seen below, Krebs also judged the anomalous results concerning odd- and even-numbered fatty acids as indications of a serious problem, but not one that he should put ahead of the one on which he had already engaged his effort. In almost all of the experiments described above, Krebs noticed other "effects" that I have not included in the discussion. If he had paused to examine each one of them, he could not have sustained any stable line of investigation.

The experiment of September 30 brought Krebs to a nodal point within one of the subproblems of the broader line of investigation he had been following for several months. Having shown to his satisfaction that under certain specified conditions glucose inhibits the oxidative formation of acetoacetic acid, he was ready to turn to further subproblems with which this small resolution put him in contact. On October 4 he took up simultaneously three further sublines suggested by the progress he had made. One of these lines was to test whether glycogen acted in the same way that glucose did. A second direction may have been to begin to close in on the locus of the action of glucose by testing again its effects on the late stages of the oxidative sequence that began with butyric acid. The third subline was to examine how the formation of acetoacetic acid from caprylic acid was related to its formation from butyric acid and acetic acid.

Krebs did not state that his intention was to pursue these three questions, but they appear tacit in the design of the experiment itself. He compared the respiration and the rate of production of acetoacetic acid in liver slices from a starved rat, in the presence of the following additions to the medium: none, butyric acid, butyric acid + glucose, caprylic acid, caprylic acid + glucose, butyric acid + acetic acid, caprylic acid + glycogen, and acetic acid. The results were negative with respect to two of the implied questions, and paradoxical with respect to the third. Glycogen, Krebs put down, "does not act at all." That is, acetoacetic acid had formed from caprylic acid in the presence of glycogen at the same rate as in its absence. Glucose did not inhibit the formation of acetoacetic from butyric acid.[65] Krebs made no comment on this outcome, so it is unclear whether he regarded it as an indication that the action he had previously observed did not occur within the sequence of reactions connecting butyric and acetoacetic acid, but that is the most obvious interpretation of the results.

The examination of the relationship between the formation of acetoacetic acid from fatty acids and from acetic acid was inherent in the comparisons between the rate in the presence of caprylic acid, of acetic and butyric acid separately, and of the latter two compounds together. If acetic acid and butyric acid give rise to acetoacetic acid through independent processes, one would expect their effects to be additive. The rate for acetic acid + butyric acid did turn out, in fact, to be roughly the sum of the rates for the two tested separately ($Q_{acetoacetate}$ with acetate = 3.87; with butyrate, 2.99; with butyrate + acetate, 5.06). It was possible that, although independent of one another, these two processes arose from a common source at the stage of the longer-chain fatty acids. In that case one would expect the rate for caprylic acid to be a summation of the rates for butyric and acetic acid. If, on the other hand, the sequence passing through acetic acid originated from a different source, one would anticipate that the total rate of formation of acetoacetic acid from acetic + butyric acid would be greater than that from caprylic acid. The actual results fit neither of these logical possibilities, and this was obviously the aspect of the experiment that most puzzled Krebs. "Caprylic acid," he wrote, "appears to form ketone more <u>rapidly</u> than acetic acid + butyric acid!!!" ($Q_{acetoacetate}$ with caprylic acid = 7.56; with butyric + acetic acid, 5.06.)[66] That result would be difficult to reconcile with a logical interpretation of the processes involved. No matter whether acetic or butyric acid were *both* intermediates in pathways arising from caprylic acid or only one of them was, the basic axiom of intermediary metabolism required that either the two together, or the one that was an intermediate, give rise to the end product at a rate equal to the overall reaction.

The initial steps that Krebs took beyond his point of resolution of September 30 thus led him not to further openings, but to new obstacles. The only results of the experiment of October 4 that raised no difficulties were simple confirmations that glucose inhibited "somewhat" the formation of ketone from caprylic acid and from the "intrinsic metabolism" of the tissue.[67]

Grappling first with the most puzzling of the problems raised by the experiment, Krebs repeated on the next day the part of it comparing the rate of formation of acetoacetic acid from caprylic acid with that from butyric + acetic acid. He modified the conditions by using liver tissue from a nourished, rather than a starved rat. The new result only reemphasized the anomalous situation ($Q_{acetoacetate}$ with caprylate = 5.05; with acetate + butyrate, 3.13), and he again recorded the outcome with exclamation marks: "<u>Again</u> caprylate more acetoacetate than acetate + butyrate!!" At the same time he carried out a large scale experiment on the formation of acetoacetic acid from caprylic acid, isolating the product in the form of its dinitrophenylhydrazone and identifying it specifically as acetoacetic acid.[68]

When a focused attack on well-delimited aspects of a phenomenon fails, one may respond by reverting to a more diffuse exploration of conditions affecting the overall phenomenon. That is what Krebs appeared to do in the wake of his two unsuccessful efforts to penetrate to the subprocesses in the formation of acetoacetic acid. On October 6 he tested the actions of two potential inhibitors—sodium azide and hydrocyanic acid—on the formation of acetoacetic acid from caprylic acid. At a concentration of 0.01 M the former "inhibited almost completely" the respiration and the appearance of acetoacetate. The latter "inhibits the decomposition of fats (formation of ketone) more than the respiration! Yet the inhibition of respiration in the presence of caprylic acid is also very high!!" The next day he tested the effects of aging tissue slices on the process, and found that "the total respiration and the oxidation of the fatty acids are about equally damaged" by the alteration of the tissues that took place over 24 hours. On October 9 he tested the influence of glycerol, of insulin, and of a change in the composition of the salt solution on the process. As so often when he sought to elucidate the metabolic action of insulin, that substance exerted here "no certain effect." The glycerol and the modified salt solution both decreased the rate of formation of acetoacetic acid. Over the next six days Krebs continued in similar fashion to examine conditions influencing the process, testing glycerol and insulin again, as well as further changes in the salt solution. He noted more effects, similar to the preceding ones, but uncovered no illuminating leads.   In two experiments he measured the disappearance of acetoacetic acid added to the medium.   In another he examined "which organs produce acetoacetic acid" and concluded, "Liver alone significantly.   Kidneys, trace.   Other tissues, within experimental error!" He tried two more fatty acids, oleic and undecanoic, but lost the tissue slices.   On October 17 he compared the formation of acetoacetic acid in the presence of various "C_2 and C_4" compounds—butyrate, crotonate, acetate, β-hydroxybutyrate, acetaldehyde, and acetate + acetaldehyde.   All of these compounds increased the rate, as he would already have expected from previous experiments, except for the two tests including acetaldehyde.   That compound appeared to inhibit the process.[69]

In an experiment on fatty acids he carried out on October 16, Krebs included also tricarballylic acid, one of the three tricarboxylic acids whose formulas he had sketched in his notebook in August.   This test therefore represented an isolated return to the brief search for intermediates in the decomposition of citric acid that he had initiated then.   Again he made no inroad on the problem.   He wrote, "Tricarballylic acid does not react," and quickly left the question.[70]

IV

In the midst of his main line of investigation Krebs interjected, on October 16, a measurement of the time course of the spontaneous formation of urea in liver slices.   During the following two weeks he interspersed eight further experiments on urea synthesis.   He examined the influence of the ionic composition of the media, the concentration of the carbonate-bicarbonate buffer, the ornithine and ammonia concentrations, and the temperature.[71]   The probable motive for his return to this subject was that he was starting to prepare a review of "urea formation in the animal body," in English, for the journal *Ergebnisse der Enzymforschung*, and he wished to present more systematic data than he had available concerning the conditions affecting the reaction.[72]

On November 3 and 4 Krebs examined an aspect of the formation of urea that he had not considered in his original papers on the subject.   In 1929 M. Kitigawa and his associates in Japan had isolated a new diamino acid ($C_5H_{12}O_3N_4$) from jack beans.   They called the compound "canavanine," after its source, and extracted from the bean an enzyme that split canavanine hydrolytically, yielding urea.   Krebs now tested an extract from jack beans with liver tissue slices and found that it increased the rate of urea formation.   He concluded that canavanine was "slowly" split by an enzyme present in the liver.[73]

On the same day Krebs carried out an experiment probably provoked by two papers recently published from the California Institute of Technology. There Henry Borsook and Geoffrey Keighley had repeated and confirmed Krebs and Henseleit's observations concerning the ornithine effect.   They wished, however, to reexamine "the relationship between oxygen consumption and urea synthesis."   Noting that, according to the single published experiment that reported the effects of ornithine both on respiration and the formation of urea, Krebs and Henseleit had been "unable to find, within the limits of their experimental error, any increase in oxygen consumption after ornithine was added," Borsook and Keighley set out to multiply such measurements, under the assumption that there must be an energy-yielding reaction coupled with the energy-consuming reactions that synthesize urea. They were able to find, under certain conditions, an increase in oxygen consumption in the ratio of approximately one molecule of extra oxygen for each molecule of increased urea production.[74] Krebs now tested their conclusion by making a new measurement of his own of the consumption of oxygen and formation of urea with and without ornithine (in the presence of lactate and ammonia). The results were:

	Without ornithine	With ornithine
Q_{O_2}	24.0	23.9
Q_{urea}	5.21	6.86

He wrote down "Unfortunately too little ornithine effect!! Nevertheless, no increase in respiration!! Contrary to Borsook."[75] The smallness of the ornithine effect would seem to make this an indecisive experiment to pit against the numerous positive results Borsook and Keighley had reported. Krebs was, however, apparently satisfied enough not to pursue the question further. The probable reason was that, even without testing their results, he did not "consider [Borsook and Keighley's] experiments conclusive."[76]

In his review article Krebs explained why he took this view of Borsook and Keighley's experiments: "They only show that an increase of the respiration takes place *after addition of ammonia.*" This increase, Krebs thought, was probably accounted for by the formation of other products to which ammonia can give rise. In the article he merely republished the same data from his earlier article that Borsook and Keighley had questioned in the first place, remarking that "the conditions of the experiment . . . were such that—if the parallelism between respiration and rate of urea synthesis claimed by Borsook is correct-the effect on the oxygen uptake should be very high indeed. Yet no change in the rate of oxygen uptake could be detected."[77] Here, as on other occasions, Krebs displayed a confidence in his own reasoning that sometimes led him to brush off impatiently suggestions of colleagues that impinged on his work. Borsook and Keighley had offered a detailed explanation for the difference between their results and the result of Krebs and Henseleit.[78] Krebs ignored this discussion, dismissing the effect they had observed as due only to the addition of ammonia. Borsook and Keighley had opened themselves to such a criticism by calculating their "$\Delta Q_{O_2}/\Delta Q_{urea}$" ratios from pairs of experiments with and without ammonia, in which the ornithine was present in the same concentration. Yet Krebs's rationale was too easy, for it did not account for the fact that it was the ratio of extra oxygen to extra *urea* that showed a pattern of near constancy in Borsook and Keighley's experiments.

During the two weeks that he occupied himself with urea formation Krebs also kept searching for footholds on the resistant acetoacetate problem. The scanning mode of attack to which he now had recourse is suggested by the title of his experiment of October 19: "Liver + diverse additions." The additions included acetate, lactate, pyruvate, acetaldehyde, acetaldehyde + pyruvate, acetaldehyde + acetate, lactate + acetate, and pyruvate + acetate. He may have chosen lactate and pyruvate because as intermediates in glycolysis they might be expected to share the antiketogenic action of glucose, and acetaldehyde because it had conspicuously reduced respiration and the formation of acetoacetate in the preceding experiment he had carried out in this line. The pairs, however, appear more likely to reflect a systematic search of the possible combinations of these substances than any hypotheses concerning their metabolic relationships. In this experiment he found "practically no acetoacetic acid !!!" in any vessels, "even though the acetate increased the respiration!! Therefore it is consumed!!" The experiment, which had been carried out on tissue slices from a "fat, old ♀ rat," thus merely raised more problems.[79] Repeating it the next day on liver tissue from a starved, but presumably younger rat, Krebs this time did obtain acetoacetic acid in each case, but, as his statement of the results make emphatic, the outcome was just as confusing:

> In starved animal large formation of acetoacetic acid from acetic acid. But why increase in respiration from acetate? Evidently consumes _more_ acetic acid (oxidation?) Strong antiketogenic action of pyruvate and lactate. Acetaldehyde _also_ acts antiketogenically. Not as expected and as previously.[80]

In the earlier experiment acetaldehyde had appeared to act as a general inhibitor, reducing both respiration and the appearance of acetoacetate. Here it reduced the latter, but not the former, a characteristic of antiketogenic substances. Pyruvate and lactate reduced the formation of acetoacetate while strongly _increasing_ the respiration.

Once again Krebs had observed effects that were striking but not clarifying. Still searching, on October 24 he tried out glucose, insulin, and butyric and caprylic acid on sarcoma tumor tissue. He only reinforced what he had already observed for liver tissue: "large effect from glucose, no effect from insulin." In this tissue butyric and caprylic acid produced no acetoacetic acid. The next day he tested the effect of pH on the action of glucose with liver tissue and found only a slight influence.[81]

A week passed before Krebs returned to these problems on November 1. Repeating his experiment on the influence of pH, he observed this time that the acetoacetate formed, both with and without glucose, at lower rates at the lower of two pHs, but that the antiketogenic action of glucose was relatively small at both pHs.[82] During the next week Krebs finished up his experiments on urea, then returned to another aspect of his acetoacetate investigation that he had left in an unsettled state one month earlier: the anomalous formation of acetoacetic acid from odd-numbered fatty acids. (See this chapter, p. 27–28).

On November 7 Krebs tested the actions of the whole series of normal fatty acids from C_2–C_9 on respiration and the formation of acetoacetate in liver tissue from a nourished rat. The results were:

	Q_{O_2}	$Q_{acetoacetic\ acid}$
Acetic acid (C_2)	-14.7	0.37
Propionic acid (C_3)	-11.7	0.25
Butyric acid (C_4)	-15.6	1.56
Valeric acid (C_5)	-13.4	0.53
Caproic acid (C_6)	-14.4	1.64
Enanthic acid (C_7)	-12.2	0.52
Caprylic acid (C_8)	-18.7	1.72
Pelargonic acid (C_9)	-13.3	0.59
Control	-10.1	0

He concluded that "odd-numbered also increase respiration, but less than even-numbered acids. Extra ketone formation also from odd-numbered, but less."[83]

There was, as the above figures show, a consistent, if modest difference between the increases in respiration caused by the odd- and even-numbered acids, and a more distinct difference between the respective rates of acetoacetate production; but this was hardly a resolution of the problem, since the odd-numbered acids were not supposed to produce an even-numbered ketone body at all. The next day Krebs repeated the experiment with liver tissue from a starved rat and reached an even less satisfactory outcome. Each of the acids yielded acetoacetic acid at higher rates than before, and the highest rate of all was from an odd-numbered acid: $Q_{acetoacetic\ acid}$ with pelargonic acid (C_9) was 7.33, compared to 6.09 with caproic acid (C_6), and 4.32 with caprylic acid (C_8). Krebs noted, "Definite formation of ketones also with odd-numbered acids," but added "(if the latter are pure)."[84]

Clearly doubting these results, Krebs pursued the possibility that the commercially prepared fatty acids he had employed were *not* pure. On November 9 he purified enanthic acid, following a method published by J.S. Lumsden in 1902. Krebs satisfied himself that the crystals he obtained were pure. By November 13 he was ready to test purified samples of all the fatty acids he had previously utilized (except for butyric acid, which he again used in commercial form). His labors did not improve the situation: $Q_{acetoacetic\ acid}$ with caproic acid (C_6) = 6.82; with enanthic (C_7) = 2.17; with caprylic (C_8), 3.72; with pelargonic (C_9), 3.75. Krebs succinctly summarized the result: "Just as with unpurified acids!"[85] It is difficult to tell whether that statement reflects frustration with his inability to confirm the known or interest in the possibility that he was on the track of something new.

Narrowing his focus, Krebs tested purified enanthic acid, in three concentrations, on November 15, and confirmed again that this odd-numbered fatty acid caused "definite ketone formation." Two days later he carried out a "comparison between acetate and enanthate in the liver," and observed to his evident surprise, "Acetate → less 3 β-ketoacid than enanthate!!" The next day he posed the question, "Can the yield of β-ketoacid be increased?" and sought to do so by adding substances that might block the decomposition of the acetoacetate formed. He tried arsenite, hydrocyanic acid, and 2,4-dinitro-phenylhydrazine. The first and last increased the apparent rate of formation modestly. He also tried using ground liver tissue, but no acetoacetate formed. Carrying out a similar experiment the next day, he found that arsenite did not increase the yield of acetoacetate. Shifting his attention, on the 20th, to possible "poisons for the formation of β-ketoacids," he tested iodoacetate, iodine, and pyridine, but found that they did not specifically inhibit the formation of ketones. On November 23 he continued looking for other substances that might "influence" the process; tested the action of several fatty acids on dog liver tissue; and tested the effects of various salt solutions on the formation of acetoacetate from caprylic acid in rat liver. On November 28 and December 2 he tested the action of several fatty acids in mouse liver tissue, obtaining results similar to what he had already observed with rat tissue. On December 5 he utilized three fatty acids he had not previously included—capric (C_{10}), undeca-noic (C_{11}), and lauric (C_{12})—and obtained "definite formation of ketones" from each of them.[86] Then, abruptly, he left his investigation, not only of the fatty acids, but of the whole set of interrelated problems involving the intermediate stages of respiration that had been his primary research concern for over 10 months.

* * *

Since Krebs frequently shifted back and forth among ongoing research projects, we cannot infer directly from his actions that he had abandoned the investigative line he had been following in December 1933. Even the fact that he did not return to these problems for two years might not indicate that he anticipated in advance that he would drop them for a prolonged period. It would be plausible for him to have been diverted as usual to another set of problems that he had already pursued successfully in the past, without making any plans about the duration of the shift. There is no contemporary record of his attitude toward his efforts in the area of respiratory oxidations at the time he ended them. Repeatedly, however, while he examined the notebook pages containing these experiments with me in 1977, he referred to the general venture in negative terms. "I did lots of experiments," he remarked at one point, "without getting anywhere." At other times he said, "all these attempts were unsuccessful"; or, "it looks rather futile here, because nothing of this . . . was published"; "I got stuck then for quite a long time"; and "altogether that got not very far. Essentially I didn't do much more than confirm what was known or what was to be expected." Summing up his situation at the end, he said, "I always felt, and acted accordingly, that I was up against a wall, and then I tried other things."[87]

These are, of course, retrospective comments, the reactions of a man looking back at himself through layers of later events; yet their consistency, and the fact that the outcome of the investigation as it can be reconstructed from the laboratory record lends itself to the assessment he made in 1977, make it plausible that he felt similarly about it in December 1933; that he did give up on these problems, then, because he felt that he was getting nowhere; and that he decided to return to areas in which he thought he had more chance for success.

Krebs's explanations for *why* he got stuck were, however, explicitly based on hindsight. "When one looks at the development of the field and the finding of the answer to the questions raised here," he told me, "it is quite clear that one couldn't have done it in those days. Many collateral developments had to take place." The next day he said, "Retrospectively one can say that so little was known in the field that one couldn't expect that I would make progress."[88]

Similar though these two statements appear, there is a crucial distinction between them. We might readily agree that, in 1933, Krebs could not have found the complete "answers" that have since been given to the questions he asked; it is less evident that he could not have made "progress" in 1933. It would be a futile historical exercise to identify, also by hindsight, some specific clues that Krebs "missed" during the 10 months that he spent on these problems. Experimental scientists inevitably pass over many of the effects they observe. Most of these effects will turn out to have been insignificant, while a few will take on special significance in the light of later developments. (Often, in examining these notebook records with me in 1977, Krebs saw "obvious" significance in data that could not have displayed the same meaning to him in 1933.) No one can be prescient enough to seize unerringly upon the category that will later appear to have been beckoning clues. What is legitimate to ask is whether, in a more general way, he fully exploited the opportunities for progress that were available to him, or whether he gave up where further pursuit might have been worthwhile.

From the preceding account of the investigation we can see that Krebs was not pursuing a single question through these months, but a series of loosely interconnected ones. The subquestions were not strictly separate, but shifting in their relationship to one another, and at some points it appeared that two or more of them might merge if he could establish metabolic links between previously proposed sequences. Progress might have come in any one of these subproblems, or by connecting two of them more closely together. We can see that he did not press the various subproblems he took up with equal intensity, nor did he press any one of them to the limit of his experimental resources before shifting to another. He departed from his efforts to test the Thunberg-Knoop-Wieland scheme before he had either confirmed or refuted it and before he had exhausted the avenues of examination open to him. The most novel facet of his investigation, his attempt to connect citric acid with known pathways, he followed up only fitfully, testing only three hypothetically possible reactions and yielding quickly to initial negative results. In one phase of his investigation he did appear, for a while, to be making progress; that was when he observed that glucose inhibited the oxidative formation of acetoacetic acid without affecting its formation from acetate. When he encountered obstacles, however, he shifted to other aspects of the broader investigation, without exploring as thoroughly as would seem worthwhile whether the distinction he appeared to have uncovered was significant. He ended the overall investigation at a point at which it was probably too early to tell whether the formation of acetoacetate from odd-numbered fatty acids was an experimental anomaly or a clue to a hitherto unknown metabolic connection. If Krebs did not, on the whole "get very far" in these investigations, therefore, that outcome may have been due as much to his research style as to the blockades intrinsic to the state of the field. In a comment not directed at, but applicable to this situation, Krebs tacitly implied as much:

> Some people get stuck, and some people get bored in research, because they can't get any further. But there are some people—some of whom I would greatly admire—who hammer away and get into the depth of the matter . . . and sometimes they are successful. They work in one field, and gradually go very far. But that is easier, perhaps, in general chemistry or physics, but in biochemistry the situation is often terribly complex.[89]

Someone else might have hammered away more persistently at these problems than Krebs did in 1933, and might have gone further than he did. That is not to say, however, that he was unwise to turn away from them when he did; for in doing so, he was acting in accord with the strengths as well as the limitations of his own style. Krebs's characteristic approach—to try many experiments suggested by simple ideas, rather than to spend a lot of time thinking through foundations for each move he might make—was not well adapted to "hammering away" at a recalcitrant problem. Better, for *him*, to exploit the versatility of his methods by moving on to other problems, in the knowledge that he could return quickly if a new opening in the line he had left should present itself to him.

When Krebs looked back at these "unsuccessful" efforts, he did not dismiss them as useless. "Even if they did not lead to any publication of this material," he remarked in 1977, "they were, I think, of importance in shaping my way of thinking, and forming my frame of mind."[90] In an article on "The History of the Tricarboxylic Acid Cycle," he had already stated, in 1970, what point of view he thought the investigation had imparted to him. "The striking impression which these experiments left in my mind was the fact that citrate, succinate, fumarate, malate, and acetate were very readily oxidized in various tissues, with the formation of bicarbonate and CO_2, and that the rates of oxidation of these substances were compatible with the view that they played a major part in tissue respiration; but this had already been shown much earlier—in 1911—by Battelli and Stern."[91]

The selective nature of Krebs's recollection here should be clear. The impression that he described was only most "striking" in retrospect, because subsequent events made that impression a particularly important "preparation" for his later return to the problem of respiratory oxidations. What he remembered afterward that the experiments had left in his mind, however, is far from a direct account of his frame of mind in 1933. Can we recover his way of thinking about these problems during the investigation itself?

From the foregoing narrative we can see that, although Krebs was growing somewhat bolder in posing novel hypotheses about specific metabolic connections, his general way of thinking about these problems did not differ markedly from the main currents in the field at the time. As in his earlier starts in the same area, he mainly tested oxidative reaction sequences that had already been discussed widely or that had been proposed recently by others, such as the ω-oxidation theory of Verkade. There is little reason to suppose that he modeled his approach to these problems on his solution to the problem of urea synthesis. It is sometimes assumed that, having discovered the first "metabolic cycle," he immediately began to look for other analogous cycles.[92] In fact, he had not yet perceived the "circuit of ornithine" as a generalizable pattern. Even though he investigated another prominent metabolic hypothesis comprised of a closed circuit of reactions, the Thunberg-Knoop-Wieland scheme, he did not think to apply to this problem the experimental strategy by means of which he had established that the synthesis of urea operates through a closed circuit. That is, he did not test the postulated intermediates of this circuit in very small quantities, to see if they acted catalytically. Had he been predisposed to search for metabolic cycles at this time, he could hardly have overlooked employing the approach that had been responsible for his most auspicious success. Evidently, in 1933, Krebs's "frame of mind" concerning the intermediate pathways of respiration was still open ended.

V

The similarity between Krebs's conceptions about the intermediary processes of respiration and the views of his contemporaries would lead us to expect that the experimental investigations he carried on in this area might also resemble closely what some others were doing. When he had first thought of applying the methods of tissue slices and manometry to intermediary metabolism, while working for Warburg, Krebs had identified an unexploited opportunity. That idea was not so subtle, however, that it would be likely in the long run to escape the attention of all other investigators interested in the field. During the 1920s, Thunberg's methylene blue method overshadowed the approach of measuring respiratory rates in isolated tissue that he and Federico Battelli and Lina Stern had earlier introduced (see Vol. I, pp. 20–29). By the end of the decade, however, numerous investigators were measuring respiratory rates in tissue slices or suspensions, utilizing either Warburg manometers or Barcroft differential manometers. It was nearly inevitable that someone else would perceive the same general opportunity that had occurred to the young assistant to Otto Warburg. By the time Krebs could implement his plan, another biochemist, almost identical in age, had already adopted the same approach.

Juda Hirsch Quastel came to Hopkins's laboratory in Cambridge in 1921. While there, he pioneered in the application of current views of intermediary metabolism to bacteria. Adopting Thunberg's methylene blue methods, Quastel was able to confirm that the oxidations of the succinic acid series took place in these microorganisms. A probing, independent thinker, Quastel developed an ingenious theory that metabolic reactions are catalyzed at "active sites" on cellular membranes. In 1929 he left Cambridge to take a job as biochemist at the Cardiff City Mental Hospital. The study of bacterial metabolism did not appear appropriate in such a setting, so he decided to shift his research to the brain. In practice that meant transferring his interests in intermediary metabolism to brain and nervous tissue.[93]

In 1930 Quastel published a paper on "Biological Oxidations in the Succinic Acid Series." In it he examined whether the oxidation of succinic and fumaric acid proceeds entirely through *l*-malic acid or whether there may be alternative paths. To answer the question, he relied mainly on quantitative criteria. Measuring the uptake of oxygen by suspensions of brain and muscle tissue, as well as bacteria, in the presence of these compounds, he applied with special clarity the basic postulate that we have seen appear in more or less rigorous form in intermediary metabolism from the first decade of the century on. For example, he argued, "it follows that if fumaric acid can only be oxidized *via l*-malic acid, the velocity of oxidation of fumaric acid cannot be greater than that of *l*-malic acid for equivalent saturation concentrations of the substrates. A distinctly higher rate of oxidation for fumarate would indicate a separate line of oxidation." Quastel showed that succinate, fumarate, and malate are oxidized at rates "of the same order," consistent with the view that malate is an intermediate in the oxidation of the first two compounds. On the other hand, aspartic acid, an amino acid with a structure similar to that of the succinic acid series, was oxidized much more rapidly, indicating that it undergoes a direct oxidation, without being transformed to malic acid. Quastel also invoked criteria of additivity, when he utilized two substrates in combination to determine whether the two formed parts of the same line of oxidations. He found, for example, that "the oxygen uptake of a mixture of succinate and fumarate is the average of that due to succinate alone," consistent with the view that "the main source of oxidation of succinate is through fumarate. Were there a different means of oxidizing succinate, it would have been anticipated that the addition of fumarate would have increased the rate of oxidation found in the presence of succinate alone." In addition to these quantitative considerations, Quastel utilized inhibitors, such as malonate, which he had shown to block specifically the conversion of succinate to fumarate, in support of his conclusions.[94]

As the foregoing examples suggest, Quastel was approaching questions in the field of biological oxidation in 1930, with reasoning and with methods similar to those Krebs brought to bear on such questions a little later. During the time that Krebs was beginning his independent investigative career, and concentrating his attention on urea synthesis, deamination, and related problems in the metabolism of nitrogenous compounds, Quastel fixed his attention on the special area of oxidations in brain tissue.[95] In 1933, however, the investigative pathways of the two men converged onto a remarkably similar set of experiments. During the time that Krebs was taking up the general question of intermediates in respiration, Quastel was taking up the oxidation of fatty acids. In order to study them under conditions in which they were consumed at a high rate, he shifted from brain to liver tissue. At the same time, he switched from

Barcroft to Warburg manometers and adopted Warburg's tissue slice method. Quastel intended to study "fatty acid oxidation, and the various factors influencing it, in isolated organs in a strictly quantitative manner." He measured the oxygen uptake, expressed in the Warburg manner as Q_{O_2}, for a series of normal fatty acids, from formic (C_1) to octoic acid (C_8). He, too, paid particular attention to the oxidation of butyric acid. In addition to measuring the respiration, he measured the rate of formation of acetoacetic acid. To do so, he adapted, independently of Krebs, the aniline method that Ostern had developed for Krebs for oxaloacetic acid, to the estimation of acetoacetic acid. Like Krebs, Quastel tested the effects of glucose and glycogen on these oxidations; in contrast to Krebs, Quastel found that glucose exerted no effect, whereas glycogen reduced the production of acetoacetic acid. Also like Krebs, Quastel found that the rate of formation of acetoacetic acid from butyric acid was less than would be expected from the rate of increased respiration caused by the butyric acid, and he reasoned in a very similar manner that "it was conceivable that some of the butyric acid was oxidized to succinic acid." If so, he thought, the succinic acid should be oxidized in turn to 1-malic acid, which ought to accumulate and be detectable. Finding none, he gave up his hypothesis, just as Krebs gave up the same hypothesis after reaching an analogous negative result. In one respect Quastel's investigation ran more smoothly than the corresponding investigation of Krebs, for Quastel did not encounter the anomalous behavior of odd-numbered fatty acids that Krebs did. In Quastel's hands the even-numbered fatty acids gave rise to acetoacetic acid, but "those fatty acids with an odd-number of C atoms produce little or no acetone bodies."[96]

Quastel presented the "substance" of these results at the Third International Congress for Experimental Cytology, in August 1933, and submitted a paper on the subject to the *Biochemical Journal* in September. It is most likely that he and Krebs each undertook the investigation of fatty acid oxidation without knowing that the other was proceeding along nearly the same track. Although Krebs and Quastel were aware of each other's previous publications on other topics, they had not met and were not in personal communication.[97] As the congress at which Quastel read his paper took place in Cambridge, as noted above, Krebs may well have heard it there. The paper itself must have appeared around December, and may have had something to do with Krebs's decision then not to continue his investigation of fatty acid oxidation.

Krebs looked back at his own work on this problem as unsuccessful, because it led to no publication. By that standard, Quastel was more successful; yet Quastel too had failed to solve one of the central problems that was blocking Krebs. Quastel's results did not add up to a consistent view of the process by which fatty acids are oxidized to the stage of acetoacetic acid. Like Krebs, he had run into evidence against the generally accepted view that butyric acid is an intermediate in this process and had not been able to provide either an alternative link between the longer fatty acids and acetoacetic acid or an alternative direction for the oxidation of butyric acid. Quastel chose to publish his results as empirical findings. He stated that certain of his results (in this case concerning the effects of glycogen) were "contrary to current ideas,"[98] but let the contradictions stand as they occurred. Krebs undoubtedly could, if he had wished, have published a paper of similar nature. That he did not do so does not necessarily imply that his experimental standards were generally higher than those of Quastel, but suggests that he placed higher importance on achieving a coherent explanation. He was more likely to be dissatisfied with results that did not fit into an overall physiological picture. Quastel showed himself here, as in other papers, more inclined to publish results that challenged the existing picture, even when he could not replace it. He was, however, himself not necessarily satisfied with the state of affairs as he presented it in this paper. The problem of how butyric acid is related to acetoacetic acid in fatty acid oxidations became, in fact, the focus of his ongoing investigation.

2

Laboratory Life in Cambridge

Life in the Cambridge Biochemical Laboratory suited Hans Krebs perfectly. He found the people friendly and helpful, the environment stimulating. To him there was a spirit of fairness and tolerance that touched him all the more deeply because he had experienced the opposite during his last months in Germany. In the laboratory intense scientific activity was coupled with an atmosphere of casual forbearance. Divergent opinions were freely expressed. There appeared to be little formality or hierarchy, yet the laboratory ran smoothly. When there were decisions to be made concerning the operation or teaching responsibilities of the department, Hopkins called a meeting of the staff or consulted informally around a staircase. After he had gathered the sense of the group, he took the necessary actions, unostentatiously but confidently. His style of leadership inspired harmony and mutual trust.[1]

In Freiburg the academic ethos had been generally encouraging to Krebs's research, but he had had few colleagues with interests that closely paralleled his. In Cambridge he was surrounded by biochemists involved in areas related enough to his own work so that he could discuss his ideas and his problems with them in detail. Next to his own laboratory bench, and therefore in daily conversation with him, was Norman W. Pirie (known always as Bill Pirie), who had been at Cambridge, first as a student and later on the staff of the Department, since 1925. Pirie had worked closely with Hopkins on the chemical preparation of glutathione, a compound whose widespread presence in cells and its properties had long challenged Hopkins to attempt to elucidate its metabolic functions. At the time Krebs arrived, Pirie was beginning to turn his attention toward the chemistry of viruses. Malcolm Dixon, easygoing and good humored, but a meticulous experimentalist, concentrated his research on intracellular enzymes. Ernest Baldwin seemed to Krebs an indifferent investigator, but a brilliant teacher, who articulated the view of "dynamic biochemistry" considered characteristic of the approach of the Hopkins school. Joseph Needham, then developing the fields of comparative and embryological biochemistry, possessed an encyclopedic knowledge, but Krebs noticed that Needham spent most of his time writing, and little in the laboratory, and Krebs probably learned more in conversation from Needham's wife, Dorothy Moyle Needham. Dorothy Needham had done important research in muscle metabolism, some of it touching closely upon Krebs's recent interests in respiratory oxidations.[2]

Of all of the researchers in the Cambridge laboratory, Krebs was most impressed with Marjory Stephenson, whom he remembered later as "scientifically . . . the best of the whole lot." Introduced to the metabolism of bacteria in 1924 by Juda Quastel, Stephenson had become the international leader in that field. Her 1930 monograph *Bacterial Metabolism* did much to define the subject. Krebs found her very imaginative, a skillful, hard worker, a good judge of people, friendly, and extremely helpful. He got on particularly well with her. Just "across the lawn" in the Molteno Institute was David Keilin. Like everyone else who knew him, Krebs quickly learned that Keilin was warm, generous, and lovable, always interested in the welfare of younger scientists, never too busy to discuss matters of interest.[3]

Into this environment, so different in some ways from anything he had previously known, Krebs fit easily. The reputation he brought with him undoubtedly eased his entrance, but his own adaptability, his eagerness to absorb the ways of his new home, were equally important. Those who welcomed him into their midst generally found him shy and quiet, but friendly. To them he was extraordinarily methodical. They worked hard, but sometimes kept casual schedules. He arrived at exactly the same time every day, and the others could set their watches by the time at which he began to clean up his Warburg manometers in the afternoon. Just as he wasted no time during the working day, so he never returned to the laboratory after his working day was over.

Having arrived with about 30 manometers and associated vessels, Krebs had far more "glassware" than anyone else in the laboratory. One younger investigator joked that this was unfair competition. In view of the current international conferences seeking to limit naval armaments, he said, there ought also to be a limit imposed on the number of manometer vessels an individual experimenter could use. But Krebs was generous with his "private fleet" of manometers. He lent them readily to others who wished to do experiments with Warburg manometers in place of the Barcroft differential manometers customarily used for metabolic work in the Cambridge laboratory. Krebs was in one sense a refugee, and in another sense somewhat privileged. That his privileges created no resentment in the laboratory was mainly due to his personality. Unlike one or two German refugee scientists who had preceded him to Cambridge, he was open and democratic. He discussed his own research freely, consulted others, and quickly became as well liked as he was respected. Outside the laboratory there were a few protests at a foreigner being given a research position, at a time when many British citizens were unemployed, but inside the laboratory Krebs encountered only good will.[4]

Some of the Cambridge biochemists were active in political and social issues, and conversations in the laboratory often ranged well beyond professional topics. For Krebs, this was a new experience. Sometimes he appeared to disapprove. To Pirie he would shake his head and say that they ought to stick to science. Those in the laboratory who did not know him well thought that Krebs lived only for his science, but others—particularly women in the laboratory—sensed that he had a much broader range of interests. Antoinette Pirie, for example, noticed that he adapted more readily than the other refugees to the English way of life, and she perceived that he was intensely interested in coming to understand the new community that he had entered. If that was not apparent to everyone, it was probably because Krebs organized his life so that his extrascientific concerns did not intrude upon his experimental activity. Eager to learn about English culture, he set out to catch up by reading the children's literature that he had missed by not being raised there. On the advice of Marjory Stephenson, he began with *Alice in Wonderland*.[5]

In Cambridge, Krebs's innate shyness was undoubtedly heightened by the fact that he had to learn to converse in a new language. He had learned English in a formal sense long before, but had until then used it only during brief visits to England and the United States. At first he thought in German and then translated. Because the case endings in the German language make the order of subject and object less critical than in English, he sometimes found that his English sentences made no sense to his colleagues. Surrounded every day by English laboratory conversation, he soon began to think habitually in English; but like many others in similar positions, he only slowly mastered the many idioms of the language, and he noticed that people did not correct his mistakes except when he specifically invited them to do so. As methodical in this as in everything else he did, he carried a little notebook in which he put down words or phrases that were new to him. He also began to seek assistance in writing, preparing for the time when he would publish in English. Meanwhile, he still kept his daily laboratory record in the German nomenclature and phrases that came more naturally to him.[6]

Those in Cambridge who saw Krebs schedule his laboratory activity rigidly by the clock tended to attribute such habits to his Germanic character. They may not have appreciated that here, for the first time since he had become an independent investigator, he was free to do so. In Freiburg, as in Altona, he had been forced to fit his experimental work around a partially unpredictable daily routine dictated by his medical duties. What he had learned there was how to use every available bit of his limited time effectively. In his new situation, with research as his full-time activity, wasting no time meant for him beginning every day exactly on time, carrying out a set of experiments each morning and afternoon as regularly as was possible, and finishing up promptly. Although not at his bench in the evenings, he was still busy reducing his data and planning his work for the next day. Undoubtedly the high value placed on reliability and punctuality in the German middle-class culture in which he had grown up played some part in his approach, as did the fact that the Warburg manometric methods lent themselves to such regularity. Perhaps he also reverted in part to the research style he had learned in Warburg's laboratory. Krebs did not perceive his manner of working as tightly organized, however. In his view he was simply making sure that he used all his time. His research habits were, in fact, a peculiar blend of regularity and flexibility—regular in the way that he scheduled his time, flexible in the way that he planned his experiments only one day ahead and shifted readily from one problem to another.

Proponents of the doctrine that institutions shape the personal interests of the scientists who work within them would expect that the move from Freiburg to the "Cambridge context"[7] must have altered the nature of the research that Krebs carried out in his new setting. He had entered a very different scientific and cultural environment. Formerly a clinician carrying out research part-time in a German Department of Medicine, he was now a full-time scientist working in a British Department of Biochemistry uniquely oriented around biological, rather than medical concerns.[8] Were such substantial institutional changes reflected in the nature of his research activity?

We can notice immediately a change in Krebs's pace. In Freiburg he had been able to carry out personal experiments sometimes four or five days a week, but often there were fewer, when other duties pressed on his time. In Cambridge he averaged five or six days per week, with infrequent interruptions. We can also notice that in Cambridge Krebs tended more often than in Freiburg to prepare his own chemicals, or to devise for himself modifications of standard analytical methods. In part he may have done so because he had more time to do so, in part because he was thrown more on his own resources. He did not, at first, have a group of younger workers around him like the one he had built up in his last year at Freiburg. He may, however, simply have been gaining experience at these tasks that gradually compensated in part for his slight formal training in organic chemistry.

Such changes, while real, were superficial. Were there also deeper influences on the direction of his work? If, as he has recalled, "the stimulating Cambridge intellectual climate helped me with the clarification of my ideas,"[9] those stimuli must also have *given* him some ideas that he could use in his research. Were we able to recover the lost conversations that took place daily around the laboratory benches, we would undoubtedly be able to identify particular suggestions, coming from some of his new colleagues, that Krebs put into practice in individual experiments. On the whole, however, the most obvious characteristic we can observe in the research Krebs carried out after coming to Cambridge is how much it resembled what he had been doing before he came. He picked up where he had left off, and it would be difficult to identify systematic differences in the intellectual or methodological style of his investigative path before and after his move. His work had acquired a momentum that enabled it to be moved from one institutional context to another without major deflection. Krebs's research flourished in Cambridge, because his interests were compatible with those of the Hopkins school. His approach to intermediary metabolism harmonized well with Hopkins's vision of biochemical processes, and that is undoubtedly one reason that Hopkins was eager to bring Krebs to Cambridge. It was not, however, the nourishing soil of Cambridge that defined the research program Krebs transplanted there. That program was shaped by all of the prior ideas, aims, experience, and equipment that the scientist himself carried with him.

As one of the early refugee scientists to obtain a position in England, Krebs repeatedly received requests from other displaced German scientists to help them find places to continue their work. For most of them he could do little more than to offer comforting words and pass on what information he had about other people to contact.[10] For his former medical colleague at Freiburg, Hans Weil, however, Krebs was able to do more.

Like Krebs, Weil was the son of a physician. His father maintained a thriving practice in internal medicine in Stuttgart, where Hans was born in 1905. Educated in a classical *Gymnasium*, Weil came to believe that art, literature, and philosophy were "the only occupations worthy of the human mind," and he resisted his father's expectation that he follow him into medicine. Reading the psychiatrist Ernst Kretschmer's influential book on personality types persuaded Weil, however, that medicine was not incompatible with these cultural values, and he enrolled in medical school in Basel, in 1924. He completed his M.D. in Heidelberg in 1928, then returned to Stuttgart to do his year of clinical work in the city hospital. There he entered the service of Thannhauser, shortly before Thannhauser left for Freiburg. Weil remained in Heidelberg, in charge of two wards, until Thannhauser's successor arrived nine months later, then followed Thannhauser to Freiburg, where he assumed responsibility for a tuberculosis ward.[11]

Weil's father wished him to obtain an academic title so that he would have access to a hospital appointment. In order to do so in Thannhauser's department, Weil saw, he would have to acquire a thorough training in chemistry. In the summer of 1931 Weil began taking chemistry courses. At just about that time Hans Krebs arrived in Freiburg, and Weil turned over his ward to Krebs. For his organic chemistry course Weil had to prepare a number of compounds, and he told Krebs at one point that if Krebs ever needed any special compound he would make it for him as one of these exercises. Krebs asked for and received a batch of α-ketoglutaric acid.[12]

Weil also belonged, like Krebs, to a prosperous, well-assimilated Jewish family, and he too experienced personally very little of the anti-Semitism conspicuous in postwar Germany. When the Nazis took power, Weil was studying for his final examinations in chemistry, which he took in May 1933. It was obvious to him that under the present conditions he had no hope of obtaining a university appointment in Germany. A young doctor from Geneva who was at the time working in the Freiburg clinic suggested to Weil that he would be able to work as an unpaid assistant in the Cantonal Hospital in Geneva.[13] Weil took up the suggestion, but began at the same time to look for more favorable opportunities. He visited Berlin to inquire whether he could enter Warburg's laboratory. In June, he wrote Krebs that he hoped to find a place in England, and requested a recommendation. Soon after he arrived in Geneva, Weil realized that his prospects for doing serious scientific work there were poor. The laboratory in which he worked was sparsely equipped, and it was oriented toward limited clinical applications of biochemistry. The chief was rarely present, and little activity took place there. In August, Weil reported to Krebs his discouragement about these circumstances and asked if he might "come to you for a couple of months." "Pardon me for this clumsy surprise,"

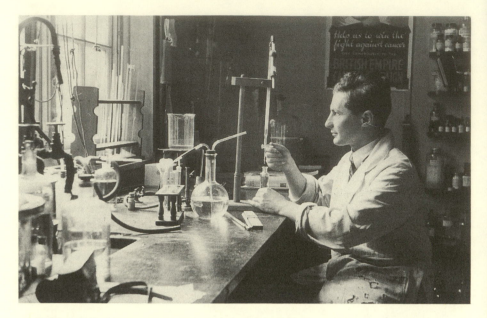

Figure 2-1 Hans Weil-Malherbe (at the Cancer Research Laboratory; Royal Victory Infirmary, Newcastle-on-Tyne) in 1937. Photo courtesy of Hans Weil-Malherbe.

he continued, "but I have the feeling that, although I may now perhaps have adequate preparatory training, I do not yet have the special knowledge necessary to work independently."[14]

Krebs discussed the matter with Hopkins, who assured him that Weil would be welcome to come to Cambridge. Then he wrote back to Geneva, "I can use a person like you, and I think there would be a possibility for a career for you in England."[15] He could, of course, offer no paid position, only a place in which to work with him. The modest Weil replied in late September, "Naturally I will come with tremendous pleasure, and whenever you think is appropriate. . . . I do not really know how I have deserved for you to accept me in this way. In any case, I thank you with all my heart. But do not go to too much trouble for me, I do not yet know whether I can actually be useful to you." Weil began studying English in earnest, obtained financial support for the venture from his parents, and arrived in Cambridge at the beginning of November. He became an "external" member of the Biochemistry Department, under Krebs's supervision.[16]

Krebs had been able to arrange for Weil to have laboratory space adjacent to his own benches. When Weil came to the laboratory, the first thing Krebs did was to give him some scientific papers to read. Then he showed him how to write an experimental protocol and organize a laboratory notebook. Weil next had to learn to calibrate a series of Warburg manometers. From then on these became Weil's manometers to work with. During the following weeks Weil learned the methods of manometry and tissue slices, as well as the equally useful technique of preparing a cup of cocoa on a Bunsen burner. Soon he was ready to begin assisting Krebs on his current research projects. During these weeks Weil was impressed that Krebs took his responsibilities as a teacher very seriously. "He never asked you to do something that he hadn't explained completely." Sharing a corner of the laboratory, and the comfort of being able to talk in their native language, Krebs and Weil worked together closely during the next months.[17]

Not long after Weil joined him, Krebs received a reminder that his own situation in Cambridge was precarious. From Paris, Robert Lambert, who had arranged the Rockefeller support that had enabled Krebs to enter Hopkins's laboratory, wrote on December 3, 1933:

> Dear Doctor Krebs,
> In a recent conversation with Prof. Warburg in Berlin the question of your plans for the future was mentioned. Professor Warburg shared my opinion that, unless you were assured of a post at Cambridge, you should consider carefully any good offer to go to America in view of the greater possibilities there for a young man of your training and ability.

After referring to the possibility that Joseph Aub would renew the offer he had made during the previous summer for Krebs to come to Boston, Lambert finished,

> I think you understand from our talk last spring that there is no likelihood of any further contribution from the Foundation to Cambridge University in your behalf. Professor Warburg thought I ought to make this clear to you.[18]

Lambert sent this warning, not because of any loss of confidence in Krebs, but because the number of "deposed scholars" that the Rockefeller Foundation was attempting to support had by then grown so large that the funds designated for that purpose were nearing exhaustion. Two weeks later Lambert wrote Aub,

> I hope very much that you can find it possible to renew your offer to K[rebs]. It is out of the question, of course, for the Foundation to continue to support such young men. Furthermore, there is already a much larger number of German exiles in England than can possibly be absorbed, whereas there is, or ought to be plenty of room in America for those in the Krebs category.[19]

It is not clear whether further support for Krebs was in fact "out of the question" or whether the foundation was merely pressing those who appeared to have opportunities for other support to pursue such alternatives. A memorandum to A.V. Hill, dated October 23, had given a far more favorable prospect: "Re

Krebs there is no reason, with our present policy, why an extension of our aid over a second year and possibly over a third year should not be sympathetically considered. We have a special interest in his field."[20] Whatever the actual situation may have been, Krebs knew only what Lambert had written him. He replied, on December 15,

> I thank you very much for your letter of December 3rd. I am very glad that you made the situation quite clear. Of course I shall do my best to find a place where I could continue my work. It is very doubtful whether there will be a possibility at Cambridge. So I would be very glad if Dr. Aub would repeat his proposal next year.[21]

I

On the same day that he wrote Lambert that he understood how uncertain his chances of remaining in Cambridge were, and 10 days after he had broken off his work on the intermediates of carbohydrate and fatty acid oxidation, Krebs and Hans Weil began an investigation of the intermediates in the decomposition of uric acid. The new work was related in two ways to his own earlier interests, for the process was viewed as an alternative pathway of urea formation and was connected also with the synthesis of uric acid that Theodor Benzinger had studied under his direction. The decomposition of uric acid was, however, itself a well-defined problem that had been investigated extensively by others.

Krebs may have been motivated in part by the need to make a fresh start after the lines of investigation he had been following bogged down. He was probably also provoked to this work in part, however, by an unpleasant affair involving two investigators at the University of Erlangen, Werner Schuler and Wilhelm Reindel. In May 1933, at a time when Benzinger's investigation of uric acid synthesis had been completed, but was not yet published, Schuler published a paper on the synthesis of uric acid in birds which reached almost identical conclusions.[22] To the workers in Freiburg it appeared obvious that the inspiration for Schuler's work had been the brief mention of Benzinger and Krebs's findings that Thannhauser had made during a lecture the previous fall. When Benzinger and Krebs finally published their work in the *Klinische Wochenschrift* in August 1933, therefore, they referred to Schuler's results as a confirmation of their own.[23] In September, Schuler and Reindel published a longer paper in *Hoppe-Seyler's Zeitschrift*, in which they rejected such a characterization on the grounds that they themselves had been the first to publish on the subject.[24] Thannhauser became upset about the situation, and wrote Krebs to say that it would be necessary to repudiate their claims. He offered to do so himself, in order to "protect" Krebs, but thought it would be better if Krebs wished to do it.[25] Krebs wrote Thannhauser back that in his "first indignation" he had considered publishing such a refutation, but that on calmer reflection he had decided not to. It had become clear to him that Schuler was "first a plagiarist, and second, an idiot." A public discussion would only "elevate him to a level that he does not deserve." "For such considerations," Krebs continued,

> I have put the manuscript back into my drawer. Recently I have taken up the uric acid problem again with Weil, and I think that a few publications will relegate Schuler to silence and oblivion.[26]

Characteristically, Krebs began his new investigation of the decomposition of uric acid by testing the conclusions that earlier investigators had drawn.

During the first decade of the century, Wilhelm Wiechowski established that most mammals are able to degrade uric acid to allantoin:

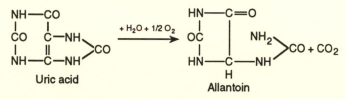

In Wiechowski's experiments allantoin appeared to be excreted without further change. Other experimenters concluded, however, that a further decomposition yielded products such as glyoxylic acid, glycine, and urea.[27]

Among their pioneering studies of respiration in isolated tissues, Battelli and Stern included the conversion of uric acid to allantoin in an investigation they reported in 1909. They found that the reaction could be readily reproduced in tissue extracts, and named the enzyme they assumed to be present in their extracts "uricase." They estimated the quantity of the enzyme in various tissues by measuring the increase in oxygen consumption and carbon dioxide formation brought about by adding uric acid to extracts made from the tissues. When they calculated the respiratory quotients in these experiments they found that many of their results approached the theoretical value for the above equation, CO_2/O_2 = 2. Some of their ratios were, however, considerably lower, indicating that more oxygen had been consumed than required for the conversion of uric acid to allantoin. To explain these anomalies, they inferred that under some circumstances the oxidation of uric acid could "activate" a secondary oxidation of other "easily oxidizable substances" contained in the tissues.[28] Battelli and Stern's work on this problem remained fundamental to further investigation of the action of uricase, but this particular observation was overlooked by most later writers.

In 1925 Stanislaw Przylecki published the results of investigations on the "degradation of uric acid in vertebrates" which extended, and improved upon, the earlier observations on the subject. Employing experiments both on intact animals and on suspensions of tissue ground with sand, Przylecki examined the process in all classes of vertebrates. He showed in particular that amphibians and fish decompose uric acid, as do all mammals—except for those few, including humans, which cannot degrade the compound—through a series of steps that include allantoin. Przylecki formulated the pathway schematically:

Uric acid → hypothetical compound → allantoin
unknown steps → urea, NH₃ and oxalic acid

As discussed in Volume I (pp. 387–388), birds and reptiles were already known to synthesize uric acid from urea. Przylecki examined the question whether this synthesis was the reverse of the process through which fish, amphibians, and mammals decomposed uric acid. Finding that birds cannot produce uric acid from allantoin, he inferred that the two pathways were distinct. For the synthetic process he accepted the theory of Hugo Wiener, that two urea atoms join with a three-carbon chain, producing dialuric acid as an intermediate.[29]

During the next few years a number of investigators turned their attention to the unknown steps in the decomposition of uric acid. In 1927 Richard Fosse isolated from plant juices a "new natural principle." Because the compound was structurally related to allantoin, Fosse named it allantoic acid. Upon heating, plant juices containing allantoic acid readily yielded glyoxylic acid and urea, and he concluded that physiologically allantoic acid yields these compounds by hydrolysis:[30]

$$NH_2.CO.NH{\diagdown} \atop NH_2.CO.NH{\diagup}CH.COOH + H_2O = {} = 2\,NH_2.CO.NH_2 + CH(OH)_2.COOH$$

Allantoic acid   Urea   Glyoxylic acid

In 1929, Fosse and his associates at the Museum of Natural History in Paris discovered in plants a "new enzyme" that hydrolyzed allantoin to give allantoic acid. Shortly afterward they obtained the same reaction in a glycerine extract of frog tissue, and concluded that allantoic acid was the "unknown step" in the scheme of uric acid decomposition as formulated by Przylecki (and that the final products were therefore urea and glyoxylic acid rather than urea, NH₃, and oxalic acid). They gave the name "allantoinase" to the enzyme assumed to convert allantoin to allantoic acid, by a "simple fixation of water." "Thus," they asserted in 1930, when they extended their observations to liver tissue from several animals, "there is established a new, unsuspected mode of degradation of uric acid in the liver of certain animals:"[31]

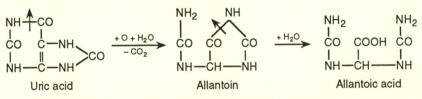

Uric acid   Allantoin   Allantoic acid

In the meantime, Kurt Felix at the University of Munich was concentrating on the first portion of the degradation pathway, the conversion of uric acid to allantoin. He sought to characterize further the properties of Battelli and Stern's "uricase," using extracts of swine liver tissue and dried powders made by treating the extracts with acetone. Felix and his associates found that the optimum pH for the absorption of oxygen differed from that for the splitting off of carbon dioxide, and therefore divided the reaction into two "phases," catalyzed by separate enzymes. They considered, but experimentally refuted, the possibility that the intermediate step connecting the two phases was uroxanic acid. Later, Werner Schuler, who was a student of Felix, pursued the question of what the intermediate might be. By 1933 Schuler had produced suggestive, but not decisive, evidence that the compound in question was hydroxy-acetylene-diureido-carboxylic acid. The first part of the pathway could then be postulated to be as follows:[32]

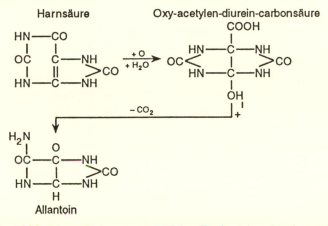

Early in 1933, Hans Kleinman and Heinz Bork claimed to have shown, with more highly purified enzyme preparations than others had used, that uric acid gives rise, via hydroxy-acetylene-diureido-carboxylic acid and hydroxy-acetylene-diurea, by hydrolysis to allantoin and by oxidation to oxalic acid and urea.[33] The consensus at the time Krebs entered this area, however, favored the clearly formulated sequence of four reactions (uric acid → hydroxy-acetylene-diureido-carboxylic acid → allantoin → allantoic acid → glyoxylic acid + urea) emerging from the cumulative investigations especially of Wiechowski, Battelli and Stern, Przylecki, Fosse, Felix, and Schuler. Krebs himself accepted this scheme as the organizing framework for his investigation.

From mid-December 1933, until late April 1934, Krebs directed nearly all of his own research, and that of Hans Weil, to the subject of uric acid decomposition. The two men apparently worked together occasionally on the same experiments, but more often they carried out individually sets of closely related experiments. There is no indication that Krebs divided the question into subproblems of which he pursued some while Weil worked on others; his notebook includes examples of all of the basic types of experiment involved in the investigation. Some of Weil's experiments were also recorded by Krebs in his notebook. It is evident, however, that Weil must have carried out many more experiments of which no direct record has survived. It is therefore not possible to reconstruct reliably the day-to-day course of the investigation. If we assume that the experiments entered in Krebs's notebook are representative of the general stages of the research, however, we can sketch in several phases in its development.

The first recorded experiment on the topic, on December 15, was performed by Weil. It was very similar to what earlier workers had done, differing mainly in the use of Krebs's manometric urease method for estimating the urea formed. The immediate subject of the experiment was "urea formation from allantoin." Weil prepared an extract of Hungarian frog liver tissue ground with sand. He added 1 mg of allantoin to each of seven vessels containing this extract, together with a buffer solution. To three of the vessels he added varied quantities of glycine. In each of the extracts substantial quantities of urea formed, in contrast to controls with extract alone and with allantoin alone, in which little or no urea formed. Calculating that the quantity of urea "preformed" in 1 mg of allantoin was equivalent to 142 cm^3 of CO_2 evolved in the urease method, and that in three of the vessels (two of which contained no glycine), more than 200 cm^3 evolved, Krebs noted, "Therefore, more than 1 equivalent split off!!"[34] Although the experiment generally confirmed the accepted view that frog liver contains an enzyme catalyzing the conversion of allantoin to urea, it was problematic in that more urea had appeared than could be accounted for on the normal assumption that one molecule of allantoin yields two molecules of urea.

During the remaining working week in December, Krebs tested slices from various frog organs, to establish which tissues can form urea from uric acid. Only liver and kidney slices performed the decomposition. With liver extracts he examined further the cleavage of allantoin and also of allantinoic acid to form urea. On December 22 he was able to record that from both compounds the extract yielded, in 20 hours, "about 90% of the expected" quantity of urea.[35] With this satisfying confirmation of the generally accepted pathway, Krebs ended his research for the eventful year of 1933.

II

Even as he established a new life in England, Krebs kept in touch with his old life through a network of correspondence, particularly with friends and associates from Altona and Freiburg. Some of them had also been forced out of Germany. Franz Bielschowsky had gone to Holland, where he was able to work in a laboratory temporarily. Bielschowsky wrote Krebs regularly about his plans, his work, and the whereabouts of other former colleagues. In the fall of 1933 he visited Krebs in England. Shortly afterward he reported that he had been invited to go to Spain, in January 1934, as chief of the division of metabolic diseases in a clinic.[36] Siegfried Thannhauser remained as head of the Medical Clinic in Freiburg, but he wrote Krebs that he did not know "how long will I be able to defend myself in my position?" Legally spared from dismissal as a Jew, because of his record of wartime military service, Thannhauser was nevertheless anathema to the Nazis, because his prominent position gave him an "influence in the community." Systematically they set out to undermine him, but Thannhauser managed his clinic so ably and so prudently that they found it difficult to fabricate an excuse to deprive him of his post.[37]

Hermann Blaschko remained in London during the fall of 1933. In October he suffered a recurrence of symptoms of his tuberculosis. His aunt, Helene Nauheim, was worried enough about him to write Krebs asking him to come to consult with them. Blaschko himself wrote Krebs frequently between October and December, embarrassed about troubling Krebs, but obviously depressed about his condition, as well as about the worsening illness of his sister in Europe. Krebs spent his first Christmas holiday away from home with Blaschko and his relatives in London.[38]

Sometime late in 1933 or early in 1934, Krebs attained a special landmark: the completion of his first scientific paper written in English. Entitled "Urea Formation in the Animal Body," it was not a report of new research, but a review article—ironically for the German journal *Ergebnisse der Enzymforschung*—based largely upon his earlier work in Germany. It had probably taken him a long time to write it, and he had had to seek help from others to put it into good English. Some sections of the article were nearly direct translations of passages from his previous German papers on urea synthesis. The phrase that he had used to describe his theory in German—*Kreislauf des Ornithins*—he translated as "cycle of ornithine."[39] The word "cycle" was not the only, or even the most obvious choice, for the literal meaning of *Kreislauf* was closer to "circulation." Nor did Krebs select that word himself, but found when he came to Cambridge that British biochemists were already using it for his term *Kreislauf*.[40] Soon the theory was called simply the "ornithine cycle."

Krebs spent much of January 1934 examining the chemical properties of allantoin, allantoic acid, and related compounds. Following methods published by Fosse and by Leopold Cerceredo, who had investigated purine and pyrimidine metabolism extensively in dogs, Krebs devised procedures for determining quantitatively allantoin, allantoic acid, and uric acid by means of the urea they yield on alkaline and acid hydrolysis. As usual he adapted such methods to his manometric urease method for urea. Employing these methods, he studied the time course of the spontaneous cleavage of allantoic acid at various hydrogen ion concentrations. Meanwhile, he continued experiments on the metabolic formation of urea from uric acid and from allantoin, utilizing sometimes tissue slices and sometimes extracts.[41] On January 9 he not only measured the urea yielded by allantoin in frog liver extract, but determined quantitatively the glyoxylic acid formed, by precipitating it as a dinitrophenylhydrazone. The proportions of the two products were "almost theoretical," but he suspected that this overly perfect outcome was an "accident." Subsequent experiments performed by Weil must have confirmed it, however, for Krebs reported afterward that they had found glyoxylic acid "quantitatively" in such experiments.[42]

On January 15 Krebs tested allantoin in frog liver extract in an atmosphere of hydrogen, as well as in oxygen. The quantity of urea produced under the two conditions was "exactly the same." This result, which he confirmed in subsequent experiments, spoke against a conclusion that Przylecki had drawn, that allantoin is decomposed in an oxidative process. Allantoinase and allantoicase were, Krebs inferred, not oxidative, but hydrolyzing enzymes that can operate anaerobically. During the same set of experiments he found that the oxidation of uric acid was complete at a time at which only one-fourth of the theoretical yield of urea had appeared. He concluded that "uricase acts much more rapidly than does allantoicase."[43]

On January 24 Krebs examined the formation of urea in frog liver slices in the presence of parabanic and dialuric acid. The most likely reason he did so was to test the specificity of allantoicase by seeing to what extent the liver formed urea from other ureido compounds. In this experiment both compounds exerted a noticeable effect (with no addition, Q_{urea} = 0.30; with parabanic acid, 2.70; with dialuric acid, 1.74). During the following days he prepared other related compounds, including hydantoic, isobarbituric, oxaluric, and allantoxanic acid, and carbonyldiurea. Weil must have tested these, as well as several other compounds. The experiment of January 24, Krebs noted, was "parallel with Weil's experiment." In the end, Krebs concluded that, besides uric acid, allantoin, and allantoic acid, only two other substances "split off urea enzymatically in the frog liver." These were parabanic and oxaluric acid.[44]

Through most of February Krebs and Weil continued their investigation along the same general lines.[45] On February 26 they were scheduled to speak at the weekly Biochemical Tea Club. Each Monday afternoon at this informal gathering, some member of the department presented the latest results of his or her work. Normally lively discussions took place, in which Hopkins presided and often asked probing questions or provided illuminating insights. When the announcement of the series for the Lent Term had been typed up in January, Krebs and Weil had put down that they would speak on the "Metabolism of Pyrimidines and Purines."[46] No details of their presentation are recorded, but it must have been mainly a progress report on the first two months of their research on uric acid decomposition.

Just after Krebs and Weil spoke at the tea club, their investigation took a new turn. On March 3 Weil carried out a set of experiments entitled "Uricase-End Values." He compared the oxygen consumed in a fresh frog liver extract to which he had added an exactly measured quantity of uric acid with that in the extract without uric acid, maintaining a pH of around 7.9–8.0 with a veronal buffer. In the same set of experiments he measured the oxygen consumption of frog liver slices in the presence of uric acid and allantoin. For the extract vessels he continued the measurements for 20 hours, until there was no further change. The difference between the oxygen consumed in the extracts with and without uric acid was 322 cm³ Krebs calculated that, according to the assumed oxidation reaction for uric acid, in which one atom of oxygen is consumed for each molecule of uric acid, the total quantity of uric acid present ought only to have used 224 cm³ of oxygen. That is, there was "44% too much" oxygen consumed. The situation in the tissue slice experiment was also anomalous. Krebs put down in his notebook, as the results of Weil's experiment:

Fresh extract: Too much O_2 taken up.
Slices: Almost 1 molecule taken up [with uric acid] instead of 1 atom.
No extra uptake with allantoin!![47]

The result with allantoin was probably unproblematic, for it would by now have confirmed Krebs's earlier finding that allantoin is decomposed anaerobically. The results for uric acid appear, however, from Krebs's repeated use of the phrase "too much," to have been not only problematic, but unexpected.

This result ought not to have come as a surprise to Krebs, because it was in keeping with what Battelli and Stern had reported 35 years before. In the majority of their experiments with uricase the oxygen consumption fit the view that one atom oxidizes each atom of uric acid; but under some conditions more oxygen was consumed. In spite of the way in which Krebs described his result, therefore, it is possible that he was looking for it. On the other hand, like other investigators in the intervening years, Krebs may not have been attentive to this particular aspect of Battelli and Stern's paper until he encountered the same phenomenon. In any case, the question quickly became of central interest to him.

In an experiment carried out on March 7 "together with Weil," on "the influence of salts on uric acid oxidation in slices," Krebs again compared the observed and calculated oxygen consumption, and found this time "only 1 O taken up;" but in a similar experiment that Weil performed on the next day, under one of the conditions tested (23°C, and in the presence of glucose), there were "almost 2 Molecules O taken up." Krebs wondered if bacteria might have been the cause.[48]   On March 9 Krebs tested the effects of uric acid and of hydroxy-acetylene-diureido-carboxylic acid on the oxygen consumption of frog liver slices. The latter substance he prepared by following the procedures of Schuler, who, as we have seen, had provided evidence that this compound is the intermediate step in the conversion of uric acid to allantoin. The result he recorded was

With uric acid 1 molecule O_2 uptake. The hydroxy-acetylene [diureido carboxylic acid] experiment is uncertain, since too much O_2 was taken up, and the extra uptake began late.

The first statement, concerning uric acid, probably contains a slip. According to Krebs's calculation, the amount of extra oxygen consumed was slightly less than the theoretical quantity for the oxidation of uric acid. He must therefore have intended to write "1 atom O extra uptake." His calculation for hydroxy-acetylene-diureido-carboxylic acid showed that the theoretical quantity was 224 cm³, whereas the measured quantity was 262 cm³, "and goes on!!"[49] that is, oxygen was still being consumed when he ended the experiment. He apparently expected that, like uric acid, the intermediate compound should have taken up just one atom of oxygen per molecule. His reasoning itself seems anomalous, for, according to the pathway formulated by Felix and Schuler (see above, p. 53), oxygen is consumed only in the formation of hydroxy-acetylene-diureido-carboxylic acid from uric acid. In its further decomposition CO_2 is given off, but no more oxygen taken up. It ought, therefore, to have been surprising that the presence of that compound caused *any* increase in the consumption of oxygen.

If we assume that the experiments beginning on March 3 were the first ones in which Krebs noticed the excess consumption of oxygen in the decomposition of uric acid, and in particular that the experiment of March 9 was the first to include hydroxy-acetylene-diureido-carboxylic acid, then he must, within the next two days, have had a substantial flash of insight. On March 11, in a letter to Thannhauser, Krebs wrote that he had "found" a new alternative pathway for uric acid decomposition, the rationale for which must have derived in part from these experiments.

> The work is making very good progress. Weil is a valuable and
> pleasant co-worker. Aside from fatty acid decomposition, we have
> recently occupied ourselves mainly with the biological decomposition
> of uric acid. Specifically, amphibians and frogs possess (Przylecki
> 1929) uricolytic enzymes, which release the entire nitrogen of uric
> acid as urea. The pathway of decomposition, or, better, one pathway,
> has essentially been found by Fosse. It passes by way of hydroxy--
> acetylene-diureido-carboxylic acid, allantoin, and allantoic acid, to
> urea and glyoxylic acid. We have now found, surprisingly [*überrasch-
> ende Weise*], that there exists a second, preferred path of decomposi-
> tion in living cells, in that the hydroxy-acetylene-diureido-carboxylic
> acid is further oxidized, and yields urea and parabanic acid. Parabanic
> acid decomposes into oxalic acid and urea.[50]

Krebs then drew the reaction scheme. It is not reproduced on the surviving carbon copy of his letter but, from a further elaborated scheme that he subsequently included in a manuscript on the subject, we can (by omitting the steps not described in the above paragraph), reconstruct what he probably sketched out:[51]

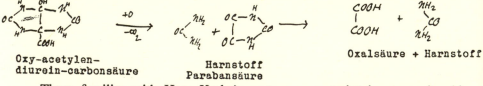

Oxy-acetylen-   Harnstoff   Oxalsäure + Harnstoff
diurein-carbonsäure   Parabansäure

Those familiar with Hans Krebs's customary restraint in expressing his feelings about scientific matters may recognize in this paragraph a tone of enthusiasm, if not outright excitement. One can sense that he felt something quite important had turned up. Each of his earlier scientific successes had emerged from a "surprising" result; it seems evident that in describing this new surprise to his old chief he anticipated that it, too, would lead him to a significant discovery—to another hitherto unknown metabolic pathway.

A difficulty with the preceding interpretation of the events leading Krebs to this position is that the experimental results described above do not provide a solid demonstration that the second pathway of uric acid decomposition "exists." All that can be found in Krebs's laboratory notebook up until this date are indications that uric acid sometimes consumes more oxygen than is required to convert it to allantoin; a single, uncertain indication that hydroxy-acetylene-diureido-carboxylic acid may be oxidized (a result that may have taken on greater significance for Krebs if he came to see, during the intervening two days, that according to the generally accepted pathway that compound should not be further oxidized); and one experiment showing that parabanic acid gives rise to urea in frog liver slices. Perhaps Krebs based his case on extensive additional experiments by Weil, not recorded in his own notebooks, and the picture, as he depicted it here, had emerged more gradually in his mind. Subsequent events, however, fit the view that he did not have more substantial evidence at hand at this point; that he may well have seen, between these suggestive results, connections which fit together so well that he was, when he wrote Thannhauser, persuaded that further work would surely confirm the existence of the second pathway.

In the next paragraph of his letter Krebs wrote with equal enthusiasm about further implications he expected to emerge from his current work:

> I think that knowledge of biological uricolysis provides important leads [*Fingerzeige*] for the reverse process, the synthesis; not, of course, in the sense that the synthesis is exactly the reverse of the decomposition (which definitely does not follow), but perhaps to the extent that one ought to investigate whether the synthesis does not begin with a C_2-compound (instead of the formerly accepted C_3-compound). That is one of the next points we shall take up in our research program. . . . It will probably take some time before we publish. I would, therefore ask you for the present not to make our results further known.[52]

Perhaps Krebs included the request for discretion because of his earlier displeasure when Thannhauser had divulged prematurely Benzinger's results on uric acid synthesis. That he nevertheless confided his latest ideas and plans to Thannhauser implies that, on the whole, there existed firm bonds of mutual trust and intellectual respect between them. Krebs began his letter, "*Sehr verehrten, lieber Herr Professor,*" a mixed salutation that blended the proper respect due to a senior with the warmth of a more personal form.

Whether or not Krebs had from the beginning viewed his investigation of uric acid decomposition as a means to the end of elucidating the synthesis of uric acid, it is clear from his letter that he was now seeking ways to follow up on the results of Benzinger's work in Freiburg. Krebs may have held back on such plans earlier, to allow Benzinger the opportunity to do so himself. During the preceding October, Benzinger had written from Königsberg, concerning the "uric acid problem"—"If you do not yourself intend for the time being to do further experiments with pigeons, I would very much like to take up these trails again." Benzinger's chief had no interest in his pursuing such research, however, and

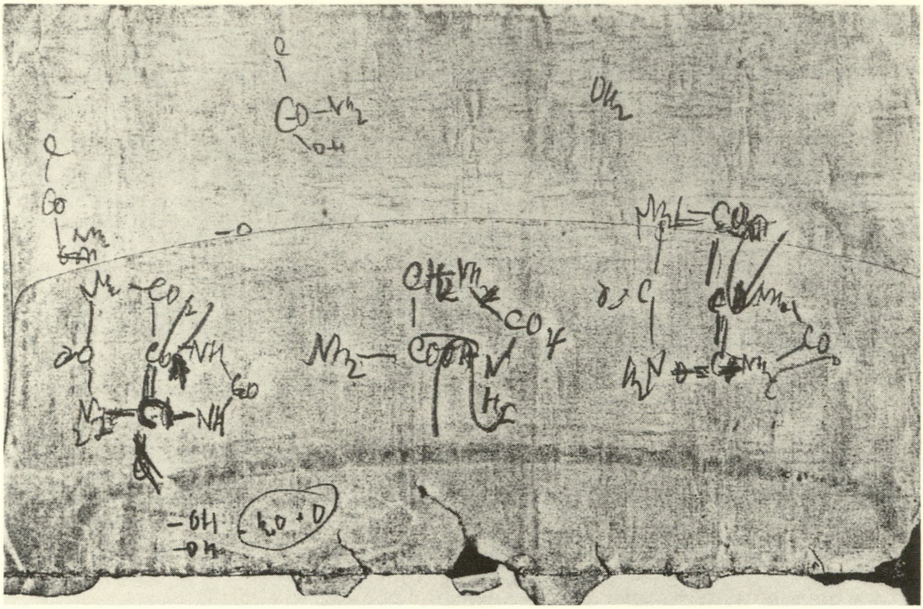

Figure 2-2 Envelope addressed to Hans Krebs, with chemical formulas sketched by him.

at the end of the year he was still unable to do anything about his plans.[53]   By the spring of 1934, therefore, Krebs would have felt free to pursue this aspect of the subject himself.   Benzinger had shown, we may recall, that neither the tartronic acid postulated by Wiener, nor any other plausible three-carbon compound, was a precursor of uric acid in birds.   What Krebs now saw was that if the products of uric acid decomposition included two-carbon compounds—glyoxylic and oxalic acid—then these compounds might also be involved in the reverse process.   Przylecki's conclusion that the synthesis and decomposition of uric acid followed separate pathways was no longer cogent, because he had based that view on the Wiener theory of synthesis now disproven by Benzinger's work.   A two-carbon compound did not lend itself as readily to a hypothetical reaction sequence as a three-carbon intermediate had done (for the Wiener scheme see Vol. I, p. 388), and Krebs probably had no specific scheme in mind.   What he perceived was, as his letter suggests, probably no more than a simple "lead" for an investigative program.   That was how he normally proceeded.

When Krebs reasoned about possible metabolic pathways, he habitually did so with paper in hand, jotting down formulas or pieces of formulas, to help him visualize what chemical transformations might take place.   He utilized for this purpose whatever scraps of paper might be at hand, and most of them quickly disappeared.   There is, however, a rare surviving trace of his thinking with pencil in hand, dating from just this period.   On the back of an envelope addressed to him, and postmarked January 25, 1934, Krebs sketched the following (see Fig. 2-2, previous page).[54] The formulas are of uric acid and related compounds. I have not been able to identify a specific set of reactions that he might have had in mind here, but it seems clear, in a general way, that he was contemplating possible steps in a decomposition pathway for uric acid. The envelope provides a nice glimpse of the way he thought about such problems—how he tried to picture the possible changes by circling atoms that might be removed together. When he discussed metabolic reactions with me many years later, he almost invariably reached for the nearest piece of paper and sketched out similar formulas. Not only did he visualize chemical compounds and reactions very much as they are depicted on paper, but he could not think about them for very long without representing his thoughts on paper.

At the time Krebs wrote Thannhauser, his mood was buoyed by more than his optimism about the current state of his research. The cloud that had hovered over his situation at Cambridge was also lifting. When he had received the news from Lambert in December that he could not expect a renewal of his Rockefeller support, he had showed the letter to Hopkins. Hopkins, who expressed publicly in a lecture in London in January "how glad he was that such a distinguished biologist was now in his laboratory,"[55] quickly set about trying to find other means to keep Krebs in Cambridge. By March Hopkins's efforts were beginning to bear fruit. Krebs reported to Thannhauser,

> Things continue to go splendidly for me here. The uncertainty about my future is also gradually disappearing. I hope, with great confidence, to receive a regular position here as an instructor after the expiration of the Rockefeller year. By good fortune, a place in the Institute as "Demonstrator in Biochemistry" has become vacant, and Hopkins is doing everything to procure the post for me. The decision rests with the Faculty, but Hopkins tells me, in this regard, "I don't think there is any doubt." It would be the first time that a German emigrant received a regular position. There are 2–3 dozen German scientists here; however, none of them yet occupies an official position. They receive, without exception, stipends arranged on an ad hoc basis. My case has therefore brought the University many headaches over considerations of principle.[56]

A week later Lambert visited Cambridge and spoke with Krebs. Afterward Lambert noted in his diary that Krebs

> thanks me for a letter I wrote him some months ago in which I gave warning that the RF would not support him a second year as we felt certain he could get a place in America. (Repetition of J. Aub's offer was expected.) K. says my letter brought matters to a head in Cambridge and that Hopkins has told him it is practically certain he will be given a regular demonstratorship. The appointment will be made in May. K. is naturally quite happy.[57]

Krebs had every reason to be happy. His professional prospects had never looked brighter. He had earned those prospects; but he had also been remarkably fortunate in those who protected, nurtured, and promoted his career in the face of so many unfavorable external circumstances.

III

During the two weeks following his letter to Thannhauser, Krebs concentrated nearly all of his attention in the laboratory on the question of the excess oxygen taken up during the decomposition of uric acid in frog liver tissue. On March 12, he tested liver extracts with and without octylalcohol added. The former rates of oxygen uptake being much higher, he again suspected bacterial contamination. On March 15, utilizing large liver slices at a pH of 7.4, he confirmed that there was "certainly more than 1 O taken up, especially with glucose added." With Hungarian frog liver slices at pH 7.6, he obtained on the next day a similar result. On the 15th he tested the influence of pH, varied by using different bicarbonate concentrations, and found that "with a somewhat more alkaline solution 2 atoms O per molecule of uric acid are taken up (in fact, somewhat more)." On the 19th he got another similar result. The next day he tested the effects of glucose again, but observed "too little O_2-uptake," probably, he thought, because he had used either too much glucose or too much tissue. The pattern of surplus oxygen consumption returned on the 23rd, when, with slices, he measured more than twice the quantity of O_2 consumption required to oxidize uric acid to urea according to the expected reaction pathway. The extra oxidation thus appeared by this time to be well confirmed. In his notebook he recorded "2 O taken up. 1 rapidly. 2 slowly. urea appears very late." The result thus appeared suggestive of the possibility that the first and second atoms of oxygen were taken up in different reactions. There is, however, no further indication of whether or not Krebs believed that he had acquired support for his view that there might be a second pathway of uric acid decomposition in which hydroxyacetylene-diureido-carboxylic acid is further oxidized. After performing one more such experiment, on the 24th, in which "almost 2 atoms" of oxygen were taken up, he interrupted his research for 15 days.[58]

At about this time Krebs received from his friend Ernst Frankel, who had gone to South Africa, a light-hearted report on how far his reputation had now spread. "Imagine yourself," Frankel wrote on March 17,

> as the famous Krebs from Cambridge, whom I have mentioned repeatedly here. Yes, you, noble Baron von Hildesheim, you are slowly but surely becoming known even here, in the neighborhood of the South Pole. Recently Carr spoke of your work [in his lecture] as though you were Hopkins himself. I have naturally given anyone who knows so much as what urea is, hour-long discourses about you and your work.[59]

Just as his future at Cambridge seemed nearly settled, Krebs was suddenly confronted with new opportunities to go elsewhere. On March 19, Joseph Aub wrote him, "Would you be interested in a possible position here at the Huntington Hospital next year? I wrote to you last year and you suggested you might be interested in the possibility next year." Aub still could not promise a position, but his comment "we should be delighted to have you here working with us" implies that the chances were good. "Inasmuch as the time is growing short," he added, "and I will have to find out whether I can arrange the finances if you would like to come, may I suggest that you cable your answer collect to me here?" Aub did not expect a "final answer from you until I can make a more definite offer," but he did not wish to go to the effort to raise the funds unless Krebs was interested.[60]

Four days later, Judah Magnes, chancellor of the Hebrew University in Jerusalem, opened for Krebs the prospect of moving in a very different direction:

Dear Dr. Krebs,
The Hebrew University has the promise of a sum of money for the creation of an Institute which shall deal with research in cancer problems.

After enumerating four sections that were planned within the new institute, Magnes wrote,

The purpose of this letter is to ask you if you would come to the Hebrew University as the head of the Division in Cell Chemistry and Metabolism.
Your name has been given by Prof. Otto Warburg and others, and we are very anxious to have you come here. . . . We hope that you are sufficiently interested in the building-up of the Hebrew University, especially in these days of anti-Semitism, to wish to come here. Although we are not able to promise anything luxurious, we hope that it will be possible to create an effective even though modest Institute.[61]

Krebs's responses to the two proposals reaching him almost simultaneously were notably different. To Aub he cabled on April 6, "Obtain position here. Krebs"; and on the same day he wrote,

My situation here is at present this: The grant of the Rockefeller Foundation expires in July. Prof. Hopkins told me several weeks ago that he would like to keep me here and he asked me whether I would accept a post as demonstrator in biochemistry. I answered, of course, in the affirmative. But this affair is not settled yet (and therefore I had not written to you about it). It will be decided by the University early in May. Prof. Hopkins told me that it looks very hopeful. I shall let you know the final decision as soon as possible.

On the one hand I feel very sorry that I should not come to Boston and work in your place, but you will certainly understand, that I would accept the demonstratorship since it is a permanent post which also offers good possibilities for the future.[62]

To Magnes, on the other hand, he had written three days earlier,

I thank you very much for your letter. I have always greatly sympathized with the Hebrew University and I am really fascinated by the idea that I could take part in the building-up of the University.

You will certainly understand that I would like to know some details about the institute before giving the final decision. Therefore I would be very grateful to you if you would let me know something about the allowances for equipment, assistance, current expenses and my personal salary. And when will the Institute be ready for work?

I am anxiously waiting for your reply and I hope that things will be satisfactorily settled at an early date.[63]

Thus, at the very time that he was declining an offer to go to the United States on the grounds that he was ready to accept in Cambridge a position that promised to be permanent, Krebs also appeared ready to accept an offer to leave Cambridge in order to play a leading role in the formation of a new institute in Palestine.

Returning on April 9 to his investigation of uric acid decomposition, Krebs carried on for the next five days experiments on conditions that might influence the "extra" uptake of oxygen. Varying the alkalinity, he confirmed that, with extracts, a pH above 7.4 favored the uptake of more than one atom of oxygen per molecule of uric acid consumed. In extracts from two mammalian livers, swine and rat, there was also somewhat more than the calculated quantity. He tested the uricase preparation Battelli and Stern had used, and obtained "18% more than 1 O."[64] All of these results showed that "the quantity of oxygen taken up per molecule of uric acid depends on the origin and treatment of the enzyme material," but did not elucidate Krebs's hypothesis that a second oxidative pathway of uric acid decomposition exists.

On April 14, Krebs devised a way to test his view directly. If there were a second pathway, then the decomposition of uric acid in liver extract should yield less allantoin than indicated by the primary reaction sequence: uric acid → hydroxy-acetylene-diureido-carboxylic acid → allantoin. Utilizing swine liver extract under conditions in which he had previously found that two atoms of oxygen were taken up per mole of uric acid consumed (he mistakenly stated in his notebook "2 mol. O_2 uptake"), he analyzed the solution afterward for allantoin. The amount formed, according to the titration method he used, was "quantitative"—that is, almost exactly the quantity calculated for the total consumption of the uric acid present at the beginning. None of the products of his hypothetical second reaction path—he tested for oxalic and parabanic acid—was detectable. Whether or not Krebs abandoned his hypothesis immediately in the wake of this outcome, he was soon afterward ready to interpret the result as showing "that the extra oxygen taken up is not applied to the further oxidation of uric acid or allantoin."[65]

The demise of his own theory forced Krebs back to the only other interpretation of the extra oxygen uptake that appeared possible to him. That was the original conclusion of Battelli and Stern, that there is an oxidation "coupled" with the oxidation of uric acid, such that half of the oxygen serves to oxidize the uric acid, while the other half is transferred to some other substrate. On April 17 Krebs tested some possible substrates—glucose, butyric acid, and dl-alanine-by comparing the quantities of oxygen taken up by frog liver extracts with uric acid alone and in the presence of these additions. Of these three substances, he noted, "only dl-alanine has perhaps an effect (deamination??)." It is most likely that Weil was at this time testing several other substances, including hexosemonophosphate, methionine, dihydroxyphenylalanine, and lactic acid, with similarly negative or questionable results.[66]

By this time Krebs had probably completed the manuscript for a paper entitled "Investigations of the Uricolytic Enzymes (Uricase, Allantoinase, Allantoicase)," jointly authored by himself and Weil.[67] He intended the paper to be his contribution to a volume of articles scheduled to be published in Russia in 1935 to celebrate 30 years of scientific activity by Lina Stern. Considering that the most salient experimental finding Krebs had to offer was the verification of the extra oxidation first reported by Battelli and Stern 25 years earlier, his subject was quite appropriate to a publication in her honor. In his paper Krebs acknowledged that "our experiments to find the substances [oxidized in the coupled oxidation] have up until now been unsuccessful."[68]

In the paper Krebs also described the attempt to identify a "second path of uric acid decomposition in which the intermediate products would be parabanic and oxaluric acid." He had now expanded this hypothetical pathway to the following scheme:

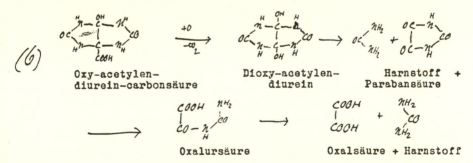

"All experiments to prove the formation of oxalic acid or parabanic acid in the frog liver," however, "yielded an unambiguous negative result, whereas glyoxylic acid [one of the products of the generally accepted primary pathway] was found quantitatively. We have therefore given up the idea of a second pathway of uric acid decomposition by way of parabanic acid."[69] On March 10 Krebs had written Thannhauser in the belief that he had "found" that second pathway. It had therefore taken about a month, for what had then appeared found, to be lost.

Unable to confirm either of the possible "meanings" of the extra oxygen consumption, Krebs and Weil had little novel to present in their paper. The 20-page typewritten manuscript provided mainly supporting evidence for the reaction scheme, developed through the work of other biochemists, that had been Krebs's starting point. Although he eventually received page proofs of his article, the volume dedicated to Lina Stern never appeared, and he did not publish these results anywhere else.

* * *

In 1977 I asked Krebs whether there were any times after the publication of his papers on urea synthesis when he wondered whether he would be able to sustain the standard set by that discovery. He replied, "No, this ornithine paper, which seemed to me straightforward and simple, gave me confidence." When I pursued the question by asking if there were subsequent periods in which things did not go so well, or were slow, he answered,

> Yes, there were, but it didn't disturb me, because my outlook was that one cannot expect that every piece of work is an outstanding discovery; but one can expect that every piece of work leads to some sound, publishable, new information.
>
> And one has to be satisfied with producing just that sort of thing. And if you do that continuously, you may produce sometimes something which is outstanding and not just sound, but is of special value. I know there are some people who feel they must keep up a certain standard. Of course, I felt also I must keep up a standard of effort; but not necessarily expect a standard of achievement, because that includes an element of luck.
>
> And I felt I had enough ideas and had the necessary methodological skill to remain productive. So I was never worried by inferiority complexes.[70]

For a short time in the spring of 1934 Krebs thought that the work on uric acid decomposition would be an outstanding discovery, but it turned out barely to reach the category of "sound, publishable, new information." That was the most he had to show for his first nine months of research in Cambridge. His recollection that such periods never worried him was undoubtedly colored in part by the retrospective knowledge that "in the following years work went extremely well."[71] We cannot be certain that he was in 1934 as totally serene as he later portrayed himself. Nevertheless, there is no contemporary record indicating that he ever became unduly anxious about his research progress.

* * *

In view of the negative description in Krebs's manuscript of his effort to identify the substrates involved in an oxidation coupled with that of uric acid, it is surprising to find in his notebook, for April 18, the following experiment: "Rat liver + phenylenediamine + uricase. following Keilin." Measuring the oxygen uptake of frog liver extracts in the presence of uric acid, with and without phenylenediamine, Krebs observed that with uric acid alone the end quantity was 229 cm³—slightly more than the theoretical 224 cm³ required to oxidize uric acid in accordance with its primary decomposition pathway—whereas with uric acid and phenylenediamine the quantity was 491 cm³ (both of these values were obtained by subtracting from the measured values those obtained in control experiments without uric acid). "With p-phenylenediamine," Krebs wrote at the bottom of the page, "it is therefore somewhat <u>more</u> than doubled." On the next page he noted that glyoxylic acid, methionine, and casein digested with pepsin caused "<u>no</u> extra uptake of oxygen."[72]

On the surface of the matter it would appear that Krebs may here have detected one of the possible coupled substrate oxidations, yet he neither revised his manuscript to mention it nor carried out further experiments on the subject. (There is no record of whether or not Weil did.) Why did Krebs and Weil apparently discontinue the search just when they had obtained the first sign of a potential successful outcome? The contemporary record gives no clue; but from later recollections of Krebs and of Weil, the probable explanation emerges. To my question, asked in 1977, was "your interest in coupled oxidations . . . to any degree stimulated by [David] Keilin's experiments," Krebs answered, "Yes . . . I think he . . . made the observation . . . that the oxidation of uric acid can be coupled with ethanol."[73] Weil recalled more vividly (after reading the above account in 1986), that "it was Keilin who suggested to us that the phenomenon was due to a 'coupled oxidation' of the type observed with peroxidases." The discussion occurred on one of the rare occasions in which Keilin walked across from the Molteno Institute to visit Krebs at his workbench. These recollections are not specific enough to link Keilin's suggestion directly with the above experiment; but the phrase in the title of the experiment "following Keilin" makes such a connection plausible, and an acknowledgment at the end of the manuscript on uricolytic enzymes thanking Keilin "for many suggestions" implies that Keilin provided Krebs with significant guidance about the problem of coupled oxidations. Perhaps he even suggested the use of phenylenediamine (although it was a substance known since the work of Battelli and Stern to be readily oxidizable in tissues). As Krebs became aware, however, that Keilin was himself deeply involved in the study of coupled oxidations, he decided not to pursue an investigation that might encroach on the interests of a senior colleague whose warmth and generosity he valued. He told Weil that "there is no point in going on with this. This is Keilin's territory and I wouldn't like to impinge on that." Weil was greatly disappointed, because it meant that all the research they had carried out over the past six months was going nowhere and would have to be abandoned.[74]

IV

While Krebs flourished in the open environment of Cambridge, his old chief in Freiburg struggled to survive in the oppressive environment of Nazi Germany. After various efforts to compromise Thannhauser through allegations of financial irregularities and anti-Nazi remarks had failed, the regime found a way to dislodge him through a legal provision that any civil servant could be required to exchange his post for an equivalent one at another place. Told that he must give up his professorship at Freiburg for a position as assistant in a clinic at Heidelberg, Thannhauser tried to defend himself through a provision in the law that the new post must be equal to the old one. In Heidelberg he would have no authority and no professional opportunities corresponding to those of his present position. When that argument was ignored, he resorted to another provision of the law, that he had the right to request retirement in lieu of the position offered. He thwarted an attempt to rule that he had not applied for retirement within the time limit provided by that law, by arguing that he could not make such a decision until he had been informed of the conditions of the Heidelberg position. On this point his tenacious defense prevailed, and he was able to retire with the pension due to him from his current position. He was also able to retain some of his laboratory apparatus, by persuading the authorities that it had been supplied to him personally through his Rockefeller grant.[75] On April 21, 1934, Thannhauser wrote to Krebs that he had been placed on leave and explained some of the circumstances. He had made arrangements to continue to see his own patients in a private clinic and to carry on some scientific research in a private laboratory placed at his disposal. He realized that these arrangements were only a temporary expedient, however, and when he wrote Krebs he was in Paris to discuss with Rockefeller officials the possibility that a position could be found for him in the United States.[76] Franz Bielschowsky commented in a letter to Krebs in June that Thannhauser's dismissal had come "one year, almost to the day, after their own."[77]

    Krebs's own father was also feeling the adverse effects of National Socialist policies. The Nazis were intent on driving Jewish practitioners from positions of influence and reducing their numbers to a figure proportional to the overall proportion of the Jewish population in Germany. An order of the Reich Ministry of Labor in April 1933 had excluded non-Aryan physicians from public and private health-insurance practice, leaving them access only to private patients. Georg Krebs lost his official position as a general throat surgeon at the municipal hospital. Jewish doctors could not be dismissed from their own private practices, but propaganda campaigns exhorted Germans to avoid them.[78] Despite the great good will that he had enjoyed among his patients, Georg Krebs found by the spring of 1934 that fewer and fewer of them were coming to his office. Early in April he complained in a letter to Hans:

> My practice has gone particularly badly since Easter. You can well-imagine the grounds for that. For me this is doubly dark. In the first place, because, as a practitioner and technician I feel, in spite of

my 67 years, quite at my peak; second, because, as an old German patriot, I feel the defamation especially strongly.[79]

The elder Krebs had additional, personal reasons to be depressed. At the beginning of March Wolf had married Lotte Rittmann. His father was opposed to the marriage, a situation that led to strains between the two. Two weeks after that event Hans received a letter from his sister, Lise Daniel, who stated that, "you cannot imagine <u>how</u> unhappy father is."[80] Earlier, Hans too had come under a cloud, because his father came to feel that his eldest son, who had earlier opposed his own remarriage, remained cool toward his second wife, Maria, and their daughter Gisela. A letter from Hans in February, assuring his father that he felt he belonged to the family, brought his harassed father "great relief and joy."[81] In March Lise wrote Hans that their father's "only joys are his little daughter and his big son."[82] When Hans wrote home on April 3, mentioning the offer from Jerusalem, his father responded enthusiastically:

> You really do have, as you once put it, an "international value," when you have once again received an offer from another country. The family is proud of you.[83]

Despite his pride in his son's reputation, Georg Krebs had reservations about the post itself. He went on,

> Does the situation in Jerusalem involve an English or a Jewish institution? If it is the latter, wouldn't you experience difficulties because of your non-Jewishness [*Nichtjudentum*]?[84]

Thus, in defiance of all that had happened in Germany, the staunch old advocate of assimilation still viewed the son whom he had raised outside the established Jewish religion as a non-Jew. He remained, in fact, a confirmed opponent of Zionism. Nevertheless, he trusted that the son who had managed his life so successfully up until now would make the "right decision."[85]

During April the prospect grew quickly that Krebs would depart Cambridge for Palestine. Magnes treated Krebs's initial reply to the offer as an acceptance in principle. "I was very happy to receive your telegram," Magnes wrote from Jerusalem on April 5,

> The fact that we shall have you here is good news for all of us. We shall look forward to the letter you are writing and I sincerely hope that there may be no hindrance to the conclusion of a satisfactory agreement between us.
>
> . . .
>
> Our negotiations with the anonymous donors of our fund are practically completed, and it is our great desire to begin the preparations for the new work as soon as possible.[86]

Krebs clearly did not regard the matter, as Magnes did, as essentially decided, but he responded, on April 14, with almost equal enthusiasm: "The more I think about my possible going to Palestine the more I feel fascinated about it. I hope for the opportunity of meeting you and discussing the plans with you in the near future." On the 17th Magnes sent him a tentative budget for the institute and stated that "if you find the £30 monthly salary too little, we are ready to raise the salary up to £40 a month." He added, "We are anxious to have the budget worked out so that everyone is satisfied." There was no reason, he thought, why Krebs "should not come to Jerusalem as soon after the middle of June as you are ready."[87]

On April 29 Krebs replied to Magnes's latest information about the situation, "In its general outlines I perfectly agree with the scheme and I am glad to accept definitely provided that the following points can be satisfactorily settled." He asked first for the £40 salary that Magnes had already said was possible. Moreover, he could not come as early as June, because "the University of Cambridge is just creating a post for me here and I have promised Professor Hopkins several months ago to hold that post at least for one year. Secondly, I think it is not necessary and desirable that I stay in Jerusalem before I could actually start working in the new institute." A temporary visit during the late summer would be sufficient, he thought, to take his part in the planning. The budget for building and equipment seemed to him adequate, "provided that the plans are worked out carefully." When Magnes received this letter he considered the matter settled. "We are all delighted," he wrote on May 9, that "you definitely accept our invitation to come to Jerusalem." Although he understood Krebs's obligation to Hopkins, he hoped that Hopkins "might be willing to let you go a little earlier," and he advised Krebs in any case to come to Palestine temporarily in June.[88]

Krebs's decision to accept this position appears precipitous. He made it at a time when he was almost certain to be appointed in Cambridge to a post which would enable him to continue his research in a laboratory where he was very happy, where he enjoyed the firm support of one of the most distinguished biochemists in the world, and where he regarded his future prospects as very good. All of this he planned to give up in order to take a position offered by a man whom he had never met, in a land he had never seen, within an institute that had not yet been built. He did so with little detailed knowledge of the responsibilities his own position would entail or of the physical conditions and the organizational structure of the proposed institution. How did Krebs come to make such a leap?

Did the prospect of moving from demonstrator to chief of a division of a research institute at so early a stage in his career appeal to Krebs's ambition and his vanity? Or did he feel an obligation to participate in the development of a Jewish homeland? Or was he uncertain, in spite of Hopkins's support, that he could remain indefinitely in England? Undoubtedly all three of these factors had some influence; but none of them seems compelling. Krebs was not without ambition or vanity, but his was not the vaulting kind that seeks satisfaction in rapid career advances. He was, first of all, a dedicated scientist who viewed posts primarily in terms of the opportunities they provided for him to pursue his research. There was little in this offer that would appear on the surface to make it likely that he could carry on his investigative work more effectively in Jerusalem than in Cambridge. Although his forced departure from Germany had taught Krebs the importance of his Jewish identity, he had never been intensely interested in Zionism. He was sympathetic to the formation of a Jewish homeland, but, unlike his friend David Nachmansohn, he was not devoted to it. Concerning his chances for staying in England, the only tangible threat was that he held a temporary visa, so that he had to apply each year for permission to stay another year. It was always possible that some year he would be refused, but the very fact that he was soon to be the only German refugee to hold an official academic position in England ought to have given him a favorable estimate of his chances.

All in all, it is hard to attribute Krebs's choice to considered calculations on his part. Rather, he seems to have decided on an impulse. It is evident from his letters to Magnes that the sudden, unexpected prospect of participating in the founding of a new institute in Palestine was exciting to him. Twice he used the word "fascinated" to describe his attitude toward it. The habitually circumspect young scientist appears to have been swept up in a mood of enthusiasm for an adventure in a land that was remote, but with which he could identify. Perhaps there was within him, after all, something of a romantic spirit.

During the last week of April, and through May, Krebs pursued his experimental work with somewhat less than his usual regularity. During that period there were, in addition to weekends, two gaps of six and nine days respectively in his notebook record (the second of these may have been due to his attendance at a physiological meeting). During this period he carried out a number of experiments on the oxygen consumption of fetal and newborn goat tissues for Joseph Barcroft, in connection with Barcroft's studies of the physiological changes that occur during birth. Barcroft performed the experiments on living, anesthetized animals. In order to collect the tissues for his part of the collaboration, Krebs came frequently to Barcroft's laboratory and waited while Barcroft carried out the *in vivo* respiration measurements. Krebs came to know the distinguished physiologist, then nearing retirement, quite well, and to appreciate particularly his whimsical sense of humor. In his own investigative trail, Krebs continued with experiments on uricase activity. He examined whether the decomposition of uric acid influenced the formation of ketone bodies from butyric and other fatty acids. He also tested the effects of glutathione on both the decomposition of uric acid and the formation of ketone bodies. He was probably induced to include glutathione in his study because of the interest of Hopkins and Pirie in that substance as a presumed respiratory catalyst. Krebs found that glutathione did appear to increase the respiration of liver slices and extracts catalytically, but it had no effect on the two specific processes he was studying. At the end of May he appeared to have uncovered no new experimental openings to exploit.[89]

In the meantime, Krebs's relationships with Palestine had become complicated. On May 6 he saw Dr. Redcliffe Salaman, a British biologist whom he had come to know well, and probably mentioned his plans. Salaman thought that Krebs should consult Chaim Weizmann about them. The Russian-born Weizmann, the most prominent leader of the international Zionist movement, was also an active chemist. A naturalized British citizen since 1910, Weizmann had become a reader in biochemistry at the University of Manchester in 1914. Interested since the beginning of the century in the project of establishing a university in Palestine, he had been one of the principal founders of the Hebrew University, which he dedicated in 1918, and which was inaugurated in 1925. At that time he was elected with Albert Einstein as joint president of the university board. Salaman had good reason, therefore, to think that Krebs ought to talk to Weizmann. The next day Salaman wrote Krebs that Weizmann would come to Salaman's home at Barley, 12 miles south of Cambridge, on Saturday May 12, and that "he would very much like to see you." Salaman sent his car to pick Krebs up and drive him to the meeting.[90]

Weizmann could stay at Barley for only about two hours, so the meeting must have been relatively short. Krebs has provided a brief account of it in his autobiography:

> Dr. Weizmann . . . began by asking me to describe to him my work on the ornithine cycle of urea synthesis, and then we discussed what kind of possibilities there might be for me in Palestine. The idea emerged that a small group of refugee scientists might join one of the research centres, perhaps the Daniel Sieff Institute.[91]

Krebs's memory of this event was obviously selective. It excluded all recollection of the fact that at the time of this meeting he had already accepted a position in the proposed cancer institute of the Hebrew University. Weizmann must, in fact, have suggested the Sieff Institute as a possible alternative to this move. In addition he may well have advised Krebs to be more skeptical than he had been up until then in his negotiations with Judah Magnes, for Weizmann had been at odds with Magnes for the previous six years. In Weizmann's opinion, Magnes was trying to establish too many insignificant institutions and degrees at the university, when he should be concentrating its resources on a few basic research institutes that could attain international stature. Frustrated by his inability to deflect Magnes from his course, and by the refusal of the university to provide facilities for his own chemical research, Weizmann decided in 1932 to found his own research institute in Palestine. Raising an endowment from the Sieff family, he was able to inaugurate the Daniel Sieff Research Institute in April 1934,[92] almost exactly at the time that Krebs was accepting the terms offered by Magnes to come to the Hebrew University.

Either just before or after his meeting with Weizmann, Krebs discussed his plans with David Nachmansohn, who was visiting England, perhaps also in connection with the meeting of the Physiological Society. With his deep Zionist convictions, Nachmansohn was very interested in the prospect that Krebs would go to Palestine. A friend of Weizmann for many years, Nachmansohn then stopped off in London on his way back to Paris, to consult further with

Weizmann on behalf of Krebs. On May 17, Nachmansohn wrote Krebs that he had seen Weizmann and discussed the matter with him very thoroughly. "The situation," he reported,

> is rather complicated, and you must proceed cautiously. Weizmann thinks very little of the people who now direct the University. He fears that you will be led there in a false direction.

Nachmansohn raised a number of critical questions concerning those with whom Krebs had been in contact about the position, said that Weizmann wished to discuss matters further with Krebs, and outlined Weizmann's own plans for his Sieff Institute. Weizmann envisaged "a great scientific research center in Palestine." Although he did not wish to compete directly with the university, his institute would provide far more favorable resources than could be expected from Magnes's ventures. Nachmansohn emphasized that he himself was as interested in these questions as Krebs, hinting that Weizmann might be interested in inviting them both, as part of a group of emigrés who would work in the institute. In addition to these incentives, Nachmansohn added a more somber motivation for moving to Palestine: "Weizmann is very pessimistic, and does not believe that the Jews can maintain themselves in France and England."[93]

These developments must have given Krebs serious pause. There is no record of his immediate reaction, but he could not have avoided the sobering realization that he had acted too hastily, too much on his own, with too little knowledge of the actual situation in Palestine. While he pondered what to do next, news of some of what had happened quickly became public. On May 22 his friend Walter Auerbach wrote from Edinburgh that he had read in the *Jewish Chronicle* that Salaman had announced, during a meeting of the Committee for the Assistance of Jewish Physicians in England, that Krebs—"one of the most eminent scientists in his branch"—had been recruited for the University in Jerusalem. Auerbach wanted to hear from Krebs himself whether the report was accurate, even as he congratulated him in case it was.[94]

Whatever the long-range future held for him, Krebs's short-term future in Cambridge was ratified early in May with his official appointment as University Demonstrator in Biochemistry. This news too spread rapidly, even before the official public announcements appeared. On May 13, his ever-supportive friend Bruno Mendel wrote Krebs from Holland, where Mendel had settled after leaving Germany,

> Yesterday, at the lectures by Warburg and Keilin in Utrecht, I heard such happy news about you that my wish to congratulate you has overcome all obstacles. If I understood correctly, you have received—what no other foreigner has yet enjoyed—an official position in a biochemical laboratory![95]

Krebs relayed the news to his family by sending them a copy of the *Times* containing the announcement of his post. His father quickly wrote back,

First of all, heartiest best wishes for being named demonstrator. Once
again I am very proud and happy over your well-earned success. And
Maria is no less! Just yesterday we had read in the preceding issue of
the Times that several of the "guest investigators" had carried out
outstanding scientific research, and we had thought keenly about you.
Today comes the information about your appointment, which lifts you
far beyond the circle of "guests." Please write in greater detail about
what the post of demonstrator means, and, further, whether your
Jerusalem plans are canceled through the Cambridge appointment.
Personally, I would prefer it if you lived in England, rather than in
Palestine. But that, of course, is less important than your own
future.[96]

Georg Krebs went on to say that he and Maria planned to visit Hans in
Cambridge or London during the summer. Hans Krebs must have been pleased
that his progress was such a consolation to his father in difficult times. He must
also have taken some satisfaction in the fact that the father who had once
dominated him, and had doubted his ability to become a first-rate scientist, was
now acknowledging that the son had succeeded beyond his dreams.

Among other things, Krebs's new status meant that he became a participant
in the staff meetings of the Biochemistry Department, chaired by Hopkins in the
Common Room. Although such meetings had sometimes in the past revolved
around relatively small issues—such as who should be permitted to use the
balances, or what to do about bicycles left leaning against the building, or
whether to hold more frequent staff meetings—there was at just this time a
weighty matter for discussion. During the previous year biochemistry had been
made for the first time a subject for Part I of the Natural Sciences Tripos. The
department had now to design a new course to prepare students for this
requirement, to decide when it could be ready to begin teaching it, what subjects
to cover and in what depth, and for what lectures various members of the
department would be responsible. There was also debate over the question of
what degree of preparation students must have in order to be introduced
successfully to biochemistry. Initially it had been assumed that the course would
not begin until the spring of 1935. When it became clear that the Governing
Board of Biology expected it to commence in October 1934, the department
realized that "no time is to be lost in settling the programme."[97]

During the staff meeting of May 23, the first one at which Krebs was
recorded as present, the minutes record that "the discussion of the teaching of
biochemistry Part I was resumed and was adjourned until 9.00 AM on Wed.
May 30." Krebs also attended that meeting, when

Discussion of the schedules for the new Part I was resumed. The
appointment of the periods was concluded. Dr. [Eric] Holmes asked
that estimates of the apparatus necessary for a class of 30 students
should be handed to him by each member responsible for a section or
sections of the teaching, not later than July 20.[98]

At the conclusion of this meeting Hopkins was able to propose to the Directors of Studies in Natural Sciences that "the first Course in the half subject Biochemistry, for Part I of the Natural Sciences Tripos, shall commence in October 1934." He added that he believed students should not take the course in their first or second year unless they had attained specified qualifications in chemistry, biology, and physiology. He forewarned the directors that "the course for the first year has been based on the assumption that not more than 30 men will attend, and it will not be possible to reorganize it at the last moment."[99] In keeping with the style of Hopkins's leadership, the department had planned the course as democratically as such things can be managed. Several members made proposals, which were discussed until a consensus could be reached. In the schedule agreed upon, Krebs was assigned lectures on colloids, osmosis, enzymes, and tissue metabolism.[100] Thus, although given his own specialty area to teach, he was also required to deal with some topics that were outside that area. For the first time in his career he had to prepare to take part in the pedagogical process of initiating students into the basic principles of his field. He had been deprived of a similar opportunity a year before, when he was dismissed just as he was about to begin teaching as a *Privatdozent* in Freiburg. Now he would have to learn to communicate the language of biochemistry in English, rather than in German.

The salary that Krebs obtained in his new position as demonstrator was less even than he had received the previous year on his Rockefeller grant. He now drew only £150 "basal pay" and a "fellowship allowance" of £100, for a total of £250, compared to the £300 he had had before. The other demonstrator, Bill Pirie, with one year of service in the position, had a total salary of £300, whereas the four lecturers earned from £350 to £700, depending on their seniority.[101]

Despite his modest income, Krebs felt ready to enjoy a few amenities. He decided that it was time to have a motorcar. That posed the difficulty that he did not know how to drive. He asked Vernon Booth if Booth would teach him to drive a car. When Booth said yes, they went out together and found a used, rather weathered Austin 12 roadster, which Krebs brought for £22. Krebs had some trouble learning to let up the clutch and put down the throttle smoothly, as well as in estimating how far he was from the left-hand side of the road. Soon, however, he had mastered the art sufficiently to set out by himself on a tour in the surrounding countryside. The first such venture must have taken place in early June, for Booth sent him a postcard on the 18th saying, "Glad you managed your solo journey so well. I hope you will see that the radiator is kept filled. If the water continues to disappear tell Mr. Hallen (when you see him about the clutch)."[102] Krebs was thus expanding his horizons in the summer of 1934 in the smaller, as well as the larger concerns of his life.

V

Sometime before the end of May, Krebs had written his sister Lise about his offer to go to Jerusalem. In reply Lise apparently expressed great surprise that he would consider emigrating to Palestine.   Hans in turn wrote a letter (unfortunately now lost) explaining, in far more detail than he usually wrote about himself, why he was interested in the position.   In it he must have emphasized his attraction to the concept of a Jewish homeland.   On June 7 Lise wrote another letter which illuminates her brother's attitude as well as her own:

> Your letter of yesterday so pleased and interested me, that I shall answer it immediately.   Contrary to your usual habits, it contains much about you that is personal.   For the first time you allow something more than your outward life to be seen, so that I was truly moved.
>
> I am sorry if the expression of my feelings about your offer from Jerusalem somewhat shocked you.   But since you have always been, up until now, reluctant to speak about everything that concerns you, I had no way to anticipate that the Jewish "interests" generally interested and moved you.   I thought that for you there were only pure human-spiritual things, and believed that you rejected nationalistic and racial points of view.   But we too [that is, Lise and her husband, Adolph Daniel] are not so remote from the Zionist question as you imagine.   For years, we have had Zionist friends (Gluckmanns, Baums, etc.), and we have often spoken, read, and heard lectures about Palestine.   We have never rejected Zionism, but—unfortunately—I can see no solution of the Jewish question there, and am therefore not whole-heartedly for it.   I find it attractive that for many young people—and not the worst of them—it signifies an ideal through which they can give content to their lives.   And even if it is only a utopia, it still does no harm.   Is not everything for which humans struggle—progress through ethics, science, and art—perhaps illusion and utopia!?[103]

Lise and Hans were both struggling to reconcile their family heritage with the devastating effects of the Nazi takeover on the beliefs they had cherished.   Lise had probably given more sustained thought to these difficult questions than had her brother.   It is possible that Hans had long harbored views such as those he must have expressed in his letter, without letting his family or friends know about them.   Other letters confirm Lise's statement that he was characteristically reticent about his inner feelings.   More likely, however, these views represented a recent conversion, brought about by a combination of the radical change the Nazis had forced on his life, the sudden opportunity created by Magnes's enticing offer, and perhaps the strong Zionist convictions of his old friend David Nachmansohn.

In early June Krebs seemed, in his research, somewhat at a loss for new experimental horizons.   He tried a number of variations on, and recombinations of, types of tests he had previously performed, as though casting around again for a trail to follow.   Having incorporated into his study of uric acid decomposition tests of the action of glutathione and connections with the formation of ketone bodies, he now left out the former, and continued to examine the effects of glutathione on the latter, as well as on other metabolic processes. In this inconspicuous manner the uricase investigation on which he had placed such high hopes only two months earlier faded out of his research program. Glutathione lasted for only two experiments longer. On June 1 and 2 he tested its action on the respiration of liver tissue in the presence of fatty acids, glucose, lactate, and pyruvate, without discovering any effects worth following up. Then he centered his attention for a while on the fatty acids alone, measuring their effects on the respiration of various tissues. These included—in addition to liver—yeast and the common intestinal bacterium *B. coli*. The single experiment on the latter he probably tried because he could easily obtain the material for it from a student working with Marjory Stephenson. In the midst of these experiments he also carried out two more measurements of the respiration of fetal and newborn goats for Barcroft.[104]

At this point Krebs found a stimulus for his research in the latest work of Albert Szent-Györgyi. With his assistant B. Gözsy, Szent-Györgyi had just published an article "On the Mechanism of the Principal Respiration of the Pigeon Breast Muscle." Although a continuation of an investigative pathway that he had been following since 1925, Szent-Györgyi's most recent contribution contained significant novelties, both of method and of concept. During the preceding years he and his associates in Hungary had extracted from swine heart muscle a partially purified substance, probably a nucleotide, that when added to washed ground tissue could restore a portion of the respiratory activity lost by the washing. Szent-Györgyi believed that he had identified a respiratory coenzyme. It was, he showed, connected particularly with the oxidation of lactic acid. Szent-Györgyi now set out to confirm and extend his earlier conclusions by studying respiration in pieces of tissue that could be suspended in a medium in a manometer cup, but that retained their cellular structure. He attained this objective by putting the tissue through a Latapie grinder with its second disk removed, so that it reduced the tissue only to "small pieces." Although he did not say so in his paper, he probably chose pigeon breast muscle because, as a flight muscle, it could be expected to have an exceptionally high rate of respiration. With this preparation he showed that washing the tissue suspension 20 times in water reduced its respiration to a low level, and that by adding the extract of swine heart muscle containing his coenzyme, he could restore the level to that of unwashed muscle tissue.[105]

Szent-Györgyi believed that there must be a second respiratory "cofactor" to act as a "hydrogen transporter" in the process. He suspected that cofactor to be succinic acid. His reasons, as he gave them in this paper, were that the existence of the exceptionally active, stable enzyme, known since the early work of Battelli and Stern to oxidize succinic acid in isolated tissues, would be "incomprehensible" if succinate were merely a chance substrate. Succinic acid was oxidized with "enormous rapidity." "It may appear strange," he acknowledged, "to attribute to a substance as simple as succinic acid the role of a catalyst," but that role could be justified theoretically in terms of its unique molecular structure. To justify it experimentally, he showed that the respiration of pigeon breast muscle suspensions, increased by 100 to 600 percent in the presence of succinic acid. To demonstrate that this effect did not result from a complete oxidation of succinic acid, he showed that the product of the oxidation succinic acid was known to undergo in tissues, fumaric acid, could also greatly increase the respiration, even when present in very small quantity. Adding to several suspensions, respectively 1/10, 1/20, 1/40, and 1/80 Mol. of fumaric acid, he found that the amount of respiratory oxygen uptake induced by the smallest portion must have consumed all of the fumaric acid added, if it were oxidized in the process; yet at the end of the experiment he could recover as much fumaric acid as he had put in at the beginning. It must, therefore, have acted catalytically.[106]

Szent-Györgyi interpreted his results as supporting the concept that succinic and fumaric acid form a catalytic hydrogen-transport system. Succinic oxidase oxidizes succinic acid to fumaric acid, which is reduced again to succinic acid in removing hydrogen from foodstuffs. Szent-Györgyi's view of the role of succinic acid can be traced back to his efforts since the mid-1920s to reconcile Wieland and Warburg's conflicting conceptions of the mechanism of cellular oxidations. He had long sought what he now saw in succinic acid, a substance that "would play the role of a hydrogen transporter, inserted between the dehydrases and the foodstuffs on the one hand, and oxygen on the other hand."[107] In identifying succinic and fumaric acid with that role he departed from the prevailing view, associated particularly with the Thunberg-Knoop-Wieland scheme, that succinic and fumaric acid were critical steps along the main pathway of substrate oxidations. To Szent-Györgyi these substances were not metabolites, but catalytic agents that enabled other substances—the "foodstuffs" [Nahrstoffe]—to undergo respiratory oxidations.

The issue of Hoppe-Seyler's Zeitschrift containing Gözsy and Szent—Györgyi's article went to press on May 24, 1934. Krebs would therefore probably not have been able to see it in the Cambridge library until around mid-June. He would undoubtedly have read it at once, with great interest, for he both admired Szent-Györgyi and regarded him as a personal friend. Most likely, he decided immediately to examine Szent-Györgyi's conclusions with an experiment of his own. On June 19, he headed an experiment in his notebook:

> Influence of succinic acid on respiration of liver and muscle (in presence of lactate) (Compare the work [that has] just appeared of Szent-Györgyi, H.S.)

As the title indicates, for the purposes of this experiment Krebs treated succinic acid from the point of view of Szent-Györgyi. To test the "influence" of succinic acid was to think of it as a catalyst, whereas lactate was the presumed substrate undergoing respiratory oxidation. A further clue that Krebs here adopted Szent-Györgyi's perspective is a notation regarding the quantities of succinate he added. They were equivalent, he put down, to certain quantities of "succinate → fumarate." He carried out two parallel sets of three experiments each on rat liver slices and pieces of rat muscle tissue. The latter were "thick," because he did not succeed in an attempt to make slices in the usual way with a razor blade. All of the vessels contained lactate, whereas for each tissue he utilized three concentrations of succinate. The results were:

Liver				Muscle			
Concentration of succinate:				Concentration of succinate:			
0	3	11.9	47.6	0	3	11.9	47.6
Q_{O_2} 20.7	20.6	21.8	22.7	1.95	2.52	1.77	1.83

He commented:

> Liver: no effect of succinic acid on Q_{O_2} in presence of lactate
> Muscle: Smallest concentration increases!!!?? Larger not.
> Error? Rapid sinking of respiration in muscle.[108]

Krebs thus obtained no effects corresponding to the dramatic ones that Gözsy and Szent-Györgyi had reported; but he had, of course, not duplicated their system. In place of the suspension of pigeon breast muscle they used he utilized liver slices and muscle pieces that were too thick for optimal experiments. His remarks indicate that he regarded the muscle portion of the experiment as unsatisfactory. Instead, however, of making further efforts to obtain the specific effect of succinic acid on lactic acid oxidation that the Hungarian group had studied, Krebs examined in his next experiment whether succinic acid might influence other oxidative processes, including particularly that of ketone body formation with and without succinate in rat liver slices in the presence of caprylic acid, glucose, and butyric acid. This time he found "no influence from succinate on the metabolism, although a slight increase in Q_{O_2} everywhere."[109]

Krebs pursued Szent-Györgyi's conception of the catalytic influence of succinic acid no further at this time. How did he view the negative outcome of his two attempts? It is hard to imagine that he could have regarded them as a refutation of Szent-Györgyi's theory, for he had not attempted to repeat the experiments on the same material they had used. He would have been quite aware that, if he had not been able to duplicate the phenomenon, it was most likely because his system did not provide experimental conditions appropriate to display the effect. On the other hand, Szent-Györgyi's conception of the role of the succinate-fumarate system would appear to have such broad potential consequences, that if Krebs were truly testing it he could hardly have been satisfied with the two experiments he had performed. Most likely he did not conceive of his experiments as tests of their results, but was merely trying out the possibility that he could utilize the effect they had found in his experimental system. If so, he could profitably extend this "succinate effect" to other problems in which he was interested. If not, it only meant that, for the time being, he could not link up their results with his own investigative pathway. The brevity of his effort suggests that he did not fully appreciate the scope of the challenge to the standard view of respiratory oxidation that Szent-Györgyi's position implied.

Szent-Györgyi's finding that succinic and fumaric acid act catalytically on respiratory oxidations could have been fitted smoothly into mainstream thought on the subject by applying to it reasoning analogous to that which Krebs had used in his investigation of urea synthesis. If one viewed these two acids as participating, like ornithine, in a closed circle of reactions that continuously regenerated them, then one could use Szent-Györgyi's results very nicely to support either the Thunberg-Knoop-Wieland scheme or the variation proposed by Erich Toenniessen and E. Brinkmann. There is no indication that Krebs made such a connection; but if he did, it did not strike him forcefully enough to induce him to follow it up. It may well be that in the summer of 1934 he was simply not prepared to reenter this field. After having devoted nine months to it with no significant success, his mind might still have been in a refractory state about immersing himself again in an investigative morass.

Simultaneously with the second of the above two experiments, Krebs carried out another set that was evidently also influenced by Szent-Györgyi's recent investigations. These involved the effects of *Kochsaft*—that is, extracts made by cooking a tissue in water-from swine heart muscle and rabbit liver on the respiration and formation of keto acids in these tissues. It was from such extracts from swine heart muscle that Szent-Györgyi had obtained his respiratory "coenzyme." Krebs utilized *Kochsaft* without treating it further, as Szent-Györgyi had, to obtain a fraction containing the coenzyme, but he was undoubtedly testing for the effects of the coenzyme. In the case of the swine heart, which formed very good slices, the *Kochsaft* caused an "enormous increase" in the respiration. (This experiment also included one additional test of the influence of succinate on respiration in the presence of lactate, but the two substances together did not increase the rate more than did lactate alone.) With rabbit liver tissue, caprylic acid in the presence of *Kochsaft* produced a "strong increase" in respiration, but none in the formation of α-keto acids.[110] Whether these marked but general effects might have led Krebs anywhere beyond a vague confirmation of Szent-Györgyi's claims for his coenzyme is uncertain, because at this point he was diverted by the impact of another new publication from the laboratory of Szent-Györgyi.

Early in 1934 Erno Annau investigated the formation of ketone bodies, particularly acetoacetic acid, from pyruvic acid. The reaction could be carried out chemically by means of peroxide, an agent which was thought to "imitate biological oxidations," but the yield was small. Annau found that by adding ammonia to the reaction he could increase the quantity obtained. He interpreted this result as due to the "condensing action" of NH_3, and inferred that the reaction must proceed through an intermediate polymerization of two molecules of pyruvic acid to form parapyruvic acid. Annau was then able to reproduce the effect of ammonia physiologically, adding NH_4Cl to suspensions of liver tissue. In the presence of NH_4Cl the ketone bodies formed from pyruvic acid at a rate four to five times higher than without it. He found, however, that NH_4Cl exerted no effect on the formation of ketone bodies from fatty acids. This was in accord with his expectations, since only pyruvic acid required the condensation step to produce acetoacetic acid.[111]

Annau's article was placed in the issue of *Hoppe-Seyler's Zeitschrift* following the one in which the article of Gözsy and Szent-Györgyi described above appeared. The new issue went to press just nine days after the previous one, on June 2. Krebs very likely saw it, therefore, shortly after he had carried out the preceding experiments. The next experiment that he carried out, after a three-day pause, on June 25, appears clearly to have been inspired by Annau's results, even though there is no explicit mention of the article in the laboratory notebook. The title of the experiment was "Formation of Acetoacetic Acid in Heart *Kochsaft*. Influence of NH_4Cl."[112]

Only a portion of this experiment concerned NH_4Cl, the remainder being a continuation of Krebs's examination of the influence of *Kochsaft*. The latter increased the respiration of liver tissue from a nourished rat, without increasing the formation of β-keto acids. NH_4Cl, on the other hand, "increases the formation of β-keto acids enormously." Q_{ket} for the control = 0.40; with NH_4Cl and no other addition, = 2.88.[113]

This striking result prompted Krebs to concentrate all his attention on the subject of "the influence of NH_4Cl on ketosis"—which was also the title of the next experiment he carried out, on June 26. On liver slices from a starved rat this time, he tested NH_4Cl in four concentrations, comparing the rate of respiration and keto acid formation to controls and slices to which caprylic or butyric acid were added. There was "everywhere increase through NH_4Cl, but the influence of the NH_3 concentration is unsteady." On the 27th he narrowed his focus to the specific process that Annau had studied, "NH_4, pyruvic acid, and the formation of β-keto acid in the liver." In fact, he tested not only the action of NH_4Cl on the formation of keto acids from pyruvate, but also from caprylate, perhaps for comparison. His result strongly confirmed that of Annau. With liver tissue from a starved rat there was an "enormous effect of NH_3 on pyruvic acid $\rightarrow$ β-keto acid." There appeared to be no increase in the "spontaneous" formation of keto acids caused by either pyruvic acid or by NH_4Cl alone. Caprylic acid alone caused a rate equal to that attained by NH_4Cl + pyruvic acid, but NH_4Cl brought no further addition. Krebs made no comment on that aspect of his results, an indication that it was the ammonia effect that had captured his interest. On the next day, this time with tissue from a nourished rat, he compared the action of NH_4Cl on the formation of keto acids from pyruvic acid with its effect on the formation from acetic acid. Again in keeping with Annau's view, he found that "NH_3 increases keto acid formation from pyruvic acid, [but] not from acetic acid."[114]

During the last week of June, Krebs had finally found, with the help of Szent-Györgyi and his associates, an experimental effect that appeared to him promising enough to pursue intensely. It remained to be seen, however, whether he could go beyond supporting evidence for what they had already found.

Throughout June Krebs was as preoccupied with what to do about Palestine as with the immediate progress of his research. Even before the complicating events of May he had begun to think more closely about what facilities he would need in his division of the cancer institute in order to carry on his work effectively. He began looking up the cost of equipment and planning the layout of his laboratory. The further he proceeded, the more dubious he became that the funds available according to Magnes's earlier letters would suffice to support an institute with several divisions.[115]   At the same time, Nachmansohn, who was keeping in touch with Weizmann in Paris, kept sending Krebs stronger suggestions that Weizmann wanted him to come eventually to his own new Sieff Institute, and sowing further doubts about Magnes's actions.   On June 24 Nachmansohn reported that Weizmann hoped that Krebs would travel with him to Palestine in October, and urged him to visit Weizmann in London as soon as possible.   "How would it strike you," he wrote tantalizingly,

> to head a division in Rehovot [Weizmann's institute]?   This is naturally preliminary, not yet a formal offer.   I am merely of the opinion that you do not place value on being at a university, and that you would be able to work at least as well in a pure research institute, if the conditions were favorable enough.   I have mentioned this (as my personal opinion) to Weizmann.   Weizmann asked me what you would need.

All of this could not be settled in less than a year, Nachmansohn acknowledged, so that Krebs must first "decide the question of the cancer institute."[116]

It was just that question which kept Krebs in such a quandary, during June, that he did not answer either Nachmansohn's letters[117] or those that Magnes had written him in May concerning details such as what name he would like for his division.   Apparently Krebs did not even inform Magnes of his decision to postpone until fall the trip to Palestine that Magnes had advised him to make in June.   Finally, on July 1, Krebs wrote Magnes,

> Please pardon me for not having answered your last two letters until today.   During the last weeks my thoughts have been thoroughly occupied with the problem of the construction of the Institute, and it has taken me some time to clarify for myself a series of things.   On closer consideration I have had some doubts about the possibility of building the Institute in the way that I had previously had in mind. Please allow me to explain to you my doubts.

The first point Krebs raised was that he was uncertain that the total sum available for building and equipping the institute left enough over for what would be required to equip his division.   He wished to know if there would be at least £500 for the latter purpose.   Second, he wished to clarify the relationship between the leaders of the three divisions and to know who would direct the institute itself.   It was very important, especially in the beginning, to know who would have the authority to allocate resources to the divisions.   Finally, although he thanked Magnes for offering funds to come to Palestine, he believed that "it would be more purposeful if I wait until the plans for the Institute have taken clearer form, before making the trip."[118]

Krebs had not resolved his question. Clearly, however, he had become uncomfortable with his position and was seeking a way to extricate himself from the commitment he had made, while leaving room for further negotiations.

Krebs wrote this letter while his father and stepmother were visiting him in Cambridge. Because the father had been permitted to take only 10 marks out of Germany, the son had to pay his hotel bills and give him spending money. For Hans it was the first opportunity to make good on his offer of eight years earlier to help his father financially if he ever needed it (see Vol. I, p. 128). The necessity to accept his son's support even for a short time, however, made the elder Krebs uncomfortable. Hans showed his father around the laboratory, but did not think his father knew enough chemistry to understand his research, and therefore made no effort to explain it. They talked more about the situation in Germany, which Georg Krebs found discouraging but not hopeless. Hans hoped that his parents would make an extended stay, but his father soon became restless to return to his patients, and they left after little more than a week. Nevertheless, his father afterward wrote a note referring to "the beautiful days in Cambridge." Undoubtedly Hans had conferred with his parents about the difficulties that had arisen over the Palestine situation, for his father sent him from Belgium a copy of a Paris newspaper carrying "the false report that you are going to the Hebrew University in Jerusalem." Georg and Maria had traveled from England to Belgium to join Lise and her husband on holiday. Lise had hoped that Hans would accompany their parents, but he disappointed her, probably because he was unwilling to interrupt his research before his own holiday was scheduled to begin.[119]

During the first week of July Krebs extended his investigation of the effects of NH_4Cl on the formation of keto acids. Under a variety of conditions he confirmed that NH_4 increases the rate at which acetoacetic acid forms with pyruvic acid. On July 9 he began to measure the rate of consumption of pyruvic acid along with that of the formation of keto acids. On the 11th he widened the investigation again to compare the effects of NH_4Cl on the formation of keto acids from butyric, capronic, and acetic acids as well as pyruvic acid. He used liver slices from a nourished rat. This time keto acids formed at a higher rate in the presence of NH_4Cl, not only with pyruvic, but with each of the acids:

		Pyruvic	Butyric	Capronic	Acetic
$Q_{keto\ acids}$	without NH_4Cl	1.45	1.58	2.76	1.06
$Q_{keto\ acids}$	with NH_4Cl	4.82	4.30	4.06	2.15

Krebs wrote, "Influence of NH_3 everywhere." Now he had the first strong evidence that the ketogenic action of NH_3 was broader than Annau had suspected. On that encouraging note Krebs closed his laboratory books for the summer vacation.[120]

VI

In the late spring or early summer, a New Zealander, Norman Edson, had come to Cambridge on a postdoctoral fellowship. He wished to do a Ph.D. in biochemistry, and the senior staff of the department suggested that he work with Krebs. Edson was a pleasant man who liked to crack jokes, and Krebs got on well with him. The little subgroup of Krebs and Weil expanded to three, and the daily conversation in their corner of the laboratory gradually shifted from German into English. Krebs turned over to Edson the further investigation of the effects of ammonium chloride on ketogenesis.[121]

During the summer, Weil suffered a major accident. While he was using a Bunsen burner to heat a flask filled with alcohol, during a lipid extraction, the flask cracked and went up in flames. His clothes caught fire, and he was burned seriously enough so that he had to spend two months recovering in a nursing home.[122]

Another new arrival in Cambridge in 1934 was Krebs's long-time friend Hermann Blaschko. A.V. Hill, whom Blaschko had been assisting in his efforts to resettle German refugees in England, sponsored Blaschko's move to Cambridge to pursue a Ph.D. degree. Blaschko was offered the choice of working in Hopkins's laboratory or that of Joseph Barcroft. Because he felt that Hopkins was already overloaded with refugees, he chose Barcroft, in spite of the fact that Barcroft was a physiologist rather than a biochemist. Krebs found a room for Blaschko in his own pension on Hills Road, but both of them soon afterward moved to other places. Whether or not he helped in other ways, as Blaschko suspected, to facilitate arrangements for his coming, Krebs was undoubtedly pleased about it. During their long earlier association he had found Blaschko among the brightest and best informed of his colleagues, one to whom he himself could turn for constructive advice. Krebs believed that Blaschko had very high scientific potential that he had so far been prevented from realizing by his illnesses and the external obstacles that had blocked his progress.[123]

In July Allan Elliott called on Krebs to consult with him about some research that Elliott had recently completed. Elliott had been a student in the Cambridge Biochemical Laboratory from 1927 to 1933. While there he assisted Malcolm Dixon, who, with Keilin, was working out some improvements in manometric methods for measuring tissue respiration. Dixon and Keilin continued to use Barcroft differential manometers in preference to the Warburg single manometers. In 1932 they published a method that allowed one to measure the consumption of oxygen, the formation of carbonic acid, and the production of total acid in a single experiment. Their technique enabled one to "make accurate determinations of the respiratory quotient . . . on one slice of animal tissue." Under Dixon's tutelage Elliott learned these methods.[124]

Shortly after Krebs arrived in Cambridge, Elliott left to take a job in a cancer research laboratory in Philadelphia. There he had the use of a well-equipped laboratory, but worked in relative isolation from a director with whom he had little rapport. Impressed, as others had been in Cambridge, with the experimental successes that Krebs had had with tissue slice methods, Elliott decided to apply such methods to other metabolic processes, and became adept in the techniques. During his first year in Philadelphia he focused on the metabolism of lactic and pyruvic acid in rabbit kidney cortex tissue. He utilized

Figure 2-3 Hans and George Krebs in front of Garden House Hotel, Cambridge, July 1, 1934.

Dixon and Keilin's manometric method, together with chemical methods for estimating changes in the quantity of lactic and pyruvic acid in the medium and of glycogen in the slices. He measured the rates of respiration of the tissues, their respiratory quotients, and changes in the quantities of pyruvic, lactic, and total acid content in the presence of lactic, pyruvic, succinic, malic, fumaric, oxaloacetic, acetic, and formic acid. All of the "complex series of observations" he had made, he concluded, could be "explained satisfactorily" by a scheme that Dorothy Needham had already applied to lactic acid oxidation in muscle (see following page). Since, however, he had found "no sign of rapid oxidation of formate, the hydrolytic breakdown of the hypothetical α α'-diketo-adipic acid apparently does not occur and this section of the scheme must be left more indefinite."[125] Elliott's scheme was, in fact, another variation on the old Thunberg-Knoop-Wieland scheme, most closely resembling the one proposed in 1930 by Toenniessen and Brinkmann (see Vol I, pp. 31–33).

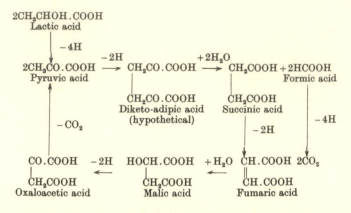

In June 1934, Elliott wrote Krebs that he would come to Cambridge on his holiday, during July. "I have," he added, "worked very hard here, under good laboratory conditions, but with no one to discuss my own work with or what is appearing in the journals. So I hope to have lots of opportunity to talk with you. And I will be glad if you will read my papers in proof or in typescript and tell me what you think of them."[126]   Apparently Elliott brought with him the typescript of the paper reporting the above results, for it was submitted to the office of the Cambridge University Press, the publisher of the *Biochemical Journal*, on July 21, soon after he planned to arrive in Cambridge.

There is no direct record of how much opportunity Elliott had to talk with Krebs—if Krebs's habits then were similar to his later ones, he could be quite ungenerous with his time when visitors came—but he went over the manuscript carefully enough to pick out a technical discrepancy with significant implications. After Elliott had returned to Philadelphia he wrote back, "As a result of our discussion in Cambridge I have added the following appendix to the proof":

> Dr. H.A. Krebs had pointed out that in the case of the substances shown in Table VII [that is, succinate, fumarate, malate, and oxaloacetate] the total oxygen uptake is not sufficient to account for complete oxidation of the acids disappearing. Also the R.Q. is in some cases higher than complete oxidation could account for. It would seem that there is another course of metabolism of these substances which also involves disappearance of carboxylic acid. A similar observation was made by Ashford and Holmes (1931) in their study of the oxidation of lactate in brain tissue.

"If you have any objection whatever to our mentioning this fact which comes also in your own work," Elliott wrote, "or to our using your name, you have our authority to go to the University Press and change or eliminate the paragraph."[127]

In the paper referred to in Elliott's footnote, Charles Ames Ashford and Eric Gordon Holmes had found that the oxygen consumed when they added

Figure 2-4 Hermann Blaschko (left) in Cambridge, with Hans Sachs.

lactic acid to brain tissue was not sufficient to oxidize completely the acid added. After ruling out several possible intermediates that might be expected to result from a partial oxidation, and making several indecisive efforts to trace the substance, Holmes and Ashford had inferred that "the larger part of the lactic acid which disappears when brain tissue is shaken with lactate in the presence of oxygen is transformed into some substance which has so far eluded identification."[128] Krebs must have drawn Elliott's attention to their conclusion, pointed out that Elliott's results pointed toward an analogous conclusion regarding succinic, fumaric, malic, and oxaloacetic acid, and informed him that he had obtained similar results during his investigation of the same problems during the preceding year.

There is no contemporary record of what Krebs did think of Elliott's work as a whole. When I brought this paper to Krebs's attention, in 1977, he commented that Elliott was "quite a neat worker," but that this investigation "didn't really add very much." Elliott found "as I had observed also but not published, that these substances are readily oxidized. But the link was missing."[129] Despite the retrospective quality of these remarks, made long after the "link" had been found, they fit well with what we would expect Krebs's immediate reaction in 1934 to have been. It would have appeared to him that Elliott had done no more than he himself had done—tested the standard list of suspected intermediates and confirmed that they are rapidly oxidized. Krebs had not thought his own results worth publishing.

Krebs's assessment was not unjustified. Far from building up the general picture of carbohydrate oxidation, Elliott had only added doubts concerning one of the current schemes at just that portion of it—the synthetic steps—constituting the most crucial and least substantiated "link."

Elliott probably had no illusion that he was making a grand contribution. A skillful research craftsman, he reported the small details of his experimental techniques with loving care. His discussion of the way in which his observations fit into the Toenniessen-Brinkmann reaction scheme was, in contrast, notably cursory. Elliott enjoyed his science, but he "was never one of those who lived only for research." [130] He was probably more content than Krebs to do work that led only to "some sound, publishable, new information." [131]

We cannot tell whether Krebs expressed his opinion of Elliott's contribution candidly, or tactfully limited himself in their discussion to technical points. In any case, if he thought that Elliott had added little to the overall picture of oxidative carbohydrate metabolism, he was in the summer of 1934 no further along himself. He had tried and failed to advance the state of the problem. If, as appears likely, the statement in Elliott's footnote—"It would seem that there is another course of metabolism of these substances which also involves disappearance of carboxylic acid"—was based on Krebs's suggestion, then the vagueness of the idea probably reflects well the current vagueness of Krebs's view of the situation. For him, as for Ashford and Holmes, if there did exist courses of oxidative carbohydrate metabolism other than the sequences incorporated into the schemes of Toenniessen and Brinkmann, or Thunberg, Knoop, and Wieland, then those courses had all "so far eluded identification."

VII

As he wound up his research on July 11 and began his holiday, Hans Krebs had much on his mind. Besides the question of what direction to take when he resumed his investigations, and the vexing matter of Palestine, he was being pressed by his father and sister to try to help his brother Wolf find a position in England. They were deeply concerned that in Nazi Germany there could be no future for Wolf commensurate with his promising talent and auspicious early achievements as an electrical engineer. He had so far survived the dismissals of Jews from the General Electric Company in Berlin, where he worked, but his family was convinced that he could never advance in his career unless he left Germany. [132]

For now, however, Hans probably put these concerns in the back of his mind. He looked forward to a very special holiday, for he planned to spend it with a special friend from Germany whom he had not seen for nearly three years. During the year he had spent at Altona, he had become particularly close to a nurse there named Katherina Holsten (see Vol. I, p. 237). After Hans left

Altona, he and "Katrina", as he called her, occasionally exchanged letters. At the beginning of 1934, Hans wrote Katrina, after he had allowed their correspondence to lapse for longer than usual. Very pleased to hear from him again, Katrina quickly replied, recalled how nice it had been in earlier times, and asked, "When can we see each other once again?" She had sometimes wondered, she added, if she might be able to come to England during one of her holidays. She doubted if anything would come of such a plan, but Hans took the suggestion seriously and wrote a succinct note encouraging her to come. In May, Katrina replied in a touching letter that explained why she could not come. Partly it was because of practical obstacles, but more important, she wrote, was that during their time in Altona she had come to care for him so much that it had been much more difficult for her after he departed than he had ever suspected. She had yearned to be "much more to you, and wanted to help you" more than he would permit her. It had taken her a long time to get over her feelings of loss. If she were to come now to England, they would surely have a very good time together, but afterward she would again be alone and would again have to overcome the conflicts within her.[133]

Hans must have been moved by Katrina's letter. He wrote back acknowledging that he had sometimes been thoughtless toward her in Altona, and urged her to come to England, to find out whether they could still tolerate one another. Perhaps, he teased, she would be cured of her feelings about him—particularly since he had gained weight since his days in Altona. His letter caused Katrina "to throw all of my long-considered and firm decisions overboard." She agreed to come, if he could arrange his holiday to coincide with hers. Asking what he planned to do during his holiday, she stated that she would prefer "much sun, much wind and water, and few people." Hans replied that he now possessed a car, and proposed that they go on a motoring and camping tour of Britain. Katrina was enchanted with this plan.[134]

Katrina arrived by steamer at Grimsby, at the mouth of the Humber, early on the morning of July 16. Hans met her there, and during the next three weeks they drove through much of England and Scotland in his dauntless Austin roadster. They saw Edinburgh and Loch Ness. In the Lake District they pitched their tent in the rain at Grasmere. They visited their mutual friend from Altona, Walter Auerbach. Stopping by Cambridge, Hans introduced Katrina to some of his colleagues and showed her where he worked and lived. For 23 days they pursued a free and happy nomadic life. On August 8 Katrina took the train back to Grimsby and boarded the *Duisburg* for Germany. Before sailing she wrote Hans, "Are you glad to be rid of me, or are you also a little bit sad?" She herself was both happy and sad. On the homeward voyage she experienced again the sense of loss she had predicted she would, and wrote to Hans that she would "very definitely come again."[135] Meanwhile, Hans Krebs had returned to his laboratory and immersed himself again in his work.

On July 16, Judah Magnes wrote to Krebs to answer the questions respecting the prospective Institute in Jerusalem that Krebs had raised in his letter of July 1.

1. We are prepared to equip your laboratory as you desire. We rely, of course, upon you to be as moderate as possible in your requirements. We should like the money at our disposal to go as long a way as possible. Would you be good enough to send me a list of the requirements for the equipment of your Section? Perhaps we shall be able to get this or that as a gift from America or elsewhere.

Regarding the direction of the institute, Magnes proposed that each section head be "in full charge of that Section," that a colleagues committee decide policies "affecting the Laboratories generally," and that because of his "age and experience," Prof. Ludwig Halberstädter, the head of the intended Radiology Division, be chairman of the committee. Remarking pointedly that "it was these and other matters that we wished to discuss with you. . . . on the spot [in June]," Magnes added that "the Board of Governors of the University is to meet in Zurich on August 13th. Could you be at Zurich around that time?"[136]

These answers did nothing to satisfy Krebs. Had he retained his initial trust in Magnes, he might have been able to accept general assurances that he would have what he needed; but in view of the opinions of Weizmann and Nachmansohn about the chancellor of the Hebrew University, he must have perceived them as mere evasions. Again he delayed answering, perhaps struggling with the decision he was approaching. Finally, on August 11—four days after returning to the laboratory, and only two days before the meeting of the Board of Governors in Zurich that Magnes had asked him to attend—he composed the following:

I thank you for your letter of 16 July. I must confess to you openly that I cannot infer from your statements what will support financially the building of the chemical division of the planned institute. In view of the given shortage of funds I consider it better if the Institute were at first set up without a chemical division. The equipment of a modern physiological-chemical division requires—even with modest claims and limitation to a narrow field of work—far greater means than are envisioned in the previous plan, for its construction as well as its maintenance. The newly founded Institute in Rehoboth ought to provide a measure, even if the Cancer Institute is to be on a completely different scale. After long consideration, I cannot overcome my doubts about the workability of the plan, and I must therefore withdraw my earlier acceptance.

After assuring Magnes of his "undiminished" interest in the University in Jerusalem, Krebs concluded:

I must explain to you why I have only come to this decision so late. When you raised the question of my coming, several months ago, I was not well enough oriented, either about general conditions in Palestine, or about the financial side of the Institute. I have only over the course of time been able to inform myself more precisely, and I hope that you will understand my decision, and that there will be no inconveniences to you arising from my present decline of the offer.[137]

Hans Krebs could not have found it easy to write this letter. It is hardly surprising that his autobiography passes over this phase of his relations with Palestine. The embarrassment must have been acute; not so much because he had been misled, but because he had been credulous enough to be so easily misled.

3

Progress Under Pressure

In early August Hans Krebs returned from his idyllic holiday not only to his work, and to his hard decision about Jerusalem, but to an urgent family problem. His father and sister had pressed him, in addition to helping Wolf find a job in England, to write a "brotherly" letter about the importance of getting out of Germany. Hans carried out this family responsibility in a manner that was not merely brotherly, but big brotherly. His letter included what Hans regarded as "gently phrased" admonitions concerning a "few deficiencies" in Wolf's behavior, in particular the assertion that Wolf was not acting "seriously" enough in pursuit of possibilities for emigration. Not unexpectedly, but apparently to the surprise of Hans, Wolf wrote back a "very remarkable" letter defending himself from Hans's criticisms. Hans was upset enough to write Lise and his father to complain about Wolf's response and to say that he would give up on further efforts for Wolf as "pointless."[1] Now it was the turn of his father and older sister to admonish Hans.

On August 15, Lise wrote Hans that she regretted very much that "your brotherly and friendly help should have failed merely because you had not handled the matter in the correct form." She believed, however, that the fault for the tone of the exchange lay on both sides. Wolf too had complained to her. To him it appeared that Hans viewed the course that Wolf's life should take "only from his own orientation." Lise felt that Hans had "smartly reprimanded the 'little' brother" in a way that was reminiscent of how the older brother had sometimes treated the younger one when they had been small boys.[2] Georg Krebs too intervened, as he probably had often done in the past to protect his gentle younger son from the strong will of his older son. In his "birthday greeting" to Hans, Georg wrote on August 22,

> The second half of your letter, in the matter of Wolf, has pleased me less. I cannot read into Wolf's letter the attacks you find in it. Wolf seems to me to believe himself to be on the defensive. With what right I cannot judge, because I do not know what you wrote him.

That a few aggressive points can be found in his letter you do not have to take as tragic. . . .

In any case, I ask you sincerely not to give up your efforts to find a prospective position for Wolf. . . . In Germany he has no chance to find a position commensurate with his ability. And who knows whether he will be able to hold his present one?[3]

Wolf also wrote a birthday letter to Hans. He was friendly but firm, saying that he could only be open with his own brother in defending himself against complaints and advice that were "very wounding to me, and were, in my opinion, for the most part unjustified." Moreover, until Lise and Maria told him about the strenuous and time-consuming efforts Hans had made on his behalf, Wolf had not been aware that Hans had done anything at all for him in England. Wolf thanked Hans for those efforts, even though the outcome apparently had been merely that there was little prospect for Wolf to obtain in England a salary better than his existing one in Germany. If Hans should find anything promising for him, Wolf added, he would come to England to discuss matters; otherwise he would not, because of the expense involved. In closing, Wolf reported that in his work at General Electric (Allgemeine Elektrizitätsgesellschaft, or AEG) he had recently entered a new field in the analysis of electric motors, a field in which there were few experts.[4]

These exchanges reveal much about the dynamics of the Krebs family and the effects upon them of the stressful circumstances under which they lived. The brotherly clash reflects accurately the respective temperaments and perspectives of the two brothers. To Hans, who had earlier seen for himself no choice but to leave Germany, who had acted decisively to do so, and who had prospered by doing so, it was self-evident that his brother must do the same. Willful and impatient with his brother's more placid personality, Hans apparently felt that he must take a strong stand to spur his brother to the kind of action he himself would have taken in the same position; but Wolf was different. He remained nearly as shy as both of them had been in their youths and lacked the relentless drive of his older brother. Moreover, Hans failed to realize how well Wolf was doing where he was. That Wolf had invented improvements for electrical machines valuable to AEG did not impress Hans. Not appreciating a fundamental difference between the forms of advancement and recognition in academic science and in industrial development, Hans had sometimes warned Wolf, "Unless you publish things, nobody will know about" your contributions.[5]

Georg Krebs and his daughter Lise probably overestimated, for quite different reasons from Hans, the need to help Wolf. As the "baby" of the family, Wolf had been held from childhood in special affection. Where Hans had experienced his father as critical and demanding, Wolf experienced the same father as loving and admiring, and felt with him a closeness that eluded Hans. Lise too had lavished particular care and attention on Wolf, not only as a child, but in his years as a student; but both the father and the seven-year-older sister who often acted as a surrogate mother sometimes still failed to perceive that

Wolf had become an independent adult, and they tried to extend to him more protection than he needed.[6]

On his own Wolf was, in fact, solving design problems for AEG with remarkable success during just that time that his family feared for his future. Early in 1933 he discovered an elegant method for rapidly braking an AC electric motor by feeding a DC current into it. By causing the DC current to enter the coils at the "star points," at which the voltages of the three-phase alternating current just canceled one another, he could avoid damaging the DC rectifier and achieve a solution so simple that it required the addition of only two extra terminals to the standard motor. Soon afterward he solved a major problem that provided his firm a significant competitive edge in an important segment of the electric motor market.[7]

For a broad range of sizes and applications the dominant form of electric motor by the 1930s was the three-phase AC induction motor. Simple in basic design, these motors typically ran at approximately constant speed under widely varying loads. There were, however, many applications for which it would be advantageous to be able to run these machines at two or three different speeds.

In the induction AC motor the steady-state RPM was determined by the frequency of the alternating current input to the motor and the number of magnetic poles built into its stator. The frequency of the current was fixed by the generating source, so that the only way to vary the speed of the motor, without adding complex auxiliary control equipment, was to vary the number of poles in the motor. Because the magnetic fields were produced by electric currents carried through wire coils wrapped into grooves in the stator, it was possible in principle to change the number of poles by switching the connections between individual coils. In practice, however, this problem had been solved in a commercially effective manner only for the special case of speed ratios such as 1:2. The applications of the motors often required much closer ratios, such as 3:4 or 5:6. A number of designs for such motors had been patented, but none of them had proved successful in practice. Motors that achieved such ratios were usually constructed with two separate sets of windings, only one of which was in operation at a given speed. This was, in effect, building two separate motors into the same frame, a workable but inefficient arrangement.

One day, the research and development department at AEG asked Wolf if he could find a better way to obtain fractional speed ratios in induction motors. The Siemens firm had recently introduced a three-speed motor that was suitable mainly for certain specialized uses, such as driving machine tools.

By studying all of the technical literature on the subject that he could find, Wolf tried to understand why the various proposed designs for changing the number of poles by switching the connections between windings had not led to a practical solution. What he realized was that when the windings were connected so as to give the second speed, the waveforms for the variation in time of the magnetic fields deviated from the ideal sinusoidal curve, and that they differed in adjacent poles in such a manner as to produce strong harmonics. These harmonics reduced the performance of the motors and caused excess noise.

Searching for a way to eliminate particularly the lower orders of these harmonics, Wolf had the idea that if he could introduce into the stator two sets of winding parts, displaced from one another by nearly 180 electrical degrees, and with currents running in opposite directions, all the even-numbered harmonics would cancel out. Beginning with two coils, he systematically built up a complete winding that fulfilled this condition. Mathematically he could easily show that the efficiency of his design was sufficiently greater than that of a motor with separate windings, so that one could attain the same maximum performance within a motor frame smaller than those in the standard AEG "model" line.

When he had worked out this solution theoretically, Wolf sent instructions to the machine shop to construct a prototype motor by rewinding a standard frame according to his design. The first motor was immediately successful. So obvious was it that the design was generally applicable to induction motors that AEG quickly began to accept orders for two-speed motors of all sizes and ratings. By the time Wolf wrote a paper on the "New Reversible Pole Winding" for the company journal, "AEG-Mitteilungen", in 1934, he could report that "on the basis of its advantages the new winding has already found successful application in numerous reversible pole AEG motors, even in those of large power capacity."[8]

* * *

During a conversation in 1988 about this work, Wolf Krebs recalled,

> I found, when I was pondering any of those problems, perhaps for a week or a fortnight, or longer, and then suddenly, I found a solution. . .it upset my emotional side to such an extent that I usually spent a half sleepless night....If it's a problem which evades one for a long time, when one then finds the way and the path which leads one to the right solution, it makes it very exciting.[9]

Wolf thought that, in comparison to the complex problems with which his brother Hans dealt, his own were "fairly straightforward." To me, however, it appears that Wolf displayed creative investigative ability comparable to that for which Hans was winning international recognition in biochemistry. But they applied their respective abilities in different worlds. For Wolf the rewards for success were recognition mainly within the firm, some notice in competitive firms, opportunities for further interesting assignments, and the award of patents.

Perhaps these two talented brothers never realized how much they shared. Neither believed he could understand enough about the other's field to learn about the work the other did. When Wolf discussed with me his engineering achievements of the early 1930s, he still believed that Hans had taken little interest in them, if he even were aware of them. In 1979 Hans Krebs had shown me, with evident pride, reprints of several of Wolf's company publications on the design of "squirrel cage" electric motors, admitting also that because of his own limited mathematical background, they did not "make much sense to me."[10] There were many barriers to communication between these two brothers, who were, I believe, in some deep ways more alike than they knew.

* * *

This suggestion, made here primarily in the context of Hans and Wolf Krebs's activities as creative individuals within their respective professions, extends also to other aspects of their personalities. Although each of them, and those who knew both of them, tended to stress their differences, I sensed in my conversations with them striking resemblances in their manner of speaking, use of language, facial expressions, sense of humor, and a broad range of attitudes. Some of these similarities were undoubtedly due to common cultural experiences and transitions. Many of my conversations with Wolf took place after Hans' death, and there were some moments in which the resemblance to conversations with Hans seemed almost uncanny.

* * *

Hans Krebs's conviction that Wolf must make more serious efforts to get out of Germany was not entirely due to differences between his temperament and that of his brother, but in part also to their respective positions outside and inside of that troubled country. Judging the current situation mainly from reports in British newspapers, supplemented by dramatic news and propaganda clippings he received from contacts in Germany, Hans tended to exaggerate the impact on ordinary life of the actions and verbal intimidation directed against Jews. Very pessimistic about future developments there, he told every German with whom he came in communication not to stay as long as he or she could, but to leave as soon as possible.[11] For many Jews in Germany in the summer of 1934, however, it appeared less urgent to leave than it had when the Nazis first came to power. The early anti-Semitic demonstrations and boycotts of Jewish shops had aroused such unfavorable foreign reactions that Hitler restrained further violent actions against the Jews, and the situation seemed to settle down. For a large proportion of those not subject to the civil service law, life went on in a nearly normal way. Despite the determined effort of the Nazi government and its local party leaders to instill its hatred of Jews into the German population, many "Aryan" Germans were indifferent to that perverse message and continued to maintain ordinary business and professional contacts with Jews. Although Jews had to face daily the unabated propaganda campaign directed against them on the radio and in the press, Wolf had the feeling that, except for those who were already converted, most people did not seem to be listening—or at least avoided the appearance of listening.[12]

In his own professional world Wolf noticed some worrisome signs. There was a young man in the drafting room wearing an SS uniform, who appeared to have no function except to observe things he might report to the SS. Once when some documents from a file Wolf was examining fell out because the file folder had worn through, this alert observer reported that Wolf had carried out sabotage. The director of Wolf's department, who was also Jewish, was nervous enough to use a mirror to keep an eye on what was going on around him. These were, however, at the time small incidents. The draftsman who converted Wolf's designs into drawings sometimes appeared in an SA uniform, but they nevertheless continued to get along very well with one another. Wolf and his wife lived in a well-furnished apartment in Berlin, and he even bought a sailboat, in which they frequently spent happy Sundays on the *Mittelsee*, one of the many lakes near the city. Like many others in 1934, Wolf thought that he could survive the present troubles, that the regime would not last indefinitely, and that things might improve.[13]

In the meantime, Lise and her husband were meeting rebuffs that were leading them to conclude that they themselves would eventually have to leave Germany. In Düsseldorf, where the Daniels had moved from Berlin, Lise found no opportunities to continue her nursing career. In an attempt to prepare for other employment, she applied to take a nutrition course at the municipal hospital, but was informed that "pure Aryan descent" was a prerequisite for enrollment. For this and other reasons, her husband became persuaded that "in 1 or 2 years—if we are still living—we shall be settled in P[alestine]."[14] Time, the accumulation of unpleasant experiences, and the example of several friends who had departed for Palestine were gradually overpowering Lise's conviction that Zionism could not hold the answer to the Jewish Question.

I

We have seen already that holiday breaks from his weekly laboratory routine sometimes stimulated Krebs to alter the investigative direction in which he had previously been moving. This effect was especially strong in the case of his summer holiday of 1934. It is evident from his subsequent actions that when he resumed experimentation on August 7, he had come to a decision to shift away from those problems over which he had labored with so little overt success since his arrival in Cambridge. This time he did not have to seek new starting points in the contemporary literature of the field. He found them instead in leads, arising from his own earlier work in Freiburg, that he had allowed to lie fallow for more than a year.

During the investigation of the oxidative deamination of amino acids that he had pursued through much of his last year in Germany, Krebs had carried out most of the experiments with tissue slices. He had also, however, prepared a tissue extract in which some of the amino acids were rapidly deaminated to their corresponding α-keto acids, the latter undergoing no further decomposition. The extracts therefore contained, he claimed, an enzyme, or enzyme system, which he designated "amino acid oxydo-deaminase."[15] In August 1934, he began his new round of research by comparing the deamination activity of various cell-free kidney preparations. The title of the first of this series of experiments, "Amino-acid Oxydase," suggests that, having previously demonstrated the oxidative deamination reaction itself in isolated tissue, he was now moving on to characterize the enzyme which catalyzes this reaction.

From August 7 to 17, Krebs tested extracts made from ground swine or sheep kidney tissue, varying the proportions of water, the pH, the time of extraction, and other factors. He also prepared dried powders from acetone extracts. He found, among other things, that fresh aqueous extracts were more active than those prepared from acetone powders, that alcohol destroyed the enzyme, and that the addition of pyrophosphate or glutathione had no influence on the activity, whereas trypsin inhibited it. Somewhat to his surprise, since he had earlier maintained that amino acids are deaminated mainly in the kidneys, he found on the 14th, that for the case of acetone powders from swine tissue, a liver and a kidney preparation were equally active. On the 16th, on the other hand, a fresh acetone swine liver extract was much less active than the corresponding kidney extract. In all these experiments he employed racemic alanine as a standard substrate for comparing the activities of the various preparations. By the 20th, apparently satisfied that an extract prepared by mixing dried sheep kidney with water, centrifuging, and neutralizing, was suitable to his purpose, he tested it on 12 different amino acids, including both racemic and optically active ones.[16]

On August 22 Krebs switched abruptly to another aspect of his earlier deamination investigation, to a question that he had just identified when his research was interrupted. The question had arisen from some special, unexpected properties of the two dicarboxylic amino acids, aspartic and glutamic acid. In the first days of April 1932, he had observed that when either of these acids was deaminated in kidney slices, the rate of formation of NH_3 was unusually low, but that the rate increased in the presence of cyanide or arsenite, the two poisons he used to block the further decomposition of the products of deamination. He had further observed that, unlike other amino acids, these two cause NH_3 added to the medium to disappear. He had been forced to leave unanswered "the question, which nitrogen compounds are formed in the unpoisoned kidney cells from the ammonia and the amino acids." It was evidently this question that he was now ready to take up again. In his first new experiment on the subject he compared the rates of oxygen uptake and ammonia formation in sheep kidney slices with glutamic acid, glutamic acid + HCN, and aspartic acid to that of controls with no additions and with HCN alone. The most marked effect was an "enormous increase in respiration with the dicarboxylic amino acids!!" There was also "a large percentage (500–650%) increase in NH_3-formation, but less than would be expected from the rise in respiration!!" The level of respiration rose, in the absence of HCN, from $Q_{O_2} = 11.2$ to $Q_{O_2} = 30.0$ for glutamic acid and $Q_{O_2} = 26.7$ for aspartic acid. The Q_{NH_3} for the control was 0.60; for HCN alone, 1.97; for glutamic acid alone, 2.79; for glutamic acid + HCN, 2.10; and for aspartic acid alone, 3.60. Thus the absolute increases were much less than predicted on the assumption that the deamination reaction produces NH_3 in the same molar quantities that it consumes oxygen. From his previous experiments he would have expected that equivalence to hold for the case of glutamic acid + HCN in relation to the case of

HCN alone. Moreover, he noted that "in general the readings are too high."[17] The experiment thus provided no immediate clarification of the situation, and Krebs turned the next day to a different question.

The experiment that Krebs performed on August 23 was entitled "Do 2 Enzymes Exist: Comparisons of *d* and *l* Leucine under Various Conditions." He established four sets of "conditions," comparing the rates of respiration and NH_3 formation of the two isomers and a control in kidney slices, in kidney slices in the presence of octylalcohol, in liver slices, and in a kidney extract. In each of the kidney preparations the "nonnatural" *d*-leucine greatly increased the rate of NH_3 formation, while the "natural" *l*-leucine raised it little or none. In the kidney slices the same was true for the rates of respiration. In the liver neither isomer had much effect. In spite of these sharp differences, Krebs did not regard the experiments as successful. It did show that "there is little deamination by *d*-leucine in the liver"; that "octyl alcohol does not inhibit the deamination of *d*-leucine"; and that the "extract does not inhibit *d*-leucine deamination"; but, since "*l*-[leucine] is not clearly attacked either in slices or in the extract," it "is an unsuitable substance for testing the question."[18] Far more significant than this unsatisfactory outcome was the very posing of the clear, simple question that the experiment was intended to test. What Krebs meant by it was, is there a separate enzyme for the deamination of each of the two optical isomers of the amino acids? He had long since discovered that in kidney tissue the "nonnatural" isomers are deaminated at far higher rates than the "natural" ones. In his past discussions of the topic, however, he had referred only to an "amino acid deaminase," implying that a single enzyme catalyzes the reaction for both isomers. To ask whether the rate differences were a clue that there might be two enzymes was an important clarification of the issue. The caption for this experiment is the earliest record of Krebs asking that question. We may wonder, therefore, whether it represents the point at which he first perceived the question that way. Quite possibly it does. The fact that he had headed his initial experiment of the new series on deamination "amino acid oxidase" can be interpreted as an indication that only two weeks earlier he had still been thinking in terms of one enzyme system. It is risky, however, to read too much into simple captions. On the whole I think it equally plausible that it was because he had thought of the question sometime before August 7, that he went back to the old deamination investigation, and in particular that he began with the preparation of extracts suitable to the study of the enzymes involved.

The next day Krebs carried out an experiment on "glutamine and liver extract." The description is short enough to quote in full:

> 15 mg. glutamine in 2 ccm. Ringer's does not decompose substantially in 1 hour at 32°.
> + 1.0 of liver extract from yesterday (2 liver + $8H_2O$) no increased formation of NH_3 in 60'.[19]

Once again, the negative result of the experiment is less significant than its subject. Krebs was here examining whether glutamine would decompose in the liver extract, but his selection of the substance to examine its enzymatic reactions at this time indicates that he had already in mind the likely candidates for the nitrogen compounds formed in tissues from the two dicarboxylic amino acids. Glutamine is the amide of glutamic acid:

$$
\begin{array}{ll}
\text{COOH} & \text{CO—NH}_2 \\
| & | \\
\text{CH}_2 & \text{CH}_2 \\
| & | \\
\text{CH}_2 & \text{CH}_2 \\
| & | \\
\text{NH}_2\ \text{CHCOOH} & \text{NH}_2\ \text{CHCOOH} \\
\text{Glutamic acid} & \text{Glutamine}
\end{array}
$$

Glutamine was little known in animal metabolism, but had been found widely distributed in plant matter and was known to give rise to glutamic acid by hydrolysis, releasing NH_3. The homologous amide of aspartic acid, asparagine, had been more extensively investigated. The enzyme asparaginase, catalyzing the hydrolysis of asparagine to aspartic acid, was found both in plants, especially yeast, and in animal tissue, particularly in the liver. Several of those who investigated the characteristics of asparaginase, such as W.F. Geddes and Andrew Hunter in Toronto, had raised the question whether this enzyme might also catalyze the hydrolysis of glutamine.[20] Glutamine and asparagine were therefore fairly obvious first choices for Krebs to test for an answer to his question: What substances are formed when aspartic acid and glutamic acid absorb NH_3 in tissue slices?

From his subsequent activity we can infer that by this time, if not sooner, Krebs had resolved to make the two questions embedded in the preceding experiments the central thrust of his research effort. He pursued the two lines of investigation, sometimes alternately, sometimes with aspects of both problems dealt with in the same experiments. During the early weeks he gave more priority to the reactions of the dicarboxylic amino acids than to the deaminase enzymes. In contrast to some of the earlier phases of his experimental career, he concentrated steadily and persistently along an experimental path guided by these questions. It appears evident that he maintained, as strong hypotheses, that there *were* two distinct amino acid oxidases for *d* and *l* isomers and that glutamic and aspartic acid *do* give rise respectively to glutamine and asparagine in animal tissues. This situation gave a somewhat different cast to his experimental style than to that of the preceding year. There was relatively little scanning for new effects, and few digressions. His investigative problem was to gather the evidence that would support or refute two well-specified hypotheses.

In the first phase of his investigation Krebs sought mainly to confirm the results of the small number of experiments carried out in Freiburg from which he had then inferred that the dicarboxylic acids form a nitrogenous compound in animal tissues. Initially he tried to repeat the observations showing that in unpoisoned tissue these acids produce little NH_3, but that in the presence of arsenite or cyanide the quantity increases, and the proportion relative to that of oxygen consumed approaches the 2:1 ratio predicted by the oxidative deamination equation. Things did not go smoothly. In one experiment the NH_3 measurements were unusable, because the cyanide solution was old, and contained much NH_3. In another the inhibiting action of the cyanide "quickly disappeared." In a third, in which he compared the inhibiting effects of different concentrations of cyanide, he unfortunately omitted a control containing the amino acid and the tissue without cyanide.[21]

Krebs interjected into this rather sorry series, on August 30, a single experiment on the "washing out of the d-system." In Freiburg he had already found out that he could extract some of the amino oxidase activity from tissue slices, without destroying the cell structure, simply by shaking slices for an hour in physiological saline solution. Returning to this technique, he shook kidney slices in phosphate Ringer's solution for various lengths of time, transferring some of them at different intervals to fresh solution. He then apparently tested the remaining activity of the slices on the "natural" isomer, l-valine. From his results he calculated that in the first hour "22% washed out," and after the second hour, 32 percent. The experiment is not easy to interpret from the elliptic notebook record. It appears contradictory to assume that what is washed out is the d-amino oxidase system, yet to measure that quantity by the remaining activity of the l-amino oxidase system in the tissue. At any rate it appears that whereas what he had once regarded as the extraction of a single oxidase system, he was now treating as a selective extraction of the d-oxidase system.[22]

Returning to the dicarboxylic acid series on August 31, Krebs had another failure in an experiment with glutamic acid, guinea pig slices, and arsenite. The rates of oxygen consumption were so anomalous that he suspected the arsenite had either been old or put into the wrong flasks.[23]

It is probably too easy to pass judgments from afar on experiments for which the notebook records provide only partial accounts. Perhaps Krebs had a string of unusually bad luck during August; but it nevertheless appears that at least some of these failures could have been avoided by more meticulous experimental design and execution. Just as he characteristically tested out simple ideas that he might have spent more time thinking through in advance, so he seems occasionally to have displayed a tendency toward hasty experimental planning. In both respects he relied on the ease with which he could multiply and vary his experiments, and on the dogged pace at which he did so, to make up for the deficiencies that sometimes crept into his performance on individual experiments.   These characteristics of his research style were mutually interdependent, for it was probably in part because he insisted on carrying out at least a full set of experiments every working day that the time he had available to plan each set was restricted to a single evening.

Continuing with similar experiments in early September, Krebs tested, on the 3rd, aspartic acid in rat kidney tissue, in the presence of two concentrations of arsenite. He included a control without arsenite and three controls without the aspartic acid. This time no immediately evident experimental flaws turned up, but the results were nevertheless puzzling. From the measurements of the oxygen consumption and NH$_3$ formation, he calculated the ratios of their rates:

Without arsenite $\qquad \dfrac{Q_{O_2}}{Q_{NH_3}} = \dfrac{26}{10.8\text{-}2} = 2.96$

0.05 arsenite $\qquad\qquad\qquad \dfrac{3.7}{4.3\text{-}1.8} = 1.48$

0.001 arsenite $\qquad\qquad\qquad \dfrac{6.7}{4.47\text{-}2.66} = 3.70$

He commented on these figures, "It is not very straightforward (as earlier with rabbit). Take less arsenite." The "earlier" experiments with rabbit slices refer to two sets of experiments similar to this one that he had carried out in Freiburg. Then he had obtained, in the absence of arsenite, much higher Q_{O_2}/Q_{NH_3} ratios. The addition of 0.05 arsenite had lowered the oxygen consumption and increased the yield of NH$_3$. These results had formed the starting point for Krebs's inference that the dicarboxylic amino acids form with NH$_3$ an ammonia compound, absorbing all or most of the NH$_3$ which an ordinary deamination reaction would otherwise release. The arsenite he had interpreted as blocking the further reaction. Although Q_{O_2}/Q_{NH_3} did not approach the 1:2 ratio predicted by the deamination equation, the change brought about by the arsenite was, in general, strongly in the direction required by this interpretation. No such clear-cut change occurred in the new results. Moreover, the changes in the ratios that did occur represented mainly the larger inhibition of the respiration, for the absolute rates of formation of NH$_3$ with arsenite were actually lower than without it. The new result therefore added no further support to Krebs's original evidence for the reaction he sought to demonstrate.[24]

One other salient feature of the comparison that Krebs drew between this result and the older ones is that he differentiated the rat kidney tissue used here from the rabbit tissue earlier employed, but did not note that the earlier experiments had utilized glutamic acid instead of aspartic acid. This was in keeping with his expectation that he was dealing with the corresponding reactions of two closely related homologous compounds.

In the first four weeks of his renewed experimental campaign on the dicarboxylic amino acids, Krebs thus appeared not only to have made little progress beyond the point he had reached in Germany, but even to have lost some ground. His luck began to turn in another experiment, carried out on the same day as the above, on the "cleavage of glutamine." Comparing the formation of NH$_3$ in rat kidney and small intestine slices in the presence of glutamine, and with controls excluding respectively the tissue and the glutamine, he obtained for kidney tissue Q_{NH_3} = 24.4, in contrast to Q_{NH_3} = 0.9 for the intestine. "Kidney cortex," he concluded, "is enormously active as 'glutaminase'."[25]

Armed now with suggestive evidence that the reverse of one of the reactions he sought occurs in kidney tissue, Krebs pressed forward to resolve the difficulties obstructing his latest efforts to repeat the observations from which he had initially inferred the existence of the reactions themselves. On September 4 he measured the oxygen uptake and ammonia formation in rat kidney slices with the natural isomers of aspartic and glutamic acid, with and without arsenite. Contrary to the advice he had given himself for improving the preceding experiment of the same type, he used more arsenite (0.05 M), rather than less. The results were unhelpful. When added without arsenite, both substrates raised the rate of NH_3 formation above that of the control. The arsenite, on the other hand, decreased the rate of NH_3 formation in the presence of both substrates. These effects were both opposite to those required for his hypothesis.[26]

Two successive experiments had turned out contradictory to what Krebs would have expected on the basis of his hypothesis, but he probably did not even entertain the possibility that they constituted a refutation of the hypothesis. To have accepted that experimental verdict would have been essentially to end the investigation. He assumed instead that he had not yet attained the experimental conditions necessary to display consistently the effects he sought. The next day he tried again, limiting himself this time to glutamic acid, switching from rat to guinea pig tissue, and now implementing his recommendation to employ smaller quantities of arsenite. This time he obtained just what he needed to consolidate his case:

	No substrate		5×10^{-2} M d-glutamic acid	
	Q_{O_2}	Q_{NH_3}	Q_{O_2}	Q_{NH_3}
No arsenite	16.8	1.20	25.7	≈0
2×10^{-4} M arsenite	8.53	1.18	10.2	2.30
4×10^{-5} M arsenite	?	?	19.2	≈0

According to these data d-glutamic acid (here designating, in the old notation, the dextrorotatory "natural" isomer) in the absence of arsenite increased the respiration, but suppressed the formation of ammonia, as it ought to if it was combining with ammonia to form glutamine. The action of the smaller concentration of arsenite was too weak to alter the situation, but the higher concentration caused more NH_3 to form than in the tissue without arsenite, in accord with his view that the glutamic acid was now releasing NH_3 in deamination and that the arsenite was preventing the formation of glutamine.[27]

Krebs wrote down as the ratio for the measured Q_{O_2}/Q_{NH_3}, in the case of glutamic acid and 2×10^{-4} M arsenite, 4.44:1. He noted, "Calculated, 4.5."[28] The two were in very good agreement. To reach this "calculated" ratio he must have added a figure for oxygen based on the theoretical ratio of the oxidative deamination reaction to a measured figure representing the oxygen consumed in other cellular processes and divided by the measured Q_{NH_3}. I have not been able to arrive at this number, however, from the data shown in the notebook. At any rate this experiment was probably very convincing to Krebs. (His confidence in this particular experiment is reflected in the fact that he later used the data from it as the basis for the published version of the reasoning outlined above.)[29] He had, at least for this stage of the investigation, and for glutamic acid, adequately reinforced the first phase of his argument.

This experiment included, in addition to kidney slices, corresponding measurements with liver slices and four other tissues. There was no "substantial" NH_3 formation in any of them, and Krebs concluded "d-glutamic acid is transformed almost exclusively in the kidneys." The substrate did cause an "increase in respiration in muscle!" he noted, but without a corresponding rise in the rate for NH_3.[30]

On September 7, Krebs moved on to repeat the second phase of his original Freiburg investigation, testing the effects of ammonia added to the medium of the slices in the presence of glutamic acid. He measured the oxygen consumption and NH_3 formation with glutamic acid alone, with glutamic acid + arsenite, with glutamic acid + NH_4Cl, and with a set of three controls excluding the glutamic acid. The results generally favored his hypothesis. With glutamic acid alone the quantity of NH_3 did not change; in the presence of arsenite, glutamic acid yielded NH_3; in the presence of NH_4Cl + glutamic acid, NH_3 was consumed. The measurements for the tests with added NH_4Cl were questionable, however, so that the experiment was unreliable. In the corresponding experiment with liver slices, Krebs measured the formation of urea instead of NH_3, and found that glutamic acid + NH_4Cl increased the rate, from $Q = 1.27$ to 1.92.[31]

As far as the record reveals, Krebs did not consider any alternatives to asparagine and glutamine as the compounds he expected to be formed by the combination of ammonia, respectively, with aspartic and glutamic acid. The reason was, no doubt, that these were the simplest, most direct reactions imaginable for the situation. He did, however, write down in his notebook at this time a page of "paper chemistry" which shows that he was asking himself whether glutamic acid might also undergo other reactions (see Fig. 3-1, next page).[32] There is no indication of the source of his interest in such a reaction, and he did not follow it up experimentally, but he did keep in mind that proline might play some role in the processes he was studying.

After carrying out experiments on the influence of the concentration of glutamic acid, and of other factors, on the respiration of kidney tissue, Krebs repeated, on September 11, the test of the effect of glutamic acid on NH_3 added to guinea pig tissue slices. This time he reached the desired outcome:

Enormous disappearance of NH_3 with glutamic acid.
No   "   " "   without "   "   .

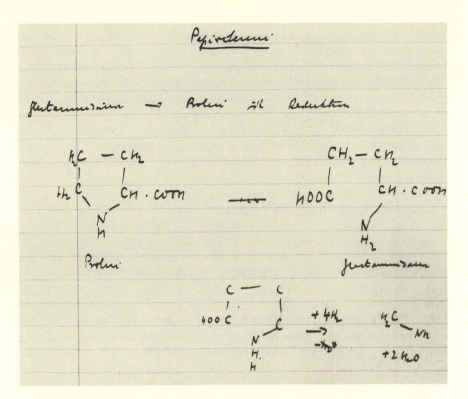

Figure 3-1 Krebs laboratory notebook No "11," p. 123.

(The figures from which he drew this conclusion were, apparently, Q_{NH_3} + glutamic acid = -5.22; Q_{NH_3} without glutamic acid ≈ 0. Krebs must, however, have placed erroneously in the column giving the latter result an arrow indicating the presence of glutamic acid.)[33]

So far in his new investigation Krebs had managed not much more than to confirm, for one of the two dicarboxylic amino acids, the evidence that had led him 15 months earlier to infer that these acids combine with NH_3 to form a nitrogenous compound. It remained now for him to extend that evidence to aspartic acid, and to establish experimentally that these nitrogenous compounds were the amides of the acids. On September 12 he incorporated both of these objectives into one set of experiments.

To detect the formation of the acid amides, Krebs relied on a well-known method for distinguishing ammonia released by amides from that released by amino compounds. Amides were hydrolyzed by heating in a dilute acid solution. To utilize this test, Krebs set up the experiments in the usual way with tissue slices in the presence and absence of glutamic and aspartic acid. After determining the change in the quantity of free NH_3 in each medium in the usual way with the Parnas-Heller apparatus, he placed each medium in HCl solution and heated it for four hours over a water bath. Then he neutralized the solutions and measured the free NH_3 again. Any newly formed NH_3 represented that released from acid amides produced during the experiment.[34] Because the test did not identify the specific acid amide present, it remained as an inference from his hypothesis that the acid amide produced in an experiment employing glutamic acid must be glutamine, and that appearing in an experiment using aspartic acid must be asparagine.

For glutamic acid, the first results Krebs obtained were decisive. With no NH_4Cl added to the medium containing guinea pig kidney slices, in the presence of glutamic acid, amide-N formed at the rate $Q = +6.76$. With NH_4Cl added, in duplicate experiments,

$$\begin{array}{lll} Q_{NH_3} & = -16.1 & -13.0 \\ Q_{amide-N} & = +21.0 & +17.2 \end{array}$$

Thus, in each experiment the slice formed somewhat more amide nitrogen than it consumed of added nitrogen. The difference could be explained as due to NH_3 arising from deamination of some of the glutamic acid. The experiment thus strongly supported Krebs's hypothesis that kidney tissue converts glutamic acid and ammonia to glutamine. (Eventually he included this result too in the published data supporting that conclusion.)[35] Aspartic acid, on the other hand, performed badly. It did not consume added ammonia, so Krebs did not even measure the amide-N.[36]

After performing this critical experiment, Krebs left Cambridge for a long weekend with Bruno Mendel in Holland. He must have done so with a very good feeling about the state of his research. There was much left to be done in order to consolidate the results obtained thus far; but he could travel with reasonable assurance that he had just discovered a hitherto unknown metabolic reaction.

II

With his trusted friend Mendel, Krebs had much to discuss besides his latest scientific progress. No sooner had he settled the Palestinian question for himself in August by withdrawing his acceptance of the offer from Judah Magnes, than the situation took a sudden new turn, and he found himself again drawn in. From Redcliffe Salaman he learned that the conference in Zurich of the Board of Governors of Hebrew University, on August 13, had drastically altered the circumstances surrounding the planned Institute. Magnes had been rebuked and his authority reduced. The planning for the Institute had been removed from his control and turned over to a commission consisting of Salaman, Weizmann, Magnes, and Emanuel Libman, a well-known New York City clinician. The previous scheme for the institute had been given up, and several of the scientists intended to work in it, about whom Krebs had had doubts, had either been excluded from it or their roles diminished. Salaman had asked Krebs to suggest how the £40,000 allocated for the institute should be employed if Krebs should become chief of the institute, or, in case he did not want to be, what he would otherwise recommend.[37]

All of this, Krebs wrote David Nachmansohn on August 24, "is gratifying beyond expectation. . . . I am at the moment very much occupied with this difficult matter [of preparing the memorandum]. Salaman will be away travelling for 4–5 weeks, so I have that much time."[38]

The central point on which Krebs intended to insist was that the institution should not be named, as presently planned, a cancer institute. It should be designated more generally as a "Medical Research or something like that" Institute. To be required year after year to publish research on cancer would be an enormous hindrance. "There are many cancer institutes," he wrote Nachmansohn, "in which nothing at all is brought forth. Almost all of the important discoveries in the carcinoma field have been made—in connection with other investigations—not in cancer institutes." Krebs was adopting here a position derived in part from similar views upon which he had often heard Otto Warburg insist.[39]

Despite these positive developments, Krebs was no longer in a mood to leap at the prospect laid before him. "The most difficult question," he confided to Nachmansohn,

> is naturally the personal question. I am not yet entirely clear in my own mind, whether I am the right person; and I have also just told Salaman that I now regard myself as committed for 3 years in Cambridge. What do you think about this [?] Have you any interest in and desire to build an Institute? Or whom would you otherwise suggest [?][40]

Nachmansohn would certainly understand, Krebs added, that under these circumstances he could not consult Weizmann, either about the present situation or about the alternative of a possible position in Weizmann's planned institute. It would not be fair to Salaman, and might arouse mistrust. Finally, he would no longer consider traveling to Palestine during the current year, but hoped to do so in the following year.[41]

Not surprisingly, Nachmansohn was excited over this news. "My personal view," he wrote back, is

> that it is for you a very great, perhaps unique opportunity. You would not only be a director of a university institute at a very young age, but the determining man in the medical-scientific faculty ... And I am fully certain that a person like you, with the very great possibilities that such a position offers you, would feel very happy in a land where, to be sure, everything is still beginning and becoming, but that is full of unheard-of vitality, of affirmation and joy in life.[42]

Nachmansohn was confident that additional funds would be found to supplement the proposed £40,000 budget for the institute, so that financial support would be no problem for Krebs. He was flattered that Krebs would suggest him for the position if he himself decided against it, but sought to dispel Krebs's doubts that he was himself the person to do it. "You are absolutely the right man to make a great thing out of this. Besides you I can see no one." Moreover, Nachmansohn thought Krebs should not delay beyond the following summer to take up the position. The next three to four years would be decisive for the development of scientific institutions in Palestine, and if he arrived too late, he would have less influence. Nachmansohn could not see why Krebs felt he had an obligation to stay in Cambridge for three years. It was generally understood that anyone accepting an academic appointment is free to leave before the end of the specified term if another institution makes an offer constituting a promotion. Nachmansohn also believed that Krebs was far too optimistic about his long-term prospects in England.

> whether England will offer you such great opportunities later on, I do not know. That Hopkins is doing everything to keep you now, I understand. I would have done that too. But I fear that if you remain in Cambridge, you will not always be spared from bitter disappointment. Even in Germany we did not consider possible what has happened. To be sure, I do not believe that there will be a Hitler in England. But against how much resistance will you have to struggle? And will you not in the end still remain an "alien"? Talk sometimes about it with people who are not, like Hopkins, personally interested in you. And do not forget that the generation of Hopkins and [A.V.] Hill, who, besides the greatness of their personalities, grew up in an age of real and genuine liberalism, will soon not be in a position to make these decisions.[43]

Krebs was not the sort of person to be swayed easily by the opinions even of close friends, where they differed from his own. Clearly he was not persuaded to take so dim a view of his future in England as Nachmansohn represented it. Nevertheless, he could not entirely disregard such warnings. He was well aware that there existed in England resentment at a foreigner being given an academic position at a time when chronic high unemployment made it difficult for English citizens to find jobs. Hopkins had, in fact, received some protests over the appointment. Moreover, Krebs was given permission to stay in England for only one year at a time, and had to keep in mind that some year his application to extend his visa might be turned down. No matter how auspicious his immediate situation might appear to be, he could not forget that his future there remained uncertain.[44]

Balanced against such concerns was the fact that the longer Krebs remained in Cambridge, the more attached he became to the Cambridge "milieu." He was particularly happy in the biochemistry laboratory itself. Much later he expressed eloquently what it had meant to him:

It was in Hopkins's laboratory where I saw for the first time at close quarters some of the characteristics of what is sometimes referred to as 'The British way of life.' The Cambridge laboratory included people of many dispositions, convictions, and abilities. I saw them argue without quarreling, quarrel without suspecting, suspect without abusing, criticize without vilifying or ridiculing, and praise without flattering.

· · ·

To me, as to many others, life in Hopkins's laboratory was a great education, a model of how true and lasting strength can be achieved in a team of scientists.[45]

Krebs may not have been able to define the environment of the Cambridge Biochemical Laboratory as crisply while he was there as he did here while reflecting back on it over a quarter of a century, but he obviously thrived in it, scientifically and personally. Some of the diverse people in it became his good friends. He went on walks with Eric and Barbara Holmes, sometimes had dinner with Bill and Tony Pirie. Surrounding the immediate circle of the laboratory were important contacts with leading scientists in neighboring laboratories, including Keilin, Barcroft, and Francis Roughton. These men too became stimulating colleagues and friends. Beyond the scientific connections lay the rich cultural life of Cambridge. Even for a person who had to be as frugal of both his time and money as Krebs did, there was much to enjoy. He found the theater excellent and inexpensive. He went swimming, hiking, and cycling, and participated in the favorite Cambridge recreation of punting along the backs. He felt very much at home.[46]

Krebs's satisfaction with Cambridge owed much to Hopkins himself. To the gratitude Krebs felt for Hopkins's firm support and welcome was added growing admiration and affection. Just as Warburg had served as his model for rigorous scientific investigation, Hopkins now became his model for leadership of a scientific group. It was Hopkins who encouraged and nurtured the diversity that Krebs saw as the greatest strength of the Cambridge school. "Hopkins," he recalled later, "was the central figure, beloved and respected as a natural leader, exercising leadership from within and not from above, utterly humble, modest, gentle, but by no means weak."[47]

For those who came to work and learn in his laboratory, Hopkins created opportunities, but he did not provide individuals with detailed supervision in finding or solving their research problems. That, he believed, had to come from within. This style of Hopkins's leadership was forcefully impressed on Krebs one day when he came to talk with Hopkins about a young Ph.D. student who was working near his bench in the laboratory. The student was having difficulty, and asked Krebs for advice. Krebs thought that the student was on the wrong track, and believed that it was his responsibility to point out that the problem was not very good, that to pursue it was a waste of time. Hopkins dissuaded him, saying that young people should not be directed too closely.

Figure 3-2 Hans Krebs punting in Cambridge: accompanied by the daughter of Murray Luck.

This was the stage at which promising investigators separated themselves out from those who would never get very far.[48]

Krebs learned little about research methods from Hopkins. He considered Hopkins's methods rather old-fashioned, in comparison to those he had brought with him from Warburg; but Hopkins's ideas about biochemistry seemed to Krebs thoroughly modern. He felt that Hopkins thought about biochemistry in much the same way that he himself did. Although Krebs remained somewhat overawed by Hopkins, he found him easily approachable, discussed scientific questions frequently with him, and benefited from the breadth of Hopkins's view. When Krebs told Salaman and Nachmansohn that he was committed to Cambridge for three years, he was expressing not just a contractual obligation, but a personal one. Hopkins had made it clear how much he valued Krebs's presence in the Cambridge laboratory. Krebs would not find it easy to disappoint this strong, gentle man toward whom he felt a deep attachment.[49]

Pressed between his happiness in England and loyalty to Hopkins on the one hand, and heady opportunities and expectations for him in Palestine on the other; between the general humanistic-intellectual values which had been instilled in him early and the special concerns of the Jewish people that had lately impinged so starkly on his life; between the view of his family that he would be best off in England and the opposite view of his friend Nachmansohn, Krebs must have been eager to talk his situation over with Bruno Mendel. From Mendel, whose counsel he had often sought ever since they had consulted each other daily about their first scientific experiences in Berlin, Krebs could be confident that he would receive opinions independent both of the enthusiastically Zionist views of Nachmansohn and of the adamantly anti-Zionist convictions of his father.

Mendel met Krebs in Amsterdam and drove him to his house in a small, nearby town. Krebs stayed with Mendel and his wife for two days. There is no record of their activities and conversations, except that both Krebs and the Mendels enjoyed the visit very much.[50]

The assertion of a three-year commitment to Cambridge undoubtedly did not arise solely out of Krebs's sense of duty to Hopkins. It was also convenient to his own purposes, in protecting him from being pushed too quickly toward a decision for or against Palestine. Having once experienced the discomfort of backing out after deciding too quickly, he was probably now determined to move cautiously. It was too early to judge whether the new plans and hopes of those who wanted him for the institute in Jerusalem would materialize. In spite of his disclaimer to Nachmansohn, he was probably by now well enough aware of his strategic importance to these plans, so that he could afford to temporize. By stressing the claims of Cambridge on him, he improved his bargaining position. In this way he could maintain the option of eventually accepting the position, without eroding his present position. Unlike Nachmansohn, who saw in the new developments the chance of a lifetime for Krebs, Krebs himself probably now perceived in them mainly an alternative to which he could turn if he were to encounter future barriers in England.[51]

III

In his personal life Hans Krebs also faced conflicting pressures, and expectations he could not immediately fulfill. Katrina had returned to Germany sad, lonely, and yearning for the happiness she had recently known. When she thought back on their days together, she wrote Hans in late August, "It is as though I open a door and look into something wondrously beautiful. Then the door closes, and everything becomes colorless."[52] She was not content, however, simply to stare at a closed door, and she pressed Hans in further letters to make clear what he thought about their future. "It can never again be as it was for us before the holiday," she wrote. She had not written him for four years, and then finally come to him, "only in order to experience a one-time pleasure, but to show you how I feel about you."[53] Her situation was particularly difficult, because, except for one or two trusted friends, she could not let anyone in Altona know about their relationship. That was undoubtedly in part due to its uncertainty, but also because she could not, in Nazi Germany, afford to reveal that she maintained ties with a Jewish refugee. Consequently she felt isolated mentally from those in her everyday working world, even as she was isolated physically from Hans.[54]

In the absence of the letters that Krebs wrote to Katrina, we can only partially reconstruct his position. From references in her letters it appears that he must have been seriously interested in her, that he at least entertained the possibility that they would eventually marry; but also that he was either not certain enough of his feelings toward her or of how matters could be practically managed to make a definite commitment to her. In one letter he wrote that he thought "really very much" about their future.[55] In another letter he pleased her greatly by reporting that he had moved to larger rooms in a boarding house near the edge of Cambridge, and that he had given consideration to the question of whether there would be space enough in them for her; but it is also evident that he had expressed doubts about whether he was ready to settle down with one person, had pointed out that he was so busy that he would have to leave her alone much of the time, and raised other difficulties.[56] He promised to write regularly, but usually kept her waiting days or weeks for a letter that more often than not turned out to be brief. It is likely that Krebs was in a genuine quandary over Katrina, as well as over the question of marriage in general. He was now far from the terribly shy young physician who had come to Berlin 10 years earlier. He had found that his quiet good humor and gently affectionate manner exerted an easy attraction. He had formed numerous attachments with women, ranging from "old friends," such as Ursula Bucky, to a number of romantic involvements. More than once he had come near to the point of becoming engaged. That none of these engagements had materialized was partly due to outward circumstances, particularly the severe economic obstacles that forced many people of his generation to postpone that step year after year. Partly, however, his own personality may have been a factor. The women he attracted wanted to help him and to become more deeply involved in his life than he would permit. The playful charm that he displayed in times of relaxation he did not allow to intrude on the single-mindedness with which he pursued his scientific life. When Katrina wrote, just after their holiday "Only do not forget these days too quickly in the bustle of work,"[57] she showed that she knew him well.

It is also evident that Krebs had come to enjoy his bachelorhood sufficiently so that he was reluctant to give up the freedom that it allowed him. Yet he must also have had in mind that at the age of 34 it was time for him to consider forming a permanent tie. All of these things must have been much on his mind in the fall of 1934. His professional uncertainties were probably not unrelated to his personal uncertainties. If Krebs were to consider marriage, he faced financial hurdles that were still formidable. His salary in Cambridge was just enough to support his modest needs as a single person. There are some hints that he was growing impatient with the stringent circumstances that he had so long endured. Since the previous spring his father had been sending him monthly sums of money, probably mainly from Hans's own bank account in Freiburg. The restrictions on sending German currency abroad were growing ever tighter, however, and by fall his father could send him no more than 10 marks per month.[58] When Krebs thought about the possibility of going to Palestine as the head of a research institute, therefore, one of the considerations that probably weighed heavily on him was the opportunity it could provide him finally to support a wife. Katrina let him know that if he decided for Palestine she would come to him there. She too thought, however, that England suited him better.[59]

IV

After his trip to Holland, Krebs continued his investigation of the dicarboxylic amino acids. During the last two weeks in September he performed mainly variations on the experiments he had previously carried out. He explored what conditions would best bring out the effects he had already observed, he refined some of his analytical procedures, and he looked for the processes he had observed primarily in guinea pig kidney slices in the kidneys of other animals, as well as in the tissues of other organs. Not all of the experiments worked out as expected. In rat kidney slices, on September 18, glutamic acid caused no consumption of NH_3 in the presence of added NH_4Cl. Some experiments raised new questions. For example, the fact that in rat liver tissue glutamic acid appeared to inhibit somewhat the formation of urea led Krebs to ask whether that might be an effect of the formation of glutamine (the glutamic acid presumably competing for the NH_3 available for the synthesis of urea). In guinea pig liver, on the other hand, he observed that "glutamic acid increases the formation of urea without a corresponding increase in NH_3 consumption. Clearly no glutamine formation. Repeat, but measure the amide-N!" Krebs was less interested in following up new leads such as these than he might have been at some other time. The urgent task before him was to accumulate further evidence for what he had already found.[60]

On September 22 Krebs tested glutamic acid with guinea pig kidney in the presence of NH_4Cl, and in an atmosphere of nitrogen, and observed, "No consumption of NH_3 anaerobically." This result fit his expectations. The formation of the amide was a synthetic reaction, requiring energy, and therefore cellular respiration as a source for the energy. The next day he tested whether the reverse reaction would occur anaerobically. Both glutamine and asparagine yielded NH_3 in liver and kidney slices in the absence of oxygen. The result was flawed, however, by another one of those essential details that Krebs sometimes neglected in his drive to experiment incessantly. "Unfortunately," he realized afterward, the experiment "lacks an anaerobic tissue slice without additions."[61]

On the same day Krebs moved on to a new phase of his investigation, examining "glutaminase and asparaginase" in extracts of kidney and liver tissue. Previous investigators had studied the action of asparaginase mainly in yeast extracts, although Geddes and Hunter had made a few preliminary tests with calf liver in 1928 which indicated that the enzyme is present also in animal tissue. They tested the yeast enzyme on a number of acid amides other than asparagine and found that the only other one hydrolyzed was glutamine. They suggested that the action of asparaginase "is limited, as far as is known, to the amides of the two amino acids, asparagine and glutamine."[62] More recently, Wolfgang Grassmann and Otto Mayr had come to a different conclusion. Their yeast asparaginase preparations did not act on glutamine. Fresh autolyzed yeast did, however, produce a "weak but definite cleavage of this amide." Grassmann and Mayr thought it likely that there existed a "special glutamine-splitting enzyme; but that requires further confirmation."[63] It was apparently Krebs's intention now to explore further the possibility that there were two enzymes.

To prepare extracts he placed one part of each tissue into three parts of water, ground with sand, and centrifuged. A phosphate buffer maintained pH 7.4. He measured the rates of formation of NH_3 in each extract in the presence of glutamine and of asparagine. (The unit weight of tissue for the calculation of the Q values he based on the dry weight of the tissue contained in the volume of extract employed.)

	Liver	Kidney
Glutamine	Q_{NH_3} = 4.48	1.70
Asparagine	Q_{NH_3} = 14.1	1.30

The main significance he saw in these results was not that the liver extract was more active than the kidney extract, but that "in the liver asparagine is more effective, in the kidney glutamine!!!" This result was clearly suggestive to Krebs that the enzyme responsible for the hydrolysis of asparagine is distributed in the tissues differently from that acting similarly on glutamine.[64]

To bring out this distinction more fully, Krebs repeated the experiment on September 24, testing each extract at four different hydrogen ion concentrations. Because the previous extracts had been too concentrated, he also made them this time with 10 parts of water to one of tissue, before adding the buffers. The optimal rates occurred at pH 8. Under these conditions the contrast between the relative rates of formation of ammonia from the two amides was sharp:

$$\text{In kidneys, pH 8} \qquad \frac{\text{asparagine}}{\text{glutamine}} \quad \frac{2.6}{7.8} \; = \; 0.33$$

$$\text{In liver, pH 8} \qquad \frac{\text{asparagine}}{\text{glutamine}} \quad \frac{75}{15.5} \; = \; 4.84$$

A further distinction between glutaminase and asparaginase was that the activity of the former was sharply dependent upon the degree of the alkalinity of the extract, whereas the pH curve of the latter was "very flat." Krebs concluded confidently, "Therefore glutaminase and asparaginase [are] different!"[65]

During the next three days Krebs repeated successively, with rat, sheep, and rabbit kidney and liver slices, the measurements of oxygen consumption, NH_3 and amide-N formation that had, in the experiment performed just before he left for Holland, provided the strongest evidence for his hypothesis that the dicarboxylic amino acids combine with ammonia to form acid amides. He carried out the new experiments with both aspartic and glutamic acid. As in earlier experiments, the results for glutamic acid proved more convincing. The difference between them appeared particularly clear in the experiment of September 28. In rabbit kidney slices, with NH_4Cl added, the comparative results were:

	Glutamic acid	Aspartic acid
Q_{NH_3}	-12.0	-0.5
$Q_{amide-N}$	+13.9	+4.06

He commented, "NH_3-consumption is certain only in the kidneys with glutamic acid. In the liver there is no effect of glutamic acid . . . Little amide formation with aspartic acid."[66]

Over the course of the investigation he had pursued since August, the scope of the problem that Krebs was studying was gradually shifting. He had begun with the view that the two dicarboxylic amino acids shared a property that distinguished them from other amino acids. They combined with ammonia in kidney tissue to form their respective acid amides. He tended to favor glutamic acid in his experiments, because it displayed the effects through which he had drawn that inference more markedly. Now he was coming to see more distinctions than similarities between the physiological properties of the two acids and their corresponding amides. The enzymes hydrolyzing the amides were different, and were concentrated in different tissues. Glutamic acid appeared now to demonstrate very clearly the reaction that he wished to establish, whereas the situation with respect to aspartic acid appeared increasingly doubtful. Krebs directed his attention more and more toward the more promising of the two cases. He did not rule out the possibility that aspartic acid might act similarly under different conditions, as he had once ruled out the possibility that lysine might exhibit the ornithine effect. He merely narrowed his focus. Instead of exploring a generic reaction, common to a class of at least two amino acids, he was now essentially investigating the synthesis of glutamine.

Despite the fact that in an earlier experiment he had described kidney slices as a very active "glutaminase," Krebs may not have regarded the hydrolysis of glutamine as a necessarily physiological reaction. In the early stages of his investigation he may have entertained the view that glutaminase normally catalyzes the synthesis of glutamine, and that it is only in extracts, when the cell structure is destroyed, that the enzyme acts in reverse, carrying out the decomposition reaction, which requires no energy. That may explain his response to an experiment that he carried out along with the ones just described.

Glutamine added to one of the kidney slices raised the oxygen consumption from $Q = 13.8$ to 21.6. This apparently surprised him. He wrote, "glutamine increases the respiration!! Do the kidneys consume glutamine?"[67]

New questions were emerging, but on the whole the investigation was going very well. Near the end of September, Krebs wrote Katrina. She had not heard from him for more than a month. He explained to her that he had been unusually busy with his research. He had made important progress, and was anxious to finish up what he was doing before he had to begin the lectures he was scheduled to give in the new biochemistry course. Anticipating that he would probably have to interrupt his experiments then, Krebs must have hoped that he could complete his investigation of the synthesis of glutamine during the five weeks he had left. Katrina was sympathetic. She understood, she wrote back, that at such a time he would "not like to be diverted by letters to golden pheasants."[68]

In his investigation of the synthesis of glutamine, Krebs had relied on a general test for amide nitrogen. That the acid amide formed in the presence of glutamic acid was glutamine and not some other acid amide was based on the assumption of simplicity; that is, the amide that glutamic acid was most likely to form was the one whose molecular structure most closely resembled it. Recently, however, Albert Chibnall and Roland Westall had published a paper on "The Estimation of Glutamine in the Presence of Asparagine," which suggested a simple modification of the amide nitrogen test that could distinguish between these two amides. Glutamine, they discovered, had the anomalous property of yielding all of its nitrogen in the Van Slyke method for amino-N determination, whereas asparagine yielded only half of its nitrogen under the same conditions. The reason, they found, is that glutamine is less stable and is very easily hydrolyzed. By examining more closely the conditions under which the hydrolysis took place, they found that in a solution of the two amides made lightly acidic and heated for a few minutes, almost all of the glutamine, but very little of the asparagine, decomposed. A determination of the NH_3 found at this point therefore became a measure of the glutamine present. Heating for a longer time hydrolyzed the asparagine, after which the NH_3 it yielded could be determined.[69]

Chibnall supplied Krebs with some of the chemicals he used in his glutamine investigation, and it is therefore possible that he himself drew Krebs's attention to this method, or Krebs might have run across it in the literature in his usual way. At any rate, when he tested the "acid-splitting of glutamine" on September 25, he listed "Chibnall" as the reference for the method, along with an article by E. Thierfelder and Hans von Cramm, who had observed the unusual property of glutamine in 1919. In a dilute sulfuric acid solution heated to 100°C, Krebs found, the glutamine was "in 4 minutes completely split," whereas asparagine was only 25 percent decomposed after five minutes. With further refinements of his method, Krebs was able to determine the quantity of glutamine present to an accuracy of 1 to 2 percent.[70]

The availability of a more positive means to identify glutamine opened a new vista in Krebs's investigation. He could begin to look for other compounds that might give rise to glutamine, and thereby attach the glutamic acid → glutamine reaction to a longer metabolic sequence. That is clearly what he had in mind on October 1, when he tested "proline, α-ketoglutaric acid and glutamic acid, [all] + NH_3," with rabbit kidney slices. He had previously, as we have noted, pondered a reaction in which glutamic acid might be reduced to proline; but here he was thinking of the reverse possibility, that proline is oxidized to glutamic acid, which would in turn be converted to glutamine. Similarly, he evidently included α-ketoglutaric acid with the expectation that it might combine with NH_3 to give glutamic acid. He obtained the following results:

	Q_{NH_3}	$Q_{amide-N}$
Control	+1.26	+0.39
Proline	-2.54	+2.64
α-Ketoglutaric acid	-1.38	+1.80
Glutamic acid	-12.5	+14.9

For an initial experiment on the subject, these were dramatically positive readings, and Krebs reflected in his statement of the results his sense that they fit remarkably well with the two possibilities he was exploring:[71]

> Proline and ketoglutaric acid give glutamine!! With proline, exactly in accord with the NH_3 consumption. With ketoglutaric acid only half of the NH_3 consumption should yield amide-N [because the other half is employed in forming glutamic acid]. The control value,

$$+1.26, \text{ plus } NH_3 \text{ consumed}, 1.38 = \frac{2.64}{1.80} \frac{NH_3 \text{ consumption}}{\text{amide formation}}$$

The reasoning through which Krebs inferred that the proline results exactly fit the assumed chemical reactions does not appear consistent with that through which he fitted the ketoglutaric acid results to the corresponding reactions. For proline he treated the direct measurement of the disappearance of NH_3 as the quantity actually consumed, because that figure was almost exactly equal to the measured value for the formation of amide nitrogen. For ketoglutaric acid, however, such a direct comparison would not work because it showed less NH_3 consumed than amide nitrogen formed instead of the 2:1 ratio predicted by the reaction sequence. He fixed the situation up by adding the NH_3 formed in the control to that which disappeared in the ketoglutaric acid test, to give a total NH_3 consumption which yielded a ratio roughly approximating the expected one.

Such an inconsistency within a developed set of interpretations would constitute a serious flaw. Here it is only an interesting example of the fact that the initial interpretive responses of an experimenter in the midst of an exploratory investigation are apt to be provisional, sketchy, and unintegrated. David Gooding has recently suggested that such first stages in the process of interpretation might be called "construals," to distinguish them from later stages in which we would expect a well-formulated, coherent interpretation.   When an investigator has only "construed" his observations, Gooding writes, "formal notions of evidential support (or disconfirmation) do not yet apply."[72]

Obviously encouraged by the outcome of this experiment, Krebs moved on to try to fill in another step in the reaction sequence that appeared to be emerging.   If proline gave rise to glutamine by way of glutamic acid, then chemical considerations would suggest a further intermediate, pyrrolidone carboxylic acid, according to the equation:[73]

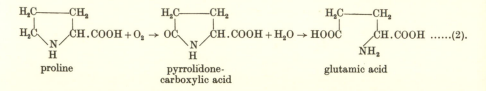

proline   pyrrolidone-   glutamic acid
                                 carboxylic acid

On October 3, he tested proline and pyrrolidone carboxylic acid in rabbit kidney slices. The results, in the presence of NH_4Cl, again accorded with the chemical prediction:

	Q_{NH_3}	$Q_{amide-N}$
Proline	-2.06	+2.63
Pyrrolidone carboxylic acid	-1.92	+2.64

Krebs's statement of the result suggests that he recorded it, not without some sign of excitement:

> l-proline and pyrrolidone carboxylic acid → amide-N!!   Both equally
> rapid, that means that the splitting of pyrrolidone is the rate-limiting
> reaction.[74]

Although the first statement appears as a summary of the immediate empirical result, it is evident from the second statement that Krebs construed the experiment as showing that the reaction sequence proline → pyrrolidone carboxylic acid → glutamic acid → glutamine occurred in the tissue. Thus, in two quick experiments it appeared that he had achieved a major expansion in the scope of his discovery. In place of the single reaction glutamic acid → glutamine, he now had the initial evidence for a metabolic pathway composed of four reactions, and a second branch connecting α-ketoglutaric acid with it.

Seeking further evidence for the new pathway, on October 6 Krebs placed pyrrolidone carboxylic acid in a kidney extract, and after 14 hours measured the quantity of amino-N in it, in comparison to that in an extract without the acid and the acid without the extract. If pyrrolidone carboxylic acid were converted to glutamic acid, as the pathway predicted, then there should be an increase in the amino-N in the extract, because the reaction would break the pyrrolidone ring, opening up an α-amino group (see above equation). To Krebs's surprise, there was "no formation of amino-N!!"[75] Thus he hit the first snag in this new branch of his investigative trail that had appeared to open up so quickly. One could not expect to establish a case in just three experiments, however, and it is unlikely that Krebs was ready to retreat from the reaction scheme that he had in mind.

Shortly after Krebs had carried out these experiments, Hans Weil returned to the laboratory after the long absence caused by his injury. After recovering sufficiently to travel, Weil had returned to Germany with his parents, where he stayed for several weeks. Rosanne Malherbe, with whom he had formed a friendship in Geneva that had deepened through the many letters they exchanged afterward, visited him there, and they decided to be married. After stopping in Geneva to arrange for the forthcoming event to take place in England, Weil came back to Cambridge, in early October, to "begin again with the work." Krebs thought that Weil was now ready to work independently. He gave Weil the project of following up these promising preliminary results on proline and pyrrolidone carboxylic acid.[76]

* * *

In 1977 Krebs remembered the origins of his investigation of the conversion of proline to glutamine quite differently from the preceding account. I had shown him a copy of the paper that he and Weil subsequently published on the subject, and he commented, "That was the beginning of what later was confirmed in many details, that this is the pathway of degradation of proline.

FLH: Now I wondered how that could have originated. It seems conceivable that you could have predicted from paper chemistry . . . the conversion of proline to glutamic acid. Do you think that you did that?

HAK: Well, I generally don't speculate too much. I tested systematically which compounds produce glutamine, for which I had a very easy test. Hydrolysis for five minutes in sulfuric acid. That produced ammonia. . . . And that is essentially the evidence that proline produces glutamine. And then on chemical grounds that could easily be understood. In general, experience has shown that . . . expecting too much on the basis of purely organic chemical predictions is risky. But it is always worthwhile to think about it.

FLH: But you mean it wasn't even sufficiently helpful to be useful in guiding experiments?

HAK: No. I had just checked systematically what could give rise to glutamine, and proline did.[77]

This conversation bears an unmistakable resemblance to those in which Krebs recalled that in his investigation of urea synthesis he systematically tested the effects of various substances until he came across ornithine (see Vol. I, pp.

283–286). There too he denied that he was guided to ornithine by a chemical prediction. In the present case, however, the laboratory record provides more compelling evidence to the contrary. In the first experiment designed to test what substances, other than glutamic acid, could give rise to glutamine, he chose just two—proline and α-ketoglutaric acid—that organic chemistry predicted might be linked through glutamic acid to glutamine. That first experiment was very successful. No "systematic" testing had been necessary to come across proline.

From the evidence that Krebs remembered the events leading to the test of proline inaccurately, can we infer that he probably also remembered the events leading to the ornithine effect inaccurately? Do the two examples taken together indicate that he systematically viewed himself as a more inductively oriented investigator than he was? Did he suppress in his own mind the extent to which he had been guided by chemical predictions because some of them proved later to have been incorrect? The comparison is suggestive, but it is easy to be misled by it. To expect too much from historical parallels is as risky as to expect too much from paper chemistry.

The difference between Krebs's recollection in this case and what can be inferred from his laboratory record does not argue for the general unreliability of the scientist's memory, but for its selectivity. If Krebs's memory of the mental events underlying these experiments appears inaccurate, his recall of the operational events through which he had shown, 43 years earlier, that proline gives rise to glutamine was impeccable.

Comparing the cases of ornithine and proline, I cannot help wondering whether, if I could show him my reconstruction of the latter situation, as I did show him my account of the former situation, he might have been persuaded by the stronger evidence available regarding the proline test, where he was not persuaded by the circumstantial argument I gave regarding ornithine. That I can no longer do so is another among the many reasons I have to regret that my collaboration with the subject of this story ended too soon.

V

Near the end of September 1934, Krebs encountered, in *Hoppe-Seyler's Zeitschrift*, an article that he found irritating. Efim London, the Russian physiologist who had written him in January 1933 that he planned to test Krebs's conclusions concerning the synthesis of urea, under "normal conditions" (see Vol. I, pp. 395–396) had finally reported his results. The paper began,

> On the basis of his numerous and difficult investigations in liver slices in Ringer's solution H.A. Krebs has put forth a harmonious and logically reasoned scheme, according to which the physiological formation of urea from ammonia is supposed to occur in the liver.

After outlining Krebs's "scheme," London went on,

If one has available a dog with a cannula in the portal vein and another in the hepatic vein, one can pursue whether the aforementioned scheme will prove true in the normal process of urea formation from ammonia in the liver.[78]

The technique that London and his associates used was to inject ornithine, citrulline, and arginine into the portal blood and to look for changes in the concentration of urea in samples of blood removed from the hepatic vein. Ornithine did cause the urea content of the hepatic blood to rise above that of the controls, but not consistently more so than ammonium chloride did. London took samples of the hepatic blood at several time intervals after the injection. He argued that, if ornithine acted as a catalyst in the way that Krebs claimed, it ought to "influence only the tempo of the conversion of a certain quantity of ammonia into urea, but not the yield." Judged by this criterion, he found that ornithine did not have the predicted effect. Moreover, when he compared the effects of injecting ammonium chloride alone with ornithine and ammonium chloride, he observed in one case only a slightly greater increase in the urea formed "in the presence of exogenous ornithine," and in another case "no difference between the presence and absence of exogenous ornithine." A further count against Krebs's scheme was that the injection of ornithine or citrulline brought no rise in the quantity of arginine in the hepatic vein blood, in spite of the fact that "if arginine formed an intermediate, through which ammonia must obligatorily pass in its transformation to urea, then in all of our experiments with ammonium chloride, ornithine, and citrulline we ought to have discovered definite variations in the quantitative content of arginine in the hepatic blood, since intermediate processes are always reflected, one way or another, in the composition of the blood."[79]

London concluded, "Up to the present time, angiostomosis investigations give no reason to believe that the formation of urea in the liver occurs under physiological conditions in the way that Krebs represents it on the basis of his in vitro investigation."[80] Elsewhere in his article London expressed in more forceful language his attitude toward the type of experiments on which Krebs had based his theory. "If ornithine [without NH_4Cl] in vitro causes no noticeable increase in the urea (Krebs), what takes place in a dead reagent vessel with dying tissue is nevertheless not determinative for the living organism."[81]

When Krebs's loyal former chief Siegfried Thannhauser saw London's article, he indignantly wrote Krebs:

I read now in the latest issue of *Hoppe-Seyler* the work of London and his colleagues, in which London thinks that he has disproven your work on the mechanism of urea synthesis. It is an unbelievable effrontery to investigate a reaction mechanism with such a research procedure. It is just as if one wished to disprove the role of iron in respiration with injections of iron, or of the dehydrogenase enzyme with injections of methylene blue. Please reply in the strongest terms. I shall write to the fellow. How could arginine or ornithine increase the formation of urea or of NH_3 contained in arginine in experiments with <u>whole</u> organs, since there is already present in intact organs sufficient ornithine for the formation of a very large quantity of urea.[82]

Thannhauser's letter is undated, so that it is unclear whether or not Krebs received it before taking his own action in response to London. Undoubtedly he did not need Thannhauser's support to work up his annoyance over the matter, for even on his own he had little patience with what he regarded as unwarranted criticisms of his work. On October 7, which could not have been more than a few days after he had seen the article, he sent off a reply to Franz Knoop to publish in *Hoppe-Seyler*.

In his tightly controlled note Krebs made five points concerning London's assertions. The first three dealt with particular features of the data of his investigation: 1. London's conclusions were not consistent with his results. From his published data, Krebs picked out two cases in which ornithine injected together with ammonia caused a large increase in the quantity of urea in the blood. 2. The fact that London obtained no increase in the rate of urea formation when he added citrulline, without ammonia, was in complete accord with Krebs's own results. 3. The fact that no additional arginine appeared in the hepatic blood merely meant that arginine is decomposed more rapidly than it is formed, and does not accumulate as an intermediate product. The last two points were more general:

> 4. Our experiments consisted of the measurement of the rates of urea synthesis under defined conditions. London, on the other hand, measured no rates of reaction, but the concentrations of urea in the blood, that is, a resultant of 1. rates of diffusion, 2. rates of excretion, 3. rates of reaction. Because of the disturbances due to diffusion and excretions, London's procedure is very unfavorable when the question concerns the measurement of reaction rates. Only very rough changes can be detected by the method, and I find it surprising that under such conditions London finds any ornithine effect. That speaks for an enormous action of ornithine under his conditions.
>
> 5. London makes a distinction between our 'in vitro' experiments and his 'physiological conditions.' No one will maintain that tissue slices in vitro behave in every respect like intact organs. But if one finds an effect such as the ornithine effect in vitro, I would not like to assume that this effect is not present in the intact organ, but is only newly created through the action of the glassware, the razor blade, or other 'injuries.'[83]

Krebs mailed a copy of his note to London on October 1, together with a tactful covering letter. "I have studied your work in <u>Hoppe-Seyler</u>," he wrote, "with the greatest interest. I must say, unfortunately, that I cannot agree with your theoretical exposition." In the note he enclosed, he said he had concentrated on only a few of the many points he might have made, and gave as a further example that the cause of the difference between their results "is not that our slices are dying, but that you have in your system the kidneys, which attack and deaminate the ornithine. I also observe no urea formation from ornithine in vitro, if I employ liver + kidney slices together."[84]

"Perhaps," he added, "it will be possible in the future to settle some of our differences of opinion through personal contacts."[85] This diplomatic gesture revealed less of Krebs's feelings about the matter than did the letter he sent to Knoop with his manuscript, complaining that an article such as London's would be published without editorial correction. Krebs's letter is not extant, but its tone can be inferred from Knoop's defensive reply:

> On this matter I would remark that the Russians suffer greatly from a lack of the literature. Perhaps Mr. London was only able to read hurriedly once through your work. That I do not know. But it is difficult for us publishers to correct the persuasions of such old gentlemen. Moreover, it is not the task of the publisher to influence decisively points over which differences of opinion are possible.

Knoop assured Krebs that his note would be published quickly, expressed pleasure about his position in Cambridge, and said, "I scarcely need to tell you that I have deeply regretted your departure from Germany."[86]

London quickly composed an "answer" to Krebs's note, which he too sent to Knoop, along with a copy to Krebs, on October 11. In it he defended briefly his own conclusions, conceded that some of the points Krebs raised required further study, and stated in milder tones his general point of view:

> The assertions of Krebs concerning the connections between the formation and decomposition of urea in vitro are undoubtedly of the greatest interest. But can we apply them immediately to the living body?
>
> . . . .
>
> When one speaks of the distinction between experiments with tissue slices and intact organs under physiological conditions, it is self-evident that one does not have in mind the influence of the glassware and razor blade, but is considering that the cells in slices are no longer nourished normally, and are no longer under nervous and humoral control.[87]

On October 23 Krebs replied again to London. Thanking him for his "friendly letter," he added a few more comments on London's results, in particular that London's measurements of arginine in the blood were conditioned by the slow diffusion of that substance into and out of the cells. "In urea synthesis, however, the arginine originates in the cells themselves, and does not need first to diffuse into them. In this regard there is in your experiments also a different 'dynamic' prevailing than under physiological conditions."[88] Cleverly Krebs was turning London's own general argument against him.

Despite the cordiality with which Krebs and London sought to soften their differences, neither persuaded the other to modify his position. What separated them was more than a debate over the interpretation of specific results. The young biochemist and the older physiologist brought to the problem irreconcilable general perspectives.

The disparity between Krebs's point of view and that of London reflected the combined effects of a disciplinary and a generational gap. A veteran of the era in which feeding experiments on intact animals or perfusion experiments with isolated organs appeared to provide the only general access to intermediary metabolism, London had spent many years refining a method that promised to combine the advantages of both. The "angiostomosis" technique would enable one to trace partial metabolic processes thought to be carried out in separate organs, while preserving these organs in their normal conditions within intact animals. During the 1920s, when London worked out his ingenious surgical procedures, the complications inherent in the conditions which his method demanded for the chemical analyses were not drastically worse than the analytical problems created by the existing methods. When Krebs came to intermediary metabolism in the 1930s, however, he brought with him a tissue slice method that provided new standards for quantitative precision and for control of the experimental conditions. With these improved standards London's older methods could not compete.

The disagreements between Krebs and London were also manifestations of the fact that the investigation of intermediary metabolism had long been carried out at the intersection of the scientific disciplines of chemistry and physiology. Scientists approaching the subject from one or the other of the fields to which it in part belonged have repeatedly differed in their priorities and perspectives, because each believed that the essential phenomena could best be displayed at the level of analysis which the methods of their own field could provide. Since the mid-nineteenth century the polar positions had been, on the one hand, that metabolic phenomena can be deduced from the properties and reactions of chemical substances isolated from organisms, and, on the other, that they must be traced as they occur within healthy living animals. London's insistence that Krebs's "scheme" must be tested under "normal physiological conditions" was an echo of the assertions of Claude Bernard that physiology must set the problems and chemistry could serve only as a subordinate tool in the study of the phenomena of life.[89] The broad distinction which had then existed between the test-tube methods of the organic chemists and the vivisection experiments of the physiologist had, however, become progressively narrower and more indistinct with the emergence of new techniques intermediate in character. The chemical changes which occurred in perfused organs were clearly produced by biologically organized material, but not everyone accepted that isolated organs functioned in the same manner as when situated within the intact animal. The processes reproducible in tissue extracts and pulp, following the example of Eduard Buchner's cell-free fermentation, were less "physiological" still, but they were something more than ordinary chemical reactions. Within the range of approaches becoming available in the early twentieth century, London upheld the tradition which sought to preserve physiological conditions in the highest degree possible, even at the expense of chemical precision. His operative techniques skillfully minimized the disturbances to which the animals were subjected, and by the time he utilized the canulas for injecting substances and extracting blood samples, the animals had long recovered from even the minor effects of the surgery. The advent of the tissue slice technique further blurred the old distinction between in vitro and in vivo experiments. If one viewed the individual cell as the basic functional unit of living organisms, then "surviving" cells, placed in media closely resembling their normal surrounding fluids, ought to function "physiologically." If, however, one considered even isolated organs as subjected to unphysiological conditions, then tiny bits of tissue floating in saline solution in a manometer flask would seem still further removed from the appropriate conditions for studying vital phenomena. Thus, to London, Krebs had proposed a mere "reagent glass scheme," because he had not worked with "living organisms," but merely in a "dead" vessel with "dying" tissue. Krebs did not contest with London that the tissue slices were "in vitro." Nevertheless, he had placed the critical dividing line between phenomena bound to life and those separable from it at a different level, between intact cells and biological materials in which the cell structure had been disrupted. Because in the preparation of the slices only the very small proportion of cells lying on the surfaces was damaged, the tissues functioned, he believed, essentially as intact tissues would. "Slices and tissue pulp," he had written in 1932, "are in this sense fundamentally different."[90]

Krebs's point of view was symbolized aptly, perhaps unconsciously, in the titles of some of the papers he published in these years, including the one in question in his controversy with London. He had called it "Investigations of the Formation of Urea in the Animal Body." He had not literally carried out investigations "in the animal body," but he made no distinction between processes he had studied in tissues removed from the body and those that occurred within it.

One person who perceived clearly that the approach Krebs had introduced did not merely shift the operational boundary between in vitro and in vivo experiments—that it pointed instead toward the eventual reduction of the dichotomy to a graduated spectrum—was Hopkins. In his Royal Society lecture of November 1932, Hopkins had concluded his laudatory summary of Krebs's investigation of urea synthesis with a forecast:

We can proceed, then, from the study of tissue extracts in which it is easy to deal with the kinetics of isolated reactions, each determined by its appropriate catalyst, to studies of other tissue extracts, made with discrimination, in which the progress of a variety of reactions retains not a little of the organisation which characterised them during life, and thence to other studies in which we follow the kinetics of reactions controlled by the intact and still living tissues or cells. Thus and otherwise has biochemistry escaped from the dilemma voiced in earlier dogma, namely, that since chemical methods must at the very moment of their application convert the living into the dead, they can do nothing to elucidate the dynamic events of life.[91]

Hopkins had lived through much of the long period in which the dilemma to which he alluded had appeared to limit the scope of the biological conclusions which could be drawn from chemical methods. He was therefore in a position to comprehend more quickly and fully than many others why it was that Krebs's discovery of the ornithine cycle was the herald of a new era in biochemistry.

Krebs was naturally not able to detach himself sufficiently from the immediate situation—the fact that his most important scientific contribution had been subjected to what he regarded as a wholly unjustifiable attack—to view the controversy between himself and London as an outgrowth of the long-term historical trends and traditional positions outlined in the preceding paragraphs. His courteously phrased replies to London only partially veiled the fact that, like Thannhauser, Krebs regarded London's approach as an egregious effrontery. The fifth point in his published note, stating that he "would not like to assume" that the ornithine effect was created by the action of the glassware or the razor blade was, he recalled in 1978, "a ridiculing comment—but perhaps it didn't quite get through."[92]

VI

In the laboratory, while Hans Weil took on the proline extension of the pathway to glutamine in October, Krebs assiduously gathered further information concerning the glutamic acid-glutamine reaction itself. He tested the effects of glucose and of various metals on the rate of the reaction. He strengthened his evidence for the separateness of asparaginase and glutaminase. He tested more tissues to find out which ones, in addition to kidneys, might synthesize glutamine.[93]

On October 8 Krebs wrote his father that he had been too busy to write sooner, because of his work. Buoyed probably by the latest development regarding proline, as well as by the generally favorable course of his work on the synthesis of glutamine, he reported that he had made a major advance in his research, but he did not give any details. His father wrote back, on October 20,

> Whenever I receive no letters from you for a long time, I try to console myself with the thought that you are deeply immersed in satisfying, productive work.   That seems to be the case also right now. I have read with pleasure in your letter of the 8th that you have made a considerable step forward.   I would like to have heard somewhat further particulars about it; but I know that you regard your father as too [...] for that.   Well, my grandchildren will take my revenge on you![94]

Perhaps it was convenient for Hans to assume that his father would not understand what he was doing. That saved him the trouble of having to explain it.

Despite his continued progress, it must have become evident to Krebs by mid-October that he would not be able to complete his glutamine investigation, and certainly not finish a paper on it, before his teaching responsibilities overtook him. Relaxing his focus a bit, he interspersed a few exploratory experiments on amino acid deamination with his ongoing main line of investigation. On October 26 he broke off his laboratory work in order to cope with his pedagogical assignment.

For his part in the newly designed biochemistry course, Krebs had to give three lectures a week for three weeks, on the topics allocated to him the previous spring. (see above, p. 75). This was the first time in his career that he lectured on basic aspects of his field to undergraduate students. He prepared his lectures carefully, writing them out fully (about whether he wrote them in German or in English there is no record), so as to be certain that he had organized the essential information without unnecessary digressions. Then he presented his lectures without reading his texts. In addition he directed laboratory exercises. The teaching staff had decided that in the new course the students should not merely carry out simple test-tube chemistry as in the past, but work with the advanced experimental apparatus used in biochemical research. Krebs's task was to teach the use of Warburg manometers. The class turned out to be much larger than expected—120 students in place of the 30 that had been planned for in the spring—so it became necessary to hold duplicate sessions in the laboratory. For all of his disciplined habits, Krebs found himself hard-pressed to keep up during this short but intense period.[95] In answer to a letter explaining why he had been too preoccupied to write for more than a month, Katrina wrote Hans on November 23, "It is no easy matter to teach so many students so often. I am only pleased that this mountain of work will soon be behind you, since you have, after all, also a small secondary profession as an investigator!"[96]

The whole-hearted effort that Krebs put into his new role was amply rewarding. In spite of his newness to the language, he was able to "deliver most effective lectures in English," and quickly impressed his colleagues as a lucid, stimulating lecturer. Hopkins thought him altogether a "most admirable teacher." Krebs himself thought that the new biochemistry course in which he participated was a great success.[97]

Among the pressures with which Krebs had to contend during these busy months, the question of Palestine continued to loom large. Concerned about how to meet Redcliffe Salaman's request that he draw up a plan for the proposed institute, Krebs had turned to others for help. On September 26 he wrote Daniel O'Brien, an assistant director of the Rockefeller Foundation for the Medical Sciences. He had been asked unofficially, he said, "What I would advise to do with a gift of 40,000 pounds for the purpose of a cancer or medical research institute in Jerusalem" and "What I would feel about organizing the institute and being the head of it." He sought an opportunity to talk with O'Brien, since "you are certainly the best expert I could think of and I know very little about the financial side and the conditions in Palestine as to scientific research." O'Brien replied from New York that he would be happy to discuss the matter during a visit to Cambridge that he planned for early October.[98]

Meanwhile Nachmansohn was pressing Krebs to see Weizmann before the latter left for six months in Palestine. That was "very important," he wrote, regardless of whether or not anything further developed with respect to Weizmann's own institute, because of Weizmann's "broad horizons and great experience." Nachmansohn himself discussed the situation with Weizmann in Paris on September 20 or 21. He reported that Weizmann believed that if Krebs were to accept the position as head of the institute in Jerusalem he must be free to use the £40,000 without restriction and to organize the institute according to his own views. Following Nachmansohn's advice, Krebs made an appointment to meet Weizmann in London on the evening of October 4. Krebs probably again found Weizmann impressive, but the details of their conversation are not recorded.[99]

Two weeks later Krebs was once again jolted by a sudden change in the situation. Salaman informed him that Magnes had sent a telegram from Jerusalem stating that the appointments of two persons as heads of sections of the institute, which the Board of Governors had opposed in Zurich in August, "were already definitive and unalterable," and that the commission of Salaman, Libman, Weizmann, and Magnes appointed in Zurich was empowered to "regulate only the internal affairs of the Institute." Salaman and Libman regarded Magnes's action as a "crass falsification of the Zurich agreement." Krebs communicated this news to Nachmansohn on November 4. "The preparations for Jerusalem," he wrote, "have completely broken down, on grounds that have made me very skeptical about the whole affair." He believed that Magnes had no authority for what he had done and that the other three members of the commission would be able to expose him. "Nevertheless, the matter has given me a great shock. In Palestine I would be forced to work closely with Magnes. He controls the money. I am very pessimistic about working together with such an untrustworthy man."[100]

At this point Krebs appeared to hold out some hope that the commission would be able to check Magnes. During the next three weeks, however, he made up his mind about the "whole affair." When Weizmann sent him a "telegraphic invitation" from Palestine (presumably to continue discussions, or perhaps to visit Palestine), Krebs replied on November 25, "I must unfortunately inform you that the development of matters respecting the Cancer Institute has come to a point at which I can no longer take an interest in the establishment of this Institute."[101]

Krebs wrote this letter just as he was finishing up his teaching in Cambridge. Under the circumstances it was probably just as well that he was so fully occupied in the classroom. The language in which this reticent, even-tempered man conveyed his feelings to Nachmansohn and Weizmann reveals that to have had to come to such a conclusion for the second time within three months was a bitter disappointment.

VII

Early in November, Krebs received a letter from his parents, who were vacationing in an Italian villa near Florence. His father informed Hans of some additions to a list of his investments that he had given him in Cambridge. "I hope," Georg Krebs wrote,

> that this small precautionary measure of giving you such information is superfluous. It can, however, do no harm. So long as the National Socialists endure in Germany, anti-Semitism will remain their slogan. As the notorious [Julius] Streicher . . . has said, the racial question is the central issue for past, present, and future. In the economic connection this makes itself felt especially in the smaller cities, where . . . the population feels itself under surveillance. One must wish that there will still in general be people, such as justices, clergy, etc., who will appear in the waiting room. I have gradually come to terms with the reduction of my practice, and am frivolous enough to diminish it further by taking the present trip.[102]

Maria added a page to the letter. All of the people in the pension at which they were staying, except for two, were "non-Aryan," she reported. She urged Hans to speak again to the director of General Electric in England about possibilities for Wolf there. "All of us," she added,

> will perhaps have to look to foreign lands for our future. To be sure, there is outrage everywhere in Germany, and it is said that things cannot go on in this way; but, for the present one can scarcely think of a change. Father will, of course, scold me for being a pessimist, but I fear that things will get worse. That does not mean that I am despondent. In the meantime we go on enjoying our lives.[103]

Lise, who was with her parents in Italy, wrote a separate letter, and expressed the greatest sense of urgency. In the event that war should occur, she thought, the situation for Jews in Germany would become desperate [*schlimm*]. Since last spring the "danger of war has become much greater. In every country—even here—people are talking about it. For that reason, I ask you to use your influence for Wolf as soon as possible."[104]

During a year in which Hitler had engineered a massacre of his own SA leadership, had assumed the powers of the president of Germany after Hindenburg's death, declared himself Führer, and begun a military buildup; at a time when thousands of Germans were already experiencing concentration camps,[105] Georg Krebs may appear to have been excessively short-sighted to measure the threat of National Socialism by the size of his medical practice. Looking back from a distance, however, we too easily characterize such a period in terms of portentous events rather than by the conditions directly affecting the majority of the people who were merely trying to go on with their lives.

Whether or not the feelings of Krebs's parents were typical of German Jews in their position in late 1934, their views are illuminating. Things were bad, and they might get worse; but for now one could tolerate them. Opinions about the future were guesses. An optimist, such as Hans's father, could hope that the Nazis would not last indefinitely, and that with their eventual departure anti-Semitism would cease to be a virulent force. His wife and daughter had equally strong grounds for pessimism. In England the son also had ambivalent feelings. As an optimist, he too hoped that the Nazis would not last long. He believed that the Allies would not tolerate developments in Germany that threatened to undo the victory they had won in the world war. In his more defiant moods he mocked the Nazi slogans, and he and his fellow refugees joked that they would return to Germany as soon as the Thousand Year Reich was over. The darker side of Krebs's temperament fixed on the horrors of the Nazi regime. He clipped and saved newspaper stories documenting the tortured logic of Nazi policies and their brutal acts, especially those directed against Jews. These he hoped to use to help persuade those around him in England of the menace the Nazis presented. He was discouraged by the strength of pacifism in England, symbolized in the resolution passed at the Oxford Union in 1933 that "this House will not fight for King and Country." One reason that he was probably unsympathetic toward the political conversations that took place in the Cambridge laboratory was that he believed the Cambridge scientists involved in the peace movement, and those who advocated collective security in place of national defense, were unrealistic in the face of what Hitler might do. Those to whom he talked were sympathetic about what had happened to him in Germany, but believed that his personal experiences caused him to exaggerate the seriousness of the situation.[106]

These deeply divided feelings must have intensified Krebs's uncertainty over his own future. On the one hand, he sought to assimilate himself to the English way of life, to make it his own; but since he could not be certain that he would be allowed permanently to do so, he looked toward Palestine and a very different identification with a vision of a future Jewish homeland. If, however, in the end, "this nonsense could not possibly go on"[107] in Germany, then perhaps sooner or later he would be able to resume his life and career in his own homeland. No more than his father was Hans Krebs ready in 1934 to accept that the Germany he had earlier known had been irreversibly absorbed into the repellent order of National Socialism.

In the meantime, another member of Krebs's German circle was making up his mind that survival was not enough and that his future lay in foreign lands. In the same letter in which he discussed E.S. London, Thannhauser wrote Krebs that,

> I have been invited to teach internal medicine at New England Medical Center (Tufts College Medical School) and to establish a scientific research group. Although my material situation here is good, and the possibilities for work that I have created for myself satisfy me, the spiritual [*seelische*] situation is so corrosive that I am considering

seriously accepting the call to Boston. In any case I shall look into the matter. What do you think about it?[108]

This was a decision that Thannhauser approached with a heavy heart. He had loved the homeland of his younger days, had been deeply involved in its artistic and cultural heritage, and believed in "the old spirit of German idealism." He had not given in easily to the catastrophe that had befallen him and his country. He had borne himself well, and defended himself ably under pressure, and when he had been driven from his official post he had resourcefully found ways to carry on both his private medical practice and his research. He had, however, endured long enough. Thannhauser accepted the offer, went to Boston, and prospered there.[109]

4

New Moves

On November 20, 1934, Hans Krebs and Marjory Stephenson left the biochemistry laboratory with Hans Weil and Rosanne Malherbe, accompanied them to the Registry Office around the corner, and acted as witnesses to their civil wedding. Then they all returned to the laboratory for a sherry party. Afterward the couple took the train to London to have their marriage confirmed in a Swiss Reformed Church. When the pastor found that they had not yet received their marriage certificate through the mail, however, he refused to perform the ceremony. Hans and Rosanne had to live together without the blessing of the church until the certificate arrived and they could return to London and present themselves again in the same church.[1]

His appointment as a demonstrator required that Hans Krebs become a member of the university, by receiving a Cambridge master's degree. This formality was handled in a leisurely fashion. Appointed in May 1934, Krebs was not notified until November 22, by the vice-chancellor, that "the Council of the Senate are willing to propose your name to the University for the degree of M.A." With Krebs's agreement to this step, the degree would be voted at an open meeting of the senate on Saturday, December 1. Concerned that the meeting could become the occasion for another expression of opposition to the appointment of a refugee to a British academic post at a time of unemployment among British scientists, Hopkins made sure that the senior members of his department were present to vote in favor of the degree. Nothing of the kind happened, and on December 1, Hans Adolf Krebs, Doctor of Medicine from the University of Hamburg, became also Master of Arts from the University of Cambridge.[2]

I

When he received his degree, Krebs had just returned from the teaching laboratory to his research bench, to resume the lines of investigation he had broken off one month earlier. Although he had, in the weeks preceding that break, mainly sought to consolidate the evidence he had previously gathered in support of the synthesis of glutamine in animal tissues, he had, in the last days, encountered new complexities. On October 23, while testing the effects of glutamic acid in several guinea pig tissue slices, he had found that in heart muscle that acid increased the respiration, but without causing any distinct formation of amide-nitrogen in the tissue.[3] This result must have caused him to consider reactions, other than the formation of glutamine, that glutamic acid might undergo, and to look up a paper on that subject published by Dorothy Needham in 1930. Needham had shown that in minced pigeon breast muscle glutamic and aspartic acid give rise aerobically to succinic acid, while increasing the respiration. Only part of the acid added could be accounted for in this way, but there was no increase in the quantity of nitrogen present in the form of amide-N, ammonia, or amino-N. Needham had suggested a possible mechanism for this reaction, involving transfer of the amino-N to form other amino acids and succinic acid, and viewed such a reaction as a possible alternative to the well-known hypothesis of Thunberg that succinic acid derives from the condensation of two molecules of acetic acid.[4] Since Needham worked in the Cambridge laboratory, she may personally have drawn Krebs's attention to the possible relationship between her prior work and his result. Using pigeon breast muscle, Krebs reinforced Needham's observation that glutamic and aspartic acid increased the respiration without forming amide-N. There was, however, an additional complication, for "this time" both acids caused "a slight increase in NH_3 formation," rather than the expected decrease. In the same experiment he tested the effects of a very different substance, adenylic acid—both by itself and in combination with glutamic and aspartic acid. It is not clear what he had initially in mind in testing adenylic acid in this way. That substance was, however, regarded as "the source of the ammonia which is produced by active muscle,"[5] and in his experiment it raised the rate of formation of ammonia to such an extent that he asked himself the question, "Is the adenylic acid totally decomposed?"[6]

Of the two leads that these experiments appeared to open up, Krebs chose, in the one day remaining before his long interruption, to explore further the actions of adenylic acid. This time he examined its effects in comparison to, and in combination with, glutamine in guinea pig brain and heart tissue. The outcome that most attracted his attention was that "in the brain the increase in respiration through glutamine and through adenylic acid are additive!!" (This result appears, from the data, to have held only roughly, for the later portions of the experiment—e.g., during the last 20 minutes, Q_{O_2} with glutamine was 10.5; with adenylic acid, 6.58; with the two together, 14.5.) When he afterward measured the glutamine present at the end of the experiment, he found that in each tissue glutamine had disappeared, and that "the disappearance is exactly stoichiometrically increased by the addition of adenylic acid!!!!"[7] It is not clear whether he had made any preliminary interpretations of these results at the time he broke off the work. The additive effects of glutamic and adenylic acid would appear to indicate that the two substances exerted independent effects on the respiration. The stoichiometric effect of adenylic acid on the disappearance of glutamine, on the other hand, suggested some kind of strong connection between them; but, as his four exclamation points imply, the nature of such a connection was not immediately apparent.

Krebs must have viewed the effect of adenylic acid on the reaction of glutamine as the presentation of a problem to follow up after his teaching stint; but the shape of the problem was vaguely defined, since there were no obvious chemical grounds relating the disappearance of adenylic acid to that of glutamine. In his first experiment after the break, on November 26, he sought to identify the problem more clearly, by testing inosinic acid with glutamine, this time in rabbit brain tissue. Inosinic acid (hypoxanthine-pentose-phosphoric acid) was known to be the product of the deamination of adenylic acid in muscle tissue. The principal result of his experiment, he noted, was that "more glutamine-N disappeared with inosinic acid, especially with inosinic acid + pyrophosphate. . . . Determine adenylic acid!" Unlike the last result encountered prior to the break, this one could be construed, as Krebs's remark indicates, in chemical terms as a possible transfer of the amide nitrogen of glutamine to inosinic acid. If so, an expected product would be adenylic acid, and he suggested to himself that next time he should check such a possibility by determining whether adenylic acid formed. The situation, he wrote down, "remains to be investigated more exactly." Thus a potential new extension of his study of the synthesis of glutamine appeared to be opening up—a clue concerning the possible further fate of the glutamine formed in the metabolic reaction he had previously discovered. At this point, however, the opening must not have looked promising enough to lead him on. The next day he turned back instead to the main lines of his two ongoing investigations. He tested "the metabolism of [swine] retina with glutamic acid" and the deamination of four amino acids in a kidney extract that he had prepared earlier.[8] Through the rest of November and December, Krebs moved back and forth between experiments on glutamine synthesis and amino acid deamination, with some experiments appearing to merge elements of both investigations. Aspartic acid, which had disappeared from his examination of the synthesis of acid amides, reappeared prominently in his deamination investigation as the "natural" (l-series) isomer most convenient to study, because it reacted more rapidly than any other amino acid in that series. Along the way a few more complexities emerged. Attempting to extend the demonstration of glutamine synthesis to other animals, Krebs found, to his evident surprise, that glutamic acid did not form glutamine in cat or dog kidney tissue. This result did not threaten the positive results he had obtained in guinea pig and rabbit kidney, but did cause him to begin to realize that the system that synthesizes glutamine must not be uniformly distributed in the kidneys of mammals.[9]

During December Krebs became increasingly interested in the various ways in which the reactions he was studying may be inhibited by compounds related to those undergoing the reactions. He began, on the 3rd and 4th, by testing whether amino acids from the "unnatural" d-series might inhibit the deamination of those of the natural series. Neither d-glutamic acid nor d-alanine had such an effect on the deamination of l-valine. On the 20th he posed a question that was

both more specific and more general: "Do decomposition products inhibit [?]."
That day he found that "the oxidation of *dl*-alanine in an acetone powder extract.
. . is *not* inhibited by . . . NH₄Cl and . . . pyruvic acid." The next day he
tested "the inhibition of glutamine cleavage by glutamic acid" in guinea pig
kidney extract. Here the outcome differed dramatically from the previous ones.
0.025 μl of 0.2 M glutamate reduced the Q_{NH_3} from 1.24 to 0 and transformed
the $Q_{amide-N}$ from -1.04 to +0.43. Two higher concentrations exerted even
stronger effects on the amide nitrogen. Such a result could, on the surface, be
explained as due to an equilibrium between glutamine and glutamic acid; but to
judge by the caption of the experiment, Krebs probably did not entertain that
view. Whether he anticipated the result or it came as a surprise, he most likely
treated it from the first as a "peculiar" phenomenon—that is, a specific inhibition
of the enzymic cleavage of glutamine through the presence of its decomposition
product.[10]

About the same time, Krebs became interested in another phenomenon,
suggested perhaps by effects he had noticed during his specific investigative line,
but that he took to be of general significance. This was the effect of the
dilution of the medium on metabolic reactions. On December 12 he tested "the
influence of dilution on kidney respiration," using kidney slices. He compared
the rate for slices transferred directly into the manometer vessels with those
shaken first in Ringer's solution. Apparently he assumed that the shaking would
reduce the concentrations of metabolites with the tissue slices. The result was
apparently inconclusive. Near the end of the month he took up, as a more
suitable process on which to study dilution effects, the fermentation of glucose
in yeast juice. He compared the fermentation rate in a standard yeast extract
with that in two others diluted respectively with one part and three parts of
water. For the latter two cases the rates were much lower than for the undiluted
extract (during the last 30 minutes the manometric pressure changes were
undiluted, +585; +1 part H₂O, +29; + 3 parts H₂O, +11). Krebs wrote,
"Strong inhibition through dilution. Repeat with controls for each dilution."[11]

In his deamination experiments Krebs continued to gather data on the rates
of deamination of various amino acids, natural and unnatural, in slices and in
extracts, and appeared to encounter no new surprises. He did not take an
extended holiday break. Aside from Christmas and the following two days, he
worked straight through to the end of the year 1934.[12]

II

In late December Krebs received a letter from Juda Quastel, with whom he
had only recently become acquainted. Quastel understood, he wrote, that Krebs
had declined the post of head of the new Cancer Research Laboratory of the
Hebrew University. "I have," he went on, "been approached (quite unofficially)
with a view to taking over this job but before I commit myself in any way I
would very much like to have your views and opinions." Although it would be
a big step for him to give up his present permanent post, for "a post in Palestine
of which I know nothing . . . , I am a convinced Zionist and could willingly

live, I think, in Palestine, if my scientific future there were secure, and if my domestic life were comfortable." He would be "deeply grateful," if Krebs could write him confidentially, "why you were not prepared to take over the post?"[13]

Krebs was not too busy to write Quastel an immediate and lengthy reply. Quastel had addressed his letter on December 19. On December 22, Krebs began, "I shall willingly tell you my experience." Outlining the events which had begun with Judah Magnes's letter to him in April and ended with his own second decline of the post in November, Krebs went on to give full vent to his feelings about the outcome. "I was most annoyed about" the matter. He had decided to decline definitely because "I have lost all confidence in the management of the Hebrew University that is to say in the personality of Dr. Magnes. As you probably know the University is not one of the bright spots in the Palestinian development. . . . In my opinion Magnes alone is responsible for those developments." After going on for a page and a half about how Magnes had mismanaged the situation at hand, Krebs concluded, "I would not go out to Palestine as long as he rules in a key position." In a final paragraph, Krebs wrote:

> This is the explanation I can give you for my refusal. I should add a few words about my attitude toward Zionism. I have never been an active Zionist but I have always furthered the movement whenever I could, especially after those events in Germany. I would be only too glad to help in Palestine in building up a university, but I believe that in this case enthusiasm cannot replace a sound material and spiritual basis and the history of the Palestinian University does not convince me of the contrary. I believe in Zionism and I would consider it a privilege if I could work in Palestine, but I do not think that I can render any useful service to the cause under the given circumstances.[14]

Krebs did not send this letter. A week later he composed a shorter, more restrained version of it. After his summary of the events, he wrote this time:

> I then declined definitely, and not only on account of the financial difficulties—they might perhaps be overcome—but I lost confidence in the sound development of the Hebrew University. From many conversations which I had with Dr. Salaman and others I gathered that the management of the Hebrew University is unsatisfactory, and I believe that under the present circumstances scientific work on biochemical lines would be very difficult in Palestine. I rather prefer to be demonstrator in Cambridge with a poor salary but good conditions for working to a good salary, an independent department but very doubtful possibilities for research. I would be very glad, on the other hand if I could help to build up something in Palestine but I doubt that my presence there would be of any use under the present conditions.

This is of course a very personal opinion. Perhaps I have too little enthusiasm for the cause to take the risks. The decisive factor for me was finally that Hopkins and other people here encouraged me very much to stay in Cambridge.[15]

This letter he did send off. Quastel answered by return mail that "your letter has helped me a good deal." He had already decided that the £40,000 provided was too little for the aims of the institute, and Krebs had reinforced his own inclinations.[16]

The contrasts between these drafts of Krebs's letter run deeper than the difference between a first, impulsive reaction and a subsequent, carefully controlled response. In the first letter he had written eloquently of his belief in Zionism, of the privilege it would be to work in Palestine, but suggested that prudence had overruled his "enthusiasm." In the second letter he wrote that he had "perhaps too little enthusiasm for the cause to take the risks." The emotion he expressed in this version was reserved for his present situation, where good working conditions, Hopkins, and the other people around him overruled his material disadvantages.

Which of these contradictory views represented Krebs's real feelings? I suspect that both of them did, and that this outwardly imperturbable man was torn between conflicting visions of himself, his future, and his priorities. If he believed in Zionism it was not with the steadfast conviction of the long-time believer, but the fluctuating feelings of the recent adherent. He could in one week believe that he was controlling his enthusiasm with his practicality, and the next week doubt the strength of his own enthusiasm. His longer-standing commitment to his science was not in doubt, but even here one can sense a more muted contest between his ideals and his practical temperament. In retrospect Krebs took pride in the frugal life he had led. His reference here to his "poor salary" has a rather different tone. It was not something with which he was content, but with which he was willing to reconcile himself in order to sustain his scientific productivity and to remain in surroundings in which he was otherwise happy. In expressing his "preference" he was making a virtue of necessity. That he could give two such different explanations for his decision from one week to the next—and, I believe, mean both of them—is perhaps a hint of the intensity of his inner struggle.

Whether despite, or because of, such emotional pressures, Krebs pursued his research in early January 1935 with an energy that was extraordinary even for him. He carried out experiments on 13 of the first 16 days of that month, and on four of those days he completed three full sets of experiments. Mainly he followed out questions that had arisen during December. During the first week he concentrated on the cleavage of glutamine, particularly on the inhibiting action of glutamic acid on that process that he had previously observed. On the 4th, he found that glutamine added to sheep kidney brei disappeared at first, but that after 150 minutes there was no further change in the quantity of NH_3 released, or of amide-N remaining. "Therefore," he noted, "the cleavage stops!!" Although he did not put down any reasons for the stoppage, it is reasonable to assume, from his prior observation of the effect of glutamic acid, that he inferred that in this case glutamic acid arising from the reaction itself halted the further progress of the reaction. In a second, similar experiment carried out that day he found that the formation of ammonia came to a halt when only 106 of the 156 mm^3 of NH_3 preformed in the glutamine had been released. Finally, on the same day, in an experiment entitled "Inhibition (of Glutaminase) through Glutamic Acid," he compared the rate of glutamine cleavage in swine extract with and without glutamic acid present, and concluded that there was an "enormous inhibition (85%)." The next day he followed the time course of the cleavage of glutamine in swine kidney extract, from which he could show that the reaction began rapidly, but gradually tapered off, coming essentially to a halt when a little less than half of the glutamine had been consumed. Simultaneously he tested comparatively the effects of d-glutamic, dl-glutamic, l-glutamic, α-ketoglutaric, and l-aspartic acid on the process. The first three each reduced the quantity of NH_3 formed after 40 minutes by 88, 80, and 91 percent respectively, whereas the last two had no effect. Emphatically he commented, "Inhibition through l and d-glutamic acid!!" Lastly on that busy day, he examined the "inhibition of asparaginase," that is, he tested whether aspartic acid would reduce the cleavage of asparaginase in guinea pig liver. There was, he observed, "no definite inhibition."[17]

Although some mopping-up work remained to be done, Krebs had, in this group of experiments compressed into two days, acquired the essential evidence he needed to characterize the action of glutamic acid on glutaminase. He had established that both optical isomers of glutamic acid were equally active as inhibitors and that the action was probably specific to these two compounds. He had shown that the phenomenon was most likely peculiar to glutaminase—or at least that there was no corresponding action involving the similar enzyme asparaginase. It is not clear whether he had in mind at this point an explanation for an action shared equally by two isomers, only one of which would arise in the course of the reaction that both isomers inhibited. Nevertheless, in remarkably short order he had identified within an ongoing investigation of one phenomenon an unusual subphenomenon, and carried out an experimental subinvestigation that neatly analyzed the latter. If we think of this episode as an investigative unit contained within the larger investigative line concerned with the synthesis of glutamine, an investigation that had branched out in turn from a broader investigation of deamination that Krebs identified by the titles of his publications as parts of a still larger investigation of the "Metabolism of Amino Acids," then we can begin to appreciate the complexity of the organization of a scientist's personal research pathway. That metaphor is inadequate to express the manner in which the "pathway" is composed of components of different degrees of magnitude, occupying different time scales, interwoven within one another. Only exceptionally are subinvestigations like this one bounded so compactly in time that there is a simple correspondence between their temporal relationships to, and their conceptual relationships as subproblems within, larger problems.

In the remaining experiments he carried out during these intense working days of January, Krebs pursued three distinct sublines, not in a temporal order, but "simultaneously." Sometimes he literally carried out experiments belonging to two or three of the lines at once. Otherwise, although he alternated between them from one day to the next, his mental continuity linked successive experiments of similar type rather than those performed successively.

One of the three lines he followed up was the unexplained effect of adenylic acid on the action of glutamine, and its possible relation to inosinic acid. The first of the experiments, on January 7, tested adenylic acid in combination with d-glutamic acid in guinea pig tissue slices, and was perhaps intended to explore whether adenylic acid might supply NH_3 for the formation of glutamine. The $Q_{amide-N}$ for this combination (8.60) was higher than for d-glutamic acid alone (6.36), but Krebs did not comment on the significance of the result. On January 12 he tested adenylic acid, in the presence of glucose, with and without glutamic acid, in guinea pig brain slices. Here his only comment was that, unlike in an earlier experiment without glucose, "here no NH_3 from adenylic acid." Two experiments utilizing inosinic acid, one on January 14, the other on January 16, are difficult to interpret in detail. One of the questions he asked was whether inosinic acid, the deamination product of adenylic acid, would, in the presence of ammonia, combine with it to yield adenylic acid. The answer, at least in guinea pig brain tissue, appeared to be no. All of the NH_3 added could be accounted for as such, or as glutamine for the cases in which glutamic acid was also present.[18] Krebs pursued these questions no further. In contrast to the way he had been able to turn his observation of the inhibiting effect of glutamic acid rapidly into a coherent subinvestigation, his effort to do the same for the initially observed effect of adenylic acid quickly dissipated.

The second thread Krebs followed in January was his interest in "dilution effects." In two cases out of three he was able to obtain "large" to "enormous" effects on the rate of deamination of aspartic acid in a sheep kidney brei produced with a Latapie grinder. He was also able to repeat the enormous effect of dilution on glucose fermentation in yeast juice.[19]

Krebs devoted the majority of the experiments during these 16 days to examining the basic phenomenon of glutamine synthesis in other tissues. Increasingly it became clear that the reaction was especially prominent in brain and retina tissues and that here it was not distributed selectively by species, but existed in all the animals whose tissues he examined. The character of the process appeared to vary, however, from one type of tissue to another. In swine retina, for example, glutamic acid inhibited the respiration, but slightly increased the rate of anaerobic glycolysis. In rat brain "differently from swine retina," glutamic acid inhibited glycolysis but had "no definite effect on respiration."[20] By now Krebs must have come to appreciate that there were numerous subtle differences in the metabolic character of the same reaction from place to place in the animal body, and from animal to animal, and perhaps that there was more than one type of "glutaminase."

Following this spurt of intense research activity Krebs performed no experiments for 10 days. A reference in a letter from Katrina indicates that during this time he drove to Birmingham in his car, but there appears to be no record of his purpose in going there. He was probably not gone for the whole time, for he was scheduled to speak at the weekly departmental Biochemical Tea Club on January 22, on the topic of "Amino Acid Metabolism." It was probably then that Krebs discussed his results concerning glutamine synthesis. Following the usual custom, he spoke informally for about half an hour, and there was a discussion afterward. Hopkins undoubtedly supplied pertinent comments relating the specifics Krebs had presented to a broader point of view. By this time Krebs had already written much of the paper he intended to publish after he had completed his investigation, and he found the "pleasant" criticisms he received from his colleagues helpful in rewriting his draft.[21]

Meanwhile, it was becoming clear that Krebs's definitive rejection of the directorship of the proposed cancer institute in Jerusalem had not ended, but only introduced a new phase in, the efforts to recruit him for the future of science in Palestine. During January Nachmansohn wrote that he had again discussed the situation with Weizmann, who had asked him if Krebs was, in a fundamental sense, prepared to go to Palestine. Nachmansohn had answered that he had the impression that Krebs was not "in principle disinclined to go," if he could be assured of an appropriate position and good working conditions. It turned out that Krebs had been mistaken about one of the deceptive actions he had attributed to Magnes, and besides, Magnes's days in authority "are numbered." More important than such reflections on previous difficulties, however, was that Weizmann himself intended to bring together in London a group of scientists who had expressed a willingness in principle to emigrate to Palestine. The important thing for Krebs was not specific details: "The question just now is, how far are you prepared, overall, to go to Palestine, and to contend with the other difficulties, the material conditions, since you must first create something there and it will naturally in the first 1–2 years not be so simple to work there as it is in Cambridge." Krebs replied positively to Nachmansohn's challenge and agreed to participate in Weizmann's meeting. Very pleased, Nachmansohn arranged to come to Cambridge on February 9, in order to talk the "fundamental" questions over with Krebs in preparation for the meeting in London the next day.[22]

In addition to Krebs and Nachmansohn, the group assembled by Weizmann included Paul Rothschild and Hermann Blaschko from Cambridge, two other scientists from London, and one from Jerusalem. Since his move to Cambridge during the previous year, Blaschko had been tutored in manometric techniques by Krebs, and was now carrying out metabolic experiments in the physiology laboratory.[23] It was mainly through the efforts of his friends Krebs and Nachmansohn that Blaschko had been persuaded to take an interest in the project. At Weizmann's request, Nachmansohn began the meeting by summarizing the situation. Those present here, he said, represented diverse fields, and could become the nucleus for a biological research group in Palestine, provided that sufficient means could be found to support its work there. Weizmann replied that no further steps at the university in Jerusalem were likely to be taken until the next meeting of the governing board in August. He advised Krebs not to accept an offer at the cancer institute in the meantime, and indicated that he himself had been exploring the prospects for establishing an institute within which such a group could work. Several people had already declared their willingness to contribute to its support, but larger sums were still needed. After a discussion of the relative merits of locating the institute in Jerusalem or Rehovoth, Weizmann stated that what he needed from the group as soon as possible was a detailed plan for the construction and equipment of such an institute. He estimated that, if the project were realized, building could begin within one year. It would be very important that Krebs, the assumed likely director of the institute, go to Palestine before then to examine the whole project in detail. Krebs remarked that he intended to remain in England for two more years, but that, in accordance with Weizmann's view, he would be willing to go to Palestine for several weeks during the course of the present year. Weizmann suggested that the visit take place at the end of the year.[24]

Krebs left this meeting in a more comfortable position than he had been in for the previous nine months. He could now enjoy the prospect of one day playing a major role in the development of science in Palestine, without making any immediate commitments. It was up to Weizmann and others to attract the type of support and establish conditions that might induce him to decide for Palestine sometime in the future. The only short-term inconvenience he incurred was that, as the putative leader of the projected research group, the responsibility fell to him to draw up the detailed physical plan for the institute.[25]

Ecstatic over these developments, Nachmansohn wrote Krebs on February 17 to say how happy he was that Krebs had declared himself ready to take up the task in Palestine once the material prerequisites were given. The fact that Krebs would delay his departure for two years now appeared to be no problem. More confident than ever, after a further telephone conversation with Weizmann, that the financing of the project was under control, Nachmansohn waxed eloquent over the creation of "a powerful center of Jewish culture in Palestine."[26]

Whether or not Krebs was as committed in principle as Nachmansohn took him to be, he found it difficult in practice to commit the time necessary to carry out his part in the immediate planning. "I must confess to you," he wrote at the beginning of March,

> that I have not yet got very far with respect to the "memorandum," and would like to ask you if you would not think it correct if, in the present situation of uncertainty we limit ourselves to general considerations. The working out of a detailed plan for the Institute would require a total occupation (every day) with the problem for at least 1-2 months, with the requisition of numerous catalogues, the clarification of special cost conditions, consultations with architects, etc. But such work would appear to me at the present moment, when the overall plan is still in question, to be unrewarding.

He was finding it difficult even to compose an introduction on the necessity for biological investigation in Palestine, and hoped that Nachmansohn would take over that part for him.[27] It was clearly not just because the situation was uncertain that Krebs was reluctant to do the work involved in a detailed plan, but because he was unwilling to take that kind of time away from his research. When it came to such a choice, Krebs was very certain about his priorities.

What was for Hans Krebs the imperative to press on with his work above all else was, for his friend Katrina, the source of ever-growing anxiety. Month after month, despite her pleas, she received only short, infrequent letters from him. By mid-February she could no longer sustain her hopes for a future with him. On the 16th she opened a package from him, containing some handkerchiefs she had asked him to buy for her and only a tiny accompanying note. Lying in bed that night, with moist eyes, she wrote:

> Hans, I must write you a letter today that is perhaps somewhat uncomfortable for you. I have already written several times that for me, _all_ of my thoughts about you rest on our finally being together. Perhaps my heavy [nursing] duties in these winter months have made me a little nervous and unwell. But I am through with trips . . . Sometimes I think that you have forgot all that existed between us. . . . Since last summer you have had time enough to think it over, and to reach a conclusion—for or against a marriage between the two of us. The problem is not resolved, Hans, with a yearly holiday trip. Take the trouble to think about the situation from my point of view. I am truly homesick for you, Hans, and I must live with people, and under conditions, in which there is not only no spark of understanding, but in this matter only opposition and difficulty . . . You must not merely write me a date on which we shall see each other again, but you must tell me unequivocally what you think about the further development of our situation.[28]

There is reason to infer that Hans was not unmoved by Katrina's letter. Nevertheless, so preoccupied was he with his relentless laboratory life, that it took him a month and a half to respond.

III

The two investigations whose pursuit claimed Krebs's attention so exclusively during the spring of 1935 were both by this time well advanced. An indication of the confidence he felt in the results already attained was that, on February 15, Krebs gave talks on both subjects at the Biochemical Society meeting in London. In his presentation on the "Deamination of Amino Acids in Mammalian Tissues," he stressed the "facts" he had gathered suggesting "that the deamination of the d- and l-amino acids is due to two different enzymic systems ('d- and l-amino acid deaminases')." Because the l-system was destroyed by extraction methods that did not affect the d-system, he concluded that "the l-system [for the natural amino acids] is more complicated than the d-system." In his second paper, on the "Synthesis of Glutamine in Mammalian Tissues," he summarized his evidence that glutamine is formed from glutamic acid in kidney tissues and his reasons for inferring that "glutaminase" is not identical with asparaginase. He described the "specific" inhibition of glutaminase by glutamic acid and the distribution of the glutamine synthesis system in different tissues and animals.[29] Despite the clarity of these outcomes so far reached, Krebs did not treat either investigation as complete. Through February, March, and April he went on gathering further information about both processes at an unflagging pace. As in early January, he often carried out three, sometimes even four sets of experiments in a single day. He worked at least six days a week, sometimes including Sundays. The rate of experimentation increased still further in March when he was able to hire a "personal assistant," probably with grant funds provided by the Ella Sachs Plotz Foundation.[30]

Most of the experiments Krebs carried out during these months were straightforward extensions of the previous ones, the majority of them concerned with glutamine synthesis. In contrast to the earlier years, when he had published quickly after reaching a significant conclusion, he now appeared anxious to multiply his data before doing so. Perhaps he was in part simply becoming more thorough as his scientific style matured. In addition, however, the special character of the glutamine problem—that is, the evidence that the physiological processes involving its reactions were not alike in the different tissues in which it occurred—drove him to examine more extensively than he might otherwise have done the varied circumstances under which the synthesis and its reverse reaction took place. Beyond such considerations it is clear that he wished to penetrate past the goals of his earlier research, to do more than simply identify steps in metabolic pathways. In both of these lines of investigation he had been able to extract a cell-free enzyme preparation, that is, glutaminase and d-amino acid oxidase. This circumstance provided him with the opportunity to establish the properties of the enzymes. The approach that he followed in doing so was not novel. Following standard procedures, he determined the influence of the concentrations of the substrates, of the hydrogen ion concentration, and of the temperature. He looked for inhibitors and assessed the degree of specificity of the enzymes by testing compounds related to their normal substrates. Such experiments, though relatively routine, were time consuming, and Krebs must have felt a particular urgency about carrying them out, in order to complete two investigations whose central problems he had already solved. Having, moreover, published no research papers during the nearly two years since his arrival in Cambridge, he must have been more eager than usual to bring two successful investigations to a close. By early March he must have felt that he was engaged in a final push, after which he might be able to relax a bit. After a long silence, he wrote Lise that he was working very hard, but that he expected to "be somewhat more free in the future," and he invited his sister to visit him in England.[31]

Even though he was probably concerned mainly with wrapping up his current investigations, Krebs encountered at this point some observations which led him to reflect upon the significance of one of the general principles underlying the methods he customarily applied in studies of intermediary metabolism. That was the meaning of the "additive" effects when two potential metabolic substrates are added to a tissue medium in combination. His interest in the implications of "summation" seems to have surfaced over the course of a group of experiments encompassing diverse objectives that he began near the end of February. On the 26th he had tested the influence of glucose on the respiration and the formation of NH_3 in kidney slices in the presence of aspartic or glutamic acid. What he was most likely looking for was a manifestation of the "sparing action" of carbohydrates on the metabolism of amino acids. The result did not come out as predicted according to this concept. "Glucose sustains a constant respiration for a long time," he noted, but "inhibits little the formation of NH_3!!"[32]

The next day Krebs examined the influence of pH on the deamination of *l*-aspartic acid in rat kidney slices, "with *dl*-lactate." The fact that he included the same concentration of lactate in each of the runs at different hydrogen ion concentrations implies that he was probably not testing for specific effects of lactate, but perhaps employing it, as in his old experiments on urea synthesis, to enhance the general metabolic rates of the tissue. In the event, however, there was a "strong inhibition through lactate." Since there was no control excluding lactate, Krebs must have drawn this conclusion from the fact that the rates of formation of NH_3 were all very low compared, for example, to the results of the previous day in the presence of glucose.[33]

On the same day he carried out another set of experiments in which he might have been testing several possible relationships at once, without necessarily distinguishing them sharply in his mind. The title was "Keto Acids, + NH_3 Formation from Amino Acids." Again using rat kidney slices, he measured the Q_{O_2} and Q_{NH_3} with the following additions:

> Aspartic acid
> Aspartic + pyruvic acid
> Aspartic + lactic acid
> Pyruvic acid
> Lactic acid
> Glutamic acid
> Glutamic + α-ketoglutaric acid
> α-Ketoglutaric acid

The comparison of these rates for aspartic acid with and without lactic acid can be viewed as a follow-up on the observation that lactate had inhibited NH_3 formation in the preceding experiment, and the pyruvate might have been included because of its close relationship to lactate. On the other hand, the title, and the correspondence between the sets aspartic acid + pyruvic acid and glutamic acid + α-ketoglutaric acid, suggest that one of his aims was to explore whether keto acids might absorb NH_3 released through the deamination of amino acids. The results were also ambiguous. The respiration increased from Q_{O_2} = 21 to between 37 and 42 for the additions, whether the latter were separate or in combination. Q_{NH_3} for aspartic acid alone was 12.6, for aspartic acid + pyruvate 4.56, for aspartic acid + lactate, 4.21. From that comparison alone one might infer that pyruvate and lactate were capable of absorbing the NH_3 yielded by the aspartate; yet, as Krebs noted in the only aspect of these results on which he recorded a remark, there was "no definite NH_3 consumption!!! with keto acid or lactate."[34] That is, pyruvate, lactate, and α-ketoglutarate tested by themselves did not absorb NH_3. At this point the situation must have appeared mainly puzzling.

On the 28th Krebs carried out an experiment apparently intended to sort out some of the complexities posed by the preceding one. Avoiding the keto acids, perhaps to obviate their possible absorption of NH_3, he tested the effects of lactic and succinic acid on the deamination of l-aspartic acid. He complicated the situation, however, by testing the influence of arsenite, in three concentrations, on the effect of the combination of lactic and aspartic acid, and he tested malonic acid with and without aspartic acid. Here all of the additions inhibited the formation of NH_3 in comparison to aspartic acid used alone, but arsenite reduced the inhibiting effect of lactic acid.[35] He did not comment on these results, and they do not appear to have resolved his problems.

As he reflected on these experiments, Krebs must have noticed that, while he had fixed his attention mainly upon the rates of NH_3 formation, an anomalous pattern was showing up in the rates of respiration. The amino acids and the nonnitrogenous metabolites he had been adding to the medium of the kidney slices each increased markedly the rate of oxygen consumption over the controls; but combinations of the two types raised it little or no higher than either one alone did, contrary to what one would expect of two substances, both of which were readily oxidizable. It occurred to him that the maximum total rate of oxygen consumption might be limited by the rate at which the substrates could diffuse into the tissues. To test this possibility he carried out the next set of experiments, on March 1, at 25°C rather than the usual 37°C. At this temperature, he assumed, the tissues would respire more slowly, so that diffusion could not be the limiting factor. Again simplifying the experiment in one direction while expanding it in another, he employed only lactate as the second substrate, but tested it in combination with dl-alanine as well as with l-aspartic acid. At the same time he included the combination of l-glutamic acid and α-ketoglutaric acid that he had also tried in the first experiment of the series. As in the previous experiments, lactic acid reduced the rate of formation of NH_3 from aspartic acid, and α-ketoglutaric acid reduced that from glutamic acid. In the latter case arsenite partially restored the rate at which NH_3 was released. These were not the effects, however, in which Krebs was at the moment most interested. Rather, the key question, as his notation "Summation?" at the bottom of the page indicates, was whether the effects on the respiration of the pairs of substances were equal to the total of the effects of the individual substances alone. Apparently he wrote this query down after recording the main results in the usual form, and then added, with a different pen, a line of calculated Q_{O_2} average's in order to bring out more clearly whether or not they were. The figures were:

Control	6.3
l-Aspartic acid	12.3
dl-Lactic acid	13.3
l-Aspartic + lactic acid	17.1
dl-Alanine	21.9
dl-Alanine + lactic acid	22.5
Control	8.05
d-Glutamic acid	11.6
α-Ketoglutaric acid	11.8
d-Glutamic + α-ketoglutaric acid	12.3

The answer to his question was, therefore, not unambiguous. In the case of aspartic and lactic acid, there appeared to be additive effects, but they amounted to less than full summation. With lactic acid and alanine, and glutamic and α-ketoglutaric acid, there was little or no summation.[36]

Going back over the complexities of the situation, Krebs decided to check more fully the possibility that the reduced formation of NH_3 in the presence of the paired substrates was due to absorption by the second substrate of NH_3 formed by the first. Adding *dl*-lactic, pyruvic, and α-ketoglutaric acid, each singly, to rat kidney slices in the presence of NH_4Cl, he found that each absorbed some NH_3, but at rates too low to account for the reductions in the formation of NH_3 that these compounds had caused when added in combination with the amino acids. This experiment satisfied him that he could eliminate the explanation in question.[37]

Next Krebs pushed forward his effort to find conditions under which the effects of combinations of amino acids with other substrates might be additive. Lowering the temperature further, to 21°C, he repeated his comparisons of the effects of combinations of lactic acid with aspartic acid and alanine to the individual effects of these compounds. The title of his experiment suggests nicely the way in which a question that had arisen a few days earlier only as a subordinate methodological criterion was beginning to rise to a more prominent level of significance. In the previous experiment of this type he had written "Summation?" at the bottom of the page. The new experiment was headed:

Summation Effect
l-Aspartic acid + lactic acid
dl-Alanine + lactic acid

The question was now the main topic of the experiment. It was still posed only within a specific experimental context, but the form of the heading implies that in Krebs's mind there was hovering an inchoate feeling that this was a potential subject for an experimental inquiry concerning a general class of "summation effects," of which the cases listed were examples.[38]

Although the question was now sharply posed, the answer remained indistinct. With aspartic and lactic acid together the respiration was higher than with either alone, but not fully additive (Q_{O_2} for aspartic acid, 8.65; for lactic acid, 11.25; for both, 14.40). With alanine, lactic acid exerted no additional effect. In a second set of experiments Krebs employed aspartic acid as the amino acid, with α-ketoglutaric acid as the other substrate, and carried out comparative tests at 21°C and 37.5°C. These results too were inconclusive. He observed only "slight summation," but since there was also only a "small effect of aspartic acid" itself, he could conclude little more than that the experiment was unilluminating. On March 5 he extended his tests of summation effects to other substrates, including the pairs aspartic acid + oxaloacetic acid, and glutamic acid + α-ketoglutaric acid but again found that at 37°C the combinations did not yield markedly higher rates of respiration than did each compound alone. At 20°C aspartic acid was not oxidized at all.[39]

By this time Krebs must have come to believe, in spite of the occasional additive effects that he had observed, that the absence of consistent "summation" was not due to some secondary cause such as limits on the rates of diffusion. Rather, in the presence of one another within the cells, the two substrates involved were not metabolized as rapidly as either one would be by itself. Taken together with the experiment showing that the nonnitrogenous substrates he was using lowered the rate of formation of NH_3 by some means other than absorbing the NH_3, these results suggested that the second substrates must act by *inhibiting* the deamination of the amino acids. They did so, however, while maintaining, and sometimes even increasing, the rates of respiration. At this point, if not sooner, Krebs probably inferred that the effect of the second substrate could best be explained as a competition with the amino acids for the available oxygen. If so, then by inhibiting the oxidation of the second substrates he ought to be able to restore the oxidative deamination reactions to rates approaching those yielded by the amino acids by themselves. From his previous work he had available a convenient method to achieve that aim, in the agent arsenite, a specific inhibitor of the oxidation of keto acids. This was the likely rationale behind the next experiment that he carried out, on March 7, on "amino acids + keto acids + As_2O_3." Again he tested the pairs *d*-glutamic + α-ketoglutaric acid and *l*-aspartic + oxaloacetic acid, but now comparing the rates of formation of NH_3 for these pairs with and without arsenite. The results were not dramatic, since the values were quite low throughout; but in the case of the first pair arsenite suggestively raised the level of Q_{NH_3} from 0 to 1.13 (compared to a control value of 0.63). For the other pair there was no marked effect.[40]

Krebs now had the outlines of an explanation linking the various effects he had observed, but did not yet possess adequate data fully to confirm his view. On March 8 he carried out an experiment intended perhaps to lead him toward better conditions in which to pursue the question. Using guinea pig in place of rat kidney tissue, he measured the disappearance of NH_3 with "various substrates," including lactic, pyruvic, α-ketoglutaric, succinic, and nicotinic acid. He found "very slight NH_3 consumption, even with ketoglutaric acid." Then, whether he was for the time being satisfied with, or frustrated with, the state of this problem—or merely diverted—he turned to other aspects of his ongoing investigation. After performing two sets of experiments comparing the deamination of aspartic acid in slices and extracts, he concentrated on the synthesis of glutamine and the properties of glutaminase for the next three weeks.[41]

No matter how busy he was during this period, Krebs must have worked from mid-February on with the question of what to do about Katrina much on his mind. By late March he had finally thought enough about their future to write the letter for which she had asked. From her reply to that letter we can infer that, if he did not decide definitely in favor of marriage, he came close to it. Apparently he wrote that the main obstacle to that step was his poor salary. He told her that he was trying for some "financial improvement" at Cambridge, but that he did not know when he would be able to obtain a decision. He must also have asked her how quickly she could leave her post in Altona in the event that he was successful.[42]

By the time she opened Hans's letter, Katrina had "almost given up hope" that she would ever hear from him again, but after she read it she was joyful and satisfied. She resolved to stop chastising him for the troubles he so often caused her. "I will also promise," she wrote on April 1, "not to be so influenced by the bad traits in your character—they are more difficult to endure from a distance than from close by." She estimated that she could hold out in Altona for as much as a year, and could depart on relatively short notice. She asked him if it would be possible to manage a modest household on his present income, and was ready once again to plan a summer holiday in England.[43]

IV

On that same first day of April, in Cambridge, Krebs began a new week in the laboratory by returning to the subject of "summation in slices." He limited himself this time to one pair of substances, dl-alanine and dl-lactic acid, with rat kidney slices. Placing the slices directly into the manometer vessels, without prior washing in Ringer's solution, he attained exceptionally high rates for both Q_{O_2} and Q_{NH_3}. He also obtained no summation; on the contrary, there was a "high inhibition of Q_{O_2} through lactate," which also inhibited the formation of NH_3. In another experiment performed the same day, he employed cyanide to inhibit the deamination of dl-alanine. He must have been comparing the action of cyanide with that of lactate on the process, but it is not immediately obvious in what respect the two actions illuminated one another. The next day he continued this association, carrying out a set of experiments in which he employed both cyanide and dl-lactate, and added urethane as a third inhibiting agent, comparing the effects of all three upon the deamination of dl-alanine. This time there was a clear-cut difference in their effects. Cyanide caused only a "questionable inhibition" (apparently with respect to the formation of NH_3, there being a substantial decrease in the Q_{O_2} compared to alanine alone), whereas with lactic acid and alanine there was "half summation"—that is, Q_{O_2} for dl-alanine, 45.0; for dl-lactate, 35.6; for alanine + lactate, 53.0.[44]

On April 3, Krebs once more carried out an experiment on the inhibition of deamination with HCN in association with one on "summation, lactate +

dl-alanine." The latter was performed at 25°C. In this experiment the results were:

	Q_{O_2}	Q_{NH_3}
Control	9.30	1.18
Lactate	15.2	0.95
Alanine	19.7	8.00
Alanine + lactate	19.1	3.72

At the bottom he wrote "Good experiment. No summation!"[45] By this time his interpretation of the results of his experiments of the previous month on this subject had probably led Krebs to look for such a result, and this one evidently satisfied him that the lactate inhibited the formation of NH_3 because the tissue was oxidizing it in preference to the alanine. He then went on to elevate a question that had surfaced as a subordinate consideration in the earlier experiments of this series into the central topic of the next phase of his investigation.

Once or twice before during his three-year-long study of deamination, Krebs had touched experimentally on the question whether he could reverse the reaction and obtain amino acids from keto acids in tissue slices. Now his attention had again been drawn to that subject as he considered whether the decreases in the release of NH_3 in tissues in the presence of amino acids was due to consumption of part of the NH_3 by the other substrates he added. Even as he ruled out the explanation, he was apparently induced to take up the question whether these substrates would, under other conditions, combine with NH_3 to form amino acids. Perhaps too, he had in the back of his mind a challenge that Franz Knoop had made to him during the previous fall. "May I ask," Knoop had written in October 1934,

> whether, with your experimental methods you have ever attempted to find something that would signify a reversal of amino acid decomposition? I have always had the impression that this discovery of mine, which comprised the first proof of a physiological amino acid synthesis, has been insufficiently valued. Therefore my question. To be sure, I am of the opinion that, in general, damage to organs causes synthetic reactions to suffer first, much more than the decomposition ones, and I have often emphasized the significance of synthesis for the definition of the concept of life. Consequently I would doubt in advance whether a synthesis can be captured with your methods. My question in spite of that. Every contribution in this direction appears to me to be of general significance.[46]

It was characteristic of Knoop to think that the scientific contributions he had made early in his career were now largely overlooked, and his skepticism in principle that Krebs's methods were suitable for synthetic reactions overlooked the fact that the discovery of the mechanism of urea synthesis had already refuted such expectations. Nevertheless, from Knoops's remarks Krebs could know, at least, that if he elucidated the synthesis of amino acids with his methods, the senior scientists in his field would not be unimpressed.

Krebs approached the question in a straightforward manner. If any of the substrates involved did give rise to amino acids, then the quantity of NH_3 present ought to decrease, while the quantity of amino-N ought to increase corresponding-ly. On April 3 he carried out his first experiment on "amino-N formation in the liver." Using rat liver slices, with NH_4Cl added to the medium, he measured the rates of change in the quantities of nitrogen present in the two forms with no other additions, with dl-lactic acid, and with α-ketoglutaric acid added. The results were:

	Q_{NH_3}	$Q_{amino-N}$
Control	-7.30	+3.92
dl-Lactate	-8.50	+5.18
Ketoglutarate	-5.50	+4.99

This outcome was mildly suggestive, but inconclusive. Lactate appeared to increase both the disappearance of NH_3 and the formation of amino nitrogen somewhat above the control, whereas ketoglutarate reduced the former rate and left the latter unchanged. In each case the tissue consumed nearly all the NH_3 added to the medium, and in two of the three more nitrogen disappeared in the NH_3 than appeared in the amino-N.[47] The main result of the experiment was to suggest how another one like it could be improved.

The next day, when Krebs repeated the experiment, he added twice as much NH_4Cl at the beginning. Assuming that the most likely place for the missing nitrogen to appear was as urea, he added measurements of the rates of urea formation. He also added pyruvic acid to the substrates previously used. This time none of the runs exhausted the NH_3 supplied, and a more significant pattern emerged. The results were:

	Q_{NH_3}	Q_{urea}	$Q_{amino-N}$
Control	-3.82	+2.12	+2.85
dl-Lactate	-7.12	+3.74	+6.45
Pyruvate	-7.40	+1.57	+7.10
Ketoglutarate	-3.58	+2.18	+4.68
Ringer's solution without NH_4Cl	+0.37	+1.02	+4.44

Scanning these results, Krebs saw that the experiment was again partly spoiled, this time because he had not washed the tissues before. There was a "large blind value for amino-N," so that even without NH_4Cl or an added substrate, $Q_{amino-N}$ was not much smaller than the other rates of its formation. Nevertheless, he drew a conclusion: "Only with pyruvic acid large amino-N formation. Good agreement between NH_3 disappearance and amino-N formation."[48] (It is not immediately clear why he did not regard the amino-N formation from lactate as large also; on the other hand, taking into account that the urea formation was probably to be subtracted from the NH_3 consumption to match the amino-N value, there was not the "good agreement" here that was evident in the case of pyruvate.) Now Krebs had at least a promising lead to follow up with a better-executed experiment. Two days later he did so, being careful this time to wash the liver slices three times before placing them in the manometer vessels. Adding two further substrates he obtained the following:

	Q_{NH_3}	Q_{urea}	$Q_{amino-N}$
Oxaloacetate	-13.2	+2.68	+5.08
l-Malate	-5.38	+1.39	+2.62
Ketoglutarate	-8.06	+2.32	+5.88
Pyruvate	-14.8	+1.75	≈7.65
Lactate	-14.4	+2.91	+6.56
Control	-5.95	+1.39	+2.38
Solution without NH_4Cl	0	+0.55	4.00?

These results seemed to Krebs incongruous. "Everywhere," he wrote down, "too much NH_3 disappeared, compared to urea formation. The excess is *not* found as amino-N. Unclear what becomes of NH_3!!!" To emphasize the discrepancy he added an additional row of calculations of the values for "Q_{NH_3} unexplained" for each substrate. Then he summarized the procedures he had employed to determine NH_3, urea, and amino-N, perhaps so that he could later check whether they had been the source of some systematic error. Finally, at the bottom of the page he recorded some shorthand speculations about where the missing nitrogen might be: "Acetylization of amino-N? Ring-N? Creatine?"[49]

Such anomalies can serve either to induce the investigator to pursue new questions or to discourage further moves in the direction taken. The decision whether to pursue or desist is not based on logical judgments alone, but also on competing priorities. In this case, Krebs dropped the subject of amino acid synthesis, and returned the next day to glutaminase.[50] That he did so may mean that he saw no way around the obstacle. More likely it meant that he was too impatient to finish up an investigation nearly completed to allow himself to become further involved in a new one that quickly came to look complicated.

During the two weeks of April 6–18, Krebs carried out mainly experiments on glutaminase and deamination. With the former he worked particularly on the specificity of the enzyme, testing its action on compounds related to glutamine, such as isoglutamine, phenylacetyl-glutamine, and benzylglutamine. With the latter he multiplied his measurements of the action of *d*-amino acid oxidase in extracts of various tissues, especially in order to establish whether the proportions of O_2 consumed to NH_3 released were in accord with the theoretical ratio of 1:2. In most cases they were, within the expected limits of error. In one case, however, he found "too much O_2 used!!!" and wondered if there had been a "coupled oxidation!!!" From David Keilin he learned that Keilin had been able to couple the oxidation of amino acids under the action of this enzyme with the oxidation of other substrates such as ethyl alcohol. By now Krebs had probably written much of the two papers he planned to publish, and was striving primarily to fill in gaps in his data, rather than to follow up leads such as this one. He limited himself in his paper to reporting what Keilin told him about coupled oxidations.[51]

During this time Krebs also performed a few more experiments on the effects of second substrates on the deamination of l-amino acids in tissue slices. On April 7 (a Sunday), he paired succinic with aspartic acid, and obtained the following:

	Q_{O_2}	Q_{NH_3}
Control	23.5	1.34
l-Aspartic acid	36.8	9.65
Aspartic + succinic acid	39.0	3.28
	38.7	6.95

The result, he noted, was "succinate displaces aspartic acid (inhibition of NH$_3$ formation)." Almost parenthetically he added below, "No summation."[52] A comparison of this statement with his treatment of similar previous experiments shows the extent to which the explanation at which he had arrived for the fact that second substrates such as this one lowered the Q_{NH_3} without substantially raising Q_{O_2} had become self-evident to him. That the succinate must have "displaced" the aspartic acid—that is, that it was oxidized in place of the aspartic acid that would otherwise have been oxidized—was an *interpretation* of the result, but it had by now nearly merged with his statement of the result itself.

Continuing his peak working pace, Krebs recorded experiments on all but three of the first 18 days of April, including one remarkable day in which, presumably with the aid of his new technician, he performed seven full sets.[53] Then he left his laboratory for a week. That may have been the time in which he took an Easter holiday in Devonshire;[54] but no details of where he went or what he did are available. When he resumed work on the 26th, he continued along the same lines as before. A greater emphasis on deamination experiments suggests, however, that he may have felt that he was so close to completing this investigation that he now gave first priority to wrapping up its remaining loose ends. There were still a few surprises in store, such as an observation on April 27 that amino acids were not oxidized in rabbit liver or kidney slices. He wondered if rabbit tissue might contain some special inhibitor, and verified his suspicion by showing that an extract of rabbit tissue inhibited 72.5 percent of the deamination in pig kidney extract.[55] On the 29th he performed an experiment on combinations of an amino acid with other substrates that was more extensive than the previous ones of this type. Using rat kidney slices, he paired l-aspartic acid with proline, hydroxyproline, and glutamic, acetic, citric, and butyric acid.

Some of the results were like the earlier ones, but there were also some interesting variations:

> With acetate and butyrate, small (incomplete) summation!! But inhibition of the formation of NH$_3$!!
> Citrate inhibits [respiration and NH$_3$ formation].
> Proline and hydroxyproline small increase [in respiration], no inhibition of NH$_3$ formation.
> Glutamic + aspartic acid average value [the rate of respiration was midway between the rates for each alone]. No summation!![56]

These variations did not alter his conclusion that the substances earlier tested, which had inhibited the formation of NH$_3$ with no summation of respiration, did so by displacing the amino acid. The "incomplete" summation displayed with acetate and butyrate could be seen merely as a case of incomplete displacement. The other pairs tested for the first time here were probably not examples of the same kind of action he had previously observed. This experiment probably marks instead, whether as cause or as effect, the growth of Krebs's interest in the phenomenon of summation beyond the bounds of the specific experimental context in which he had been utilizing it as a methodological criterion. The same approach, he was coming to believe, could provide a broadly useful means to elucidate the relationships between other oxidizable substrates. By this time, or shortly afterward, he may have written the following passage in the paper on deamination that he was preparing for publication: "I explain the fact that no summation takes place [in the experiments on the inhibition of the oxidation of amino acids by other oxidizable substrates] by assuming that the mechanism of activation of molecular oxygen is identical when different substrates are oxidized." When two substrates share the same mechanism, they compete for utilization of the system. Following his discussion of the case at hand, Krebs moved on to the broader implication of this explanatory concept. "The method of investigating the summation offers the possibility of deciding generally whether the oxidation of various substances in cells involves entirely separate systems for each substrate."[57]

This methodological possibility was attractive enough to Krebs to induce him to interrupt the final stages of his deamination investigation to test summation effects in respiring yeast. Yeast was a "very suitable material" for the purpose, because without substrates it has a very low respiration, which can be very much increased by adding substrates to the medium. On May 3, he tried out seven substrates—lactic, butyric, acetic, aspartic, glutamic, ketoglutaric, and succinic acid—singly, on the respiration of baker's yeast at 17°C. He observed "no definite respiration with amino acids and succinate. All other substrates give readings, rising with acetate." Selecting those of the substrates—lactate, ketoglutarate, butyrate, and acetate—which had increased the respiration, he examined the next day whether they would exhibit "summation" with the same yeast when paired with glucose. The answer was clear. Q_{O_2} for the control was 6.4, with glucose 81.5. For none of the combinations of glucose with the other substrates did the rate exceed that for glucose alone. There was, "therefore, no trace of summation."[58] Inserting these results into the section of his paper on deamination in which he discussed the application of the summation method, Krebs concluded, "This [experiment on yeast] seems to prove that the systems responsible for the oxidation of various substrates have one component in common, most probably the part which activates the molecular oxygen."[59]

In addition to the "summation effect," the "dilution effect" that Krebs had observed during his deamination investigation came to have for him general implications concerning the nature of metabolic reactions. During the spring he carried out several experiments showing that the dilution of the medium could greatly reduce the rate of the processes he was studying. In particular, the "natural" *l*-amino acids, which could not be deaminated in tissue extracts, as the *d*-amino acids were, were oxidized in ground kidney brei at rates nearly as high as they were in tissue slices, provided that the suspension was concentrated. By diluting such suspensions 8- or 16-fold, he caused the reactions to be strongly inhibited. In the paper on deamination that he was writing he argued that the oxidative enzymes which cannot be extracted, and are therefore "said to be bound up with the structure of the living cell," are inactivated "not by grinding and destruction of the cell, but by the dilution of the protoplasm which necessarily accompanies extraction." To explain these results, he offered a "theory of the effect of dilution." In enzymatic reactions of this type, which include most oxidations and fermentations, he assumed, a ternary or higher-order collision determines the velocity of the reaction. The components which must be brought together may include an activated substrate, activated oxygen, and a coenzyme. It is because the probability of such a collision decreases in proportion to the dilution that the effects he had observed take place. "Ternary collisions occur extremely rarely in homogenous solutions." Therefore, "it is one of the functions of the structure of the cell to arrange the catalysts in such a way that a ternary collision reaches a certain degree of probability." Krebs concluded his brief discussion with a disclaimer: "It is not however within the scope of this paper to discuss this problem in full."[60]

* * *

In 1977 I asked Krebs if he recalled the circumstances in which these ideas had occurred to him. He replied,

> Well they certainly occurred to me whilst writing up the paper and trying to explain the simple observation that if one dilutes a tissue too much the activities disappear. If you dilute a simple chemical reaction the rate decreases in proportion to the dilution. But this is when there are binary collisions involved. But in these homogenates the fall with dilution was very much more rapid. It was merely an attempt. Essentially it has proved correct. Of course there are many factors involved in the oxidations. It didn't involve oxygen only—a catalyst plus a substrate, but also a resynthesis of ATP [from] . . .

ADP and inorganic phosphate. But it was too general an idea to be really useful. It was superseded by a more detailed analysis of the factors which control the oxygen uptake in homogenates. This was done by various people, adding cofactors like magnesium and purine nucleotides which were not available at that time.[61]

For our understanding of the genesis of his theory of the dilution effect, the only helpful portion of Krebs's discussion is the initial recollection that he had thought of it while writing his paper—a nice illustration of the connection between writing and the emergence of ideas. The rest of the passage inextricably mingles his view of the idea itself with his knowledge of later investigations of the problems to which he had addressed it. Retrospectively he could recognize that the situation proved to be far more complex than he had anticipated, and that it was elucidated by detailed analyses carried out by others. His admission that his own idea was "too general . . . to be really useful" appears also mainly as a retrospective judgment—a recognition that it had exerted no identifiable influence upon the work of those who later performed the detailed analyses of the factors involved. Finally, Krebs tacitly excused himself for not having made a more detailed analysis himself at the time. He could not have done that successfully, he believed, because the cofactors that later proved essential were not yet available.

We do not require this foreknowledge of later events in order to agree that Krebs's idea was too general to be really useful. It was so because Krebs did not attempt to develop it. He did not test whether a detailed analysis of the factors involved was then possible. What he put forth was not a theory formulated in the light of all of the relevant contemporary information available, but a reflective insight that occurred to him while he was writing, and that may have appeared in his paper in essentially the form in which it occurred.

About the "method of summation" Krebs commented similarly in 1977:

> I think it was just an ordinary logical analysis of the situation. If they are separate systems, then [the rate of oxygen consumption] would add up, provided there is enough oxygen. But if they all—as we now know—go through the same channel . . . , there would be no summation possible, provided that these rates are already maximal.[62]

Here too Krebs was viewing the question partly from the perspective of the way in which it was afterward answered, but in this case he distinguished the historical from the later situation:

> I think at that time it wasn't clear whether the pathway of oxygen consumption [for different substrates] ended up with the same catalyst, especially after it was shown that . . . the d-amino acids can react directly (through this enzyme) with molecular oxygen. But nowadays this is a matter of course. We know that it all goes through the respiratory chain.[63]

The method therefore might have offered the potential to clarify an important general question. Krebs, however, claimed very little for it in retrospect. He did not think that "my comments at that time were of any particular importance to the development of the subject."[64]   In referring to his idea as mere "comments," Krebs overlooked the fact that he had tested his idea experimentally, not only during the course of his deamination investigation, but with two experiments on yeast. Nevertheless, his assessment was basically valid. He did not develop his approach into a comprehensive investigative attack on the problem, and the question was eventually answered by other, more detailed types of analyses.

* * *

If neither Krebs's method of summation nor his theory of the dilution effect were important to the "development of the subject," they were, nevertheless, significant indications of the development of their author's point of view toward his subjects of research. We have seen that during the five years in which he had been carrying on investigations in intermediary metabolism, his immediate goals had been to identify concrete steps in metabolic pathways. He had stayed close to mainstream definitions of problems and areas of opportunity within this field.   He owed his successes to resourceful tactical deployment of the methodological and conceptual equipment he had acquired, but he did not have any underlying concept of the nature of metabolic pathways that differed markedly from those current at the time.   His ideas about summation and dilution effects represent efforts to think past the concrete problems of identifying metabolic reactions, to envision how such processes are organized within the framework of the cell. These were not particularly novel or deep ideas. His own retrospective view of them as "comments" accurately categorizes them. What they show is not that he thought profoundly about the broader questions in this field, but that he sought to orient his particular lines of investigation within a broader framework. He viewed the metabolic pathways he explored not merely as sequences of chemical reactions, but as physiological processes whose properties could be fully explained only in relation to the structure of organized cells.

The manner in which Krebs discussed his explanation of the dilution effect reveals a significant evolution of his view of the relation between cell structure and metabolic reactions. In 1932, (see Vol. I, pp. 354–356), he had described the formation of urea as "a process taking place in the living structure of the cell." He had distinguished between the reaction of arginine as "not bound to the life of the cells" and the conversion of ornithine to arginine as a reaction "tied to life." In his current discussion of the dilution effect, which he placed under the subtitle "Cell Structure and Enzyme Action," he implicitly distanced himself from expressions he had once used as though they were unproblematic. He now differentiated enzymatic reactions which can take place in extracts from those, such as cellular oxidations, which "are said" or "which have generally been considered" to be "bound up with the structure of the living cell."

Structure was still important to his distinction, but he sought explanations for "why the structure is of importance for chemical reactions in cells" that went beyond their designation as "tied to life." To explain that cell structure is necessary in order to provide an arrangement of catalysts that increases the probability of a ternary collision is to seek to understand in physical terms what he had formerly referred to merely as "the life of the cells."

Does this development signify a shift in Krebs's philosophical position with respect to "vital" and "chemical" processes? As discussed previously (Vol. I, pp. 354–356), Krebs himself retrospectively insisted that his earlier statements had been merely empirical, not philosophical views. I suggested there that he probably adopted the terms he used then from Warburg, without much independent thought about their philosophical implications. According to his own memory, later changes in the language he used amounted only to being more careful about how he "expressed matters."[65] Whether in fact he did shift his point of view, or whether he only became more aware of the issues involved than he had formerly been, is a question that Krebs would probably have dismissed as too subtle to answer.

The explanation that cell structure functions as a means to arrange catalysts so as to increase the probability of ternary collisions fits comfortably within the general biochemical point of view developed with special clarity by Frederick Gowland Hopkins. Hopkins portrayed the living cell as the locus of highly coordinated chemical events, organized in large part by the way in which the cell structures control the location of catalysts and substrates.[66] We are tempted, therefore, to identify here a clear influence of the "Cambridge milieu" on the thought of the emigré from Germany who had been working within the Hopkins sphere for the past two years. In light of Krebs's testimony that Hopkins's comments often helped him and others to see the broad connections between specific facts, it is most likely that Hopkins's presence did encourage Krebs to think in the directions represented by the theory of dilution. There is no evidence, however, that Hopkins supplied the particular form of the explanation that Krebs gave.

Krebs took an unusually long time to write up his third paper on the deamination of amino acids, because it was the first research paper he had attempted in English. He turned to a number of friends to help him express himself accurately in his recently acquired second language. Vernon Booth, and a librarian named Ann Barbara Clark were especially helpful.[67] This paper can be regarded as a third revision of the first two papers he had published on the same topic. It included much additional experimental data; but what differentiated it most importantly from its predecessors was that it was organized around the distinction implicit, but not fully elucidated, in the earlier papers, between the deamination of d- and of l-amino acids. He summed up the thrust of his advance in a single succinct paragraph:

> In this paper it will be shown that the enzymic system catalysing the deamination of the natural amino-acids is different from the system catalysing the deamination of the non-natural optical isomerides.

Kidney and liver contain (at least) two different enzymic systems responsible for [the oxidative deamination] reaction. The two systems differ in many ways. The system deaminating the natural amino-acids is destroyed by drying the tissue; it cannot be extracted; it is inhibited by octyl alcohol and by cyanide. The system deaminating the non-natural amino-acids is not destroyed by drying; it is readily soluble in water and can be extracted from fresh or dried tissue by aqueous solutions; it is not affected by octyl alcohol or by cyanide.[68]

Most of the lengthy paper consisted of the systematic presentation of the evidence for this claim and the comparative description of the properties of the two enzymic systems.

If the evidence for the existence of two systems was compelling, the meaning of the discovery was anything but clear. Inconspicuously, in the middle of his paper, Krebs acknowledged the paradox it posed: "It seems strange that an enzyme exists that deals specifically with non-natural substrates. But it may be pointed out that α-amino-acids of the d-series have been found occasionally in nature."[69]

As the weak form of the second of these sentences tacitly acknowledged, the occasional existence of d-amino acids in nature was not a convincing explanation for the widespread existence of an enzyme specific to them. Krebs was casting around for ways to account for the strange situation he had uncovered. Inside of the laboratory notebook he was keeping at the time there is a draft of a paragraph which documents one stage in his search for an explanation. The page was numbered "30," and probably belonged to an early draft of his paper on deamination. It provides a rare glimpse of him in the process of unfolding his thoughts on paper, and in this case also captures some of the little errors that he made in expressing himself in English. The following is the original form of the draft:

The difference of the l- and d-amino acid deaminases is of interest in connection with two problems.

Firstly it shows that the cell uses a complicated system in the case of the l-amino acid deaminase, although it would be certainly possible for the cell to perform an identical reaction in a simpler way—as the existence of the d-amino acid deaminase proves. It is probable that a complicated mechanism, that is a reaction proceeding in several steps, is involved when transmission of energy takes place. The l-enzyme behaves as respiration and we assume that it is capable to providing energy for the cell. The l-enzyme [sic: should be "d-enzyme"] on the other hand, is probably a scavenging reaction, removing not wanted material and converting it into body substances. The free energy of these reactions cannot be utilized by the cell.

Secondly the existence of the two enzymes shows that the cell employs different mechanisms according to the purpose of the reactions. The same reactions proceeds entirely different when it acts

– 30 –

~~The difference of the l- and d- amino acid deaminase is of~~
~~interest ~~in~~ in connection with two problems.~~
It seems of particular interest
~~Firstly it shows~~ (That the cell ~~uses~~ *employs* a complicated system

in the case of the l-amino acid deaminase, although ~~the~~ it ~~would~~

~~be certainly~~ possible for the cell the perform ~~an identical~~ *the*

reaction in a simpler way- as the existance of the d amino acid

deminase ~~shows~~ proves. ~~It is probable that a~~ *the* complicated mevhanism
that is a reaction proceeding in several steps
(is involved when transmission of energy takes place . The l-enzyme

behaves as repiration and we assume that it is capable to

providing energy for the cell. The l enzyme on the other hand,
preliminary
is probably a ~~scavanging~~ reaction, removing not wanted material
(free
and converting it into body substances. The energy of this reac-

tions cannot be utilized by the cell. *Thus the mechanism of a*
reaction in the cells depends on the physiologal role of the reaction.

~~Secondly the existence of the two enzymes shows that~~ the cell

employes ~~diferant mechanism according to the purpose of~~ the
(chemical)
~~reactions.~~ The same reactions proceeds entirely different when ~~the~~
it acts as energy giving reaction, or not. ~~Therefore~~ the investi-

gation of the d-enzyme does not give any information about the
Several
cell repspiration. ~~Many~~ other biological oxidations for instance

xanthin oxidase ,or uricase, or dyhydroxy alanine oxydase (dopa-
oxidase) are equally unintersting for the problem of respiration ,

the energy giving oxidative processes.

as energy giving reaction or not. Therefore the investigation of the
d-enzyme does not give any information about the cell respiration.
Many other biological oxidations, for instance xanthine oxidase or
uricase, or dehydroxy [phenyl] alanine oxydase (dopa-oxidase) are
equally uninteresting for the problem of respiration, the energy giving
oxidative processes.[70]

From the accompanying reproduction of the page we can see Krebs working
over these ideas, tightening the somewhat loose prose he had composed on his
typewriter, and making one significant conceptual modification by substituting
the word "preliminary" for "scavenging." He was struggling here for a
plausible interpretation of the existence of the two enzymes. He was speculating
well beyond his data. His argument was already complicated, but not yet fully

integrated. The draft leaves the impression that Krebs was putting down ideas that were still only partially framed in his mind, so that the writing was an integral part of the mental formulation.

These two paragraphs reveal a great deal about Krebs's general viewpoint. They exhibit, behind his immediate investigative goals, a driving motivation to integrate his results into a broad functional picture of the metabolic processes of the cell. Moreover, they suggest that Krebs was not equally interested in all metabolic processes; that the problem he most wanted to elucidate was the "problem of respiration, the energy giving oxidative processes." In that respect his investigation of deamination and his so far unsuccessful forays into carbohydrate and fatty acid metabolism were facets of a single larger venture.

With respect to the more immediate problem of explaining the existence of the d-enzyme, this discussion shared the weakness of the briefer passage quoted from Krebs's published paper. Why should so many animals possess a system to remove "not wanted material" that they would rarely, if ever, encounter? At any rate, the argument did not survive to the final draft of his paper, apparently because Krebs came upon another argument that he preferred. In the published version, under the heading "Interrelation between d- and l-Amino-Acid Deaminase," he wrote:

> From the experiments described, it was concluded that two different deaminating systems exist. But it must be pointed out that this difference only concerns the two systems as whole systems. It may well be that the two systems have certain components in common; for instance it may be that the l-deaminase is the d-deaminase *plus* an additional factor. The idea that the d-deaminase is a fragment of the l-amino-acid deaminase is supported by the fact that d-deaminase can be obtained only from those tissues which contain the l-system, and by the fact that the l-system cannot be separated from the d-system. Moreover it would explain the occurrence of an enzyme for which practically no substrate is found in nature. If this view is correct, the additional factor, which makes the d-enzyme into the l-enzyme, is a substance which reacts with l-amino acids and enables them to react in the same way as d-amino acids react by themselves without an auxiliary substance.[71]

By comparison with his earlier argument this one was crisp, sharply focused, tightly reasoned. It was no less speculative, but was internally far more satisfying. Here there is no digression onto problems that are more or less "interesting," only a logical analysis of the problem at hand. The previous argument had attempted to make some sense out of "the occurrence of an enzyme for which practically no substrate is found in nature," whereas the present argument obviated the difficulty by treating the *apparent* occurrence as an artifact. In the new argument Krebs was able to eliminate the disjunction between the observed phenomena and the assumption of functional integration. Where the looseness of the draft paragraphs suggests an early stage in the emergence of a line of thought, the smoothness of the published paragraph has the feel of a product of extended mental refinement.

Different as the scope and orientation of these two arguments were, they too had "certain components in common." The most salient feature they shared was the assumption that *l*-deaminase is a more complicated system than *d*-deaminase. That similarity suggests that in devising the second argument Krebs did not start from scratch, but that he somehow worked his way from the first to the second. The second is obviously not a direct *development* of the first, or one that could be derived from the first by any single transformation. A complex mental process must have intervened between them, in which elements of some ideas were added, while others were subtracted, and in which there was no orderly succession of coherent intermediate states. It was probably the sort of prolonged, convoluted process that Krebs liked to sum up in the phrase "hovering over a problem."

* * *

When I asked Krebs about this explanation in 1977, he commented:

Yes. That has proved wrong. I thought it might be an artifact coming through, but . . . in fact there are hardly any *l*-amino acid oxidases. Snake venom and special materials contain it, but in the animal most amino acids lose their nitrogen through first transamination and then the dehydrogenation of glutamate—although at one time people thought there are also *l*-amino acid oxidases.

After I had brought up a third explanation that he had proposed in 1936, and he had remarked, "There is also no evidence for this," I commented, "I suppose these explanations are examples of the persistence of your concern to find the physiological function." He answered,

Yes. I think it is justified to express an hypothesis if it can easily be tested. But it is very unsatisfactory to have no explanation at all.[72]

In some respects this conversation echoes one described in Volume I (p. 347), in which Krebs expressed the belief he had acquired from Warburg, that it is better to state a clear position that may be wrong than to hedge. Here, however, Krebs is not only justifying hypotheses that may be wrong, but expressing the need for an *explanation*. We have seen in this chapter, in connection with his investigation of deamination alone, three different explanations that he proposed, each of which proved to be either wrong or historically unimportant. They *are* important, however, for understanding Krebs's scientific style. He was, as we have seen, not the sort of scientist who pursues deeply imaginative conceptions thought out in advance, but a pragmatic exploiter of leads near at hand. It was in response to unexpected effects that he best displayed his experimental imagination; and it was in response to puzzling outcomes that he sometimes best displayed his conceptual imagination. His drive to *explain* what he observed may have led him sometimes to explanations of only passing value, but that same drive guided his sense for what was interesting and important. It was one of the sources of his creative force.

The foregoing episodes illustrate also that what Krebs meant by "having an explanation" was not to give any type of explanation, but to understand the functional significance of the chemical processes he studied. In the above conversation it was I who voiced his concern to find the physiological function, but in doing so I was referring to his own definition of his orientation as "physiological." He was not content to identify a chemical process in biological material unless he could also envision a physiological reason for the existence of that process. Without such an understanding Krebs felt that he could not "explain" the process, and that was, for him, an unsatisfactory situation.

On the morning after our discussion of the deamination paper which included the exchanges quoted above, Krebs began the conversation with the comment:

> Perhaps I might add to our discussion on amino acid oxidase that I didn't consider that work as being particularly important to the development of my study of biochemistry. I just happened to come across an enzyme which, however, turned out to be of no major physiological interest. I was interested in amino acid metabolism, but it was rather a disappointment when I realized the enzyme attacked only the nonnatural amino acids.[73]

Krebs's work on amino acid oxidase was, as the preceding section suggests, important to the development of his biochemical research in ways that were not obvious to him; but that is not to contest the validity of his own *feeling* that it was not. Nor can there be any question that he really was disappointed by the outcome. What is uncertain is the time at which he felt that disappointment. It may have been sometime during the course of the investigation itself, when he first recognized that only the *d*-amino acids could be oxidized in extracts, or it may have been at some later point when it became clear that no corresponding *l*-enzyme could be found. Perhaps there was no distinct point in time at which he was "disappointed," if the realization which induced that feeling was itself gradual.

* * *

One indication that Krebs had experienced some disappointment with the results by the time he completed his paper on the subject in May 1935, was that he told David Keilin, who wanted to examine some of the properties of the *d*-amino acid deaminase, that he did not intend to continue working on the enzyme. Keilin then undertook investigations that were in part extensions of the work that Krebs had done, as well as extensions of his own interest in coupled oxidations.[74]

One scientist's disappointment can be another's opportunity. A few weeks after Krebs had stopped working on *d*-amino acid deaminase, Otto Warburg visited Cambridge and stopped in briefly to see him in the laboratory. It was the first time Krebs had met his old chief since his lecture in Berlin in 1932. Warburg asked Krebs, "Can you show me the enzyme?" Krebs handed him a bottle and said, "Well, here is the solution." Looking at it, Warburg commented, "I notice it is yellowish," to which Krebs replied, "Well, most tissue extracts look yellowish." Warburg asked, "Why don't you try to isolate it?" His former assistant said, "Because I've only got one pair of hands, and to do this one has to use a large quantity."[75] Warburg took a sample with him. What he afterward did with the enzyme was the most important consequence of Krebs's discovery.

V

The physiological orientation that guided Krebs in his research was, as suggested above, harmonious not only with the views of Hopkins, but with the general approach associated with the Cambridge school and symbolized in Hopkins's phrase "dynamic biochemistry." Within that school, however, as within biochemistry at large, individuals differed over the relative priority to be given to physiological and to chemical considerations. During the spring of 1935 such differences surfaced in the form of a debate over the time that should be allotted to various subjects within the Part II biochemistry course. Bill Pirie, whose training and expertise fell on the chemical side of the spectrum within the department, criticized the course for treating such basic topics as the organic chemistry of sterols, sugars, and polysaccharides and the kinetics of enzymes in a perfunctory manner. He drew up for discussion a revised syllabus in which the first part of the course would be devoted to chemical preparations. Three other members of the department, Robin Hill, Dorothy Needham, and Marjory Stephenson, wrote a memorandum commenting on Pirie's proposal. Although agreeing with him "that certain subjects have hitherto been either wholly neglected or given scant attention," they did not think that his scheme would "be an improvement of the course as a whole." "The postponement of any tissue or organism," they argued,

> as well as any consideration of function till about one third through the Lent Term is, we think, calculated to disappoint those students whose interest in the subject is biological and functional rather than physical and chemical. To many students the great interest of biochemistry is the association of function with chemical happenings; for example the isolation of inosinic acid apart from any consideration of its function is a sterile exercise.[76]

Opinions were expressed freely in this democratic department. Eric Holmes wrote a memorandum criticizing the proposals of Pirie and of Hill, Needham, and Stephenson for giving too little time to "the biochemistry of the function of organisms and of living tissues."[77]

Pirie's proposals were discussed at a departmental meeting on April 30. According to the minutes,

> It was decided that no change in the present course would be needed before Oct. 1936, at earliest. The discussion proceeded on several lines. There seemed to be a certain cleavage of opinion between those who favored a predominantly chemical scheme (such as Mr. Pirie) and a more biological arrangement, roughly such as the present one.[78]

Krebs attended this meeting, but took no active role in the question. That was probably in part because he still felt a newcomer in the department, but also because he saw the debate as a superficial matter about how to present the subject.[79] That he did so is an indication that he was too preoccupied with research to probe deeply into pedagogical issues. On the surface the discussion was only a negotiation over how many hours to allot to the various topics covered in the course; but as the quoted passages suggest, underlying the negotiation were different images of the subject itself. Questions about how to teach biochemistry were reflections of questions about the nature of the discipline, and decisions about how to teach it would influence how future biochemists viewed their field. Evidently the divisions of opinion in this case were not sharp enough to be disruptive, in part because of the generally cooperative spirit within the Cambridge Biochemistry Department, in part because the various positions were taken in ways that allowed for compromise. On the other hand, the decision simply to postpone change suggests that it was not easy to achieve a consensus. That was, in microcosm, a reflection of the general state of biochemistry. It was becoming increasingly visible as an organized scientific field; but, as a report on the needs of the Cambridge department acknowledged even 10 years later, it was "not easy to define the boundaries of biochemistry as a science, for it is really the connecting link between many fields of specialized knowledge."[80]

If the pedagogical discussions that must have animated conversations in the Cambridge laboratory in April could not distract Krebs from his work, there was another sort of institutional discussion to which he was finally forced to divert attention. By the end of that month he knew that he had to do something about the memorandum he had been asked to draw up concerning the proposed research institute in Palestine. What he did was to write Nachmansohn, asking him to utilize the occasion of a Physiological Society meeting scheduled to take place on May 18 in Cambridge, to come there "so that we can finally come to a conclusion regarding the planned memorandum." If they could sit down at a table together with Blaschko, he believed, they could in a short time put together an outline. So far the only progress he had made on his own was to decide that the institute should begin on as modest a scale as possible, so that as much of the capital as possible could be preserved for the future.[81]

Nachmansohn did come to Cambridge, and the three friends must have worked out some of the questions that faced them, although there is no direct record of whether or not they were able to put together the memorandum. Krebs must have expressed concern about whether enough capital could be raised to make their future positions in Palestine secure. When Nachmansohn returned to Paris he spoke with Weizmann, and reported back to Krebs that the situation remained about the same as it had been in February. Again he sought to reassure Krebs with generally optimistic predictions about the extent to which Jews in America and elsewhere were now prepared to support such ventures in Palestine.[82]

While Krebs planned in a desultory way for the possibility of a future in Palestine that remained a remote vision, his prospects for a future in England suddenly took a new turn. As he was attending a few sessions of the British Pharmacological Society, which also met in Cambridge in May, Edward Wayne, the Professor of pharmacology at the University of Sheffield, came up to him and asked whether he might be interested in a post as a lecturer in the department. The position had previously been held by a biochemist, Douglas Harrison, who had left it to become professor of biochemistry in Belfast. When Frederick Gowland Hopkins heard that the Sheffield post was vacant, he had approached Edward Mellanby, Wayne's predecessor, who was now secretary of the Medical Research Council, to inquire whether the lectureship could be made available to Krebs. Much as he valued Krebs in Cambridge, Hopkins was well aware that he could not provide for him there a position or resources comparable to what Krebs was likely to be offered in other countries, and he was anxious that an effort be made to keep Krebs in England. Mellanby in turn told Wayne that it would be a service to Sheffield and to the country if he could get Krebs there. Wayne agreed to this proposition, even though it would have been in his own best interest to hire a younger pharmacologist who could assist him with his research. Krebs was, in fact, two years older than Wayne himself was.[83]

A small provincial university in an industrial city, Sheffield could not provide a milieu comparable to that of Cambridge. Nor would Krebs be able to work there in the midst of a group of biochemists with similar interests. All of the reasons he had given to others for his attachment to Cambridge must have come to mind as he discussed the new opportunity with Wayne. It immediately became apparent, however, that there would be two major compensations—more space in which to work and a substantially larger salary. For both professional and personal reasons, therefore, Krebs evinced a strong interest in this unexpected opening. Wayne was impressed also by the fact that for someone who had made a major biochemical discovery, Krebs seemed quite diffident about himself.[84]

By the end of May Krebs had finished his investigation of deamination and submitted his paper on that subject to the *Biochemical Journal*. Simultaneously he moved on to complete his other major current investigation, that on the synthesis of glutamine. The success of that research was already assured, and as he took up the topic again, on May 29, he probably planned merely to fill in a few more gaps in his data. He began this last phase with a test of the influence upon the rate of that synthesis, in sheep retina tissue, of the presence of glucose and of an excess of glutamic acid. The outcome, however, was not routine. When glucose, glutamic acid, and NH_4Cl were present glutamine formed, as expected, and as measured by the quantity of amide-N present at the end. The rate of formation ($Q_{amide-N}$ = 4.50) was, however, markedly lower than the rate of consumption of ammonia (Q_{NH_3} = -6.90). There was, therefore a "large deficit!"; that is, some of the NH_3 that disappeared could not be accounted for in the amide–N that appeared. Krebs construed this deficit as an indication that "besides glutamine, something else forms."[85]

Such a result could mean either that ammonia was used in some other reaction or that some of the glutamine that formed underwent a further reaction. To examine the possibility that the glucose had induced glutamine to decompose, Krebs added glutamine to sheep retina tissue in the presence and absence of glucose, and found "no great difference" in the rates at which the glutamine disappeared. Under both conditions it decomposed "in the expected order of magnitude." Turning next to the possibility that the excess of glutamic acid employed in the initial experiment was a cause of the deficit, he repeated it with "small quantities of glutamic acid," using the same kind of tissue. The result was more striking even than the first. "For every glutamic acid 1 $\underline{NH_3}$!!; Amide-N only 66–67% of NH_3 disappearance!!"[86] The fact that the glutamic acid and the NH_3 had disappeared in 1:1 proportions suggested that the deficit did not represent ammonia consumed in some reaction independent of the synthesis of glutamine. Nevertheless, only two-thirds of the expected glutamine appeared. These relationships must have led Krebs to favor the view that some of the glutamine formed subsequently disappeared. On May 30 he repeated the same type of experiment on rat retina and brain tissue, but the results contained obvious errors. The next day he made changes to eliminate the probable sources of these errors, and obtained deficits that were smaller than in the experiments on sheep retina, but nevertheless still "significant."[87]

These "deficits" must have appeared at first to Krebs as an intrusion into the smooth ending of an investigation. When he had carried out the same type of measurements on kidney tissue eight months earlier, in the experiments which had provided the basic demonstration of the synthesis of glutamine, he had regularly found somewhat more amide-N formed than NH_3 consumed (see p. 106). He had attributed the excess then to NH_3 released by the deamination of some of the glutamic acid present. Quickly, however, he now came to view this converse result in retina tissue as an opportunity to pursue the question, What happens to the glutamine arising from glutamic acid?

On May 31, Krebs approached this question from two directions in one set of experiments. Again using sheep retina tissue, he added glutamine and measured the quantities of NH_3 and amide-N present after two hours. There were only 29 mm^3 of NH_3, whereas the quantity of amide-N represented a decrease of 161 mm^3 compared to the quantity preformed in the glutamine. He construed this result to mean that the tissue had consumed glutamine by some means other than splitting it into ammonia and glutamic acid. The other part of the experiment was designed to begin the search for that other means. To this end he added glutamic acid to the same tissue: 1) with inosinic acid + NH_3, 2) with NH_4Cl, and 3) with adenylic acid. There was also glucose present, and in 1 and 3 pyrophosphate. What he apparently had in mind was the possibility that inosinic acid may react with the glutamine produced by the glutamic acid and NH_3. In such an event the inosinic acid might absorb amide nitrogen, causing a deficit in the sum of the NH_3 and amide-N larger than under the other test conditions. The resulting deficits, however, were respectively 1) 31 mm^3, 2) 106 mm^3, and 3) 37 mm^3. "Therefore," he remarked, "deficit not larger, if inosinic acid present!!!"[88]   Evidently Krebs eliminated inosinic acid from further consideration.

On the same day Krebs tested glutamine alone in pigeon retina and muscle tissue. With the retina he obtained the same kind of deficit that he had observed with sheep retina, but in the muscle there was "_no_ glutamine cleavage."[89]   At the beginning of the next week he turned back to guinea pig kidney tissues to see if he could detect a consumption of glutamine there in the presence of other substrates that might react with it. He began with glucose, lactate, pyruvate, and α-ketoglutarate. He measured the change in NH_3 and amide-N for combinations of each of these compounds with glutamine, in comparison to glutamine alone. The parallel between this experiment and those he had carried out a few weeks earlier on combinations of these substances with amino acids is evident. For a time then he had thought that one or more of these substrates paired with the amino acids might absorb their NH_3. That hypothesis had not been confirmed. Now he was raising the same question, except that he was testing whether the amide-nitrogen of glutamine might undergo a transfer that ordinary amino-nitrogen does not. Measuring the NH_3 and amide NH_3 present at the end of the experiment, and comparing the total with the quantity of nitrogen preformed in the glutamine, he inferred that there was "a small disappearance [of glutamine] with lactic, pyruvic, and ketoglutaric acid," but "not with glucose or with no additions."[90]

A "small" disappearance may appear as a positive or negative result, depending upon the expectations of the investigator and on other results with which it is compared. In this case Krebs was impressed not with the smallness of the deficit, but with the fact that there was a deficit at all, when there was none with glucose and in the control. He viewed it as a result suggestive enough to pursue. The next day, on June 4, he sharpened his focus, now posing explicitly in his notebook the question, "Does glutamine aminate keto acids?" To find out, he tested with rabbit kidney slices glutamine alone and in combination with α-ketoglutaric and oxaloacetic acid. For comparison he tested the two keto acids with NH_4Cl. Again he measured the change in NH_3 and in amide-N.[91]

We should pause, as we visualize Krebs preparing his slices, his solutions, and his manometers to carry out these measurements, to reflect on the implications that a positive experimental response to his query would hold for him. If glutamine did aminate keto acids, then he would simultaneously have found an answer to the question of what is the physiological function of glutamine synthesis and a solution to the problem of reversing the deamination of amino acids. He would have discovered not only another metabolic reaction, but an integrated process. In short, he would have expanded the important discovery he had already made into one of landmark proportions.

In each of the runs there was a "large deficit!! in glutamine." There was, that is, in each case a loss of amide-N, indicating that some glutamine had disappeared. This loss was, however, somewhat smaller in the presence of ketoglutaric, and considerably smaller with oxaloacetic acid, than with glutamine alone. Therefore, he inferred, there was "no increased amination with glutamine."[92] Whatever the reaction was that was consuming the glutamine, it was not the reaction that he had had in mind.

Two days after performing this experiment, Krebs drove to Edinburgh to attend a meeting of the Biochemical Society, taking Hermann Blaschko with him. Krebs was scheduled to speak again on the "Synthesis of Glutamine in Animal Tissues." The meeting began at 2 p.m. on Friday afternoon, in the Department of Medical Chemistry at the university, and included 15 papers. Krebs gave the ninth paper on the program, and probably had about 15 minutes to talk. He used most of his time to discuss the enzyme glutaminase, which he believed to be concerned "with synthesis alone" in the living cell, even though the thermodynamic equilibrium of the system (glutamine $\leftrightarrows$ ammonium glutamate) in aqueous solution is "completely on the side of" glutamate. Apparently he did not discuss his latest attempt to expand the scope of his research to the further reactions of glutamine, as he stated simply that "the products of the decomposition of glutamine in those tissues have not yet been found." After the remaining six talks were over, the participants went to dinner at the Royal British Hotel. Saturday was more relaxed. They traveled in private cars to a lunch in the Highlands, and returned to Edinburgh on the Forth ferry.[93]

When Krebs resumed his glutamine experiments, on June 11, he continued to measure "deficits" that pointed to the disappearance of glutamine, but he made no further efforts to identify the reactions that consumed the glutamine. Quickly bringing the investigation to a close with a few additional experiments on the inhibition of glutaminase through glutamic acid,[94] Krebs made his final revisions on the paper describing his investigation and sent it off on June 29 to the *Biochemical Journal*.

VI

Krebs opened his article, "The Synthesis of Glutamine from Glutamic Acid and Ammonia, and the Enzymic Hydrolysis of Glutamine in Animal Tissues," in seminarrative form; that is, he described how the investigation he was reporting had begun:

> The starting point of this investigation was the observation that glutamic acid behaves differently from all other α-amino acids in guinea pig and rabbit kidneys, in that although it increases the oxygen uptake more than other amino acids of the *l*-series, in most cases it actually diminishes ammonia formation.[95]

This was only in a very approximate sense a historical account. It is true that, early in a phase of his investigation of deamination that included glutamic acid—that is, within two weeks of the time he had begun it, in March 1932— Krebs had made an observation similar to the one described, and that that observation had then appeared to him to distinguish glutamic from most other amino acids. If we turn back to the experimental series including this observation, however (Vol. I, pp. 416–417; this volume p. 97), we can see that this specific observation was not literally the "starting point" of an investigation. The investigation involved both aspartic and glutamic acid, and the first suggestion that these amino acids behaved differently from the others came from the observation that the addition of arsenite caused the formation of NH_3 to increase. Experiments of the type mentioned in the above passage came as a follow-up of this lead. In his paper Krebs reversed this order, introducing the experiments with arsenite as the subsequent step in the investigation:

> However, if guinea-pig kidney is poisoned with arsenious oxide, ammonia is formed from glutamic acid. This could be explained by assuming that ammonia is formed in a primary reaction but disappears in a secondary reaction, the latter being inhibited by arsenious oxide. Experiments were therefore set up to see whether ammonia reacts in the expected way if it is added to kidney.[96]

The last sentence is, as a historical statement, roughly accurate. Not long after performing experiments with arsenite like those described in the first part of the paragraph, Krebs did try experiments in which ammonia was included in the medium, and observed, as described in the next section of his paper, the "disappearance of ammonia from kidney in the presence of glutamic acid."[97] He also carried out other, related experiments, however, about which he said nothing here. They proved afterward unessential to the logical progression of his investigation.

Even in the absence of a notebook record, we could have seen that what Krebs referred to as a "starting point" incorporated an aggregation of data that could only have been gathered during an extended period. The "point" in question could not, therefore, even in principle, have been located precisely in time.

In his discussion of the further steps of the investigation Krebs did not employ narrative phrases like "The starting point was . . ." and "Experiments were therefore set up . . . ." He shifted instead to timeless analytical forms—"If ammonium salts are added . . ."— and, in the following section on the formation of amide-nitrogen, "If the solution from which ammonia has disappeared . . ." is heated ammonia appears again, indicating the presence of acid amides. After describing the latter result, he wrote, "We may therefore conclude that the kidney tissue has converted ammonium glutamate into glutamine."[98]

In discussing Krebs's first paper on urea synthesis (Vol. I, pp. 332–333), I pointed out that it contains only a few traces of narrative form, embedded in an analytical structure bearing little relationship to the order of the investigation that had led him to the conclusions he presented. In his paper on glutamine there are, in the sections described so far, just two explicitly narrative phrases. In this case, however, perhaps in part because one of the phrases identifies the beginning of his argument with the beginning of his investigation, the two phrases give these sections the semblance of an investigative sequence. There is, in fact, a rough parallel between the order of presentation and the historical path of research. Krebs's account can be viewed as a simplified, idealized recapitulation of what appeared in retrospect as the "essential" steps through which he had arrived at the conclusion stated above. His description would probably qualify as an example of what Imre Lakatos has called a "rational reconstruction" of a historical discovery.[99]

For Krebs, the degree of resemblance between the historical and the rational investigation was not an issue. His task, in writing this paper, as in writing the paper on urea, was to present the most effective case for his conclusion. For the synthesis of glutamine the sequence of investigative steps through which he had reached that conclusion happened to require less modification to transform it into a strong argument than had been necessary in the case of the synthesis of urea.

The preceding discussion pertains to the first three sections of the glutamine paper, those in which Krebs established the basic phenomenon that he had discovered. Most of the remaining sections summarized the numerous experiments he had carried out after making that discovery, examining the characteristics of the reaction, its reverse, the properties of glutaminase, and the variations on the characteristics that he had found in different tissues. The organization of this portion of the paper was correspondingly descriptive and topical rather than logical. The action of glutaminase differed so markedly in detail, depending upon the type of tissue from which he prepared the extract containing it, that he inferred: "There are at least two types of glutaminase distinguishable by their pH optima and their inhibitions by glutamic acid ('brain type' and 'liver type')."[100]

In a section placed near the end of the paper, and entitled "The Fate of Glutamine in Brain and Retina," Krebs described a few of the experiments he had carried out near the end of his investigation, beginning with the observation of the deficits in the amide-nitrogen formed when he placed glutamic acid and NH_4Cl with these tissues, and including those using glutamine alone, from which he had inferred that "the tissue utilizes glutamine but not by splitting it into glutamic acid and ammonia." Omitting his efforts to identify the reaction in which glutamine might be consumed, either in these tissues or in kidney tissue, he concluded merely that "the nature of the products of the conversion of glutamine remains to be investigated."[101] Evidently he saw no reason to make public that he had begun such an investigation, but that he had not yet been successful.

In a final section, Krebs discussed, in very compact form, four broader questions raised by the results of his investigation. The first was:

Reversibility of glutamine synthesis.
The synthesis of glutamine in tissues can be reversed *in vitro*, for instance by changing the pH. However, two facts indicate that the synthesis of glutamine is practically not reversed in the living cell. 1) The hydrolysis occurs in practice only outside the physiological range of pH. 2) Glutamine disappears from brain or retina without forming ammonia. Thus there appears to be a cycle of ammonia in nervous tissue in which the conversion of ammonium glutamate into glutamine is one step. Nothing is known about the other stages.[102]

This passage inevitably strikes our historically oriented antennae as the first occasion, beyond his descriptions of urea synthesis, in which Krebs referred in his published writings to a cycle occurring in the living cell. (Even in his summary of the Thunberg-Wieland-Knoop scheme he had not employed a German equivalent of that term.) Do we see here a germinal stage in the emergence of his conception of the cycle as a general pattern of metabolic reactions? We must guard against the temptations of hindsight; but neither should caution lead us to dismiss the possible significance of this development. It is evident, even from this compressed statement, that the reasoning from which he had inferred that ammonia undergoes a cyclic process was 1) ammonia is consumed in a reaction that is not reversible; 2) overall there is not a net consumption of ammonia in the tissues in question; 3) the ammonia must be regenerated by a pathway different from the reversal of the synthesis of glutamine. By definition, such a sequence was a cycle. He had reasoned in the same way toward his solution of the problem of urea syntheses: 1) the arginase reaction is not reversible; 2) the ornithine produced in the arginase reaction must again form arginine; 3) there must be another pathway. The fact that Krebs followed analogous mental steps in his analysis of two different situations sharing a common feature does not necessarily show that he was beginning to conceive of the metabolic cycle as a generic phenomenon. It does suggest that he was falling into a pattern of thought that could well lead him in such a direction. Finally, we should note that Krebs did not discover an "ammonia cycle," but merely suggested that there appeared to be one waiting to be discovered.

The second general point Krebs made was about the "physiological significance of glutaminase in brain and retina." From his inference that the enzyme that he had shown to hydrolyze glutamine was "one and the same" as the enzyme that he believed to be "concerned *in vivo* with synthesis only," he speculated that "it may well be that several hydrolysing enzymes found in tissue extracts are only components of synthesizing systems. For instance hippuricase (histozym), or proteolytic and lipolytic enzymes are found in extracts of those tissues known to perform the respective syntheses *in vivo*."[103] Again we see a parallel to reasoning that he had applied in an analogous situation. In his paper on deamination he had also inferred that an enzyme catalyzing a reaction that he did not consider physiological might be a component of an enzyme system concerned with a reaction that was physiological. Again it is difficult to distinguish whether he was consciously moving toward a generic conception of artifactual enzyme reactions produced by components of enzyme systems or whether he only thought in the same way when confronted with analogous problems.

The third point Krebs made was based on some of the experiments he had carried out—about which no special point was made in my foregoing historical account—showing that if the energy-giving reactions of the tissue, that is, the respiration, or in some circumstances anaerobic glycolysis, are inhibited, the "synthesis of glutamine ceases. This makes it evident that the system which synthesizes glutamine consists of glutaminase and of an additional factor concerned with the transmission of energy. The transmission of energy results in a change in the thermodynamic equilibrium between ammonium glutamate and glutamine in favor of the latter. If this occurs the enzyme catalyses the attainment of the new equilibrium."[104] Taken together with Krebs's unpublished draft explanation for the existence of two deaminases discussed previously, this statement suggests a growing general interest in the connection between the particular metabolic reactions he was studying and the transmission of energy within cells.

Each of these three brief discussions is a further illustration of Krebs's drive to explain what he observed and to understand the physiological significance of the chemical events he uncovered. His fourth point was a tacit admission of his inability to explain or understand in such terms the central achievement of his investigation:

> *Significance of the "glutamine system."*
> Nothing definite is known at present about the physiological function of the glutamine synthesis. Certain experiments suggest a connection between the system and the energy-giving reactions: l (+)-glutamic acid is the only amino-acid which increases respiration in brain and retina: unlike other amino-acids it inhibits anaerobic lactic acid fermentation. These experiments will be described in a later paper.[105]

The experiments to which Krebs referred had been carried out, at his suggestion, by Hans Weil.[106] Given Krebs's feelings about what it is like "to have no explanation at all," he must have felt privately that the situation summarized here was far from satisfactory. From his perspective, to understand "the physiological function of glutamine synthesis" was nearly synonymous with linking that reaction to known metabolic pathways. As we have seen, in several phases of the investigation that he did not report in his paper, he had sought, but failed to find, such connections. He could only suggest in broad, vague terms that these connections must exist and that he had some experimental leads for further investigation.

The discovery of the synthesis of glutamine was thus a closely circumscribed one. Krebs had been able to add a single discrete chemical step to the repertoire of known metabolic reactions, but he had not been able to extend the reaction sequence or to integrate it into a broader process. In that respect the discovery appears less comprehensive than his discovery of the ornithine cycle. It was, however, a major achievement. To identify a new metabolic reaction was, in the early 1930s still a rare, difficult feat. The "discovery" was not a simple experimental "finding." Krebs was not in a position to isolate the synthesized substance and identify it with positive certainty as glutamine. As in the case of urea synthesis, he put together several pieces of evidence, no one of which was decisive by itself. These were that glutamic acid does not cause NH_3 in a tissue medium to increase to the extent expected from the behavior of other amino acids; that the addition of arsenite in this situation increases the formation of NH_3; that if NH_3 is added to the tissue medium along with glutamic acid, the NH_3 disappears; that there is a corresponding increase in the quantity of amide nitrogen-present; and that the way in which the amide-nitrogen was determined indicates that it is more likely to derive from glutamine than from asparagine. All of these observations could be interpreted as manifestations of a reaction in which glutamic acid combines with ammonia to yield glutamine, and nothing was inconsistent with this interpretation. The discovery was thus an inference built upon a set of indirect, but convergent observations. Krebs might have continued seeking ways to confirm his conclusion, but he probably felt no need to. It was his view that, once he had sufficient evidence to draw a conclusion with confidence that he could not be wrong, it was a waste of time to accumulate further evidence. In this, as in his discovery of the ornithine cycle, he built a case that was decisive from the combined strength of its supporting evidence, yet also lean enough to be elegant.

The discovery of the synthesis of glutamine was, despite its discrete immediate limits, significant in a broader sense to the direction of the development of the field of intermediary metabolism. As we have seen (Vol. I, pp. 1–35), the specific metabolic reaction sequences established at the time Krebs entered the field were fragments of a picture in which the blank spaces loomed larger than the areas filled in. Nevertheless, the boundaries of the picture were fixed, and the task of the investigator seemed to be to identify, according to rules that were set in advance, the steps connecting the starting and end points of the metabolic chains; that is, the reactions through which the foodstuffs entering the body were converted to the decomposition products leaving it. As Knoop put it, the goal was to leave no gaps along these pathways. Knoop pressed also for more intense investigation of synthetic reactions, but he envisioned these reactions as essentially the reversal of the decomposition reactions. Even as it became evident—as in the case of the decomposition and synthesis of uric acid—that the steps in a synthetic sequence might not be identical to those in the corresponding decomposition sequence, the one was still regarded as leading back from the end point to the starting point of the other. The synthetic pathways were those through which, from partial decomposition products, the "building blocks" of the constituents of the organism—mainly amino acids, carbohydrates, and fatty acids—were reconstituted. Investigators in the field were, therefore, in a general sense, seeking closure, to fill out a picture whose broad outlines were already situated.

The discovery of the synthesis of glutamine in animal tissues did not simply fill in an empty space in this picture. It led somewhere beyond the existing framework. The reason was that glutamine itself had no established role within that framework. It was not a prominent building block of the foodstuffs or of the animal organism, being known mainly as a substituent found in plants. Instead of closing a gap within the conventional boundaries of the field, therefore, the discovery had the potential to open those boundaries, to make room for processes not previously included in the picture.

Krebs, of course, could not foresee how far that opening might lead. Sharing the general picture sketched above, he tried to connect the synthetic reaction with it through the most direct leads possible. If, for example, it had turned out that glutamine did aminate keto acids, the synthesis would have appeared merely as a short loop facilitating one of the steps in an expected synthetic pathway. Since it did not work out that way, the opening was, for the present, left as an opening into the unknown. In contrast to the discovery of urea synthesis, which was a closed (although by no means complete) solution to a long-standing problem, the discovery of glutamine synthesis was the solution to a previously unknown problem, and the invitation to pursue a new set of problems.

When I asked Krebs what role chance had played in his discoveries, he picked out the discovery of glutamine synthesis as an example. He had, he said, been looking for the formation of NH_3, and found instead its disappearance. He invoked a favorite word of scientists—"serendipity"—to describe the situation.[107] Whether or not we classify the synthesis of glutamine as a "chance" discovery, it is clearly one of the best examples in his early career of Krebs's special ability to exploit the unexpected.

This discovery also epitomizes Krebs's scientific style. What he had set out to find was a conventional extension of a well-known theory of oxidative deamination. What he found was far more original than what he anticipated. As well as any of his investigative achievements, this one reveals how Krebs's scientific imagination, his creativity, lay not in what he formulated in advance, not in a long-range vision of the road ahead, but in his responses to what he encountered along the way.

* * *

During May and June 1935, Hans Krebs published two major research papers, the first ones he had completed since leaving Germany two years before. If we compare his research record during those two years, it appears that somehow in the second year he "hit his stride." Although he worked at the same steady pace through both years, he appears during the first year to have been floundering, by comparison to a second year marked by steady progress. It is unlikely that he consciously changed anything in his approach to scientific research. Perhaps he simply had better luck; but his successes of 1935 may also be a sign of maturation. He had come to Cambridge in the wake of a success so stunning, so early in his career, that it was not easy to match, even by its author. Whether or not he fully recognized it, he may have felt some pressure to repeat such a peak performance. As noted in an earlier chapter, Krebs did not admit, during our conversations, to any insecurity about his ability to maintain the standard set by his first great discovery; but perhaps some hint of what I have suggested here is embedded in shorthand in a remark he made in 1977. I had asked him about the period of two years in Cambridge before his first research paper came out. He answered,

> Well, there was a gap which was—you're quite right—unusually long, but then it took a fairly long time for a paper to be written and to be published. Actually I made . . . well, first in Cambridge I didn't make much progress, and I turned to other problems of glutamine and amino acid metabolism.[108]

Arrivals and Partings

By the time Hans Krebs completed his investigation of the synthesis of glutamine, Hans Weil was no longer with him in Cambridge. During the winter of 1934–35, Weil's financial position had become untenable. Tighter restrictions on sending money out of Germany sharply decreased the support his parents could send him, even as his new marriage increased his monetary needs. Believing that he could not hold out for longer than six months on the reserves he had in hand, he decided that he would have to use that time to study for the qualifying examination to practice medicine in England. Meanwhile, he began applying for paid research positions. Late in March he noticed an advertisement in *Nature* for a "research worker in cancer" at the North of England Cancer Campaign in Newcastle-on-Tyne, and showed the notice to Krebs. Despite his skepticism concerning specialized cancer institutes, Krebs thought that this might be a good situation for Weil, because the research director, Frank Dickens, was interested in metabolism. Krebs told Weil, "That may be just the thing for you. I know Dickens. He has worked for Warburg, and wants someone who knows how to handle Warburg manometers." Weil sent in his application, then left for a belated honeymoon cruise with Rosanne, in the company also of his parents. Meanwhile, Dickens had asked Krebs to recommend someone, and Krebs probably suggested that Weil was just the person for the job. After sailing to Algeria, spending a week in San Remo, and returning by another steamer to Southampton, Weil found a telegram waiting for him, requesting that he come to Newcastle for an interview on the next day. He did so, and obtained the position. Then he returned to Cambridge to finish up his work there, so that he could begin in Newcastle on June 1.[1]

I

During his last months in Cambridge, Weil finished the study Krebs had given him of the possible conversion of proline to glutamine in kidney tissue. The methods Weil employed were mainly those that Krebs was using in his investigation of the conversion of glutamic acid to glutamine. Weil was able to put together a convincing case in support of Krebs's initial experiments indicating that proline is converted by way of glutamic acid to glutamine. Proline (and hydroxyproline) increased the oxygen uptake and, like glutamic acid, gave rise to amide-nitrogen. To confirm that the "primary step" was the oxidation of proline to glutamic acid, Weil showed that amino-nitrogen is formed in the quantities expected for the reaction:

$$C_5H_9O_2N + O_2 \qquad \rightarrow \qquad C_5H_9O_4N$$
$$\text{Proline} \qquad\qquad\qquad \text{Glutamic acid}$$

Weil could not substantiate, however, that the intermediate step that Krebs had postulated, the formation of pyrrolidone carboxylic acid, really occurred, and he concluded that this compound "is not the intermediate." Weil pursued the possibility that, in addition to forming glutamine, the glutamic acid derived from proline gives rise in the kidney to α-ketoglutaric acid by deamination. By "checking its further oxidation with arsenite," a technique that Krebs had utilized often in analogous situations, Weil was able to isolate the dinitrophenylhydrazone of "pure α-ketoglutaric acid."[2]

Before his departure from Cambridge, Weil wrote up his findings in a paper that he discussed with Krebs and probably left for Krebs to make corrections. Early in June Krebs stopped to visit Weil in Newcastle on his way by car from the meeting of the Biochemical Society in Edinburgh, and they may have gone over some details then. Their jointly authored paper arrived at the offices of the *Biochemical Journal* on July 11. In their conclusion, they stated that "the metabolism of proline may . . . be formulated in the following way:"[3]

$$\text{proline} + O_2 \rightarrow \text{glutamic acid} \begin{cases} + NH_3 \rightarrow \text{glutamine} \\ + \tfrac{1}{2}O_2 \rightarrow \alpha \text{ ketoglutaric acid} + \text{ammonia} \end{cases}$$

The outcome of this investigation must have been satisfying both to Weil and to his mentor. Krebs had given Weil a well-defined problem whose solution was foreshadowed in his own preliminary experiments, but which left room for his younger collaborator to make an identifiable contribution. While confirming the "primary step" in the reaction scheme Krebs had postulated, Weil rejected the further intermediary step that Krebs had assumed. The project supplemented Krebs's own demonstration of the synthesis of glutamine, while marking a significant step in Weil's progression from apprentice to independent investigator.

As noted in the previous chapter, Weil began in Cambridge some experiments on glutamic acid in brain tissue, following up another subproblem that Krebs had come across in his study of the synthesis of glutamine. Before leaving, Weil found that glutamic acid was the only amino acid that increased the respiration of brain and retina tissue, and the only one that inhibited the anaerobic formation of lactic acid. When he arrived in Newcastle, he learned that his main research responsibility there would be to investigate the effects of radio shortwaves on the metabolism of cancer cells. This problem did not particularly interest him, and since the instrumentation for the work had not yet been set up, he decided to continue his experiments on the metabolism of glutamic acid in the brain. Using a "few milligrams of glutamine, glutamic acid, and α-ketoglutaric acid" that he confessed to Krebs that he had taken with him from "your supply" and an "idiotic" manometric apparatus that was the only one available to him, Weil made some new measurements of the effects of these substances on the respiration of guinea pig brain slices. On July 27 he wrote Krebs to report his new results. In the main they confirmed the work undertaken in Cambridge; but in one set of experiments in which he added arsenite to the medium, he noticed something which "very much surprised me." In the vessel containing glutamic acid there was "unquestionably more ammonia than in the other two, which does not agree with your findings." This result suggested to Weil that some of the glutamic acid might be deaminated, producing α-ketoglutaric acid. In view of the unsatisfactory apparatus he had employed, however, this determination might not be "entirely free of objections," so Weil wrote that he would seek to confirm his conclusion by isolating α-ketoglutaric acid. Four days later he wrote again to announce that he had succeeded.[4]

Enthused, Weil described his plans to continue the investigation, and reported that he had persuaded Dickens to purchase a Parnas apparatus so that he could measure the quantities of NH_3 formed in the presence of arsenite; but he paused to ask Krebs a delicate question: "Or do you want to do that? It belongs really more in your territory." In any case, Weil added, he wanted to see "whether, as an end product malic acid is formed, in spite of Quastel. The oxidation of glutamic acid to malic acid would yield an R.Q. of exactly 1, which is what I have, in fact, found." After raising another question related to these problems, Weil ended, "I have a vast metabolism program, and I do not know if I can do all that I have in mind," especially in view of the fact that the shortwave project would begin after the holidays.[5]

Krebs's reply has not survived, but we may surmise that he wrote Weil that he was ready to turn to other problems, and encouraged Weil to go on with his study of the metabolism of glutamic acid in the brain. Finding that the shortwave experiments occupied little of his time, Weil was able to continue this "more important" activity and to publish by the following winter a comprehensive paper on his extension of the work begun in Cambridge.[6]

Although Hans Weil had worked informally with Krebs in Cambridge, he became, in effect, the first student Krebs had after leaving Germany. Moreover, he was the first of any of Krebs's students to make the transition to independent investigator pursuing a sustained research program grounded in the training he had received from his mentor. In the preceding remarks in Weil's letter we can glimpse a type of negotiation that must, either explicitly or tacitly, accompany such a process of transition. The student has earlier undertaken to solve problems posed by his teacher, enclosed by broader problems pursued by his teacher, applying conceptual approaches and technical methods provided by his teacher. Gradually the student learns to work independently, but there remains the problem of finding within the domain defined by his teacher the scope for an investigative pathway that will diverge sufficiently from that of the teacher so as to avoid encroachment on the latter and to allow autonomy for the former student. In this case the negotiation was eased by the fact that Krebs did not take a proprietary attitude toward the problems he gave his students, but recognized the need to leave them room to follow the directions that he had put them on, and by the fact that Weil was sensitive enough to sound out his mentor's feelings before proceeding.

II

Two weeks after Weil sent his paper on proline to the *Biochemical Journal*, in July 1935, Krebs's other student, Norman Edson, also submitted the results of his first extended research effort to that journal. During the previous summer Krebs had suggested to Edson the project of following up his own extension of the discovery of Erno Annau that NH_4Cl increases the rate of formation of ketone bodies in liver tissue. At the time he turned the work over to Edson, Krebs had already obtained a preliminary indication that if the slices came from a nourished rat, the keto acids formed at a higher rate not only in the presence of pyruvic acid, as Annau claimed, but also with other acids such as capronic and acetic (see pp. 80–81, 83). Edson spent nearly a year "examining the phenomenon in greater detail." Much of this time he probably devoted to learning and refining, under Krebs's close supervision, the methods for the microdetermination of acetoacetic and β-hydroxybutyric acid. Then he examined the effects of NH_4Cl with a large number of substrates and varied nutritional conditions. His results confirmed Annau's original findings and added to them, in extension of Krebs's few experiments on the subject, that "ammonium chloride accelerates the formation of β-ketonic acids from most fatty acids in well nourished liver."[7] Edson's effort did not lead him or Krebs either to more novel results or to an explanation of the overall effect of NH_4Cl. Edson continued the investigation by examining whether amino acids were the source of the ammonia that caused these effects.[8]

On the day that Krebs probably finished up his paper on the synthesis of glutamine, he began a new series of experiments on a problem he had not pursued for more than a year: the synthesis of uric acid in birds. The question had undoubtedly been on his mind ever since Benzinger's experiments in Freiburg had left the subject in a state of uncertainty, and he may have been motivated to return to it now simply because he had completed the two major projects that had occupied him for the past year. It is likely, however, that he was provoked to this move by a new incursion into the subject of uric acid synthesis by Werner Schuler and Wilhelm Reindel. In a paper published in the issue of *Hoppe-Seyler's Zeitschrift* that appeared on June 19, Schuler and Reindel and their students claimed to have established that pigeon liver tissue synthesizes a purine compound that is subsequently oxidized to uric acid in the kidneys. They attempted also to identify the source of the carbon skeleton for this purine, but were able only to eliminate a number of potential substrates and to come up with some vague evidence for a natural "precursor" produced by the liver cells.

They identified the purine product as such by a method that was not specific, but which they asserted to be superior to existing methods. "We are," they reported, "at the present time working on the question of which purine originates in this process."[9]

Krebs may have just seen Schuler and Reindel's paper when he renewed his own attack on uric acid synthesis on June 27. If so, he would probably have been both annoyed at the pretensions and shortcomings in their work and aware that they might be on the track of a real advance. He might well have felt that he ought to move quickly to ascertain whether or not their lead was promising, before he left for his holiday.

In his first experiment Krebs measured the rates of absorption of NH_3 and of formation of uric acid in pigeon liver and kidney slices and in the two together. He was, in a sense, picking up where Benzinger had broken off in 1932. Krebs tested each tissue and the combination under four conditions—in the presence of NH_4Cl, of NH_4Cl + pyruvic acid, of NH_4Cl + acetic acid, and a control with no additions. The object of the experiment was probably twofold: to identify which of the two tissues previously shown by Benzinger to be essential in the pigeon to the overall process was the one that binds ammonia, and to "obtain information about the source of the carbon skeleton of uric acid."[10] Concerning the former question, the results pointed clearly to the liver. $Q_{NH_3}^{Liver}$ in the presence of ammonia was -7.40, in the presence of ammonia + pyruvate, -14.2; and in the presence of ammonia + acetate, -3.79. In the kidney the comparable figures were ≈ 0, -2.30, and -1.84. In confirmation of Benzinger's earlier investigation, only the combinations of liver and kidney slices produced substantial quantities of uric acid.[11]

This initial experiment provided promising leads for further exploration. The indication that ammonia is absorbed in the liver rather than the kidney was compatible with Schuler and Reindel's contention that a purine compound is synthesized there, while the increased absorption due to pyruvate suggested a possible role for the latter in supplying the carbon for the purine.

Two days later, on a Saturday, Krebs sought to characterize more fully the absorption of NH_3 in pigeon liver tissues. Comparing the process aerobically and anaerobically, he observed that the absence of oxygen reduced the rate to about one-fourth of the rate in that gas. He also attempted to reproduce the process in liver extracts. Probably in order to check the possibility that the NH_3 absorbed was utilized to form amino acids rather than a precursor of uric acid, he measured the amino-N produced in the slices and found its rate to be roughly half of the rate of absorption of NH_3 under each of the four conditions tested.[12] Although he made no comment in his notebook, this result probably fit his expectation that a part of the NH_3 absorbed entered the composition of the purine compound in question.

At the beginning of the last week before his holiday, Krebs returned briefly to glutamine synthesis, testing whether the addition of NH_3 and α-ketoglutaric acid to guinea pig liver or kidney slices would give rise to amide-N.[13] If so, such an outcome might not only extend the scheme of the metabolic reactions connected to the formation of glutamine, but advance his so far unsuccessful efforts to demonstrate the formation of amino acids from α-keto acids. Perhaps he was stimulated to look again at these questions by an article by M. Neber in the same issue of *Hoppe-Seyler's Zeitschrift* in which Schuler and Reindel's article appeared. In a discussion which made repeated reference to Krebs's work on urea synthesis, Neber reported that he had detected the formation of amino acids from pyruvic acid in liver slices. Neber's article would also have reminded Krebs that in the preliminary paper from Freiburg on the disappearance of NH_3 in the presence of glutamic acid in kidney slices, Krebs had included, without comment, a protocol showing the disappearance of "a great quantity of ammonia" in kidney slices with ammonia and α-ketoglutaric acid.[14]

Whatever his expectations, Krebs's new result was not striking enough to divert him further, and he continued with uric acid synthesis.

On Tuesday, July 2, Krebs measured the absorption of NH_3 and formation of uric acid in an acetone powder of pigeon liver and in this powder combined with chicken liver slices. On the same day he examined the effects of pyruvate, lactate, glycerophosphate, and glycine on the formation of uric acid in the presence of NH_4Cl in chicken livers (in which bird, as Benzinger had found, the whole process occurs in the liver), but made no calculations because the measurements were too small. He performed no further experiments that week until Saturday, when he again compared the rate of disappearance of NH_3 and the formation of amino-N in pigeon liver and found the latter to be less than half of the former. At the same time he measured the formation of uric acid in combinations of pigeon liver and kidney slices, allowing the vessels to sit in ice over the weekend before he determined the quantities of uric acid produced in them. In one last effort to make further progress before leaving the laboratory, Krebs measured the absorption of NH_3 and formation of amino-N in chicken liver slices in the presence of NH_4Cl and three different substrates. In each case the latter rates were sufficiently lower than the former to indicate that not all of the ammonia absorbed was consumed in forming amino acids. When he made the uric acid determinations for the experiments performed on Saturday, he obtained quotients that were large, "but irregular."[15] His time had run out, and he had established nothing conclusive. There might be other ammonia-binding mechanisms in the liver to account for the difference between that absorbed and that incorporated into amino-nitrogen. His results thus far did provide room, however, to sustain the hypothesis that at least part of the balance of the ammonia absorbed in the liver was employed in the synthesis of uric acid.

The experiments that Krebs performed on uric acid synthesis during these two weeks were probably intended as preliminary groundwork for an investigation to be carried out in collaboration with Edson, with assistance also from a newcomer named Alfred Model. A German emigrant physician who had found a position in a hospital in Belfast, Model had planned for more than a year to spend a few months working with Krebs when he finally arrived in July.[16] Krebs set for Edson and Model the problem of identifying the purine that Schuler and Reindel claimed to be synthesized in pigeon liver and oxidized to uric acid in the kidney. On chemical grounds Krebs assumed that the most likely candidates were xanthine and hypoxanthine, which can give rise to uric acid by the addition respectively of one and two atoms of oxygen to their molecules:[17]

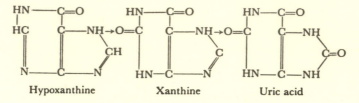

Hypoxanthine   Xanthine   Uric acid

On the morning that Krebs left the laboratory for his holiday, Edson was already able to tell him about some preliminary results indicating that "xanthine is oxidized to uric acid by the kidney, but not by the liver of the pigeon, whereas both organs of the hen are capable of bringing about the oxidation."[18]   That was just what Krebs would have predicted if xanthine were the intermediary metabolite they sought.

Late in June Krebs received a notice concerning the appointment of a lecturer in the Department of Pharmacology at the University of Sheffield, about which Edward Wayne had approached him in May.   The appointment was expected to take effect "as from the 1st October, 1935," and to afford the Lecturer "ample opportunities for prosecuting original research.   He will undertake such teaching duties as shall be prescribed by the Professor."   Before leaving Cambridge Krebs sent off his application for the lectureship.   On a single typewritten page he outlined his present position and his previous educational and research experience.   "I lost my position at Freiburg after the advent of the Hitler regime owing to my Jewish descent," he stated simply, "but obtained a Rockefeller grant for prosecuting research at Cambridge."   He enclosed a list of his publications, noted that "I have teaching experience both in lecturing and in the conduct of practical classes," and indicated that Hopkins and Keilin had agreed to his giving their names as referees.[19]

With camping equipment packed into his green Austin roadster, Krebs drove off on July 12 for Southampton.   At noon he met Katrina disembarking from the Hamburg-America steamship liner *Albert Ballin*, and they headed once again for the open road and "the beautiful clear air, the sun and wind, and the carefreeness" with which Katrina anticipated "we want to live our days" together.[20]   This time they drove to the west of England, camping through Cornwall and the spectacular coastline of Devon.   They hiked, swam, cooked potatoes, and fried eggs.   The Austin broke down only once during the three weeks they traveled in it.   All was not quite the same as the first time.   Katrina learned that even on holiday Hans was not willing to share all of his time with her.   When he read the *Times* every morning, and when he received and answered reports from his laboratory delivered to prearranged addresses on their route, she felt left out.   All in all, however, it was still a beautiful holiday for her.[21]

Back in Cambridge on the evening of July 30, Hans and Katrina had a long talk, during which Katrina expressed her feeling that they must now come to a clear choice.   She could not continue as a "holiday wife" [*Zeitfrau*].   She saw

Figure 5-1 Hans Krebs, camping in Devon.

Figure 5-2 Katrina Holsten on the heath.

her way as leading unerringly to him, but he must decide how things were to be resolved. Hans probably said that he needed more time to think it over, and that much depended on the result of his application to Sheffield. Despite the indecisive outcome of their conversation, Katrina felt that it had brought them closer. The next day Hans drove her to London on a sun-filled morning and left her at the railroad station to journey alone to Southampton. She stood watching him drive off, until his car dwindled to a green speck and she could see only the top of his head disappearing in the distance. From the *Albert Ballin* she wrote him that she had had "very much happiness" on their holiday. He had caused her some tearful moments as well, but they had served in the end to help them know and understand each other more fully. Nevertheless, she feared the loneliness and the emptiness of the days she knew to be ahead for her.[22]

III

While Krebs relaxed in the sun, his professional interests were moving forward. On July 16 Wayne wrote him from Sheffield that the applications for the post of lectureship had been considered, and that with the approval of the Dean and vice-chancellor,

> I am putting your name forward to the Faculty, Senate and Council for election. I gather that there is a good chance of your being appointed. The meetings at which the matter will be considered take place early in October. . . . I think it would be a good plan for you to come up here for a day or two in September to see the department and to meet a few people.[23]

Krebs must have received this letter in Devon and replied from there, asking about the facilities that would be available to him and suggesting that he could make the visit in August—that is, almost immediately after his holiday. On August 1 Wayne wrote back that September would be more suitable, because "everyone is away in August." Accommodatingly he added,

> I think you would like being here. There is plenty of space in the department and it is quite well-equipped. I could arrange for biochemical teaching for you or not, just as you wish. The total amount of teaching you would have to do is, however, including pharmacology, quite small.

There were also excellent clinical facilities, in case Krebs had "any problems of a clinical nature."[24]   Meanwhile, Wayne had already written to Krebs's referees, Hopkins and Keilin, and had received rapid replies. Their assessments of Krebs are worth quoting in full. Even keeping in mind that such recommendations are expected to present a favorable picture, these letters reflect convincingly the high esteem in which the two distinguished senior scientists held their younger colleague.[25]

Dear Professor Wayne,

In answer to your letter I may say at once that it would be impossible to say too much in praise of Dr. Krebs. He is not only an investigator of most exceptional ability with the widest possible knowledge of Biochemistry and the Medical Sciences generally, but also (though this may be of less importance to you) a most admirable teacher. He can deliver most effective lectures in English.

Let me be frank. Krebs is so valuable to us that if it were at all possible to persuade the University to take such steps as would enable us to retain him here, I should certainly do my best to make a move in that direction. Unfortunately it is not very likely that we can do anything adequate, but between now and the beginning of October, when, as I understand, your decision will be made, I shall at least sound the authorities on this matter. I feel it only fair to tell you this, but success is so unlikely that I think you may safely proceed without hesitation.

Yours sincerely
F.G. Hopkins

Dear Professor Wayne,

In reply to your letter of July 16, I should like to say that I know Dr. Krebs well and have been following his work for the last few years with great interest. It is hardly necessary for me to discuss his achievements in great detail; they are, in fact, well known to everybody. His scientific contributions on urea and nitrogen metabolism are of outstanding merit. He has an exceptionally wide knowledge of biochemistry and is master of a very great variety of methods, which enables him to cover in his research a very wide field of the subject. He has had considerable experience in clinical biochemistry and with the problems which may arise from contact with clinical work. He has the excellent qualities of an independent and enthusiastic research worker, and if given the necessary opportunity, is capable of organising a good school of research. I gather from the Biochemical Department where he is already in charge of teaching that he has a distinct ability and is a lucid and stimulating lecturer. He has, moreover, a very charming character and is much liked by his pupils and his colleagues in biochemical and other scientific departments. I am sure he will be a great acquisition to any scientific department which he may join, and I should like only to express the feeling, which I am sure is shared by his other colleagues, that we shall be sorry to lose him.

. . .

Yours sincerely,
D. Keilin

Not all of the above statements can be taken as objectively accurate. Krebs had mastered a repertoire of methods that enabled him to investigate one important area of biochemistry with great effectiveness, but there were broad areas of the field with which he was only superficially familiar. He had made a good start as a conscientious teacher, but was not in charge of teaching in the Cambridge Biochemistry Department. These opinions nevertheless bear witness to the powerful impression Krebs had made on those around him in this leading center within the biochemical world.

While Krebs was away, his laboratory interests were also being looked after. On July 23 Edson wrote, in his typically humorous style, to report on the progress of the uric acid investigation. Trusting that "Your Excellency is making joyful gambols under the sub-tropical sun of Somerset," Edson informed his chief that "Alfred Model and I are stumbling along and waiting for your return." They had verified that xanthine is oxidized to uric acid only in the kidney of the pigeon, but in both the kidney and the liver of the chicken. "We have also got the results for hypoxanthine and guanine and adenine." Edson did not specify what these results were, but it is clear that they were testing systematically whether purines other than xanthine might also give rise to uric acid. He went on, "We tried adding milk xanthine oxidase to the fluid after the pigeon's liver had finished, but the increase was not convincing. The oxidase preparation, however, was not satisfactory—it was an old one obtained from Booth and the defatting process reduced its activity considerably. This will be repeated with fresh oxidase before you return." This strategy, which Krebs had probably discussed with Edson before leaving, was intended to take advantage of the fact that xanthine oxidase happened to be one of the few oxidative enzymes known that were active in solution. It could easily be obtained from milk, was readily extracted from tissues, and had been studied extensively in Cambridge and elsewhere. If, therefore, pigeon liver slices synthesized xanthine, one would expect that compound to accumulate in the medium, so that a preparation of xanthine oxidase added to this fluid might produce from it uric acid.[26]

Edson had also looked up "an old paper by [Edward] Morgan, the Prof's assistant," on the distribution of xanthine oxidase in animal tissue. Noting that in birds, unlike mammals, the enzyme is "unquestionably" present in the kidney, Morgan had written in 1926, "we find a remarkable difference between species: the enzyme is certainly present in the liver of the fowl, and with equal certainty absent from that of the pigeon."[27] This statement would have struck Edson forcefully, because it fit so closely with Krebs's current view, whereas Morgan's finding that the enzyme also existed in pigeon pancreatic tissue connected with a claim in Schuler and Reindel's recent article that pancreas slices can produce uric acid from purines. "We are going to try pancreas slices," Edson wrote. "The results with brei," he continued, "have been disappointing, but we shall persevere."

Your Excellency can see that your stewards have made a little progress in your absence, and that they hope to settle this one point before your

Excellency again inspires them by your presence. Your Excellency will surely regard it as very amusing if all the protestations of the long-winded Schuler can be replaced by a few sentences.[28]

Edson's report not only reflects the shared opinion of Krebs and his students concerning his would-be rival, but gives a nice picture of their shared attitudes at an early stage in their own investigative venture. The letter shows clearly that they were aiming to build a case for a theory that was still weak, at the same time as they sought to gain information that would further delineate the theory itself. They were not simply testing with a neutral response toward positive or negative results. A "disappointing" result indicated that they should persevere until they obtained one better fitting their expectations. An "unconvincing" increase did not convince them that their hypothesis was unconvincing, but that they needed a better preparation. Despite these departures from idealized canons of scientific detachment, these were appropriate responses for the situation at hand. They illustrate what Gerald Holton has called the "suspension of disbelief" without which "work of novelty could not get past those first hurdles whose exact nature can be identified in detail only after the fact."[29] Krebs was probably quite satisfied with this report. He had confidence in Edson, whom he regarded as a well-trained, skillful, and highly intelligent investigator.[30] He undoubtedly believed that Edson and Model were stumbling along in the right direction.

Krebs also thought Edson's letter very funny.[31] That he should have, and that Edson wrote him in this vein, illuminates the personality of Krebs as much as it does the style of Edson. Not all supervisors would react kindly to such a letter from a student. Some of Krebs's fellow German emigrant scientists would not have found it amusing to be addressed "Dear Geheimrat Professor Doktor Hans Adolf Krebs," in such obvious parody of the formalities they had observed at home. Krebs would probably not have dared to write such a letter to Warburg. For all of his seriousness of purpose, however, Krebs greatly enjoyed simple humor; and he did not mind irreverence from subordinates if he liked them and trusted them.

IV

Krebs came back to work, "after summer holiday" as he put down in his notebook, on August 2. During the next five weeks he followed several lines of investigation. Although chronologically interspersed, these were distinct ventures, and it is therefore best to describe them separately.[32]

On August 7 Krebs began a series of experiments on the "disappearance of NH_3," in rat and chicken liver slices. He was probably prompted to do so by the article by Neber "on the synthesis of amino acids from ketoacids and the synthesis of urea in the liver" that appeared just before Krebs's holiday. Building upon the old feeding experiments of Knoop and the organ perfusion experiments of Embden, as well as on Krebs's recent tissue slice demonstrations of the synthesis of urea and deamination of amino acids in the liver, Neber showed that mammalian liver slices in the presence of added NH_4Cl normally produce only urea, but that when high concentrations of pyruvate are added, the slices form, in roughly equal quantities, amino-N and urea, and that the total of these two accounts for all or most of the NH_3 absorbed. Under these circumstances, he showed, ornithine does not increase the rate of urea synthesis.[33]

Out of these results Neber drew a new interpretation of the role of ornithine. "Ornithine functions therefore as the *regulator* of urea synthesis, and establishes in that way the equilibrium between the synthesis of amino acids on the one hand, and the decomposition of amino acids and formation of urea on the other hand."[34] Neber presented this interpretation as a manifestation of the view that Knoop had long maintained, that there must exist equilibria between the processes of synthesis and decomposition in intermediary metabolism. Neber's application of Knoop's approach, and his conception of the regulation of these interconnected metabolic pathways, was far-sighted for a time in which biochemists were mainly concerned simply to discover the pathways themselves.

In his experiments Neber had measured the decrease in NH_3, and the quantities of amino-N and urea formed. Krebs measured the same three quantities, but did not duplicate Neber's conditions in other respects. Krebs prepared stock solutions of butyrate, acetate, lactate, pyruvate, caprylate, succinate, glycerophosphate, and glucose. He intended, therefore, not to test the effect of pyruvate alone, but also that of other substrates. Moreover, he did not employ the high concentrations of pyruvate that Neber had used. In the first set of experiments Krebs tested each of these substrates with liver slices from a nourished rat, with NH_4Cl also added. Pyruvate caused the highest rate of amino-N formation ($Q = 8.72$, compared to 6.76 for the next highest substrate, butyrate, and 2.54 for NH_4Cl alone). Very little urea appeared. Krebs made no remarks about this experiment, but repeated it two days later using bicarbonate in place of phosphate Ringer's solution. These results were marred by the fact that there had been "too much tissue," so that the NH_3 supplied in the medium was exhausted. Moreover, the urea values were almost null, suggesting to him that borate present in the urease solution he used to determine urea had inhibited the reaction. Testing that assumption, he found that the addition of 10% sodium borate to urease does rapidly destroy the enzyme.[35]

On the 15th Krebs tried out a variation on the manometric adaptation of the Van Slyke method for determining amino-nitrogen, incorporating a large manometric flask used by Frank Dickens. The method gave accurate results when tested with a known solution of alanine. The next day he carried out another set of experiments, using the full array of substrates he had prepared, on "NH_3 disappearance" and "urea" in rat liver slices. Although "all the amino-N" determinations were "uncertain, perhaps because of incomplete removal of NH_3," the results appeared distinctive enough for Krebs to draw the conclusion that glycerol, glucose, and α- and β-glycerophosphate exert "little influence (except amino-N)." Although he did not comment on it verbally, acetate, butyrate, and caprylate exerted little influence on any of the values, whereas pyruvate stood out too sharply. Its $Q_{amino-N}$ was 53, compared to 11.6 for the control and 9.4–13.9 for the other substrates; moreover, it depressed the formation of urea markedly. Krebs put an exclamation mark after the 53 and circled it. He probably regarded it as too high to be accurate, but not to be discounted entirely. The Q_{NH_3} with pyruvate was 16.00, highest of the group, but far too little to account for the amino-N figure.[36]

On August 20 Krebs repeated the experiment once more. The results were:

	Q_{NH_3}	Q_{urea}	$Q_{amino-N}$
Control	0	0.52	2.76
+ NH$_3$	-11.45	4.42	3.70
NH$_3$ + acetate	- 9.52	3.44	5.15
NH$_3$ + butyrate	-11.1	3.96	4.35
NH$_3$ + caprylate	- 8.48	1.80	4.06
NH$_3$ + succinate	- 9.42	3.74	3.18
NH$_3$ + pyruvate	-10.4	2.81	6.96
NH$_3$ + glycerol	- 7.15	3.23	2.61
NH$_3$ + glucose	- 8.40	2.82	3.36
NH$_3$ + β-glycerophosphate	- 8.46	3.09	—
NH$_3$ + α-glycerophosphate	- 9.58	3.48	2.98
NH$_3$ + d-lactate	- 8.31	1.94	2.51 !!

Like the previous ones, this experiment was flawed. The amino-N measurements were "not reliable," he thought, because there was too much tissue, and the $Q_{amino-N}$ values were too small. He could nevertheless infer that there is a "definite deficit. Too much NH$_3$ disappeared." That is, the decrease in NH$_3$ in the majority of the runs exceeded the total of urea and amino-N produced.[37]

Despite the deficiencies in this and the preceding experiments, Krebs did not try further to improve on them. The next day he began a series of similar experiments on chicken liver, in which a subordinate aim may have been to see whether pyruvate gives rise to amino-N in the avian liver, but the main objective of which was probably a continuation of his investigation of uric acid synthesis with Edson and Model. As far as his notebook record shows, Krebs did not return to these experiments on rat liver slices during the remainder of the year; yet he reported in a review article that he wrote during that period, "The writer is able to confirm Neber's experiments for rat and bird livers, but he finds no amino nitrogen formed when pyruvate is replaced by other ketonic acids, e.g., ketoglutaric acid."[38]

The negative result for ketoglutaric acid referred probably to the single experiment Krebs had carried out before his holiday (see above, pp. 180–181). As we can see from the above data, in his last experiment pyruvate did cause amino-N to form at a somewhat higher rate than did the other substrates, and reduced the urea output below that for NH$_3$ alone, a pattern that fitted Neber's results. These effects did not stand out conspicuously, however, from the effects of some of the other substrates. If Krebs was satisfied with so cursory an examination of the situation, it may be because Neber's results looked good enough for him so that he had no strong reasons to doubt their validity. The little-known Neber appeared in fact to have made the first successful response to Knoop's challenge to demonstrate in tissue slices the synthesis of amino acids from keto acids. Having responded himself only in a desultory fashion, Krebs had lost the opportunity to fill in that piece of the picture. It would have been consistent with his general attitude if Krebs thought it would be a waste of time to seek more rigorous confirmatory evidence. Despite the high importance accorded in principle in science to the repetition of experiments in other laboratories, scientists do not in practice often devote great energy to the pure confirmation of work done by colleagues. There is little reward for such effort in comparison to that accorded original results.

The other lines of research in which Krebs was active in August and September were collaborative efforts. One of them began on August 7 with an "experiment with Miss Stephenson." The experiment dealt with the metabolism of bacteria, and represented a combining of his manometric methods with her scientific interests. For several years Marjory Stephenson had been directing much of her research on bacterial metabolism to a special phenomenon, the ability of some species of bacteria to liberate hydrogen gas. In 1931 she and Leonard Stickland had shown that certain bacteria possessed an enzyme they called hydrogenase, which enabled the organisms to oxidize gaseous hydrogen through the reaction $H_2 + X = XH_2$. Another enzyme already recognized in bacteria was formic dehydrogenase, which transferred hydrogen from formic acid to another molecule, according to the reaction $H \cdot COOH + X = XH_2 + CO_2$ (where X represents any acceptor molecule). Stephenson and Stickland showed in addition that bacteria which release gaseous hydrogen also decompose formic acid, but by a mechanism independent of either of these reactions. The existence of this third reaction, $H \cdot COOH = H_2 + CO_2$, they established in part by measuring with Barcroft manometers the rates of formation of hydrogen in bacterial suspensions. They named the new enzyme responsible for this reaction hydrogenlyase.[39]

Performing together experiments concerning hydrogenlyase was for Krebs and Stephenson mutually beneficial. In switching from differential to Warburg manometers she profited from his long experience with this method. He expanded his metabolic interests from the experimentation with the tissues of higher organisms with which he had been mainly concerned in the past to include the specialized subfield of bacterial metabolism. Among other things, he became familiar with the techniques and precautions necessary to cultivate bacteria and to prepare suspensions of "resting" bacterial cells. He probably also welcomed on more personal grounds the opportunity to join forces with a scientist whom he particularly admired. Marjory Stephenson was, in his view, not only a highly effective, imaginative investigator, but an extremely warm, friendly, and helpful colleague.[40]

The question posed in the experiment that Krebs performed with Stephenson was whether or not the action of hydrogenlyase is reversible. If so, then in an atmosphere of hydrogen and carbon dioxide, hydrogen-forming bacteria ought to absorb these gases and to produce formic acid. To test this possibility, they placed a washed suspension of *Bacterium coli* in 95% H_2 and 5% CO_2, using

Figure 5-3 Marjory Stephenson. From *Biochem. J.*, 46(1950): opp. p. 377.

as a control a suspension in 95% N_2 and 5% CO_2. A crucial experimental assumption was that if the reaction took place according to the theory, each molecule of formic acid produced would displace a molecule of carbonic acid from the medium, so that there would be no net absorption of CO_2 in the manometer. The changes in pressure, therefore, ought to provide a direct measure of the H_2 absorbed, unless there were other gases besides CO_2 exchanged.[41]

In the experiment, Krebs and Stephenson observed changes in the expected sense. The pressure decreased in the manometer containing H_2 and increased in that containing N_2. In a further control containing no bacteria, there were no pressure changes. The readings, however, were rather small. Krebs wrote down tentatively, "Therefore, some pressure change, decreasing??" A few days later they repeated the experiment with more concentrated bacterial suspensions, and obtained this time "large pressures." Again the pressure change was positive in N_2 and CO_2, indicating a release of gas, and negative in H_2 and CO_2, indicating an absorption. After 50 minutes they exchanged the N_2 in the control for H_2, and obtained decreases in pressure so large that after 50 minutes more the total drop was somewhat larger than that in the manometer containing H_2 from the beginning. Noting that these changes could not be due to O_2, since the phosphorus present in the manometer, together with the bacteria, would quickly consume any O_2 present, Krebs wrote down "Σh in 5 [containing H_2 from the start] -327!! in 4 [switched from N_2 to H_2] -359!!" The experiment thus provided auspicious preliminary evidence in favor of the reversibility of the hydrogenlyase reaction.[42]

A week later Krebs carried out another experiment on this subject, using now as the two atmospheres pure nitrogen and hydrogen gas. To eliminate CO_2 from the system he employed a phosphate Ringer's solution. As a control he ran two parallel experiments with bicarbonate Ringer's. The pressure changes were positive and approximately equal for all the vessels except the one containing H_2 and bicarbonate, in which he calculated that the CO_2 evolved from the solution " $= 5\%$." The results he put down (in English) were:

1. *Coli* produces gas in N_2 and H_2 which is not CO_2!
2. Disappearance only when $\underline{CO_2}$ + H_2 present, not N_2; pressure change falling off.[43]

The outcome was again consistent with the hypothesis that the hydrogenlyase reaction is reversible. In the absence of CO_2 one would expect the reaction to produce H_2 and CO_2, and for the reversed reaction to take place only when both CO_2 and H_2 were present. As his remarks imply, however, these results were still only suggestive, because there was no positive identification of the gas produced or absorbed respectively under these several experimental conditions.

Two weeks later Krebs recorded in his notebook an experiment by "D.D. Woods" on "*Coli* + CO_2 + H_2O." Donald Devereux Woods was a student of Marjory Stephenson, who had originally been inspired to take up the study of microorganisms by a radio broadcast that Stephenson had given in 1930 on "how microbes live or some aspects of bacterial physiology." After studying chemistry, zoology, and botany for Part I of the Natural Sciences Tripos at Cambridge, and biochemistry for Part II, Woods was awarded a grant to stay on for research, and Stephenson took him on as a Ph.D. student. Woods carried out an investigation of the mechanism of indole production by *B. coli*. Then Stephenson "sent" Woods to Krebs to learn the Warburg manometric methods.[44]

It is not clear whether it had already been decided, when Krebs carried out the initial experiments on the reversibility of hydrogenlyase with Stephenson, that they were to be continued by Woods. At any rate, by August 30 Krebs was collaborating with Woods on this project, and it was probably by then intended to become the subject for Woods's Ph.D. thesis.

In the experiment of August 30, and a second set performed the next day, Woods compared the changes in gas pressure from *B. coli* in H_2 + CO_2 and in N_2 + CO_2, as before, but with varied concentrations of the suspension and with bacteria grown respectively in media containing formate and no formate. The latter comparison was a reflection of Stephenson's interest in the concept of "adaptive enzymes." She and her collaborators had already found evidence that the hydrogenylase content of bacteria is dependent upon the content of formate in the medium on which they are grown. She believed that this phenomenon was not the result of natural selection, but of a mechanism through which the presence of the substrate served as a direct stimulus for the formation of the enzyme in the organism. Woods's results fit the expected pattern. Only the bacteria grown in formate media absorbed gas in the presence of $H_2 + CO_2$. Another experimental comparison carried out during these days, however, raised a complication. Utilizing "the principle of Warburg's two vessel method" to test whether the only gas change occurring was the uptake of H_2, he ran parallel experiments in vessels of different fluid volume (V_f). Warburg had originally devised that method in order to measure simultaneously the evolution or absorption of CO_2, a gas highly soluble in water, and another insoluble gas such as O_2. Given in this case, however, that there was assumed to be no change in the CO_2 pressure, the readings in the two vessels ought to be the same unless some third gas more soluble in the medium than the H_2 was taken up or given off. In the event, there was a difference. As Krebs recorded in his notebook, "with large V_f less negative pressure!!" Because this result was not in accord with the theory, Krebs wondered if there might have been, in the larger vessels, "damage by dilution??"[45]

Krebs entered no further experiments concerning *B. coli* in his notebook for more than a month, an indication that Woods quickly began to carry the investigation forward on his own. The work went well, subsequent results providing firm support for the view that "the formic hydrogenlyase system can be reversed."[46] Krebs formed a "very high opinion" of Woods's ability as an investigator. As he wrote a little later in a recommendation, "Mr. Woods has original ideas, is an absolutely reliable and painstaking worker and is very industrious. He used to spend all day and a good deal of the evening in the laboratory."[47] Originality, reliability, and industriousness were qualities on which Krebs placed great value, and Woods clearly measured up to his demanding standards.

A second collaboration, begun in August 1935, brought Krebs back to the ornithine cycle to look for the additional intermediate step that he thought it probably contained between ornithine and citrulline. This effort coalesced around certain incidental points of contact between Krebs's research interests and the otherwise very different research interests of his Cambridge colleague Francis Roughton.

One year older than Krebs, Roughton had come to Cambridge as a student in 1917, and remained there ever since. Becoming interested in the kinetics of rapid chemical reactions in solution, he carried out with Hamilton Hartridge during the 1920s experiments enabling them to follow for the first time the course of reactions lasting for only milliseconds. They fixed their attention particularly upon the physiologically important reactions of hemoglobin. In the early 1930s Roughton and several collaborators studied the mechanisms by which CO_2 is exchanged between the blood and the lungs. Basic to this exchange was the reaction $H_2CO_3 \rightleftarrows CO_2 + H_2O$. Because this reaction took place relatively slowly in inorganic solutions, the question arose whether there

Figure 5-4 Francis J. W. Roughton from *Biog. Mem. Fell. Roy. Soc.*, 19 (1973) 563.

was sufficient time for the carbonic acid in the blood to be converted to CO_2 while the blood passed through the lungs. Calculations made in 1928 by O.M. Henriques indicated that only a small portion of the CO_2 evolved from the lungs could be accounted for by this reaction unaided. Henriques suggested that there is either a catalyst present to accelerate the reaction or CO_2 is bound to hemoglobin in a rapid reversible reaction. Roughton explored both possibilities.[48]

In 1932 Roughton was able to isolate from ox red blood cells an enzyme that greatly increased the rate of the reaction $H_2CO_3 \rightleftarrows CO_2 + H_2O$. He named the enzyme carbonic anhydrase. He also found evidence that not all of the CO_2 is held in soluble form in the blood, and that some of it may be bound to hemoglobin. It was already known that CO_2 can combine with the NH_2 group of proteins, amino acids, and simple amines, forming a class of "carbaminocompounds" according to the general reaction

$$CO_2 + RNH_2 \rightleftarrows RNHCOOH \rightleftarrows RNHCOO^- + H^+$$

From the similarities between factors effecting the uptake of CO_2 in blood and the properties of simple carbamino compounds, Roughton and his coworkers built up enough evidence by 1934 to claim that around one-third of the CO_2 in the blood is transported in the form of a carbamino compound of CO_2 with hemoglobin.[49]

After Krebs arrived in Cambridge he came to know Roughton well, and they discussed their work with each other frequently. Although Roughton was shy, nervous, and socially awkward, with a chronic cardiac disability that forced him to follow a rather austere life, he was a person of considerable warmth to those who penetrated his facade. Krebs found him to have rather narrow scientific interests, but he pursued these interests tenaciously until he acquired a profound understanding of them. It turned out that these interests intersected at two points with Krebs's own scientific problems.[50]

In 1934 Krebs and Roughton performed some experiments together to examine the effects of carbonic anhydrase on the urease reaction that Krebs customarily employed to determine urea quantitatively. Measuring the rate at which CO_2 is liberated in the reaction, they found that without carbonic anhydrase there is at first a rapid output, but that most of the CO_2 released is subsequently reabsorbed. With carbonic anhydrase the reaction reaches the same endpoint without the initial overshoot. They concluded that the urease reaction rapidly produces gaseous carbon dioxide, which comes more slowly into equilibrium with carbonic acid, and that the enzyme increases the rate of the latter reaction so that the lag is eliminated. They did not publish their results independently, but Roughton described them briefly in 1935, "as an example of the way in which carbonic anhydrase may obstruct the diffusion away of CO_2," in a long review article on carbon dioxide transport by the blood.[51]

Sometime after this collaboration, Krebs and Roughton began to explore the idea that the first step in the synthesis of urea might consist of the formation of the carbamino compound of the δ-amino group of ornithine. Ornithine would thereby be converted to citrulline in two steps, according to the reactions:[52]

$$R \cdot CH_2(NH_2) + CO_2 \rightarrow R \cdot CH_2NH \cdot COOH$$
$$\text{ornithine} \qquad\qquad \text{δ-carbamino ornithine}$$

$$[R = COOH \cdot CH(NH_2) \cdot CH_2 \cdot CH_2 -]$$

$$R \cdot CH_2NH \cdot COOH + NH_2 \rightarrow R \cdot CH_2NH \cdot CONH_2$$
$$\text{δ-carbamino ornithine} \qquad\qquad \text{citrulline}$$

The immediate appeal that this hypothesis would hold for Krebs is self-evident. Earlier he had analyzed the reaction:

$$\text{ornithine} + 2NH_3 + CO_2 = \text{arginine} + H_2O$$

into two steps by inferring that the four molecules entering this reaction would not join together simultaneously. Now there was the possibility of carrying this reasoning a step further, dividing the reaction into three stages, each of which involved the union of only two molecules. That being so, such a scheme would resolve the metabolic pathway into plausible single intermediate steps.

* * *

What was the source of this hypothesis? At a logical level it is easy to explain it as a natural consequence of a convergence of complementary knowledge and interests held by the two men who put it together. The historical question of precisely how and where the idea originated is, however, more elusive. There being no documentary record of how the hypothesis first arose, I was naturally interested to ask Krebs about it. I did so in July, 1977:

FLH: Can you recall . . . how the idea of the collaboration first came up?
HAK: No, I don't know this. Well, the initiative came probably from Roughton, who knew my urea work, and it was probably he who clarified the analogy. But I don't know the details. One has so many day-to-day discussions, and so many ideas popping up.[53]

Despite his own hesitation about the matter, Krebs's suggestion appears entirely plausible. The analogy that Roughton would have seen was obviously between the carbaminohemoglobin reaction he had been studying and the possible carbaminoornithine reaction. The subsequent actions of Krebs and Roughton, as we shall see, support the view that the "clarification of the analogy" supplied them with their organizing framework.

In August 1978, however, I again approached the question of how the idea on which the collaboration was based originated, from a different direction, and elicited a quite different recollection of the situation:

FLH: Did you think of this through paper chemistry similar to citrulline before you began making experiments?
HAK: Yes. And as Roughton was working on carbamino compounds I contacted him about doing joint experiments. This could be verified by Roughton's papers.

Krebs then described succinctly and accurately Roughton's work on the carbamino compounds of hemoglobin. I next asked:

FLH: Do you think you might have had that idea about the carbamino ornithine for some time before you [had] . . . the opportunity to do [this work] with Roughton?
HAK: It may have cropped up in discussing the work of Roughton. . .but I don't think that he made the suggestion that it might be an intermediate. It's paper chemistry. You are right. This is a kind of paper chemistry which I formulated.[54]

We discussed the topic once more, in May 1979, after I had examined the Roughton papers (which did not document the beginning of the collaboration). This time I asked Krebs if he had had a rationale for believing that the δ-carbamino ornithine compound is formed. (I was actually referring to a late stage in the investigation, but he construed my question as applying to the original hypothesis). He replied:

> Well, as I would reconstruct this there was a problem: In which order are CO_2 and ammonia attached to ornithine to form citrulline? It was known that carbamino compounds can be formed by amino groups. So that was one possibility and the only one which on the basis of the then existing knowledge suggested itself.[55]

These successive discussions, at one-year intervals, of events that had occurred four decades earlier, provide an illuminating case study concerning the qualities of long-term memories, and the opportunities and risks that such memories pose for historical reconstruction. Each of Krebs's comments, by itself, fits satisfyingly into the framework provided by the contemporary record, yet they do not appear fully reconcilable with one another. The first attributes the origin of the idea for the collaboration to Roughton, and views the formation of carbamino ornithine as an extension of his study of carbamino hemoglobin—which it was. The second attributes the origin of the idea to Krebs and views it as an extension of the existing conceptual framework of the ornithine cycle—which it also was. The third comment does not attribute the idea to either person, but reconstructs the problem in such a way that the reasoning is most likely to have been that of Krebs.

Oral historians might discount the second of the above accounts as the product of my violation of the precept that the interviewer should not ask leading questions. It was I who appeared to plant the suggestion that the idea came out of paper chemistry, and Krebs merely agreed to it. While recognizing the general validity of such objections, I do not think that in this case my question vitiates the answer. I often asked such questions of Krebs as my confidence grew that his memory was not easily swayed by them, that he was as ready to answer "no" as "yes." The third response above illustrates Krebs's awareness that he was at this point "reconstructing" the problem that had initiated the inquiry from his general understanding of the situation, rather than directly "remembering" it as a long-past event. All in all, Krebs was circumspect enough not to pretend to remember what he could not and to resist whatever temptation might have existed to provide me with a good story.

If at first glance Krebs's responses to my questions appear to offer at least two divergent versions of the origins of this collaborative investigation, a close examination suggests that the area of incompatibility is rather small, and that they can be assimilated into a nearly coherent reconstruction. It is quite possible that Roughton initially viewed the problem through the analogy to carbamino hemoglobin, that Krebs viewed it at first as a continuation of the reasoning that had led to the existing scheme for the ornithine cycle, and that as they talked the problem over each absorbed some of the viewpoint of the other. The discrepancy between the accounts then reduces to the point of who thought of the "nascent" idea around which both men contributed supportive reasoning. Krebs's vacillation about whether Roughton "took the initiative" or whether he himself "formulated" the problem and "contacted Roughton" does not necessarily indicate that his memory failed on this point, for the point itself may be an artifact. The questions that I posed for him reflected my own presupposition at that time that there must have been a "flash of insight" out of which the idea for the collaborative investigation was generated as a discrete event. These ideas may instead have formed gradually. Krebs's reiterated disclaimer, that because he and Roughton held many discussions together it is impossible to pinpoint when this idea "popped up"—to which he added that it often still happened to him that he discussed matters with his colleagues and when it came time to write them up "no one may remember who actually initiated it"[56]—does not merely explain why he could not remember the exact circumstances in which this idea arose, or from which of the two men it came. There may well have been no single event of this kind, but repeated discussions in which one or another aspect of the idea came up until it appeared to both of them significant enough to pursue.

* * *

When Roughton and Krebs decided to pool their respective experiences in order to test their hypothesis, they must have rejected quickly the conventional strategy of obtaining some δ-carbamino ornithine and adding it to the medium of liver slices to see whether it would accelerate the synthesis of urea at least as much as ornithine did. That course was probably impractical, although salts of carbamino acids were well known and were sometimes produced as a means to determine amino acids quantitatively. These compounds were only stable under highly alkaline conditions, so that they might be expected to dissociate into the amino acid and CO_2 at physiological pH.[57] For ornithine there was the added complication that there should be two isomers, of which only the δ-carbamino compound was relevant to the postulated reaction scheme.

Instead of choosing such an approach, Krebs and Roughton attacked the problem with methods similar to those Roughton had used to characterize carbamino hemoglobin, relying on his deeper understanding of the physicochemical mechanism of such reactions. They probably decided to open their investigation by first examining the relation between the rate of formation of urea in liver slices and the concentration of the bicarbonate-CO_2 buffer in the medium. If the first step in the reaction was the formation of the carbamino compound, then according to the kinetics of the reaction one would expect the rate of synthesis to increase with the concentration of carbonic acid present. Their second method was probably measuring the dissociation constant of δ-carbamino ornithine, in order to ascertain whether or not CO_2 would combine to an appreciable extent with the δ-NH_2 group at physiological pH. This was a particularly crucial point, because the reaction was known to take place with substances containing the amino group in the form of NH_2, but not to occur when it was in the form of NH_3. Roughton may have been able to show quickly, on theoretical grounds, that a considerable portion of the ornithine should be in the form RCH_2NH_2 at pH 7.4, whereas only about 1 percent of the

α-amino group under those conditions should be present as $-NH_2$. These considerations were clearly favorable to the hypothesis.[58]

Of the two facets of the investigation outlined above, the first lent itself to Krebs's usual manometric tissue slice methods, whereas the second depended upon special methods that Roughton and his associates had recently devised in order to estimate the amount of CO_2 bound to hemoglobin in the form of carbamino compounds. Although according to Krebs's recollection he and Roughton worked at the bench together on this investigation,[59] it is evident that they divided the primary responsibility for the experiments along these methodological lines. Krebs took up his part of the task on August 23, with an experiment on "urea synthesis with CO_2/HCO_3 concentrations." He placed rat liver slices, with NH_4Cl and glucose, in solutions containing no bicarbonate ion or CO_2 (phosphate Ringer's) and in five concentrations of bicarbonate in geometric ratios in equilibrium with equivalent ratios of gaseous CO_2. For each condition he measured the rate of urea synthesis with and without ornithine. The results were:

Bicarbonate concentration	$\%CO_2$	Q_{urea}	$Q_{urea + ornithine}$
0	0	0.53	0.62
145	1.25	2.42	3.43
290	2.5	4.17	4.30
580	5.0	4.45	6.40
1160	10.0	5.46	5.45
2320	20.0	3.58	6.68

Krebs commented:

> Without ornithine: regular increase with CO_2 pressure.
> With ornithine: relatively little more.
> Nevertheless apparently as expected larger effect without ornithine than with ornithine.[60]

The meaning of his compressed, elliptical remarks is not entirely self-evident—neither that the effect without ornithine clearly was greater than with nor that it should be, if the assumed explanation for the relation was a mass-action effect of the concentration of CO_2. In any case, the "regular increase" of the rate with CO_2 concentration was generally consistent with their hypothesis. Krebs repeated the experiment the following week, employing duplicate runs using slices from a starved rather than a nourished rat and adding lactate to the medium in place of glucose. Again he obtained regular increases with increasing CO_2 concentrations, but this time the increases were less without ornithine than with it. Moreover, the duplicates were not in good agreement.[61]

Krebs plotted the results of these experiments on graph paper and drew rough freehand curves to represent the relation between CO_2 concentrations and urea synthesis, but he did not for now carry the investigation any further.

Figure 5-5 Hans Krebs in thought.

Perhaps he was too much occupied by the other investigations in which he was simultaneously involved to give this one priority. Or it may have been that these results afforded adequate preliminary support for the hypothesis and that better data of similar nature would be of little benefit without the measurements of the dissociation constant of δ-carbamino ornithine. These depended upon Roughton, and Krebs soon learned that Roughton's scientific style was quite different from his own. Where Krebs was wont to move quickly from initial idea to experimental test, Roughton characteristically spent much time carefully working out the theoretical foundation for and the design of an experiment, before he was ready to enter the laboratory. In their collaborative venture Krebs learned to adapt to Roughton's pace.[62] It is likely that he now had to wait for his more deliberate colleague to get under way. Meanwhile he had much else to do.

During the last week of August and first week of September Krebs carried out several experiments in conjunction with Edson and Model on the absorption of NH_3 in chicken liver slices.[63] As these experiments were undoubtedly associated with a larger number that his two students were carrying out at the time on the synthesis of uric acid, it would not be fruitful to attempt to interpret his individual experiments in detail.

On September 11 Krebs performed an experiment unrelated to any of the lines of investigation he had recently been pursuing. He headed it "Dismutation of Aldehyde," and began with references to three articles on the subject. Two of them were the well-known papers of Jacob Parnas and of Battelli and Stern that had opened up the field of dismutation reactions in 1910, the third an article by Heinrich Wieland and Karl Frage published in 1929.[64] As has been discussed briefly in Volume I (p. 12), the nearly simultaneous discovery by Parnas and by Battelli and Stern of what Parnas designated the "Cannizzaro reaction" exerted a formative influence on early views of intermediate metabolism. The theoretical incorporation of a dismutation of methylglyoxal into Carl Neuberg's fermentation scheme kept the aldehyde dismutation reaction in the mainstream of thought about anaerobic respiration into the 1930s. The dismutation reaction survived the demise of methylglyoxal, as Gustav Embden replaced the Neuberg scheme with a glycolytic pathway that also included reactions of this type (see Vol. I, pp. 432–433).

Dismutation reactions were also of more general interest to biochemists in the mid-1930s as they focused attention on various forms of coupled oxidation-reduction processes. Up until this time Krebs had not directed his own research effort particularly at such phenomena; but, as he recalled in conversation in 1977, "the possibility of dismutation was always on my mind, because it was widely discussed."[65]

It may well be that, as in other areas of his field, Krebs had known of the pioneer work of Parnas and of Battelli and Stern only from summaries in later review articles or *Handbuchs*. There is nothing obvious in his activity of September 1935 to explain what prompted him at this point to look up the original articles. Whatever the stimulus may have been, he was strongly impressed by the paper of Parnas. "Very good and complete," he wrote in his notebook alongside the citation. Concerning Battelli and Stern's article he commented only on the surprising coincidence that it had appeared "4 weeks later!" than the Parnas paper. About Wieland and Frage's rather slight contribution to the subject Krebs put down, with some justification, "wretched" [*kummerlich*].[66]

Krebs had reason to single out Parnas's article for private praise, for it was one of the classics of experimental reasoning in the early history of intermediary metabolism. Parnas began with a probing discussion of the puzzle of why there are in nature a great number of esters of alcohols and fatty acids in which the acid and the alcohol contain the same number of carbon atoms and the same structure. The only explanation he considered plausible was that "the acid and the alcohol originate through a rearrangement of the corresponding aldehyde," by means of a reaction of the type:

$$2R \cdot CHO = R \cdot CO \cdot O \cdot CH_2R$$

The ester so formed is usually saponified, yielding the alcohol and the acid:

$$R \cdot CO \cdot O \cdot CH_2R + H_2O = R \cdot COOH + RCH_2OH$$

Parnas then showed that the brei, press juice, and aqueous extracts of liver tissue under anaerobic conditions act upon a series of aldehydes, converting them to the corresponding alcohols and acids. Among the aldehydes for which he was able to demonstrate the reaction quantitatively were isovaleric aldehyde, n-valeric aldehyde, and n-butylaldehyde. After providing additional supporting evidence, Parnas discussed in broad terms "the biological meaning of the Cannizzaro reaction." The reaction provided "the simplest case of a reduction in the organism the mechanism of which could be fully understood." Oxidative reactions were easy to explain in terms of known chemical processes, because they were exothermic, but the only known chemical types of exothermic reductions involved free hydrogen, which is not available in organisms. There must, therefore, be a source of energy able to drive reductions through "coupled reactions." The Cannizzaro rearrangement offered "a simple system of coupled reactions in which, through the displacement of oxygen and the uptake of water a simultaneous reduction and oxidation takes place." The class of such reactions might be extended beyond those he had demonstrated, to include molecular rearrangements yielding mixed esters or acids and alcohols of differing structures.[67]

Parnas's results and his views remained as cogent in 1935 as they had been in 1910. In fact, Krebs may have felt as he read the paper that advances in the meantime made it now more feasible to exploit the lead Parnas had provided than it had been when the paper first appeared. The experiment Krebs carried out on September 11 was clearly inspired by Parnas. He chose to test, in rat liver and kidney slices, n-valeric aldehyde, one of the aldehydes that had figured prominently in Parnas's experiments. Measuring the effects of valeric aldehyde on the respiration of the tissues, he found a large inhibition in the kidneys, and a slight one in the liver. This was to be expected, since aldehyde was not metabolized in the kidneys. The crucial experiment was the anaerobic one. There he traced the effects of n-valeric aldehyde in terms of the formation of acid, measured manometrically. He found,

	Liver	Kidney
Q_{acid} (control)	8.04	2.18
Q_{acid} (+valeric aldehyde)	22.7	7.40

The outcome was obviously a striking confirmation of the results of Parnas. After listing the immediate results, Krebs wrote down:

Investigate further   various aldehydes
                       "   tissues
                       keto acids + aldehydes[68]

Although he did not pursue this plan at once—as we shall see, his research was interrupted at this point—it seems evident that he saw in this result the potential starting point for a new line of investigation. The notation to try "keto acids + aldehydes" implies that he had in mind more than a fuller confirmation of Parnas's work; that he envisioned following up Parnas's suggestion that biological dismutation rearrangements might include reactions beginning with two dissimilar molecules.

We have seen that Krebs often interjected single experiments outside the main lines of his current lines of investigation rather casually, and quickly dropped them if nothing exceptional turned up. This isolated experiment was clearly something more. Krebs has remarked that reading scientific papers often "put my mind on a certain line of thinking."[69] We are fortunate to be able to isolate in this case such a reading that gave a powerful impetus, not only to his thinking, but to his future investigative course. Even though some time was to elapse before that happened, we can infer from later events that he probably formed at this point a strong intention to move in this direction as soon as a favorable opportunity should present itself.

V

By August 1935, Krebs had been living in England for more than two years. He spoke mainly English, and he wrote his scientific papers in English, yet he still kept his private laboratory records in German. On August 30, for the first time, he recorded an experiment in his notebook entirely in English. The next experiment, performed on September 6, is mostly in English, but includes one German phrase.[70] The notebook then reverts to German, with a few later exceptions. These switches were clearly not deliberate. When I discussed with Krebs in 1978 the transition from German to English in his notebooks (referring not to this notebook, which had not yet turned up at that time, but to the subsequent one), he commented, "Well, occasionally, as it happened that I didn't know in which language I was writing or speaking, this happened when I had for some reason or another been speaking English or speaking German."[71] Perhaps it is not irrelevant that the first experiment written down in English was one performed by Donald Woods, so that the rough data that Krebs would have reduced in order to record the results may have been written in English. To have become unaware of which language he was using was a subtle mark of the progress of Krebs's *de facto* naturalization.

During these weeks of August and early September, Krebs's thoughts must have been as much on his future in England as on the present. He was apparently optimistic enough about his prospects in Sheffield to let his friends know about it. On August 19 Nachmansohn wrote him

The latest turn in your situation pleases me very much. Hearty good wishes! Perhaps the Cambridge people will on their side outbid Sheffield? Still, that would be difficult![72]

Bruno Mendel was also happy for his old friend. On August 25 he wrote

> We are very glad about the news that you may possibly emigrate from
> Cambridge to Sheffield, because if you make up your mind to move
> to this industrial city it must be because the position being offered to
> you at that university is a substantial improvement. Will you write us
> sometime with more details about the situation?[73]

Pleased though his friends were, both hinted that Krebs must have compelling
reasons to think about leaving beautiful Cambridge for a sooty industrial city,
to exchange a university with international prestige for an obscure provincial
one, or that he might use an offer from Sheffield to strengthen his position in
Cambridge. Such ideas must have occurred also to Krebs.

On September 2 Nachmansohn wrote again with news about the other, more
distant future that still beckoned to Krebs. "In the meanwhile, therefore," he
announced, "important things have happened." Blaschko had come to Paris and
he and Nachmansohn had held a short conversation with Weizmann "concerning
our institute. . . It was definitely decided that we will travel [to Palestine] in
December, independent of the question of how far the financing has progressed
by then. Weizmann believes, in fact, that he can go forward decisively in the
negotiations during September or October." Nachmansohn reported on a
number of related developments, and urged Krebs to come quickly to Paris,
while Weizmann was there.[74]

These developments undoubtedly stirred Krebs's interest, but he could not
give them the priority that Nachmansohn did. In a letter that crossed Nachman-
son's he had just written that he was unable to set a time for a visit to Paris,
because he was planning to go to Sheffield for a consultation in the middle of the
month, and the date depended upon "the other side." Moreover, he was
expecting a visit from his sister as soon as she recovered from a knee injury.
He proposed coming to Paris near the end of the month instead.[75] No sooner
had he posted this letter than he received one from Wayne asking if September
17 would "suit you to come up to Sheffield and look around the laboratories
here." Krebs replied immediately that he would be able to come then.[76]

Lise arrived in England on Friday, September 13, and stayed with her
brother for 10 days. She had decided that she could get more of his attention
if she were there over two weekends. There is no indication that she accompanied him to Sheffield during the week, but he seems to have found more time to
spare for her than for most visitors who came to see him. Hans was somewhat
"shocked" to realize how little English Lise knew. He attributed her limited
proficiency to the fact that his sister, like his brother, had failed to make the
special effort he had made to use his holidays to learn foreign languages.[77]

Hans and Lise had much to talk about, including the big question of
Palestine. During the previous spring she had put aside the idea of emigrating.
Reports from friends who had gone there were mixed. Her husband Adolf
seemed to be adjusting well to his position in Düsseldorf, and "at this point,"
she had written in March, "our life is quite enjoyable." It seemed to her that

"it would be senseless to give up this good and pleasant existence and for us 'old' people to dare a difficult experiment."[78] By fall, however, the tensions of life in Germany were again taking their toll, and Lise and Adolf were wrestling with the formidable obstacles to emigration. The most difficult problem was the impossibility of taking out of Germany enough money to live on until they could establish themselves in Palestine. Lise discussed all of this with Hans, who urged on her the necessity of getting out before their situation deteriorated.[79] Perhaps his advice was given special force by the fact that, while Lise was with him in Cambridge, Hitler's government passed in Nuremberg the notorious Law for the Protection of German Blood and German Honour, which deprived Jews of their citizenship and their legal rights.[80]

To overcome the financial difficulties, Lise and Adolf were contemplating an approach to Georg Krebs, who might be able to put at their disposal funds drawn from his foreign investments. Hans supported this idea and suggested ways in which they might overcome their father's doubts.[81] Undoubtedly Hans also discussed with Lise the possibility that he too might eventually settle in Palestine, and they probably agreed that it was hard to form a realistic picture of conditions there without ever having seen the country.[82] Hans's counsel proved very helpful to his sister in making up her mind. Afterward her husband wrote him:

> I thank you heartily for your loving reception of Lise. Her stay in England has been good for her in many respects, and, strongly impressed by the big brother, she now sees many things differently.

Lise thanked Hans "from my heart for all the love and care during my stay there."[83]

Hans drove to Sheffield, as planned, on Tuesday, September 17. The trip took about four hours. If he followed Wayne's directions, he had to enter Sheffield through the industrial district and follow the tram lines through the center of the city to reach the university. Most of the University of Sheffield was concentrated in a large brick building situated part way up one of the streets that ascended the steep hills forming the northern suburbs of the city. In the afternoon Wayne probably showed Krebs the laboratories of the Pharmacology Department, located on the second floor of this building. If he were to come, he would have available a spacious, well-lighted laboratory room across the hall from Wayne's own laboratory quarters.[84]

In the evening Krebs went to Wayne's home, met Edward's wife Nan, and stayed overnight with them. The next morning Krebs had a brief interview with the vice-chancellor and met the dean of the medical faculty and the retired but still influential professor of medicine, Sir Arthur Hall. Afterward Wayne probably discussed various matters with Krebs, cautioning him that there might be some opposition to the appointment of a foreigner, but expressing confidence that it would be approved.[85]

As Krebs drove back to Cambridge on Wednesday afternoon, he must have been very encouraged by his experience in Sheffield. He had been received in the most friendly way by everyone concerned. He realized that he would have there ample room not only for his own work, but to accommodate other people who wished to work with him. In Cambridge, on the other hand, he competed with many others for space at crowded laboratory benches. It was probably also clear to him immediately that Edward Wayne was offering to share his own resources in an extremely generous manner.[86]

While he motored home Krebs must have pondered also the personal consequences of a move to Sheffield. To leave the Cambridge milieu would not be easy. He would miss friends, he would be separated from the daily discussions with colleagues and the supportive atmosphere of a laboratory that had come to represent his ideal of scientific organization. Even in the last weeks the collaborative ventures he had begun with Marjory Stephenson and Francis Roughton had enhanced for him the advantages of working in Cambridge. On the other hand, Sheffield was only 110 miles from Cambridge, and as he made the trip back he must have reflected that it would be feasible to return often enough to keep in touch with these people.

To give up the physical beauty and cultural richness of Cambridge for the bleakness of an industrial city was another matter. It must have given Krebs pause when he drove through some of the grimmer districts. There was, however, the prospect of ample compensation in the surroundings. For someone who enjoyed hiking in beautiful settings as much as Krebs did, the high moors of the nearby Peaks District offered a strong attraction. Finally, the Waynes themselves must have made a forceful impression on him. Their warmth, their openness, their broad interests, and their capacity to make a stranger feel instantly at home must have drawn Krebs to sense that here was an environment in which he might thrive.[87]

Back in Sheffield, Edward Wayne probably wondered whether he was being too generous. He was not only committing himself to turn over to Krebs much of the research space and budget available to him, but was sacrificing his only opportunity to acquire a junior assistant in pharmacology who might have taken over a major portion of his teaching and furthered his own research interests. Wayne was doing more than his share to keep this awesomely successful young biochemist in England.[88]

Krebs went quickly to Hopkins to discuss his situation. True to his prediction, Hopkins had not been able to induce the Cambridge authorities to do anything, and he had to tell Krebs that he was not in a position to offer him a post comparable to that in Sheffield. As Krebs weighed his alternatives, he soon came to the conclusion that the space available to him in Sheffield to bring in collaborators was a decisive advantage. The prospect of doubling his salary was also very welcome after a decade of living on minimal incomes.[89] Given the personal questions he faced concerning Katrina, the financial factor may have weighed heavily on him at just this time. Krebs did not linger over his decision. Within about a week after his "exploratory" visit to Sheffield, he was able to write Wayne that he was looking forward to coming there.[90]

Professional, public, and private events seemed tightly linked in Krebs's life in September 1935. Although personally beyond the direct reach of the

Nuremberg laws, he quickly felt their impact through their effect on Katrina. These brutal statutes not only forbade marriage between German citizens and Jews, but made extramarital relations between them a serious crime.[91]

During the month since their parting, Katrina had been cheerful in spite of her loneliness in Altona. Her habitual complaints about the scarcity of his letters were light hearted. She willingly gave him permission to show the snapshots of their holiday to his sister, but warned him not to allow any other girlfriends he might have in Cambridge to see them.[92] But the Nuremberg laws suddenly turned her mood somber. On September 19 she wrote

> Dear Hans. I am waiting every day for you to write me. You will also have read the proclamations [*Ereignisse*] of September 15 in Nuremberg in their original wording. My position remains always the same, Hans, my greatest wish is, as it has been before this, that I will come to you soon, and can prove the value of what has held us together for these years. Do not leave me in the dreadful anxiety into which these days have driven me, but write me honestly your thoughts about these matters.
>
> If you lived here, you would have to recognize that now there can be only an either-or. I am absolutely clear about it, that I am willing to summon up all my courage, strength, and responsibility for you, on the other hand I also know that giving up our friendship would mean a complete extinction . . . . Will you help me?[93]

This time Hans did not delay, but answered her plea by return airmail. What he told her can only be inferred from her next letter to him. "Your quick answer," she wrote, "has consoled me very much."

> I believe that you find my reaction to things such as this latest law somewhat exaggerated; but you are sitting far from the firing-line, and it reaches you only second hand. If I know that you want the same things as I do, and you understand that we must now more than ever hold together, then all the clamor that is made about these matters over here is to me completely indifferent.

Besides assuring her of his feelings toward her, Hans must have included the news that—assuming his appointment was approved—he would go to Sheffield, a matter of considerable importance to both of them. Her spirits revived, Katrina expressed great curiosity about "the impression you have of Sheffield, your working conditions there, and your chief. And what do people say in Cambridge about your leaving there?"[94] Given the either-or message in Katrina's previous letter, Hans must have come close to a declaration that if all went well in Sheffield he would be ready for her to join him there.

The events of September crowded in on him so rapidly that, for once, even Hans Krebs could not maintain his normal relentless pace in the laboratory.

During the week that he divided between Sheffield and his sister he did no experiments. On the week after her departure he managed to get in three days at the bench. He used them mainly to do a little more work on his part of the carbamino ornithine investigation. On Tuesday, September 24, he found that increasing pressures of CO_2 increase the rate of urea synthesis in the presence of ornithine and NH_4Cl at pH 6.8. The next day he carried out a similar experiment with similar results at pH 7.4. On Thursday he examined "the effects of ions on urea synthesis." He arranged a series of liver slices in special media prepared so that each of them, except for one, lacked one or more of the normal constituents of physiological salt solution. The missing constituents were, respectively, potassium, calcium, magnesium sulfate, calcium + potassium, and all three of these. To one he added an excess of KCl. None of the single omissions lowered the rate of urea synthesis in the presence of ornithine and NH_4Cl. He wrote down, "Absence of K, Ca, $MgSO_4$ no effect: excess K increased!!"[95] The probable objective of the experiment was to test whether there were other ions beside bicarbonate whose absence prevents urea from forming. The fact that these three did not supported his view that the bicarbonate effect was specific and could be explained in terms of the hypothesis that the first step in the synthesis is the formation of carbamino ornithine.[96] This was far from a confirmation of the hypothesis, but promising enough so that it probably heightened his interest in getting on with the part of the investigation that depended on Roughton.

On Friday (September 29) Krebs left for Paris to spend the weekend with the Nachmansohns. Events in Sheffield may have preempted his immediate future, but did not lessen his interest in longer-range possibilities in Palestine. He was, in fact, highly encouraged by the recent developments about which Nachmansohn had informed him, and probably eager to talk over the situation in person. Nachmansohn and his wife were also eager to see Krebs for personal reasons, and the weekend proved to be a happy renewal of a long-standing friendship.[97]

Returning to England on Monday, Krebs stayed over at the Red Court Hotel in London in order to meet in the evening with Weizmann. There Weizmann suggested that there was no reason that Krebs should move to Palestine until the proposed institute had become a reality. Krebs made it clear to Weizmann that, although keenly interested, he could not commit himself to future participation. He would reexamine his position according to later developments. It seems evident that the better his future in England looked to him, the more he would require in the way of firm guarantees of good working conditions to draw him away. Afterward Nachmansohn agreed with him that he should not settle for anything "provisional," and assured him that he remained so crucial to the enterprise that he was in a strong position to influence the course of events.[98]

Arriving back at the laboratory at Cambridge on Tuesday, Krebs performed during the next three days several experiments connected with Donald Woods's research on the hydrogenlyase reaction in bacteria, and one or two related to urea synthesis.[99] Then he received from Wayne the news for which he must have waited impatiently:

1st October, 1935

Dear Krebs

At the meeting of the Faculty today it was decided to recommend your appointment as Lecturer in Pharmacology. The recommendation will have to come before the Senate on Friday and the Council a week on Friday before it is finally settled. I think, however, you may take it that the Vice Chancellor will see to it that the appointment is approved in these bodies, and the chances of an adverse decision are very slight indeed. I think you might begin to make your arrangements for coming up here. If you will let me have a copy of an advertisement for rooms, I will have it placed in the Sheffield Telegraph.

After discussing a few further details, Wayne added,

I think you will find people friendly and facilities good. I shall be personally delighted to have you as a colleague.[100]

Krebs was not only ready to accept this appointment, but, as Wayne's letter implies, to come to Sheffield as soon as possible. That choice ran against the awkward circumstance that he was required to give a full term's notice before leaving his post in Cambridge. Moreover, he was scheduled to take his turn teaching in the biochemistry course in one month, and could not be replaced in so short a time. Hopkins resolved these problems by releasing Krebs from his contract on condition that he return in November to fulfill his teaching assignment.[101]

Meanwhile, Wayne reported, on Friday, October 4, that the Senate had approved his appointment as expected, and he had no doubt that "it will be recognized by the Council next Friday." Krebs's advertisement for rooms appeared in the Sheffield newspaper on Friday and Saturday. On Monday he drove to Sheffield to look for his lodgings and found waiting for him one response to his notice. Mrs. F.M. Hall, of 191 Ringinglow Road, had "the accommodations you require. This is a modern house with every convenience." Her terms, "including garage, rooms, and board," were £2.2.0 per week. After spending Monday evening with the Waynes and discussing various arrangements for the move, as well as applications for money for research equipment, Krebs inspected Mrs. Hall's rooms on Tuesday and accepted her offer.[102]

In his autobiography Krebs has given strictly professional reasons for his decision to leave Cambridge for Sheffield. The factor that influenced him most was the plentiful space available to "expand my team." Supporting conditions were the increased salary and the fact that the lectureship was a "semi-permanent" appointment.[103] These were undoubtedly important considerations, although the emphasis on the opportunity to expand his team seems clearly influenced by what later did happen in Sheffield. We may still be left with the feeling that the explanation is somehow incomplete, that in view of the obvious disadvantages of giving up the "Cambridge milieu" for a small provincial university, the haste with which he made his choice and his departure remain puzzling. It was not that he had been attracted to a major post in the smaller university. In explaining to me why opposition to the appointment did not materialize, Edward Wayne remarked in 1977 that the appointment was not greatly sought after. "There wasn't a great tradition of people of ability taking anything like a lectureship. If it had been a chair, then of course there would have been a lot of opposition; but it was not a particularly attractive appointment."[104] Nor does Krebs's recollected point that the position was semipermanent provide a compelling advantage, since it appears evident that that was also true of his position as demonstrator in Cambridge. There is, I believe, sufficient circumstantial evidence to suggest that Krebs may have been impelled to his rapid decision in some part by the very private reason that the immediate financial improvement it offered would help him to resolve the pressing dilemma of his relationship with Katrina.

VI

From his new lodgings Hans quickly sent an airmail letter to Katrina sharing with her all that had taken place. His rooms were not particularly comfortable, he wrote, but otherwise everything looked very good to him in Sheffield. Then he drove back to Cambridge to organize his move.[105]

To Katrina it now seemed that the barriers that had stood between her and Hans were about to fall away. Answering his letter at once, she sympathized about the state of his newest quarters, but implied that that was a temporary inconvenience.

> I will be glad on the day when we can look for a small neat dwelling. You must also have a place to work that is neat and pleasant. Now you must first get used to the Sheffield circumstances. And if you then want to follow the advice of your new chief, you shall write me, and I will come to you as your genuine wife . . . . I am so happy Hans, that you have now really gone to Sheffield, your letter sounds so satisfied, and the outlook for later on is certainly more favorable than in Cambridge.[106]

In Cambridge Krebs rapidly assembled his belongings and prepared his research equipment for transporting. He was able to pack most of the manometers and other equipment he had brought from Freiburg into his roadster and drive off with them. By the time he reached Sheffield again, on Monday, October 14, he had learned from Wayne that the council had allocated £100 to purchase additional equipment for him over two years. Installing himself in his new laboratory quarters, Krebs set up his equipment so quickly as to astonish his new chief. It took Krebs just one week to resume his normal daily schedule of experimentation. He went about it so quietly that it appeared to Wayne at first that he had almost vanished into the background.[107]

UNIVERSITY, SHEFFIELD. 293

Figure 5-6 View of the University of Sheffield.

In Sheffield Krebs opened a new laboratory notebook, and apparently made a deliberate decision now to keep it in English. He did so with only occasional lapses into German in the early pages.

Krebs fixed his attention in Sheffield on the investigation of uric acid synthesis that Edson and Model had been carrying on under his supervision since June. Presumably Krebs devoted his personal research effort to salient or troublesome points that had come up as a result of the experiments of his students, but there was evidently no systematic division of labor involved. In the absence of a detailed record of the experiments that Edson and Model had conducted in the meantime, the experiments that Krebs now undertook cannot be placed with precision into the fine structure of the developing problem. The general situation was, however, that stronger data were still needed to confirm the preliminary results supporting their hypothesis: that pigeon liver tissue synthesizes hypoxanthine, which is oxidized in the kidney to uric acid, and that in chickens both stages can occur in the liver. Moreover, Krebs still hoped to identify the precursor that supplies the carbon skeleton for hypoxanthine, and eventually fulfill the objective he had expressed to Thannhauser 18 months before—to establish the synthetic pathway. That he placed priority on this investigation over the other problems on which he had been engaged during his last weeks in Cambridge may be connected with the fact that he, Edson, and Model were scheduled to present a paper on uric acid synthesis at the November meeting of the Biochemical Society. It would be well to be as far along as possible before then.

On the first three days of his research Krebs tested preparations of xanthine oxidase from rat liver and kidney, pigeon kidney, and cat liver on the oxidation of hypoxanthine in solution. He found in general, however, "no increased uptake of oxygen." These experiments fit into an effort that Edson and Model had begun in July to confirm that hypoxanthine is an intermediate by showing that a substance produced by pigeon liver slices is oxidized by xanthine oxidase.[108]   Apparently Krebs was searching for suitable preparations of xanthine oxidase to use. That he was still doing so in October suggests that the problem had proved more difficult than Edson suspected when he hoped to solve it before Krebs returned from his holiday.

Another technical problem that became acute during the course of the investigation was the need for a new method for the microdetermination of uric acid.   The Folin colorimetric method on which Krebs had depended since Benzinger began working on uric acid synthesis in Freiburg was unspecific and, as now became clear, "very unsatisfactory in the presence of tissue extracts." Sometime during the summer or late fall Edson must have begun, under his supervision, to try to develop another method that would utilize a reaction converting uric acid quantitatively to urea, after which the estimation could be made manometrically with the usual urease method. To carry out the conversion they turned to a method devised by Richard Fosse and his colleagues in 1930, based on the discovery of the enzymes involved in the metabolic decomposition of uric acid (see p. 52). This method was highly specific, but it took a very long time. Krebs and Edson tried to improve it, but finally abandoned the effort in favor of a nonenzymic procedure to decompose uric acid.[109] It is not clear how far along they had got in developing this method by the time Krebs went to Sheffield. Most likely they had attained something provisionally workable but not entirely satisfactory. On October 24 Krebs resumed experiments of the type he had earlier carried out on "NH_3 consumption by pigeon liver," examining this time the influence of the concentration of NH_3 and lactate.   At the same time he tested the effects of hypoxanthine on the formation of uric acid in pigeon liver, using probably for the first time a version of the new method for uric acid determinations that he and his collaborators were developing. He had, however, employed "too little tissue," so that the readings were "too small."[110]

The next week he began a new series of experiments testing systematically for substrates that might influence the rate at which NH_3 disappeared in pigeon liver slices.   On Tuesday the 29th he included 10 substances—succinate, α-glycerophosphate, pyruvate, acetate, l (+) glutamate, glucose, glycerine, glycerol, caprylate, and ornithine—and used liver slices from a pigeon starved for 24 hours.   Of these substances α-glycerophosphate, acetate, and caprylate inhibited the process, pyruvate and glycerol increased it, and the rest had "no influence."   The next day he tested a largely different list of substances with liver slices from a pigeon starved for 48 hours. Glycerol, crotonic acid, and ketoglutaric acid had "no effect or little increase," whereas there was an "enormous effect of pyruvic [Q_{NH_3} = 11.21] and oxaloacetic acid [Q_{NH_3} = 12.4]," and "almost as much with lactate [Q_{NH_3} = 9.24]." The control gave Q_{NH_3} = 2.93.[111] It is not clear whether Krebs had in mind a hypothesis concerning precursors of hypoxanthine when he carried out these experiments. In previous, less systematic experiments he had already observed that pyruvic and lactic acid accelerated the absorption of NH_3, so these two tests could have represented an effort to find out if these were specific effects or only manifestations of a more general metabolic influence. In any case, the effects of pyruvic, oxaloacetic, and lactic acid stood out so clearly in the second experiment that he might well have sensed that he was onto the trail of the precursors for which he was looking.

On the same day Krebs performed an experiment comparing the effects of xanthine and hypoxanthine on the formation of uric acid in pigeon pancreas and kidney tissue. He found the "xanthine yield about twice as much uric acid as hypoxanthine," a result compatible with the presumed intermediate position of xanthine in the oxidation of hypoxanthine to uric acid.[112]   Having barely got set up and started with his research in Sheffield, Krebs had to break off again and prepare to return to Cambridge to carry out his final teaching obligations there.[113]

Traveling the now familiar route to Cambridge on the morning of Friday, November 1, Krebs arrived at the laboratory in the afternoon to make advance preparations for teaching practical laboratory work. He was scheduled with Eric Holmes to direct student experiments on oxidative mechanisms, using "the Thunberg and Warburg techniques." For the next four weeks he was even busier than he had been the year before. In addition to the laboratory instruction he had to give 15 hours of lectures. Getting undergraduates to perform meaningful experiments with the same complicated apparatus that was used on the research front was no easy matter. When the staff had reviewed the results of the first year of teaching the new course, during the previous spring, it had "felt that the practical work was badly and sloppily done by the students." If Krebs did his part in carrying out the suggestions made for improving the situation in the fall of 1935, he had to work hard at it, not only showing how to use the apparatus, but supervising the work and marking the notes and results of each student. Moreover, the course was again so crowded that double sessions were necessary. On November 17, in a letter to Nachmansohn, Krebs wrote with understatement, "My time is rather rushed at the moment." It did not make things easier for him that in the midst of all this, and other activities in which he participated in Cambridge, he had to make several trips (for purposes that are not stated) to Sheffield.[114]

Among the other activities that Krebs was able to crowd into this month, one was undoubtedly the supervision of Edson's experiments on uric acid synthesis—Model was no longer working on the project, having left in September to take a clinical position in Bristol.[115]   Edson probably worked particularly during this time on the method for determining uric acid. On Friday, November 15, Krebs, or Edson, or both, left for London to present their work on "The Synthesis of Uric Acid in the Avian Organism" at the Biochemical Society meeting at Guy's Hospital. From an abstract published in *Chemistry and Industry* we can obtain an outline of what they reported. They began by describing their new micromethod for uric acid. "Uric acid is decomposed to urea through the following states: uric acid → allantoin → allantoic acid → urea + glyoxylic acid. The urea is determined by means of urease."[116]   This summary is too general to indicate whether Krebs and Edson were still struggling with the enzymic method for decomposing uric acid or had turned to other chemical means.

After summarizing the experiments on pigeon liver, kidney, and pancreas through which they had identified hypoxanthine as the intermediate formed in the liver, they stated that they had been able to "replace" pancreas or kidney tissue with hypoxanthine preparations from milk, using methylene blue in place of oxygen. Finally, they had found

> The rate of disappearance of ammonia in the starved liver is greatly (300–800%) increased by lactic and pyruvic acids, the latter being more effective than the former. The possibility of pyruvic acid furnishing the carbon skeleton of hypoxanthine is discussed. Hypoxanthine may be conceived as arising from pyruvic acid and two ureido radicals with elimination of water.[117]

This idea, which must have been illustrated by structural equations in the talk itself, was not really new. It was a variation on a theme dating back to the late nineteenth century: that uric acid is formed from the joining of a molecule containing its three-membered carbon chain with two molecules of urea. In the most influential version of this theory, proposed by Wiener in 1902, tartronic acid was the bearer of the carbon skeleton (see Vol. I, p. 388). Krebs was now suggesting pyruvic acid in place of tartronic acid, and adapting the general reaction mechanism to the synthesis not of uric acid directly, but its precursor hypoxanthine. From the standpoint of Krebs's own thinking the interesting feature of his view is that it was not a development in the direction in which he had been looking in the spring of 1934 when he wrote Thannhauser that the synthesis might begin with a two-carbon compound (see pp. 58–59). Rather it was a return to a position that he had earlier believed ruled out by Benzinger's experiments in Freiburg (see Vol. I, p. 391).

It is evident that the experimental results summarized in the above passage were those obtained by Krebs in Sheffield on September 29 and 30. (This identification is supported by the fact that in the paper later published the data reported on this topic are taken from the same two experiments.)[118] From the experimental record itself, however, we can see that the result which struck Krebs immediately was that lactate, pyruvate, *and oxaloacetate* greatly increased the rate of disappearance of ammonia. Why then did Krebs omit mentioning the latter (and, we may add, omit the data for oxaloacetate when he afterward published these results in detail)? It appears that this is a case in which his theoretical presuppositions influenced the significance he attached to his data. Pyruvic acid could be connected through a plausible reaction mechanism to the synthesis of hypoxanthine, and lactic acid was already closely connected metabolically to pyruvic acid. Oxaloacetic acid, however, contained a four-member carbon chain that could not so readily be fitted into this picture.

* * *

In 1978 Hans Krebs and I examined the pages in his laboratory notebook containing the experiments of September 29 and 30, 1935. At that time I had

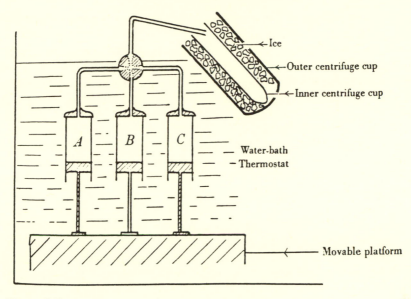

Figure 5-7 Roughton carbamino hemoglobin apparatus.

not seen the abstract of the presentation made to the Biochemical Society, and neither he nor I were aware of the reaction mechanism he had discussed then. When I read out from his statement of the results "enormous effect of pyruvic and oxaloacetic acid," he looked at the data and commented, "Yes. I thought oxaloacetic was even more. A little bit more, yes. Now this retrospectively was highly significant."[119] With the benefit of later knowledge and a changed point of view, Krebs picked out instantly from his old experiment a result whose significance was at the time he obtained it hidden from him, in part because the interpretation of the situation he then entertained excluded it from consideration.

* * *

It was probably also during these four busy weeks in Cambridge that Krebs and Roughton began their joint experiments to measure the dissociation constant for carbamino ornithine—that is, the equilibrium constant for the reaction:

$$CO_2 + \text{ornithine} \leftrightarrows \text{carbamino ornithine} + H^+$$

The objective of the measurements was to ascertain whether there would, at physiological pH, be appreciable quantities of δ-carbamino ornithine in equilibrium with ornithine, as there must be for their hypothesis that the carbamino compound is an intermediate in urea synthesis to be viable.[120] To obtain the measurements necessary in order to calculate the equilibrium constant, Krebs and Roughton planned to equilibrate ornithine solutions with gas mixtures containing CO_2, then estimate the carbamino compound in solution through a special procedure that Roughton had devised in 1934 to estimate the CO_2 bound in carbamino hemoglobin.

The experiments were carried out in Roughton's laboratory in the Physiology Department, using the same apparatus that he had previously used for carbamino hemoglobin. The principal difficulty that he had solved through a brilliant experimental design was that in the pH range of the reaction, carbamino compounds dissociated so rapidly that it would not be possible ordinarily to carry out the reactions necessary for the determination quickly enough to measure the equilibrium concentration sought. The apparatus with which he obviated the problem incorporated three syringes inverted over a common platform arranged so that when the platform was raised their plungers expelled their contents simultaneously in exactly equal quantities: In syringe A was placed a solution of ornithine previously shaken with air containing CO_2. Syringes B and C contained respectively $BaCl_2$ and NaOH. When these three solutions came together in the mixing chamber, the alkali brought the reaction to a pH of 12–13, converting all $H_2CO_3^-$ and HCO_3 present into the form of $CO_3^=$ which was precipitated as the barium salt. This left in solution, in the chilled centrifuge tube, only the CO_2 that had combined in the form of a carbamino compound. The whole process took place so quickly that there was no time for significant quantities of the compound to dissociate. The CO_2 concentration of the fluid could then be measured by the standard Van Slyke method and regarded as a measure of the CO_2 bound in carbamino ornithine.[121]

Given that the entire experimental setup belonged to Roughton, he was undoubtedly also the principal experimenter. He liked to work with other people in the laboratory, however, and Krebs spent as much time as he could spare at the bench where the experiments were carried out. As a result he came to admire the great care with which Roughton planned and conducted experiments, and he was impressed with the contrast between Roughton's research style and his own more hit-and-miss approach.[122]

There is no record of how many experiments the two collaborators managed to complete while Krebs remained in Cambridge. They were sufficient, however, to calculate an equilibrium constant favorable to their hypothesis. Adding this evidence to what he himself had acquired from his measurements of the effects of CO_2 concentration on the rate of urea synthesis, Krebs was optimistic enough about the status of carbamino ornithine in the ornithine cycle to include a brief discussion of it in an article on amino acid metabolism that he was preparing, either then or shortly afterward, for *Annual Reviews of Biochemistry*:

The rate of urea synthesis increases rapidly with increasing concentrations of the bicarbonate-CO_2-buffer of the medium. There is almost no synthesis if the carbonic acid buffer is replaced by phosphate, whereas the reaction proceeds almost normally in the absence of other ions. This specific effect of the bicarbonate buffer may be explained on the assumption that the first stage in the urea synthesis is the formation of the carbamino compound of the δ-amino group of ornithine. . . .

As references for the above results he cited "Krebs, H.A., *unpublished experiments.*" After depicting the equations that are reproduced above, he added, "This view is supported by measurements of the dissociation constant of δ-carbamino ornithine," for which he cited "Roughton, F.W.J. and Krebs, H.A., *unpublished experiments.*"[123]

As in his talk to the Biochemical Society on uric acid synthesis, Krebs chose here to make public a proposed metabolic reaction scheme for which he had only preliminary supporting evidence. Commenting in 1977 on this scheme he defended it as a "legitimate hypothesis" to "suggest."[124] On this and other occasions in our conversations he implied that there is a clear distinction between mere "suggestions" and serious claims to have discovered something. His position is perfectly sensible; yet this was not a casual suggestion. Krebs had deemed it important enough to devote half a page to it in an article for a review journal intended to cover concisely the significant developments of the preceding year in the various subfields of biochemistry.

These two instances illustrate nicely the fact that there is really not a line of demarcation between suggestions and what philosophers call "knowledge claims." Rather there is a gradation of degrees of possibility or probability that scientists habitually assign to the theoretical statements they make. The language Krebs used to present these ideas—his abstract of the paper on uric acid synthesis stated that "the possibility" of the reaction in question "is discussed," and that hypoxanthine "may be conceived of as arising" in the way described, and his treatment of carbamino ornithine stated that the results summarized "may be explained on the assumption that the first stage" is the reaction described—is typical of the cautious phrases scientists regularly use to put forth ideas without appearing to claim too much. Whether it is wise to put forth hypotheses with this degree of uncertainty at all is, of course, a matter of individual judgment. Krebs believed it to be useful to do so, and he had sufficient self-confidence not to be concerned that he might lose credibility[125] if some of his suggestions proved wrong.

VII

Hardly had Krebs transferred his apparatus and belongings from Cambridge to Sheffield than he was confronted with the unexpected opportunity to reverse that move. One of the lecturers in the Biochemistry Department had resigned to become Master of Emmanuel College, and Hopkins asked Krebs if he were interested in the position. The exact circumstances under which Hopkins communicated this question to Krebs are uncertain. In his autobiography Krebs stated that Hopkins "wrote" a "few weeks after my move."[126] A few weeks after his move, however, would put him back in Cambridge, and allusions to the question in contemporary letters also indicate that it probably came up during this period. In any case it was during his already crowded time in Cambridge that Krebs had to decide what to do. That the choice with which he was presented must have been a difficult one is self-evident from the circumstances. The timing could not have been worse.

There must have been a great deal of interest in the British scientific world about what Krebs would do. A German doctor who happened to be a mutual friend of Hans and his sister visited Lise in Düsseldorf in early November bringing greetings from Hans, and

> He told about the extremely flattering competition for you that has broken out between Sheffield and Cambridge. He hinted at a victory for Cambridge, although the decision will not be easy, since Sheffield will presumably also have much to offer.[127]

It may well be that Hans wrote back that he had not yet made that decision, for in Lise's next letter, on December 5, she asked whether it was "Sheffield or Cambridge?"[128]. By this time he had already made up his mind: in a letter to Nachmansohn from Cambridge, he wrote on November 17:

> By the way, I shall remain in Sheffield. To leave there again after so short a time would be regarded as unfair and would influence the attitudes of these people unfavorably toward German Jews.[129]

In his autobiography Krebs did not recall such considerations, describing his decision as a comparatively simple one. "I replied [to Hopkins] that I was much attracted but Sheffield had made such a major effort to meet my wishes for equipment and space that I felt morally bound to stay for a while."[130]

It was during this same hectic November that Krebs saw his younger brother for the first time in nearly three years. In the early part of 1935 Wolf had continued to do well professionally at General Electric (AEG) in Berlin, and despite the general conditions in Germany, he and his wife appeared "cheerful and contented."[131] When a vacancy occurred in a post at AEG for which he thought he was qualified, he applied for it. The technical director of the firm told him that he would be "exactly the right person for the post, but under present conditions we cannot put you into it." Then he took Wolf aside and explained that that was what he was required to say officially. He added, "I can tell you something privately that is not official. I would suggest that you try to find something *now*, outside of Germany." Because of Wolf's fine record of invention, he said, AEG would be able to supply him with very supportive references.[132]

This experience suddenly brought home to Wolf what his family had been saying all along: that he had no future in Germany. He took the advice and began looking for openings abroad. Hans was apparently able to make a contact for him with one of the directors of the General Electric Company in Birmingham, who wrote Wolf that he would like to interview him. Wolf made the trip to England in mid-November. The circumstances for the interview, however, were not ideal. Hans had booked a room for him in London above a railway station that proved so noisy he hardly slept. The next day was very foggy, trains and taxis were delayed, and when Wolf arrived in Birmingham for his interview he was probably not at his best. Hans managed to find just enough time to meet Wolf in London on Wolf's way back.[133] Wolf found the number of things Hans had to do during his own stay in England almost incomprehensible, but he was undoubtedly disappointed by the brevity of their reunion. Afterward Lise wrote Hans in her big-sisterly fashion, "but I found it very deplorable that you had so-o-o little time for him."[134]

Among the innumerable claims on Hans's attention that left him with so little to spare for his brother were the ongoing vicissitudes of the Palestine situation. The prospects were again looking less favorable than they had in September. Weizmann had made little progress toward financing the Institute that Krebs was intended to head, but had moved instead to try to bring about a reorganization within the University in Jerusalem that could eventually open up a place for it there. As a first step he was hoping to install one of the group of scientists whom he had assembled the previous spring in an advisory post in the university, from which further planning for the institute could be developed. To Nachmansohn and Krebs, Hermann Blaschko appeared to be the ideal person to fill this position. Blaschko's knowledge of biochemistry and medicine together with his administrative experience in the refugee relief organization run by A.V. Hill made him, in their estimation, very suitable for this role. Moreover, the stipends that had been raised to support Blaschko in Cambridge were likely to end, with little prospect that he could attain a paid position there. Blaschko, however, evinced little interest in this plan. So disquieted was Krebs about the consequences of Blaschko's attitude that he wrote him, just before returning from Sheffield to teach in Cambridge, an unusually sharp letter, stating that "it would be a great and irretrievable mistake on your part, not to seize" this opportunity.[135]

Blaschko nevertheless vacillated so indecisively that by mid-November both Krebs and Nachmansohn were ready to give up on his candidacy. Sometime before this, Krebs, Nachmansohn, and Blaschko had formed an agreement with Weizmann that they would travel to Palestine over the Christmas break to assess the situation firsthand. Now, however, Krebs began to think that he did not want to go through with the trip at this time. On November 17 he wrote Nachmansohn,

> I am asking myself whether, in the present state of affairs, anything can be missed if we do not travel until Easter. As you know, I am now already away from Sheffield on leave [in order to teach in Cambridge], and it would not be very pleasant for me (and my chief) if I were to be away again so soon for several weeks. In keeping with my earlier promise, however, I can make it possible to travel, if it is for a definite reason. Now I had thought that the immediate objective of the trip is to get Blaschko into place there. As far as I can see that is the only urgent point. The rest will scarcely suffer from a delay. It was originally agreed that we would travel if the existence of the

Institute was to some extent assured, or appeared to be. Now, however, there is still nothing to be seen of this Institute, and I therefore think that Weizmann cannot be offended if we postpone the trip until Easter.[136]

Nachmansohn protested that Weizmann had already been disappointed twice by the group: not only by Blaschko's lack of enthusiasm, but also by the recent sudden withdrawal of another important potential participant, Paul Rothschild. A third setback might be catastrophic. Krebs's full participation at this point was crucial for two reasons: first, because the planning would go forward much more securely if Krebs were in on it "from the start" and second, for the psychological effect on Weizmann. "You are the leader of the group," he intoned Krebs, "you have the greatest scientific authority, and Weizmann trusts you." If Krebs were to hesitate at this critical juncture, all could be lost.[137]

Unmoved by Nachmansohn's appeal, Krebs replied at the beginning of December that "I really cannot justify traveling now." Two days later he wrote Blaschko, "My presence seems to me to be desired purely on sentimental grounds." Blaschko and Nachmansohn could take care of everything that had to be done for now in Palestine. Moreover, the term did not end in Sheffield until December 18, and he could not just put off until the last minute all the preparations for his own lectures scheduled to begin on January 16. Furthermore he had to read drafts of papers that Edson and Woods had written, and "there are several other matters besides, so that every objective viewpoint speaks in favor of remaining here."[138]

Graciously accepting Krebs's decision, Nachmansohn wrote back that he would plan a trip for Easter; but before he began to make arrangements he wanted to know if "you can be absolutely certain that you will go then."[139] Nor did Weizmann appear put off when Krebs informed him that he would not go in December. From his own newly established Daniel Sieff Research Institute in Rehovoth, Weizmann wrote Krebs on December 16 that he understood the grounds for the postponement. "I shall count on your coming here [in March] and will be very pleased to be able to receive you here in Rehovoth."[140]

Perhaps Krebs's decision to postpone the Palestine trip was influenced by some good news he had recently received. On November 14 the British Home Office returned his passport with a statement that "the condition requiring the holder to leave the United Kingdom on or before the 20th August, 1936, has been canceled. There is now no time limit on his stay in this country."[141] For the first time he was free to think of his future in England as open for more than a year at a time, and he had correspondingly less need to pursue alternatives.

VIII

Settled finally in Sheffield with no obligation to leave again for three months, Krebs was free at the end of November, for the first time since late spring, to resume the uninterrupted pace of experimentation, six days per week, that he

Figure 5-8 Hans Krebs in 1935.

regarded as normal. He turned his full attention to the technical problems that stood in the way of completing the work on uric acid synthesis. The first of these was the need to improve the method for determining uric acid quantitatively.

As mentioned above, there is no record of exactly when Krebs and Edson abandoned the effort to devise a method oriented around the enzymatic oxidation of uric acid. When they turned to a chemical catalytic method instead, they drew on a paper published in 1933 by Schuler and Reindel. Seeking "model" reactions that would duplicate the enzymatic decomposition of uric acid, Schuler and Reindel had found that by shaking uric acid with the oxidizing agent MnO_2 in alkaline solution they could obtain almost theoretical yields of allantoin, an assumed intermediate step in the metabolic pathway.[142] When Krebs took up the problem of uric acid determinations again in Sheffield, his starting point was this reaction of Schuler and Reindel. Given his opinion of their work it must have been somewhat annoying to him to have to resort to another of their results, but they had provided a lead he could not ignore.

What was for Schuler and Reindel a "model" for a metabolic reaction was for Krebs a foundation for a potential microanalytical method. As he habitually did in adapting such reactions to his needs, Krebs arranged for them to take place in manometer flasks, employing very small quantities of substrate and reagents. Where Schuler and Reindel had used 0.164 g of uric acid for the protocol reported in their paper, Krebs began with fractions of a milligram. He placed the uric acid solutions in the manometer cups, with NaOH and MnO_2 solutions in side bulbs. Following the order of Schuler and Reindel's protocol, he tipped in the NaOH first, then the MnO_2. On the first set of 10 trials carried out on November 28, he included five different quantities of uric acid, each in duplicate. After 20 minutes in the manometers, during which oxygen was taken up as expected, he acidified the solution and applied his customary urease method—an operation which assumed that the decomposition of uric acid had not stopped at allantoin, but went on to form urea quantitatively. He obtained yields representing 77 to 87 percent of the original uric acid. He noted, "good duplicates! but deficit everywhere!" As "possible errors," he considered that the uric acid may have been impure, some uric acid may have been oxidized before the MnO_2 was added, or that the stirring had stopped in some of the flasks, giving uncertain readings.[143]

The next day Krebs began varying his procedures in an effort to improve the yield. He used fresh uric acid from Hoffmann-La Roche, but still found a "large deficit." He varied the quantities of MnO_2, but obtained the same end figure, leading him to infer that there was no sidereaction independent of the catalytic action. Trying again with freshly crystallized uric acid, he attained a yield of 94 percent of the theoretical figure, but regarded the situation as "still unsatisfactory." He carried out another set of measurements, increasing the alkalinity of the solution by adding more NaOH, and obtained results ranging from 90 to 95 percent of the calculated values. This time he noted that there was "generally good agreement," but it was "uncertain where loss occurs?" He wondered if it might be that the "hydroacetylenediureido acid [an intermediate postulated by Schuler and Reindel—see above, p. 52] not *decarboxylated*?"[144]

On December 2 Krebs again varied the conditions for the uric acid determination. He allowed a "longer alkaline hydrolysis," then acidified and heated the solution in order to decompose the hydroxy-acetylene-diureido acid. The yields were still consistently less than they should be according to calculation, and he thought that there should be more acid hydrolysis to assure that the decarboxylation took place. Later that day he followed up on his suggestion, acidifying the solution for five minutes at 100°C, then adding strong alkali, "(perhaps *too* strong)" he remarked. On the page following this experiment he wrote down a list of "points for uric acid determination." These included the instructions "add acid carefully to Na_2CO_3 (outside water bath)"; "Add MnO_2 before NaOH (oxidation in alkaline solution)"; "Activity of MnO_2 depends on 'state of suspension' E.q. MnO_2 powder from British Drug is completely inactive. Settles quickly, particles are less spongy. MnO_2 suspension keeps at least a week." On the 4th he performed two more sets of experiments, presumably conforming to the points he had enumerated, summarized the results so far attained, and wrote out a "full example" of the procedures for a uric acid determination. The deficits had not been eliminated, and he could only guess at their cause. By this time, however, he must have concluded that the method was nevertheless usable, because the ratio between the measured and calculated results deviated little enough from the average of about 0.92, so this figure could be used as a constant correction factor.[145]

In one week of concentrated effort Krebs had improved a method that was "unsatisfactory" at the beginning of this period, into one with which he had confidence that "0.1—2.0 mg uric acid can be determined with reasonable accuracy."[146] At this point he probably turned the method over to Edson in Cambridge to test further for interference from other substances and for specificity, while he moved on to other problems. His work on the method for uric acid is representative of a craft skill at which Krebs was almost self-taught and at which he had by this time become highly resourceful. He seldom invented original analytical methods—his knowledge of analytical chemistry was not extensive enough for that—but by trial and error, increasingly supported by accumulating experience, he readily turned basic methods devised by others into micromethods suited to his own requirements. His ability to do so was essential to the success of his research program. In discussing the uric acid method in 1978, Krebs remarked that "in all the research I ever did the elaboration of methods took a very large part of my time, and of that of my colleagues."[147]

After finishing this work on uric acid determinations, Krebs turned briefly back to an old subject, the action of d-amino acid deaminase. In a single experiment he tested two "new substrates," δ-benzoyl ornithine and δ-benzoyl lysine, together with the corresponding unsubstituted amino acids. Then he tried out a method for determining pyruvic acid based on one published recently by Claude Fromageot and Pierre Desnuelle. The principle of the method rested on the property, specific to α-keto acids, of reducing ceric ions to cerous ions. Fromageot and Desnuelle claimed that their method was quantitative, simply and rapidly performed, and adaptable either to macro or microquantities of pyruvic acid. They determined the quantity of ceric ions reduced through titration methods. Since the reaction in question

$$CH_3 CO COOH + 2Ce^{++++} + H_2O \rightarrow CH_3COOH + 2Ce^{+++} + 2H^+ + CO_2$$

yielded CO_2, Krebs saw that the method might be adaptable to ordinary manometric measurements. Testing the idea with two concentrations of pyruvate placed in manometric cups and the ceric sulfate solutions tipped in from the side bulbs, he obtained quantities of CO_2 "of expected order." There is reason to surmise that Krebs was particularly interested at this time in having available a suitable manometric micromethod for pyruvic acid because he was already contemplating a new line of investigation oriented around that crucial metabolite. He was not yet ready to move further in this direction, however, and after this single test he fixed his attention on another analytical problem central to the ongoing uric acid investigation.[148]

The identification of hypoxanthine as an intermediate in the synthesis of uric acid remained circumstantial. Krebs and his students had shown that pigeon liver produces a substance that can be oxidized to uric acid by xanthine oxidase and that hypoxanthine added to pigeon kidney tissue increases the rate of formation of uric acid. It would be desirable to confirm their conclusion by identifying hypoxanthine directly as a product of the metabolism of pigeon liver slices. To do so the collaborators needed a suitable method for the microdetermination of hypoxanthine. That is probably what Krebs had in mind on December 6, when he began a series of experiments on "hypoxanthine-determination."[149]

In the first experiment, Krebs measured the oxygen absorbed and—utilizing his newly elaborated method—the uric acid formed by the action of xanthine oxidase on hypoxanthine in simple solution, in an extract of rat liver, and in each with p-phenylenediamine added. In all cases there was a "quick oxidation of hypoxanthine," the uptake being increased by phenylenediamine. The oxygen uptake was in all cases 40 to 50 percent greater than that theoretically required to oxidize hypoxanthine to uric acid, whereas the yield of uric acid was only 50 percent of that expected in solution and 91 percent in liver extract (a figure that was questionable because of an "uncertain" blank) with "no improvement with p-phenylenediamine." Krebs made a note to "try less liver: <u>alcohol</u> instead of p-phenylenediamine."[150]

It is difficult to sort out all of Krebs's intentions in this experiment. His choice of rat liver indicates that he wanted to compare the action of xanthine oxidase alone with its action in combination with the enzyme uricase, known to convert uric acid to allantoin. Phenylenediamine was known to increase the oxygen uptake in the oxidation of uric acid through a "coupled oxidation." All this he had already encountered in the spring of 1934 working with Hans Weil on the decomposition of uric acid (see pp. 53–58, 62–67). What is not clear is how these complicating factors would help Krebs to devise a method for determining hypoxanthine, for which it may be assumed that he sought a quantitative conversion of hypoxanthine to uric acid, which he could now determine through the new micromethod.

On Monday, December 9, Krebs repeated the experiment with the changes he had planned after the first one. Again under each condition the oxygen consumed was considerably in excess of that necessary to convert hypoxanthine to uric acid. In the solution alone, and in both solution and liver extract with alcohol, much less than the calculated quantity of uric acid formed, but in the liver extract the yield was 89 percent of the theoretical. Krebs wrote down "Satisfactory with liver extract, but further oxidation of uric acid by uricase!!! Try pigeon liver extract."[151] The only surprising thing about this conclusion is that it came as a surprise to him, and that he had not used pigeon liver extract from the start.

Krebs did shift from rat liver to pigeon liver on the next day, but he also shifted the objective of the experiments. He did not use pigeon liver extract with hypoxanthine, but pigeon liver slices in a medium containing NH_4Cl and lactate. Instead of the title "Hypoxanthine Determination" that he had given the preceding experiments, he labeled the next one "Pigeon Liver + Kidney,

Pancreas, Xanthine Oxidase." He compared the rates of formation for uric acid under six conditions: with pigeon liver, kidney, and pancreas slices alone, and with combinations of liver + kidney, liver + pancreas, and liver + xanthine oxidase. The results were impressive. As he would expect, the three tissues alone each produced little uric acid. Each combination produced much more. Comparing the quantity of uric acid formed in each combination with a "calculated sum," that is, the totals for each two tissues (or, in the last case, the enzyme) alone, Krebs found large differences:

Uric acid formed in 2 hours (μl CO_2)

	Calculated sum	Measured
Liver + kidney	21.5	56.6
Liver + pancreas	12.5	35.0
Liver + xanthine oxidase	12.7	63.0

Krebs had known since Benzinger's experiments that combinations of liver and kidney can synthesize the uric acid that neither alone can produce (Schuler and Reindel had added the pancreas). That xanthine oxidase in combination with liver could do the same trick implied that xanthine oxidase can "replace these tissues." Quite satisfied with this result, Krebs wrote down "good experiment."[152]

The question of whether this experiment confirmed the identification of the intermediate conveyed from the pigeon liver to the kidney as hypoxanthine revolved around the specificity of xanthine oxidase. That enzyme acted in vitro on both hypoxanthine and xanthine. Krebs addressed himself to this question in a "large scale experiment" carried out simultaneously with the one just described. He put 303 mg of liver in the same medium in a large flask and left it at 40°C for two hours. Then he tested the "effect of xanthine oxidase" on the medium. He measured the oxygen uptake and uric acid formed in the usual way. To distinguish whether hypoxanthine or xanthine was being oxidized to uric acid in the process, he could compare the ratio between these two quantities with those required respectively for the stoichiometry of the two reactions:

1 hypoxanthine + O_2 → 1 uric acid, O_2:uric acid = 1:1
1 xanthine + ½ O_2 → 1 uric acid, O_2:uric acid = 1:2

When he calculated the experimental ratio he found it to be

"O_2:urea [15 μl. CO_2 in urease reaction = 2 x uric acid] = 1:1.70"

Dividing by two to convert the CO_2 to its stoichiometric equivalent of uric acid (which Krebs did not do in his laboratory notebook) gives a ratio of 1:0.85, or 1.17:1. That outcome would indicate that the intermediate involved was probably hypoxanthine, but there was enough of an excess oxygen consumption so that Krebs probably suspected that there had been a coupled oxidation involved.[153]

On the same day Krebs tested in another way for the presence of the "purine," whichever one might be left in the solution remaining from the same large-scale experiment. He placed two portions of it, one with and one without xanthine oxidase (and a control with xanthine oxidase in saline solution), in Thunberg tubes with methylene blue. The tube containing the solution and the enzyme decolorized the methylene blue in 3.5 minutes, the other two did not do so in 24 hours.[154]

The distinction between using xanthine oxidase as the basis for an analytical method to determine hypoxanthine and using it in experiments designed to support the view that hypoxanthine is an intermediate in uric acid synthesis is subtle enough so that Krebs himself may not have distinguished these two objectives unambiguously in the series of experiments he had begun. If he were devising an analytical method, it would depend upon the enzymatic replication in vitro of the same reaction whose occurrence in intact tissue he was seeking to confirm. That does not mean, however, that his reasoning would have been circular. The circumstantial evidence he already possessed for the existence of the metabolic reaction gave reason to suppose that a useful analytical method could be based on it; whereas the demonstration of the enzymatic reaction with known samples of hypoxanthine could lead in turn to stronger methods for identifying the metabolic reaction. Such easy interchange between means and ends in successive experiments is commonplace in scientific investigations.

From his next moves it is evident that Krebs suspected the excess oxygen uptake in his large scale experiment to be due to a particular type of coupled oxidation. It was generally thought at the time that the oxidation of xanthine to uric acid proceeds through an intermediate reaction producing hydrogen peroxide:

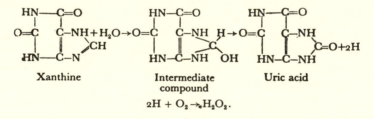

2H + O$_2$ → H$_2$O$_2$.

In the presence of the enzyme peroxidase, the hydrogen peroxide oxidizes other substrates. Hydrogen peroxide is toxic, however, and most tissues, including blood, contain an enzyme called "catalase," which removes or destroys it.[155] This reasoning lay behind the experiment that Krebs conducted on December 10 on the "determination of hypoxanthine" in pigeon blood and ground pigeon liver. The blood he regarded as a source of catalase. He placed hypoxanthine, the blood, and ground tissue in solution in manometer flasks, with controls, added xanthine oxidase from inner cups, allowed the oxidations to proceed for two hours, then compared the oxygen uptakes to the quantity necessary to oxidize the hypoxanthine present to uric acid. The hypoxanthine with ground liver and in the control solution absorbed "about calculated value" (154 and 155 μl O_2 respectively, compared to 165 μl calculated), but with both of the two concentrations of blood there was a "doubling" (308 and 321 μl). This result must have been surprising to Krebs, because he would have expected the catalase in the blood to eliminate a coupled oxidation if it were due to hydrogen peroxide. It was known, however, that hematin compounds catalyze the oxidation of some compounds by hydrogen peroxide, and he must have suspected that such a factor could be involved, for he added at the bottom of the page a calculation of the quantities of hemoglobin contained in the blood used.[156]

Pursuing the questions raised in this experiment, Krebs repeated it the next day using the same blood in various dilutions, perhaps in order to reduce a suspected hemoglobin effect. This time he found that, even in the one run which contained the same quantity of blood as the preceding experiment, there was "less additional uptake . . . than yesterday!"—"not doubling, but [only] 66% more." The next day he tested the effects of changing the concentration of the enzyme (xanthine oxidase), using more blood and boiling the blood, the latter presumably to destroy the catalase. He measured both the oxygen uptake and the uric acid formed. The oxygen uptakes were again in excess of the theoretical value. For uric acid, however, there was "most incomplete yield, the more enzyme the worse: blood increases yield." (The values ranged between 25 and 75 percent of the calculated figure.) Trying a combination of liver extract and blood (with controls including each alone), he measured only the oxygen uptakes and found that "liver + blood → final value less than 2 O_2." (That is, the oxygen uptake was 228 μl, compared to 318 μl for blood alone, and 156 μl for liver alone, the calculated quantity for a ratio of $2O_2$:1 hypoxanthine being 330 μl.) Apparently the presence of the liver somewhat inhibited a coupled oxidation due to the blood.[157]

In order to sort out these complications, Krebs performed another similar experiment on December 12, including liver, blood, the combination, and the enzyme in saline solution. He summarized the results as follows:

1. with enzyme [no blood or liver] theoretical O_2 consumption, a little more [185.5 μl, theoretical 1 O_2 = 169 μl] ≈50% uric acid formed [158.5 μl CO_2 in uricase test, compared to 330 theoretical].

2. + liver about theoretical O_2 ($1O_2$) [164 μl]: 85% of uric acid [278 μl CO_2].

3. blood doubling of O_2 [316 μl] but only 50% uric acid!!! [156 μl CO_2]

4. liver + blood almost = no blood. [204 μl O_2, 165 μl CO_2].

These results reinforced the patterns visible in the preceding experiments, and prompted him to try to give an explanation for what was happening:

H_2O_2 seems to oxidize hypoxanthine! Therefore theoretical O_2 does not mean theoretical reaction; since each H_2O_2 may destroy 1 hypoxanthine (peroxidative reaction?).   No literature found on oxidation of hypoxanthine by H_2O_2.

That is, the fact that in the case of the enzyme acting on hypoxanthine without tissue extract the oxygen consumed was in theoretical quantity for the oxidation to uric acid, but only half the theoretical quantity of uric acid formed, suggested that the other half of the O_2 absorbed might be taken up in another oxidative reaction, catalyzed by H_2O_2, whose stoichiometric ratio of O_2:hypoxanthine was the same as that for the hypoxanthine → uric acid reaction. This was, for Krebs clearly not a welcome possibility, because it threatened to interfere with his use of the stoichiometry of the hypoxanthine → uric acid reaction to identify hypoxanthine, whether as an analytic method in general or as the intermediary in uric acid synthesis. Having found no pertinent information in the literature, he could only seek further clues by varying the conditions in his own experiments. "Try," he wrote down

1) addition of <u>new</u> enzyme (is enzyme destroyed?)
2) <u>less</u> enzyme + <u>liver</u> (<u>less</u> liver)[158]

These suggestions implied a possible alternative to the explanation he had just given: that H_2O_2 reduces the yield of uric acid by destroying the xanthine oxidase rather than by catalyzing a coupled oxidation (leaving unexplained, however, the cause of the rest of the oxygen uptake). If so, one might obviate this effect by replacing the enzyme. The second point was based on the observation that in this experiment the presence of liver partially restored the theoretical uric acid yield (and perhaps, although Krebs did not indicate it, did so by somehow protecting the enzyme).

On December 14 Krebs repeated the experiment again, putting both of his suggestions to the test. In two runs he used half as much and one-tenth as much liver extract as he had employed previously, and in place of the 0.5 μl of xanthine oxidase solution utilized previously, he placed 0.2 μl in each of the two side bulbs. He tipped in the contents of one side bulb at the beginning of the run and added the "second lot of enzyme" after 80 minutes. He included also a run with washed and lysed pigeon red blood cells in saline solution and a similar one containing human red cells, but determined the uric acid only for the two liver extract experiments. In each of these the results for both O_2 and uric acid were close to theoretical (O_2 = 313 μl and 322 μl, theoretical 326; urease CO_2 = 170 μl and 166 μl, theoretical 165). He put down "good agreement with liver." In the solution prepared from pigeon blood, the oxygen consumption (306 μl) was nearly double the theoretical figure, while in that from human blood it was only a little in excess (180 μl). He did not comment on this aspect of the experiment, but he graphed the results with calculated levels for $1O_2$ and $2O_2$ marked in. His graph showed clearly that the second quantity of enzyme added to the liver extract increased the rate of O_2 consumption that was otherwise tapering off, and elevated the total to about the "$1O_2$ calculated" line. This outcome decided for him between the two explanations he had posed for his results two days before. He turned back to the page on which he had suggested that H_2O_2 oxidizes hypoxanthine and wrote above it: "Wrong, it destroys enzyme."[159]

On the same day Krebs conducted an experiment on the "coupled oxidation of amino acid oxidase," using alanine as a substrate, and in pigeon blood found "except for the first 20' exactly doubling!!"[160] It seems evident that his objective was to explore whether the doubling of the oxygen consumption he had just observed in the hypoxanthine reaction in the presence of a pigeon blood preparation was a manifestation of a more general coupled oxidation reaction attributable to some property of that type of blood.

On December 16 Krebs again carried out a "determination of hypoxanthine," similar to the series he had already identified with a title implying an objective no longer strictly applicable to the questions he was pursuing. He again used his standard liver extract in diluted form (5x diluted) and tested particularly the effects of "various quantities" of hypoxanthine. He also tested the effects of the "presence of blood," with and without liver. He completed the uric acid determinations only for the liver experiments, and obtained:

Vol. hypoxanthine sol. (30.6 mg/100 μl)	0.5	1.0	2.0	3.0
O_2 uptake (corrected for blank)	25.7	49.3	101	150
Calculated O_2	25.2	50.4	100.8	151.2
Uric acid (urease CO_2, with correction factor)	51.6	104	199	301
Calculated urease CO_2	50.4	100.8	201.6	302.4
Uric acid measured: calculated	+2%	+3%	-1%	-0.5%

He wrote down "excellent results for O_2 and uric acid with liver. Blood less satisfactory."[161]

Because the higher oxygen uptakes occurring in the experiments with blood seemed to be due to a coupled oxidation that was not occurring in the liver and that had also appeared in previous experiments to be reduced in combinations of liver extract with blood, Krebs pursued further the question of the "effect of liver in the presence of blood." Adding four different quantities of his liver suspension in geometric proportions, he observed a striking effect:

0.00 [μl] liver [suspension]	73% [O_2 uptake]
0.02	46%
0.04	23%
0.08	2%
0.16	3%

He concluded "0.08 liver is necessary! liver[:blood] 1:3." However clear the effect was, it did not make sense to him: "Liver prevents thus coupled oxidation!!" he wrote in puzzlement. "Why? Catalase is in blood and should prevent it." Noting that even without liver the excess was only 73 percent, he attributed that to "probably not enough blood." Thus, convinced that the cause of the excess uptake must be a coupled oxidation due to hydrogen peroxide, he found himself in a theoretical dilemma. His next comment "111 μl O_2 excess, how much Hb?"[162] indicates that he thought that the way out must somehow be bound up with the catalytic effects of hematin compounds.

The next day Krebs began to broaden the inquiry into this question by examining the effects of methemoglobin on "coupled oxidation with *dl*-alanine." Employing the hematin compound in two concentrations (and a blank), he obtained O_2 uptakes, 20 and 51 percent respectively, in excess of that required for the oxidative deamination reaction (compared to 2 percent for the blank). He wrote down, "excess greater, the more methemoglobin present," and made a calculation of the approximate ratio between added methemoglobin and extra oxygen consumption.[163] The lead was therefore not unpromising.

This series of experiments, begun 11 days before in order to "determine hypoxanthine," displays very well the step-by-step way in which complications arising in an investigation of one problem can envelop the investigator until he finds himself engaged in another line of investigation. Krebs had clearly by December 16 taken the first step in a study of the causes of the coupled oxidation reaction that had forced its way into his purview, but he did not take the second step. Easily diverted as he could sometimes be, this time he quickly rejected the temptation to branch out along another research trail.

* * *

In August 1978, I went over with Krebs some of the pages in his laboratory notebooks containing the experiments just described. He did not recall at a glance the meaning and context of the activity whose traces are preserved there. When we looked at the first of the experiments employing pigeon liver extract, he remarked, "I don't know what we attempted to do there."[164] As we proceeded, however, he became more definite. It is impossible to separate sharply the extent to which he was interpreting the experimental record in front of us from the extent to which it was reminding him of what he had been doing at the time he kept this record. By the time we reached the statement "Liver prevents thus coupled oxidation. Why? Catalase is in blood and should prevent it," he offered an expansive commentary:

HAK: It may either be a matter of the amount of catalase. . . . The excess oxygen indicates that the hydrogen peroxide was not decomposed, but reacted probably by using that oxygen for peroxidase; but this here produced then the right theoretical value.

FLH: Well, in the absence of the liver extract the excess was very large; but there wasn't any explanation for that, was there?

HAK: Well, I may have been satisfied with the assumption that there was enough catalase in the liver to do this, while there may not have been enough in the early samples with blood. Is it preceded by experiments with blood?

FLH: There are some, yes.

HAK: But I didn't want to be diverted from the main object here. That was merely to get a method for analysis. The object was not here a fundamental study of what was going on there, but . . . getting the stoichiometry. If one followed up every interesting observation one would be sidetracked from the main aim. And I think one has to be very careful. There are always interesting things to follow up.

When I said that one must make a judgment about when to follow a sidetrack, he acquiesced:

HAK: It is, of course, extremely important in connection with serendipity. There may be unexpected observations which don't fit in or don't lie on the main route, but which nevertheless can be of outstanding importance. So it is a matter of judgment.[165]

In this instance, as we looked back at such a judgment, hindsight did not suggest to Krebs that he had missed anything by sticking to the main route.

* * *

The main route for Krebs in December 1935 was to try to unravel the metabolic synthesis of hypoxanthine. Already on the 12th, in the midst of his efforts to develop a reliable analytical method to determine hypoxanthine, he interposed an experiment on the subject of its formation. The title of the experiment, "Pigeon Liver. Hypoxanthine Formation," implies that even though he still sought further confirmation that hypoxanthine was an intermediate in the synthesis of uric acid, he had no doubt that it really was.[166]

For this experiment Krebs employed pigeon liver slices in saline solution including NH_4CO_3. He was testing mainly the effects of lactate, pyruvate, and glucose on the rate of formation of hypoxanthine. The design indicates that he was following up the ideas he had expressed in his talk at the Biochemical Society meeting in November about the source of the carbon skeleton for hypoxanthine. To measure the rates of formation he had to rely on the same problematic analytical method with xanthine oxidase that he was still seeking to perfect. Since he was still assessing the comparative effects of blood and liver extract as a means to eliminate the interfering coupled oxidations, he included double runs, one with each of these added, for each of the substrates tested. He measured the O_2 uptake and uric acid formed in 90 minutes. The uric acid determinations failed. There were only small pressures, and he commented that it was a "rather dirty mess." There was a "distinct increase [in O_2 uptake] with lactate." The O_2 uptake was "in some cases higher" with blood, but the results were "not reliable enough. So far only qualitative value."[167]

By the time he returned to the formation of hypoxanthine in pigeon liver six days later, on December 18, Krebs had worked around the difficulties in his method for determining hypoxanthine and could apply it with confidence. He tested lactate and pyruvate as before, but now included for the first time arginine and ornithine. They represented a very different point of view from the one he had been pursuing concerning the origins of the hypoxanthine molecule. It was, as Krebs recalled in 1978, "just feasible that ornithine would be involved somehow, because the link N-C-N is in urea and is formed through the ornithine cycle. We thought it might be involved."[168]

Now that he was able to regard his analytical method as safe, he could simplify the procedure. He merely added liver extract to each cup along with xanthine oxidase, in order to suppress the coupled oxidation, measured the oxygen uptake, and calculated directly from it a set of "$Q_{hypoxanthine}$"s. The results, however, were not striking. NH_4Cl by itself gave a "slight inhibition." Lactate and pyruvate caused an increase, merely confirming what he had already known, and there was "no effect of ornithine." That was enough to "exclude that ornithine plays any . . . role" analogous to its role in the mammalian synthesis of urea.[169]

On the 19th Krebs performed another similar experiment, using liver slices from a starved pigeon, and obtained this time an "effect of pyruvate, not of lactate or glucose." He also tested the effects of KCN and octyl alcohol (together with lactate). Each of them, he found, "reduces additional formation when NH_3 is added." The next day Krebs pursued further the effects of octyl alcohol, along with that of temperature, and found that there is "still hypoxanthine formed, but not from NH_3." Since less was formed at low temperature than at 40°C, he inferred that "most of the hypoxanthine is not washed out, but formed." At the same time he tested the effects of octyl alcohol with and without NH_4Cl on the formation of hypoxanthine in a large-scale experiment. The results were "not very good altogether."[170]

On that negative note Krebs ended his research for the holiday break. Afterward he did not return to these problems. Perhaps he intended to but gave other investigations higher priority when he resumed his work. There is reason to surmise, however, that either by the time he stopped or while he was away from the bench he decided that there was no point in pushing further his attempt to work out the reaction steps for the synthesis of hypoxanthine. He could have had no way to know how far away he actually stood from an understanding of these intricate reactions, but he must have sensed intuitively that he could not reach such a goal soon. A few months later the paper that he, Edson, and Model published on the subject of uric acid synthesis stated:

> We added [various substrates] . . . expecting to obtain information about the source of the carbon skeleton of uric acid. Although lactic and pyruvic acids have a more marked effect than other substrates especially in the starved animal, it seems premature in view of the complex nature of the systems involved to draw definite conclusions from this fact.[171]

It is a "matter of judgment," not only when to desist taking a sidetrack, but how long to stick to the main route itself. We can see here, as in earlier episodes, that one of the ways in which Krebs maintained his effectiveness as an investigator was that he avoided getting bogged down. Unflagging though he was in the pursuit of experimentation in general, he also recognized instinctively when it was time to give up on a particular question.

IX

Among the "several other matters" weighing against a trip to Palestine in December to which Krebs alluded vaguely in his letter to Blaschko was one personal affair. Now that his professional position in England seemed assured, it was time to face the question of what to do about Katrina. Early in December he received a letter which again reflected her sadness when she had not had a letter from him recently. She was not "angry," for she assumed that the gap this time was due to the "much much work" he had to do during his month in Cambridge, but "it is not so simple to maintain contact for a long time through letters alone." At the end of her letter she asked, "What will become of us?"[172] Soon afterward Hans wrote her to report that he would neither travel to Palestine nor return to Cambridge, and he invited her to come to England for a few days during the holidays.

Hans's letter put Katrina into a state of excitement. At first she thought it unrealistic of him to imagine she could make arrangements for such a trip on such short notice. Despite her doubts, she wrote him on December 12, she had gone immediately to the travel bureau and found that she could book a steamship to arrive in London on December 28. Before going ahead she wanted to know what

> you think about your immediate future, and what role our conversa-
> tions of last summer play in it. You know my position quite well, and
> you know also that I am perhaps readier than you to risk breaking
> away from here.[173]

His reply must have satisfied her, for she wrote again on December 15, "I will therefore come." She would be able to get away for six days in all, leaving four days in England, and she asked whether they would stay in London or go to Sheffield. On Christmas day she wrote that she had received the tickets from him and would be arriving in London at 8:38 A.M. on Sunday the 28th. She sent photos so that he could get used to her appearance again, and described what she would be wearing.[174]

There is no record of what happened when Hans and Katrina met in London. We can only infer from later events that they did not decide to join together and from the absence of further letters that Katrina's either-or position meant that the alternative before them was to end the contacts between them. It would be possible to construct various scenarios to fill the silence that surrounds that outcome, but such events are best left to the reader's imagination.

6

The "Great Work"

At the time Hans Krebs accepted his appointment at Sheffield, he had applied successfully to the Rockefeller Foundation for an annual grant of £200. With this fund, together with the £100 that Edward Wayne had obtained for him, Krebs was able to equip his laboratory amply and to hire a technician. On January 2, 1936, Leonard Eggleston entered the Department of Pharmacology as an assistant to Krebs. A shy young boy of 16, Eggleston had left school in 1934 after completing intermediate school, but he attended evening classes at a commercial college and technical school in Sheffield to further his education in the basic sciences. He had been working for two years as an assistant in the Physics Department of the university, but had become bored with his routine tasks, which included making minor repairs on laboratory apparatus and charging batteries. Despite Eggleston's youth and inexperience, Krebs found him to be exceptionally reliable, conscientious, and loyal. Eggleston learned quickly to carry out experimental operations with the care and precision that Krebs required. He could think independently, and soon he even began to spot occasional errors that Krebs made in the calculations he performed on experimental data.[1]

During the first 10 working days of January, Krebs made only one entry in his laboratory notebook. If he was not still preoccupied with personal matters during this period, he may have spent much of his time preparing the lectures on biochemistry that he planned to begin on January 16. He was to give just one lecture per week, with no laboratory instruction, and the course was introductory. According to his later recollection, however, he organized as much of his course as possible in advance, a task which might have kept him quite busy during these two weeks.[2]

On January 12, Frederick Gowland Hopkins once again reopened the question of whether Krebs might be induced to return to Cambridge. "The filling of our vacant lectureship," he wrote, "is still worrying me to death."

Suppose we were to decide to get through this year without making the appointment and were able, next October, to offer you a post to be called 'Lecturer and Director of Research' with a salary about double that of a Lecturer, would you feel that the year's service you will then have given to Sheffield will justify you in leaving them?

It is a shame to disturb your mind again. Don't trouble to write more than just a word—just yes or no—if you feel that a decision is possible for you at this time.[3]

Coming just as Krebs was beginning to feel settled at Sheffield, this letter inevitably *was* disturbing. From the distance of 35 years he described it as having "caused me much soul-searching." He discussed it with Wayne, who consulted the dean of the faculty and assured Krebs that they would not regard it as unfair if he chose Cambridge. That left it to him to resolve the difficult question of whether Cambridge or Sheffield really offered him more. Of the factors that had earlier prompted him to move to Sheffield there remained now only the greater amount of laboratory space available to him there. "On the other hand," as he recalled in his autobiography, "to a biochemist the Cambridge environment was certainly much more stimulating and inspiring; in fact there was no other biochemist in the Sheffield area."[4] In a conversation with me he put the matter more bluntly: "I sensed [that] in my narrow field I was in an intellectual desert."[5]

While raising no obstacles to Krebs's possible departure, Wayne and the Dean held out an inducement to stay, by increasing his salary from £500 to £600, and by suggesting that a Department of Biochemistry might be established at Sheffield. Krebs's decision was, as he much later put it in muted tones, "not easily reached."[6] His quandary is reflected in a surviving preliminary draft of a letter he wrote to Hopkins on January 19:

Dear Sir Frederick

I am very glad indeed to learn from your letter that there is still some hope for my return to Cambridge. I have discussed this question with Prof. Wayne and he tells me, after having consulted with the Dean of the Faculty, Prof. Clark, that I may certainly feel justified to leave Sheffield by next October, if the post in Cambridge is a better one than the post in Sheffield.

There is, however, some difficulty in defining what is a better post. The points which you mentioned, a post to be called "Lecturer and Director of Research" with a salary about double that of a lecturer, would definitely suffice to make the position in Cambridge preferable to my present post. A complication, however, arises from the possibility of an improvement in my position here. I have tried to find out something definite about the prospect here (and this is the reason for the delay of my reply) but it was not possible to obtain any reasonably definite information. But I feel that I should mention that there is a prospect of an independent department . . . although this is only a possibility and perhaps not very probable.[7]

When he struggled over this draft, Krebs had obviously not yet been able to make up his mind what to do. It is not clear how long it took him to arrive at a decision. In his autobiography he stated that "the amount of space at my disposal, coupled with the fact that I was very happy there and loved its beautiful surroundings with the Peak District touching the edge of the city, made me decide to stay."[8]

The draft letter corroborates Krebs's retrospective description of some of the considerations that he weighed, but we can surmise that the situation was more complicated than he remembered it so much later. The idea must have entered his mind that if he were to come back to Cambridge he might be well placed to become Hopkins's eventual successor. Hopkins himself hinted at this prospect by commenting in his letter, "I feel the responsibility very greatly, because the appointment will long outlast my own stay in the Department."[9] Krebs apparently did not explore what the additional title "Director of Research" implied, but it is plausible that Hopkins anticipated entrusting Krebs with a broad responsibility that he had for many years exercised personally. As Hopkins approached his 75th birthday the question of who would be the next head of the Cambridge Biochemistry Department was by no means remote. Later in the year Krebs's father heard that Hopkins had told Thannhauser in Boston, "If I could determine my successor, it would be Krebs."[10] There must have been at least rumors to that effect circulating at the time Krebs had to make his decision. He might well have discounted them, but is unlikely to have been totally indifferent to the possibility of some day heading the laboratory he so admired. On the other hand, his esteem for the still vigorous present leader of the Cambridge school might well have inhibited him from giving much thought to calculations about the situation there after Hopkins.

By conventional standards of professional ambition, Krebs appears to have made an unlikely choice. On the surface Cambridge offered him far more scope for his career, for it would place him at the institutional center of his chosen field in his adopted land. During his two years in Cambridge he had already made a major impact there, and in the position Hopkins hoped to obtain for him his influence would be expected to grow further. Nor does the space factor appear from a distance to have been compelling. Hopkins was so anxious to bring Krebs back that it ought not to have been difficult for Krebs to negotiate for the facilities he would need. If we believe that people often construct reasons for what they want to do, then we may suspect that Krebs's general happiness in his new surroundings at Sheffield was more decisive than the spacious laboratory facilities he possessed, or the "possibility" of a future department. He may also have come to feel a loyalty to his warm and generous new "chief" that counterbalanced his gratitude to Hopkins and the stimulation of Cambridge. Whatever the subtler motivations underlying his decision may have been, the fact that Krebs felt free to choose a provincial university in which he was the sole biochemist over a prestigious university at the center of his field is a measure of the self-confidence he had attained. It is also illuminating about his personal values. Hardly an unambitious man, Krebs was nevertheless secure enough to ground his ambitions in his own inner drive and accomplishments.

He had no need to aggrandize his reputation and his opportunities through the institutions with which he associated himself. Moreover, as an emigré, he was not imbued with the English attitude that Oxford and Cambridge overshadowed all other universities. In the German university system it was not uncommon to build one's reputation at a smaller institution, and only later be called to one of the leading universities. He was not troubled when someone remarked to him, "You've gone out into the wilds."[11]

For Edward Wayne, too, there must have been some cause for soul-searching. Having done his part to keep Krebs in England at substantial sacrifice to his personal research interests, Wayne might well have been excused if he had come to the conclusion that under the new circumstances he need not go out of his way to retain Krebs in his department. On the other hand, Krebs was already an ornament to his small university. Moreover, both Wayne and his wife had quickly become fond of the attractive foreigner who had come into their circle. Although Wayne was never quite sure why Krebs did not go back, he did not regret that he stayed.[12]

Krebs has written that Wayne "gave me as much—or rather more—space in his laboratory than he could afford."[13] Trained as a pharmacologist, Wayne had research interests of his own, but had little time to pursue them, because he had assumed responsibility for a large clinical practice that absorbed most of his energy. Wayne knew enough organic and biochemistry to understand what Krebs was doing, but found that Krebs seldom discussed the details of his research with him, even though they lunched together regularly. Wayne nevertheless helped Krebs in many ways beyond the institutional support he gave him. In particular he went over the drafts of Krebs's scientific papers, improving their style and providing welcome critical judgment.[14]

Edward and Nan Wayne also provided the principal social orientation in Krebs's life during his first year in Sheffield. They invited him to dinner nearly every week. Like others who knew how to penetrate his outward reticence and win his trust, they found him approachable and engaging. Avid conversationalists themselves, the Waynes were probably better than most at drawing him into animated discussions. He seldom talked about science at their home, but was interested in a wide range of other subjects. He liked particularly to talk about literature, and often asked Nan's advice on books to read. They learned that his favorite German author was Thomas Mann. Krebs and Nan Wayne developed a warm rapport, and he sought her help in correcting his English grammar. As a Latin teacher Nan had a critical sense for the structure of language, and taught him also to improve his idiomatic English. He taught her a little German.[15]

I

The first occasion that Edward Wayne may have had to assist Krebs with his scientific writing was in the completion of a review article (mentioned in the previous chapter) on the "Metabolism of Amino Acids and Related Substances" that Krebs submitted to *Annual Review of Biochemistry* on January 17. For some of the topics covered, especially deamination, urea synthesis, and uric acid

Figure 6-1 Edward and Nan Wayne at the time of their wedding in 1934.

synthesis, Krebs relied heavily on his own investigations. He presented
prominently his hypothesis that δ-carbamino ornithine is an intermediate in the
ornithine cycle (see p. 195). For subjects such as the special metabolism of
some individual amino acids, amines, inborn errors, and clinical aspects, he
merely summarized the most recent research literature. His treatment of the
synthesis of amino acids began, as previously discussed, with the recent
experiments of Neber showing that rat liver forms amino-nitrogen when
ammonium pyruvate is added. Reporting that he had confirmed this result, but
found that "no amino nitrogen formed when pyruvate is replaced by other
ketonic acids, e.g., ketoglutaric acid," he added:

This lends support to the idea proposed by Knoop long ago that pyruvic acid may play a special role in the synthesis of amino acids in such a way that condensation between ketonic acid, ammonia, and pyruvic acid is the primary step, leading to an acetylamino acid according to the following scheme:[16]

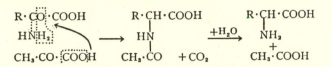

The scheme that Krebs here reproduced was based on one that had first appeared in 1910 in a paper by Franz Knoop on "The Physiological Breakdown of Acids and the Synthesis of an Amino Acid in the Animal Body." Following the same strategy that had previously led him to the β-oxidation theory for fatty acids, Knoop and a medical student named Ernst Kertess had fed a dog a phenyl-substituted amino acid—in this case, γ-phenyl-α-amino butyric acid—on the assumption that the benzene ring would hinder the complete metabolic decomposition of the molecule and permit them to recover intermediate products from the urine. Finding the levorotatory isomer of the acid unchanged, they concluded that the organism did not act "symmetrically" on the two optical isomers. They obtained in addition a substance extractable in ether that proved to be acetyl-phenyl-amino-butyric acid. Drawing, in his customary fashion, on chemical analogies to extend the significance of this result, Knoop argued that acetylation might be a more general reaction of amino acids. When they obtained the same product after feeding the corresponding α-keto acid, Knoop inferred that the acetylated amino acid might be an intermediate in a reversible reaction linking amino and ketonic acids.[17] These results "reminded" him of a reaction produced in vitro by W.K. deJong, in which

α-ketonic acids combine with one another when ammonium carbonate is added in an aqueous solution, in such a way that one molecule is reduced to an amino acid with the attachment of NH_3, at the cost of the other, which is oxidized with the loss of CO_2 to the next lower acid, and then with the loss of water is attached [to the other molecule] as an acyl residue. Pyruvic acid, for example, yields acetylalanine.[18]

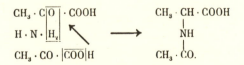

These reactions had "physiological interest," according to Knoop, because they suggested a way in which animals might synthesize amino acids through molecular rearrangements and coupled oxidations and reductions analogous to the aldehyde dismutation reaction recently discovered by Parnas.[19]

In 1925 Knoop gave up this hypothesis when he and another student found in further feeding experiments that the acetylated product in the urine was the opposite optical isomer to that predicted by the reaction scheme.[20] In 1935, however, Vincent du Vigneaud and Oliver Irish attempted to revive Knoop's theory. Knoop had abandoned it, they claimed, only because of "a misconception with regard to the spatial configuration of the acetylphenylalanine excreted after the feeding of acetyl-*dl*-phenylalanine, and that actually the data were in support of the theory." In an abstract of a paper to be presented at the annual meeting of the American Society of Biological Chemists that April, du Vigneaud and Irish gave a brief summary of further supporting evidence they had obtained by feeding *l*-phenylaminobutyric acid.[21]

Succinctly summarizing these developments in his review article, Krebs judged that "the fact that the formation of the amino acid does not require the presence of oxygen" constituted "strong support for the scheme." He saw further favorable evidence in "the fact that the formation of acetic acid, which must be a by-product according to [the] scheme [see the right side of the scheme reproduced above from his article] has actually been demonstrated by the work of Annau and Edson on acetoacetic acid formation in the presence of ammonium pyruvate. Acetic acid would be expected to appear under the given conditions as acetoacetic acid."[22] Here Krebs referred to the experiments published by Erno Annau in June 1934 that had stimulated him to take up, and then turn over to Norman Edson, the investigation of the effects of ammonia and pyruvate on ketogenesis (see pp. 80–81). The supporting evidence that Krebs could adduce was no more than suggestive. At the time he finished his review article he was, in fact, preparing to *begin* an investigation of Knoop's scheme for the synthesis of amino acids. If, as appears likely, it was through writing this review, or examining the literature in order to do so, that Krebs perceived the opportunity for a new research venture of his own, that was not a unique experience for him; as we have seen, reviews he wrote for Oppenheimer's *Handbuch* early in 1933 had similarly helped to launch him on his first investigations of fatty acid and carbohydrate metabolism (see Vol. I, pp. 397–404).

On January 6 Krebs recorded preparing *d*-acetylglutamic acid according to the procedures of Max Bergmann and Leonidas Zervas and *dl*-acetylalanine according to deJong. The latter procedure, based on the deJong reaction, he remarked, "works very well."[23] One week later he was ready to use the acetylalanine preparation to begin the experiments. His initial approach was straightforward. Using rat tissue slices (probably liver, although his protocol omits that detail), he examined the "splitting of acetyl-alanine (*dl*)," measuring comparatively, by the Van Slyke method, the amino-N formed in the presence and absence of the acetylalanine. From the reaction scheme contained in his review article we can see that the experiment was intended to test whether the second step—the hydrolysis of acetylalanine to form alanine and acetic acid—was taking place, for the appearance of extra amino-N could be interpreted as due to the amino group freed in the "splitting." He recorded no results of this first experiment. The next day he expanded on it, testing the effects of pyruvate and lactate on the process. The result was only a "small effect [of acetylalanine, and] small increase of deacetylation through pyruvate" ($Q_{amino-N}$:control, 3.56; + acetylalanine, 4.05; + pyruvate alone, 3.54; + acetylalanine and pyruvate,

4.83; + acetylalanine and lactate, 3.43).[24] These first results therefore provided no strong confirmation of the scheme, but they encouraged Krebs to continue seeking conditions that would yield more marked effects.

The following day Krebs extended the experiment further, testing both acetylalanine and acetylglutamic acid, with and without pyruvate. He tripled the concentration of the acetylamino acids over what he had previously employed. At the same time he tested a combination of pyruvate and NH_4Cl. The results were:

	$Q_{amino-N}$		$Q_{amino-N}$
Control	2.96		
+ Pyruvate	3.80	+ Acetylglutamic acid	3.50
+ Acetylalanine	5.70	+ Acetylglutamic acid and pyruvate	4.11
+ Acetylalanine and pyruvate	5.63	+ NH_4Cl and pyruvate	10.1

From the perspective of the reaction scheme in question this was a very mixed outcome. Krebs wrote down, "Some hydrolysis of acetylalanine; but more amino N formed from pyruvate + NH_4Cl; little hydrolysis of acetylglutamic acid!; no clear effect of pyruvate on deacetylation."[25] The negative result for acetylglutamic acid would suggest that, at best, the Knoop theory was not generalizable to all physiologically significant amino acids. Even with respect to acetylalanine the signs were equivocal. Although there was enough extra amino-N formed to show that "some hydrolysis" of the compound took place, the fact that it yielded much less than did pyruvate + NH_4Cl was clearly unfavorable to the view that the acetylalanine was an intermediate in an overall reaction leading from pyruvate and ammonia to alanine. That theory was, however, not a casual idea that Krebs was ready to drop at the first unsupportive turn. There were other conditions and other tissues to try.

In a similar experiment on January 17 with liver tissue taken from a well-fed rat, Krebs again compared the amino-N formed with acetylalanine to that formed from NH_4Cl + pyruvate. He omitted acetylglutamic acid, but added NH_4Cl + ketoglutarate, evidently testing the counterpart of the effect observed previously with NH_4Cl and pyruvate: that is, the possible formation of glutamic acid. The ketoglutarate result was essentially negative. Otherwise he mainly confirmed the mixed outcome of the previous experiment: "hydrolysis of acetylalanine certain, but slower than amino-formation with NH_3 + pyruvate." In a simultaneous test of acetylalanine alone in rat kidney tissue, however, he obtained a strikingly different result. $Q_{amino-N}$ was 24.4, compared to 4.50 for the control. "Enormous splitting in kidney," he commented, and added "try less acetylalanine, O_2 uptake, NH_3 formation (deamination would show splitting of d(-) alanine."[26] It appeared that the investigation had yielded its first strong lead to exploit.

Krebs first pursued his opening in a different direction from that suggested in his immediate comment on the experiment. He utilized the difference between the behavior of kidney and liver tissue to see whether acetylglutamic acid too might be split in the kidney, in this case from a rabbit. He added a twist, however, testing not for the formation of amino-N, but of amide-N in the presence of ammonia. He asked the question whether in rabbit kidney

>   ketoglutaric acid → glutamine?
>   acetylglutamic acid → glutamine?

Resting on his discovery of the synthesis of glutamine, the questions implied in both cases that if glutamine did form, glutamic acid, would be an intermediate. In addition to the amide-N he measured the change in NH_3. The outcome appeared decisive: "Amide-N formed from acetylglutamic acid [$Q_{amide-N}$ = 10.6, control, 0.44]. Quick hydrolysis of acetylglutamic acid. No glutamine formed from ketoglutarate, but NH_3 absorbed!!" Interesting though the last of these results appeared, in the context of his current investigation the first two were more central, for he now had encouraging evidence that both of the acetylamino acids he was testing behaved in kidney tissue in accord with the Knoop theory. A simultaneous similar experiment with rabbit brain tissue also yielded apparently favorable results, but they were questionable because he had omitted a control.[27]

Returning to acetylalanine, Krebs moved to strengthen his evidence for its hydrolysis in rat kidney tissue, at the same time beginning to test other tissues. On January 22 he measured the formation of amino-N in rat kidney and intestine, with and without acetylalanine. Not much happened in the intestine. For the kidney he ran three experiments extending respectively over one, two, and four hours. The $Q_{amino-N}$ values showed there was "strong splitting" (which was further increased in another experiment with octyl alcohol added). The long duration of the experiment was intended to permit the reaction to come to an end so that he could ascertain whether the tissue acted upon only one of the optical isomers of the racemic acetylalanine he employed. The total after four hours, 668 μl, was suggestively near to half of the 1308 μl equivalent to the initial acetylalanine, but Krebs did not think the result conclusive. There was "not enough tissue to prove that only half the substrate is split." Measurements of the NH_3 after two hours in a control and with acetylalanine showed that there was "no NH_3 formed from acetylalanine." The next day he carried out a similar experiment in a kidney extract and measured, after five hours, a change of 595 μl amino-N, compared to a calculated corrected value of "594 μl for 1/2 (natural form)." That close agreement convinced him that there was "half splitting"; that is, that the tissue extract was hydrolyzing only one of the two optical isomers, as would be expected of a physiological process. Finally, an experiment on "respiration with d-l acetylalanine" showed that its addition to rat liver or kidney slices had no effect on the rate of oxidation. At the end of this experiment Krebs concluded, "This proves, together with the non-formation of NH_3, that d-alanine-acetate [acetyl alanine] is split (natural stuff)."[28] That is, the two negative results eliminated the possibility that a deamination reaction, which would be oxidative and release NH_3, was occurring, thus assuring that the amino-N measurements provided a true measure of the amino-nitrogen formed in the hydrolysis of acetylalanine. Krebs's comment was not a simple summary of his results, but a strong statement. It implies that he had proven the existence in rat kidney tissue of the second step in Knoop's theoretical scheme for the synthesis of an amino acid. Accordingly, he now turned his attention to the first step, the "deJong reaction" through which Knoop had suggested that two molecules of pyruvic acid condense with one molecule of ammonia to produce acetylalanine.

For reasons that are not immediately obvious, Krebs employed rat liver slices to examine the first stage of the scheme, even though he had been able to demonstrate the second stage convincingly only in kidney tissue. He began, on January 27, by measuring the anaerobic formation of amino-N in the presence of pyruvate and NH_4Cl, using a full set of controls including aerobic and anaerobic comparisons. He also measured the change in NH_3. Because some of the controls were "lost," the only conclusion he could draw was "aerobic disappearance of NH_3 certain . . . if pyruvate is present." That result was not directly useful, since one of the crucial features of the reaction scheme was supposed to be that it did not require oxygen. On the 29th he repeated the experiment in phosphate rather than bicarbonate saline. Now he found a "definite (though small) anaerobic amino acid aerobic synthesis if pyruvate + NH_4Cl are present" ($Q_{amino-N}$ aerobic = 7.87; anaerobic, ~5.55; other controls between 2.97 and 3.84; Q_{NH_3} aerobic = -7.50; anaerobic, -3.6; controls between +0.41 and -1.42). In both this and the previous experiment he tested for the "nonnatural" isomer of the amino acid by means of the enzyme d-amino acid and oxidase that he had earlier discovered, and found "no indication" that it had formed.[29]

As his newest research venture began to unfold in promising ways, Krebs felt that he was embarking on a major project for which he needed more help. He turned to Wayne, who again responded supportively. On January 30, Wayne wrote to the chairman of the Medical Research Council, Edward Mellanby:

Dear Professor Mellanby

Krebs is finding the lack of assistance here rather a difficult problem. He has a good deal of work in progress, some of which could be delegated to a suitable person. Such an arrangement would leave him free to come down to the Infirmary with me, and as we are starting an investigation there, it is important that he should be in a position to look at a few cases. Krebs thinks that a chemist would be of most use to him and we have consulted Professor Bennett about the matter. He has recommended to us and we have interviewed a young man, W.A. Johnson. He is twenty-two years old, took a first class in chemistry, and is interested in the biochemical side.... He would be working mainly in collaboration with Krebs, but if he turned out to be capable might be given a problem to work at under supervision.

Noting that Johnson was willing to come for £200 per annum, which "is the least we could offer," Wayne inquired whether Mellanby thought the Medical Research Council would "look favourably on an application" for a grant to hire Johnson.[30]

As this letter implies, Wayne also had a reason of his own for finding help for Krebs, because (at least as Krebs remembered it) it was Wayne who wanted Krebs to work with him on a clinical study at the hospital. Their joint project was to be a comparison of the action of insulin in subjects fed glucose and levulose, with its action in subjects fed ordinary sugar. Krebs was, however, not easily diverted from his central concerns. He went to the clinic only a few times, found it impractical to combine the clinical work with his laboratory research, and ended the collaboration.[31]   Wayne did not, on this account, diminish his efforts on behalf of Krebs.

II

On January 31, Krebs carried out a second set of anaerobic experiments on the first stage of the acetylalanine reaction scheme. He did them in duplicate, with liver slices from a starved rat, in which he measured the formation of CO_2 as well as amino-N and the change in NH_3. The experiment was not entirely satisfactory; there was too much preformed CO_2, and all of the pyruvate was used up. The increase in $Q_{amino-N}$ with pyruvate and NH_4Cl was very modest compared to the previous experiment (2.55 and 2.84; controls 1.88 and 1.78). Nevertheless all the measurements came out in a sense favorable to the theory. He put down:

Result:
1) anaerobic NH_3 consumption
2) NH_4Cl + pyruvate increase Q_{CO_2}
     (would be expected if deJong's reaction occurs.)
     2 pyruvate + NH_4^+ → 1 acetylalanine + CO_2 (new acid produced by oxidation of pyruvate to acetate)
3) Amino N formed.
4) CO_2 actually newly formed, but too much preformed CO_2 present, however, good duplicates.[32]

Moving to strengthen his evidence for an increase in CO_2, Krebs measured, on February 3, the "CO_2 production in liver (and kidney) in the presence of pyruvate + NH_4Cl, ketoglutarate." As his title suggests, and as was his custom, Krebs was here making three moves at once in the logical progression of his investigation. Even as he attempted to consolidate his case for the basic reaction in liver tissue, he extended it to kidney tissue, and he began an exploration of a variant of the reaction involving two ketonic acids (pyruvate and ketoglutarate). Referring back to the reaction scheme in his review article, we can see that in the first condensation step, one of the two keto acids is always pyruvic, but the other one may be either a second molecule of pyruvic or a different ketonic acid. To design experiments incorporating several related questions at once was characteristic of Krebs's scientific style, but was also in part a merely mechanical result of his manometric methods. It was most efficient to employ a full set of manometers for each run. In this case all of the results were strongly positive with respect to the reaction scheme:

Results: small increase of CO_2 production by pyruvate (50%) little by NH_4Cl (6%)
more increase by pyruvate + NH_4Cl (94%)
ketoglutarate + pyruvate (311%)
kidney also increase!![33]

In the month that he had been engaged in this new investigation Krebs appeared to have made remarkable progress, and there is reason to believe that he felt he was onto something very significant. Although his experimental data was still thin, its overall pattern constituted a preliminary confirmation of the theory originally proposed by Knoop a quarter century earlier. To provide more definitive proof would seem to demand only further extension of the experimental line Krebs was pursuing, and to solve the long-standing problem of how amino acids are synthesized would be a major achievement.

On the same day Krebs further expanded the investigation by examining the combination of another ketonic acid, acetoacetic acid, with pyruvic acid. He measured the anaerobic formation of CO_2 in the presence of NH_4Cl + pyruvate and acetoacetate, in comparison with NH_4Cl + respectively pyruvate and acetoacetate, as well as each of the latter by itself, and a blank. He did not include a run with pyruvate + acetoacetate without NH_4Cl. The results were:

	Q_{CO_2}	Percent increase
Blank	1.51	
NH_4Cl	1.86	23
Pyruvate	3.14	109
NH_4Cl + pyruvate	5.19	244
Acetoacetate	2.25	50
NH_4Cl + acetoacetate	1.80	19
NH_4Cl + pyruvate + acetoacetate	12.5	730

The outcome was thus dramatic. Krebs put down, "enormous increase of CO_2 production with acetoacetic + pyruvic + NH_4Cl," at first sight a strong confirmation of this variant of the reaction scheme. Then, however, he asked himself, "Is NH_3 necessary?" His reason for raising this doubt was not merely that he had neglected a critical control, but that it had occurred to him that a different reaction might account for the large production of CO_2. He wrote it down:

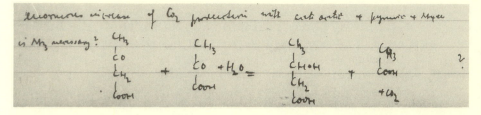

Like the reaction he had been studying, this one involved a molecular rearrangement in which an oxidation is balanced by a reduction, so that molecular oxygen is not required. Two of its products, CO_2 and acetic acid, were also the same as those of the Knoop scheme. Unlike that scheme, however, this reaction had nothing to do with the synthesis of an amino acid. The next day he probed his question by testing the "influence of NH_4Cl" on pyruvic acid + acetoacetate in rat liver and kidney. The essential outcome was "enormous CO_2 production: pyruvate + acetoacetic, no (definite) effect of ammonia on this reaction."[34] Krebs underlined "this" reaction, because the result did not apply to the reaction involving only pyruvate and NH_4Cl, for which he had already found an "influence" of the NH_4Cl. Nevertheless, a perturbation had arisen in the hitherto smooth course of his investigation. It is not evident whether he simply improvised the above equation on paper in immediate response to the possibility that pyruvic and acetoacetic acid could undergo a reaction without ammonia, or whether he drew the reaction from some other source. At any rate, the possibility of its occurrence opened up a new fork in his investigative pathway, one that was attractive enough to him to overshadow the problem with which he had been occupied.

Later the same day Krebs carried out an experiment on rat liver tissue similar to the previous ones, but in which he used his aniline citrate method to measure the change in acetoacetic acid present. The result, that there was a "larger disappearance of acetoacetic, if pyruvate present, no effect of NH_3," clearly supported his new hypothesis. Between February 5 and 12 he performed a series of experiments seeking further evidence for its occurrence. He used liver tissue from rats, pigeons, and a cat that Wayne had killed with ether. In general the results fit the reaction in question, although there were some anomalies. In a few cases there appeared to be either an effect of NH_3 on the quantity of CO_2 formed or the absence of such an effect could be attributed to ammonia present already in the tissue. In one experiment "acetoacetic + pyruvic yielded 278 μl of CO_2 more than pyruvic alone," whereas the acetoacetic acid that disappeared was equivalent only to 132 μl CO_2. "Thus," he wrote, "one acetoacetic disappearing → 2 CO_2 manometrically!!!" whereas the reaction in question "would yield one CO_2 for one acetoacetic." He thought, however, that the discrepancy could be explained: "but possibly 2 acetic → acetoacetic." In another experiment "pyruvic alone [yielded] enormous CO_2 production!!" His response to that result was to "try starved rat," presumably to eliminate metabolites in the tissue that might be reacting anaerobically with the pyruvate.[35]

On February 12 Krebs wrote to Norman Edson in Cambridge,

I am sending you a copy of our paper and I should like you to comment on it. I am sorry for the delay, but I was recently terribly interested in other work. It would be nice if we could send in the other paper at the same time. Have you done another experiment on the isolation of hypoxanthine?[36]

The first paper to which Krebs referred was probably one entitled "Micro-Determination of Uric Acid," based on the method he and Edson had worked out during the previous fall for their study of uric acid synthesis. They submitted this paper about two weeks later to the *Biochemical Journal*.[37] The other paper must have been on the investigation of uric acid synthesis itself, and their discovery that hypoxanthine was an intermediate. This paper was not ready, however, to send in "at the same time."

That Krebs had put off working on these papers because he had been "terribly interested in other work" was, for him, a strong statement. The work in question was undoubtedly the investigation that had begun with Knoop's theory of amino acid synthesis and suddenly been diverted to the study of a possible new metabolic reaction involving pyruvic and acetoacetic acid. The reasons for the intensity of his interest are not hard to guess. He had long been interested in the metabolic role of pyruvic acid without having made any significant advances concerning that prominent intermediate. Similarly, he had undertaken in the spring of 1932 to study the metabolism of acetoacetic acid, and had given up on it nine months later. Now he had a potential opportunity to link these two compounds through their participation in a new example of the dismutation reactions that were of such current interest. There was also a special reason to press ahead quickly, for he was not alone in this subcorner of the field. Juda Quastel and his associate Arnold Wheatley were moving ahead with a series of "studies in fat metabolism," and had published just three months earlier a paper on "acetoacetic acid breakdown in the kidney." Quastel and Wheatley had studied both the aerobic and anaerobic breakdown of the acid and had shown with respect to the latter that, among other substances, both glucose and pyruvate accelerated the process. They suggested, however, only that "the presence" of these substances "helps to establish more definitely the optimum conditions for the rate of breakdown of acetoacetic acid under anaerobic conditions." Quastel thought, in keeping with the general view that the ketone bodies are interconverted, that the acetoacetic acid gave rise to β-hydroxybutyric acid, but he offered no further reaction mechanism.[38] A glance at the equation Krebs had put down in his notebook shows that what he had in mind would account for this transformation by incorporating pyruvic, acetoacetic, β-hydroxybutyric, and acetic acid into a specified chemical equation. Given the long-standing unsolved puzzles about the metabolic interactions of the ketone bodies, it is no wonder that Krebs appeared excited at the prospects.

On February 14 Krebs moved to a crucial test of his hypothesis: is "acetic acid formed?" as the reaction required. Repeating the basic experimental series in liver tissue and finding the CO_2 production elevated by pyruvate + acetoacetate, he examined each medium afterward for acetic acid. After collecting steam distillates in a Parnas apparatus, acidifying and concentrating the distillates in a water bath, he applied a lanthanum reaction devised by Deodata Krüger and Erich Hirsch, which was a highly specific test for acetic and propionic acid. In this case a positive reaction would have provided "conclusive evidence" that acetic acid had formed. In each of the pertinent runs, however, he found the "lanthanum test . . . negative!!!" Although the emphasis suggests that he did take this result as something of a setback, he clearly did not treat it as a refutation of his hypothesis. One could explain it in various ways, from there being too little acetic acid for this test, to a secondary reaction removing the acetic acid. The next day he went on to test for the third of the products indicated in the reaction, β-hydroxybutyric acid, and obtained a clear positive indication that it had formed only in the presence of pyruvate and acetoacetate ($Q_{\beta\text{-hydroxybutyrate}} = 5.87$; for controls and individual substrates, 0.59 to 1.58).[39] Nevertheless, after obtaining the negative result for acetic acid, Krebs may have loosened his approach, recognizing that the phenomena he was observing were more complex than what his equation encompassed.

On February 18 Krebs performed two experiments on the formation of hypoxanthine and uric acid in chicken tissue slices, probably to help fill in a gap in the uric acid synthesis paper. He obtained "almost nothing!" Returning to his own current investigation, he measured on the 19th and 21st the effects of acetoacetate, pyruvate, and their combination on the formation of ketone bodies in rat liver slices. At the same time he examined "the influence of acetate on [the] disappearance of pyruvate" and observed no effect. It is not evident that he had in mind any specific reaction connecting acetate directly with pyruvate; perhaps he was checking the possibility he had earlier raised to explain an anomalous ratio of CO_2 formed to acetoacetate consumed—that is, that two molecules of acetic acid can give rise to one of acetoacetic. On the 21st he also measured the "effect of acetoacetic and pyruvate on glycolysis," that is, their effects on the anaerobic formation of CO_2 in the presence of glucose, and three days later he carried out an elaborate set of experiments on rat testis and brain tissue in which he measured the anaerobic changes in the quantities of pyruvate and acetoacetate present when either or both of these substances were added. In both experiments the combinations of the two substances increased the changes somewhat over what either alone did, but in both cases the most striking effects obtained were those of pyruvate itself. In the first case pyruvate caused an "enormous" rise in the CO_2; in the second there was an "enormous disappearance of pyruvic acid [anaerobically] in testis and brain."[40]

These effects of pyruvate were not appearing for the first time. From the beginning of the investigation, when pyruvate was added alone as a control on its addition with a second substance, it caused by itself marked increases in whatever aspect of the process was under scrutiny. As the action of pyruvate alone repeatedly appeared more striking than the additional action of the combination, however, the status of its effect in his mind must have begun to change. From a background control phenomenon, it was probably by this time moving toward the foreground as a potential object of investigation. Krebs must by this time have begun to view the anaerobic breakdown of pyruvate as a broader subject than the particular reactions involving that substance that he had been studying. (A nice trace of the change his viewpoint eventually underwent is that when he later published a paper on the investigation whose beginnings we have been following, he selected from the experimental series entitled "Aceto-acetic + Pyruvate" in rat testis and brain the control runs with pyruvate alone and included them in a table of results entitled "Anaerobic Disappearance of Pyruvic Acid in Animal Tissues.")[41]

It may have been such a recognition that pyruvic acid probably reacts anaerobically in a variety of ways that prompted two experiments Krebs carried out on February 25. In rabbit liver he tested the combination of β-hydroxybuty-ric and pyruvic acid, with the usual controls, but found "no particular effect." With rabbit kidney he asked, is "glutamine formed??" from a combination of pyruvate, NH_4Cl and acetoacetate. He was not merely returning to the Knoop theory of amino acid synthesis, for he posed the "theory" that the "condensation of pyruvic acid is coupled with the reduction of acetoacetate," according to the equation:

Implied are further steps in which the ketoglutarate produced in this reaction is converted through glutamic acid to glutamine. Measuring the change in NH_3 and amide-N, he found for both a "small effect in expected order!!" In brain tissue, on the other hand, there was "no definite effect on amide formation by β-hydroxybutyric and acetoacetic."[42]

As he loosened his approach in order to examine individually the phenomena that had appeared a little while earlier as linked components of a hypothetical reaction scheme, Krebs also explored other possible causes of the disappearance of acetoacetic acid. On the same day as the preceding experiments, he measured the change in that substance when he added the following to liver slices from a well-fed rat: pyruvate, pyruvate + As_2O_3, ketoglutarate, succinate, glucose, pyruvate + insulin, ketoglutarate + NH_4Cl, NH_4Cl, and NH_4Cl + pyruvate. It is doubtful that he had any specific "theories" in mind; when he carried out such heterogenous comparisons he was usually simply looking for something to "turn up." He found "no effect of pyruvate today! (freshly prepared reagents)," but an "effect of ketoglutarate and of pyruvate + NH_4Cl." "Repeat" he instructed himself. When he did so, on the following day, he observed an "effect of all substances added, not additive, ketoglutaric + pyruvic no effect of NH_4Cl!! different from yesterday." His explorations of the anaerobic disappearance of acetoacetic seemed, therefore, for the moment not to be leading anywhere.[43]

More encouraging to Krebs was a simultaneous experiment he carried out on the effects of insulin on the aerobic metabolism of acetoacetic acid. Q_{acac} without insulin equaled -3.25; with one unit of insulin, -2.46. "Insulin," he wrote down, "inhibits acetoacetic acid disappearance!!!?" Always on the lookout for specific metabolic effects of insulin, Krebs seldom encountered any. This apparent exception must have raised his hopes. Following up his lead, he tested insulin with β-hydroxybutyric acid and observed that one unit of insulin both "increases Q_{O_2}" and "increases acetoacetic from β-hydroxybutyric." Induced to pursue the sudden new opening further, on the 28th he tested insulin with β-hydroxybutyrate and with lactate on liver from a starved rat. The results, he noted, were "different in starved animal from well fed beast." In this case, among other effects, insulin increased the Q_{O_2} "if nothing added, and no effect of insulin on Q_{acac}." On the 29th he carried out two further experiments on the effects of insulin, respectively on pigeon brain and liver from another well-fed rat. In the latter case he found that "Insulin <u>increases</u> Q_{O_2} and $Q_{ketonic\ acid}$, especially from β-hydroxybutyric!!! Effect everywhere!"[44]

Despite these aerobic effects of insulin that were obviously of great interest to him, Krebs returned after the weekend to the problem of anaerobic acetoacetate metabolism from which the insulin results had diverted him for three days. Perhaps he saw that, however promising the new direction might appear, he should not follow it until other things were finished. In 1977, in response to a question about whether he generally had more things in mind than it was possible to follow up, he replied, "That was always so. I have to restrain myself not to scatter my interests, and to complete a piece of work."[45]

On March 2 Krebs tested the effects of acetoacetic acid itself on the disappearance of acetoacetic acid anaerobically in several rat tissues. The next day he examined in rat testis tissue the "reduction of acetoacetic acid [in] various substrates," including glycerophosphate, phosphoglyceric acid, lactate, pyruvate, glucose, succinate, fumarate, and α-ketoglutarate. Only in glucose did he find an increase in Q_{acac}. On the same day he tested aerobically the effects of β-hydroxybutyric acid on the formation of acetoacetic acid in various tissues. As on other occasions when the guidance of a particular theory had become uncertain, he seemed to be just "gathering more information," in the hope that a more definite lead would emerge.[46]

Even though his latest experiments appeared to blunt, rather than to sharpen the reaction schemes he had been testing, Krebs remained confident enough about them to provide a very positive summary of the work in progress for Edward Wayne to include in the application he made on March 3 for two Medical Research Council grants for Krebs. Having received an encouraging reply from Mellanby to his initial inquiry, Wayne requested £200 a year to hire William Arthur Johnson as an assistant to Krebs, and £50 for materials and apparatus "necessitated by the presence of an additional worker in the laboratory." The "problems under investigation in which he would assist," Wayne wrote, summarizing a description that Krebs probably drafted for him, "are as follows":

Preliminary experiments carried out in this Department with the tissue slice technique suggest that the primary reaction in the synthesis of amino acids from ketonic acids and ammonia is the formation of an acetyl amino acid. This is then deacetylated by a specific enzyme. The energy necessary for the synthesis is provided by the oxidation of pyruvic acid to acetic acid and the actual transmission takes place by an intramolecular arrangement of dismutation type in an intermediate compound. Other biological reductions such as that of acetoacetic to β-hydroxybutyric acid can be explained on similar principles. Here an intermediate compound is formed with pyruvic acid and this breaks down with the liberation of acetic acid and carbon dioxide. This reaction is of special interest, representing as it does a link between fat and carbohydrate metabolism, and its study promises to shed light on the problem of ketogenesis and on the metabolism of the diabetic organism.

The proposed research worker would assist in working out the details of the above schemes and in following up their applications.[47]

It is easy to see why Krebs would have been enthused about these "suggested" reactions. To find the link between fat and carbohydrate metabolism was an ambition that he had held since the spring of 1933, which he had pursued for the next nine months without success, and for which he now perceived the opening that had then eluded him. In the process he hoped also to elucidate problems concerning ketogenesis and diabetes that had eluded other investigators for nearly half a century.

* * *

The optimism with which Krebs proposed, in March 1936, that a new assistant would help him "work out the details" of a scheme whose outlines appeared already clear to him, contrasts with the skeptical view of these same ideas that Krebs took in May 1979, when they were brought to his attention. "One could put forward many other schemes...at a time when the main ingredients of the mechanisms were not yet known...[This was] just feeling one's way in complete darkness."[48] What was for the Krebs of 1977 complete darkness, was for the Krebs of early 1936 a beckoning beam of light.

* * *

In the same letter to the Medical Research Council in which he requested a total of £250 for Krebs's investigation, Wayne asked for "a grant of £50 for my own work."

Preliminary observations have shown that the rise of blood pressure which occurs in rabbits after a standard small dose of adrenaline is much greater when the vessels have been damaged by feeding with large doses of calciferol. The cause of this increased sensitivity is being investigated and substances other than calciferol are being studied. This grant is required for the purchase of animals and apparatus and expensive drugs.[49]

That Wayne requested five times as much support for Krebs as for himself may reflect a realistic assessment of the importance of their respective investigative ventures, but it is also another mark of the generous spirit in which he placed the needs of the ascending star in his department ahead of his own needs.

On the day after Wayne submitted the request for funds on his behalf, Krebs carried out one further experiment on the aerobic action of insulin, noting this time only slight effects. At the same time he tried another pair of substances, this time pyruvic and glutamic acid, on the aerobic formation of CO_2 in rat brain tissue. He undoubtedly chose this combination with the special prominence of glutamic acid in brain metabolism in mind; nevertheless, the most conspicuous result was again the control experiment without the glutamic acid. "Anaerobic CO_2 production from pyruvate!" he noted, "not influenced by glutamic acid."[50] Reinforcing similar earlier observations, this one apparently convinced Krebs that the anaerobic behavior of pyruvate added alone should become the focus of his next efforts.

At first sight an anaerobic disappearance of pyruvic acid would appear easy to explain, since it was long known to be reduced in tissues to lactic acid. The latest work on glycolysis, a development of Embden's reaction scheme recently published by Otto Meyerhof, had just established that the main pathway of glycolysis in muscle tissue ended with a dismutation reaction in which "pyruvic acid with α-glycerophosphoric acid is transformed into two molecules of lactic acid."[51] Krebs had reason to suspect, however, that the effects of pyruvic acid he was observing could not all be due to this known process. In the first place, the extra CO_2 it produced was not necessarily represented by carbonic acid displaced from the medium by lactic acid formed. More cogent for him at this point was that the effect was occurring in situations in which he would not expect there to be an adequate supply of carbohydrate to account for the level of glycolysis that would be required. In his experiment of February 21, for example, he had employed brain tissue, which contains little or no carbohydrate. He had done comparative runs with and without glucose added, and remarked that "pyruvate increases CO_2 although there is no glucose."[52]

On March 6, Krebs performed a set of experiments whose title "Testicle + Pyruvate (Anaerobically)" implies that his interest now centered on the anaerobic metabolism of that particular acid. The experiment itself, however, incorporated the usual comparisons of pyruvate alone with its combinations—in this case combinations of pyruvate with acetoacetic, malic, fumaric, and oxaloacetic acid and with glucose. The situation illustrates the way in which the objective of an experiment may be changed while its form remains nearly identical with previous ones. It seems that Krebs was not now seeking compounds that might enter into specific reactions with pyruvic acid, but testing whether the effects of pyruvate on the anaerobic formation of CO_2 were *independent* of the effects of these other compounds. The criterion he had earlier established for such independence was that increases due to two compounds in question be "additive," that is, that the sum of their separate effects equal the effect of their combination. In this case he found that acetate + pyruvate and glucose + pyruvate were both additive. The remaining compounds he did not test alone—perhaps because he did not have enough manometers. He found "no effect by NH_4Cl, malic," a questionable effect of fumaric, and an "increase by oxaloacetate!!!!"[53] This last result clearly caught Krebs's attention, but he nevertheless put it aside for now in order to hold his focus on pyruvate itself.

The next day Krebs compared the anaerobic rate of CO_2 formation in cat brain tissue in the presence of pyruvate and of glucose, but found that neither substance raised it above the level of the control. In a similar experiment with rat heart muscle immersed in pigeon muscle *Kochsaft*, both substances *lowered* the rate.[54] Finally that day he carried out a large-scale experiment on rat testis tissue to measure the quantity of pyruvic acid that disappeared anaerobically and the quantities of the products that might arise from it. Using a gasometric method of B.F. Avery and Baird Hastings, Krebs determined the lactic acid that formed, and calculated that the ratio of pyruvate disappeared:lactate = 1.86:1. After making some changes in the method he obtained a higher result for lactate and consequently a lower ratio of pyruvate:lactate = 1.56:1. "Again," he commented, "too much lactate found!!" A test for β-hydroxybutyric acid indicated that none had formed, and one for acetic acid was "unsuccessful!" There was "no recovery" even when acetic acid was added to the solution.[55]

The fact that he calculated these ratios and stated that there was "too much lactate" implies that Krebs must have had in mind a possible chemical reaction whose stoichiometric proportions he was comparing with the measured proportions. He did not put down such a reaction in his notebook, however. It is clear that neither the measurements nor his expectations fit the glycolytic conversion of pyruvate to lactate in the Embden-Meyerhof pathway, which would have yielded pyruvate:lactate = 1:2. The only clue to a reaction Krebs might have been thinking about when he performed this experiment was that underneath the first measured ratio he wrote "(3:5 = 1:1.67);"[56] that is, possibly a reaction in which three molecules of lactic acid appear while five molecules of pyruvic acid disappear. (Later on Krebs developed evidence for a specific reaction yielding both acetic and lactic acid, but the theoretical ratio for that reaction was pyruvate:lactate = 2:1, not 5:3.)[57]

At this point Krebs once again digressed from the line of investigation he had been pursuing. As had happened several times already, he did so in response to work emanating from the laboratory of Albert Szent-Györgyi in Hungary. This time Krebs's reaction was less instantaneous than it had sometimes been, for the publication that influenced his next moves had been available to him for several months. As discussed in Chapter 2 (pp. 76–78), in the summer of 1934 Szent-Györgyi had put forth the theory that succinic and fumaric acid act catalytically on respiratory oxidations, serving as a hydrogen transport system. He and his group continued to work on this question, and published a year later a collection of papers in which Szent-Györgyi provided a broad overview and his collaborators supplied most of the supporting evidence.

Szent-Györgyi identified the "special function" of the C_4 dicarboxylic acids as the link between the "Warburg-Keilin-System" (the *Atmungsferment* and the cytochromes) on the one hand and the dehydrogenation of the foodstuffs on the other. Reviewing the theory as he had earlier stated it, he described succinate as oxidized to fumarate by the hydrogen transport system, the fumarate being in turn reduced to succinate by the dehydrogenases of the foodstuffs. It was this alternating oxidation and reduction that enabled the succinate-fumarate system to act catalytically. Szent-Györgyi and his coworkers now added further supporting evidence for this conception. In place of the earlier experiments showing that fumarate in catalytic quantities can increase the respiration of pigeon breast muscle tissue, he now invoked experiments of his associate Ilona Banga indicating that fumarate does not "increase," but only "conserves" the respiratory rate. Minced tissue suspended in a phosphate medium "ordinarily loses the intensity of its respiration very rapidly." Very small quantities of fumarate added to the suspension, however, enabled the tissue to maintain the initial rate. This new evidence proved, according to Szent-Györgyi, that the fumarate was not merely activating an "artificial system," but sustaining the main respiratory process itself.[58]

The system as he had earlier presented it gave no explanation, Szent-Györgyi now acknowledged, for the function of the oxidative enzyme fumarase, and this lacuna had prompted them to examine whether "fumarate can exercise its function as a hydrogen transporter through an exchange with a higher oxidation state." After exploring various possibilities, another of the coworkers, Koloman Laki, concluded that only oxaloacetate could play this role. Further experiments on the reduction of fumarate to succinate in the presence of malonate (discovered by Quastel to inhibit specifically the oxidation of succinate) led "unexpectedly" to the conclusion that "succinate appears to be formed not from fumarate, but through an 'overreduction' from oxaloacetate, since this reduction takes place only under aerobic incubation, when the fumarate is given the opportunity to become oxidized to oxaloacetate." Szent-Györgyi accordingly formulated an enlarged scheme in which succinate is oxidized through fumarate to oxaloacetate, but the reduction proceeds directly from oxaloacetate to succinate:[59]

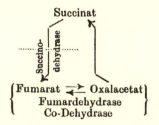

The issue of *Hoppe-Seyler's Zeitschrift* containing this complex multiauthored paper appeared at the end of September 1935. Szent-Györgyi sent Krebs a reprint, which Krebs studied carefully. He marked with two explanation points the question posed in Szent-Györgyi's introduction "What is therefore the meaning of the fact that the cells maintain the most active enzymes for succinic, fumaric, and malic acid, which obviously do not normally arise as intermediate steps in the decomposition of foodstuffs?" Krebs also put a mark by the statement that "fumarate added to muscle under anaerobic conditions is either not, or only extremely slowly reduced."[60] It is not certain that Krebs made these annotations the first time he read Szent-Györgyi's paper (when I showed them to him in 1977, he could not himself tell when he had put them in).[61] Nonetheless it is evident that Szent-Györgyi's work and his interpretation of his results strongly impressed Krebs, probably more so than the earlier version had. He did not, however, like Szent-Györgyi's new idea that oxaloacetic acid is converted directly to succinic acid by "overreduction," a concept to which he could attach no chemical meaning.[62] Evidently Krebs perceived Szent-Györgyi's latest scheme as highly significant, but not quite right; that is, it probably stimulated him to wonder if he could find a better explanation for the observation that fumarate can be reduced to succinate in the presence of malonate aerobically, but not anaerobically.

It is not clear how long it took Krebs to develop these responses to Szent-Györgyi's paper. Given his habits, he probably read it as soon as he saw it in *Hoppe-Seyler's Zeitschrift* or received the reprint. That he did not begin to act experimentally upon it until March 1936 may mean that its significance only grew gradually on him, or merely that up until then his other research projects had taken precedent. A third possibility is that he had ordered a Latapie grinder to employ Szent-Györgyi's method for mincing muscle tissue, and had to wait for its arrival.[63]

The first experimental step that Krebs took, probably on Saturday, March 7, was to try out a new method for the microdetermination of succinic acid described in a section of Szent-Györgyi's paper authored by B. Gözsy. The method "was based on the observation that washed muscle is able to oxidize aerobically only very few substances." Any such substances that a muscle suspension might contain Gözsy removed by oxidation with permanganate and extraction in ether. He then determined succinate "in a biological way with the muscle suspension, calculating the quantity of succinate from the oxygen consumption" of the tissue. To ensure the specificity of the reaction, he employed a control containing malonate, which, as Quastel had shown, inhibits the oxidation of succinate. Subtracting the oxygen consumption in this control from that in the main reaction would give the quantity due to succinate.[64] The method was ingenious, and since it offered for the first time the possibility of determining succinic acid in quantities of as little as 7 mg, its great potential usefulness in intermediary metabolism would have been obvious at once to Krebs. He followed Gözsy's procedures closely, using pigeon breast muscle minced in the Szent-Györgyi manner by means of a Latapie grinder with the second disk removed. The method appeared to him to work well. Although he obtained a measured quantity greater than the calculated quantity from his known solution of sodium succinate, he attributed the difference to the probability that the "sodium salt lost H_2O." In general, he commented, there was a "very active enzyme!" and it "does not oxidize lactic [or] pyruvic."[65]

When he resumed work on Monday, March 9, Krebs carried out simultaneously two unrelated sets of experiments. One tested the effect of β-hydroxybutyric acid on aerobic glycolysis in pigeon brain tissue, during which he observed "no considerable influence. . . !!" The other was a direct test of one of the most salient features of Szent-Györgyi's new evidence for his theory: that fumarate "conserves" respiration. Krebs did not this time duplicate all of Szent-Györgyi's procedures. He used pigeon breast muscle, but sliced it instead of mincing it, and suspended it in *Kochsaft*. He measured the respiration of the tissue by itself, with fumarate, and with pyruvate. Over a period of two hours and 20 minutes the Q_{O_2} declined roughly equally in all three cases. Krebs considered that these were, in general, "low figures," and concluded that there were "no stabilizing effects!" He had not been able to observe what Szent-Györgyi's group reported, but he evidently did not doubt the correctness of their observations. Rather he attributed his own negative result to the "old Kochsaft" he had used, and to the possibility that there was not enough fumarate present.[66]

Krebs did not try further to verify this or other aspects of Szent-Györgyi's theory at this time, probably because the succinic acid method with which their work had armed him offered a new opportunity connected with his own current investigation of the anaerobic metabolism of pyruvic acid. Among the compounds to which pyruvic acid, either alone or in combination with another compound, might give rise, succinic acid was a particularly interesting candidate, because it had already been linked to pyruvate in the Thunberg-Knoop-Wieland scheme for respiration and somewhat more directly in the Toenniessen-Brinkmann modification of that scheme (see Vol. I, pp. 22–32). Krebs apparently now asked himself whether "the conversion of pyruvic acid into succinic acid" can take place "in the absence of molecular oxygen."[67]

On March 10 Krebs applied Gözsy's method to determine whether "succinate formed?" in rat testicle tissue anaerobically in the presence of pyruvate, pyruvate + glutathione, pyruvate + ketoglutarate, and pyruvate + acetate. There was, however, "no succinic formed." The next day, simultaneously with measurements of the effects of acetate and succinate on the respiration of rat testis tissue, he again tested for the anaerobic formation of succinate, this time under the influence of pyruvate, insulin, and succinate itself. Again he found "no significant change in succinic acid, neither formed nor consumed." On the 17th he tried once more, this time with pigeon breast muscle suspension prepared according to the procedures of Szent-Györgyi. Limiting himself to the influence of pyruvate alone, he encountered his third straight negative result. Clearly not accepting this outcome as definitive, he checked the Gözsy method to see whether the malonate in the control might be disappearing through oxidation. He concluded that there was a slow oxidation, and that malonate in the concentration he was using "inhibits ca 75%."[68] In this indecisive state of affairs Krebs closed down his laboratory to prepare for his long-deferred "journey to Palestine."[69]

During the two months from the time that he resumed his research in January 1936 until he left it for the Middle East, Krebs had not followed any single line of investigation to a point of closure. Beginning with an effort to confirm an old theory of amino acid synthesis, he had abandoned that path just as he appeared to be approaching some success, in order to explore a different theoretical reaction involving one of the same compounds—pyruvic acid—implicated in the Knoop theory. From the particular dismutation reaction that had led him to take this turn, he began to branch out to explore analogous reactions and obtained suggestive evidence, but did not establish any reaction in detail. Gradually his attention shifted from specific combinations of pyruvic and another ketonic acid to the anaerobic metabolism of pyruvic acid itself as his central focus. After interrupting this line briefly to examine the work of the Szent-Györgyi school, he had just returned to the study of pyruvic acid when he broke off his research.

Although this short episode in Krebs's research life produced no completed investigations, it encompassed a significant transition in his investigative pathway. When he began it he had been working for two years almost entirely on "the metabolism of amino acids and related compounds." It was no accident that he was asked to write an article on that subject for *Annual Review of Biochemistry*, for that was the area within which he had attained all of his major experimental successes, and he was coming to be regarded sometimes as a specialist in the subfield of intermediary nitrogen metabolism. As we have seen, the nine months of concentrated effort he had put into the problems of carbohydrate and fatty acid metabolism in 1933 had yielded no significant advances. After reaching an impasse he had turned away from these problems and had not returned except for occasional brief forays. By the time that Krebs broke off his work in the middle of March of 1936, he appeared to have engaged himself for the first time since then in a major attack on these problems. We are left with the question, Did Krebs drift almost inadvertently, step by step, back into the arena he had left, or had he already made a basic decision to reenter the field, so that the steps we have followed represent only the occasions he exploited in order to implement a more general plan?

No definitive answer can be given to such a question. The day-by-day reconstruction we have followed does seem to offer an *adequate* account of the transition. If we add to the evidence embedded in his contemporary record Krebs's own reflection that he ordinarily planned his research one day at a time, the case for the first of these interpretations appears persuasive. On the other hand, in 1970 he wrote:

> between 1933 and 1935 . . . my main research ... was on other topics—the oxidation of amino acids, the properties of D-amino acid oxidase, and the biosynthesis of glutamic acid in animal tissues. The problem of the intermediary stages of respiration remained in the forefront of my mind as one of the big unsolved problems of biochemistry, and I often pondered about new experimental approaches.[70]

Krebs's recollection a quarter century later appears vague and uninformative by comparison to the chronologically precise records of his actions that we can retrieve from his notebooks. We should not, however, too readily choose the story that is more concrete and that is derivable from evidence that appears both more accessible and more immediate in time to the events in question. Krebs's "ponderings" during these years left no such easily traceable marks on the record, but they may have exerted as much influence over the directions his investigations took as did these daily interactions with experimental operations, and the direct short-term conclusions which happen to survive for our scrutiny.

III

In January Krebs had affirmed his willingness to make the trip to Palestine in March. Weizmann had invited him, Nachmansohn, and Blaschko to come at Weizmann's own personal expense, and the time appeared convenient for all concerned. Nachmansohn undertook to handle the travel arrangements, emphasizing in a letter to Krebs on January 17 that reservations must be made early, because they would be going at the peak of the tourist season.[71] He had had enough experience by now to realize that it was difficult to pin Krebs down to commitments that drew him away from his scientific preoccupations. Since Krebs's father and stepmother were planning a trip to the Mediterranean in March, Krebs wrote them in February giving them his tentative schedule, and suggested that they meet. Writing back, "I will be awfully glad to see you again," Georg Krebs altered his own travel plans so that they would intersect in Italy on Hans's way toward the Middle East.[72]

As the time for the voyage approached, a sudden rise in international tensions clouded the prospect. For some time it had appeared likely that Hitler might occupy militarily the Rhineland area of Germany, in defiance of the Locarno Treaty of 1925 which had established that territory as a demilitarized zone. Germany had already begun to rearm in violation of the Versailles Treaty, but its military strength was still small enough so that it appeared that firm policies by France and Britain could forestall such action. These two countries and their allies were, however, distracted by their inability to respond effectively to the Italian invasion of Ethiopia, and seemed unprepared to counter Hitler's increasingly bold policies. On Saturday, March 7, German troops moved swiftly into the Rhineland. France and Britain were shocked, but neither government took any immediate action.[73]

On March 9 Nachmansohn wrote Krebs that Weizmann had been too busy in Paris to see him, but had said on the telephone that "in principle nothing stood in the way of our trip. But he was very agitated about the latest political events, declared that he was 'completely disoriented' and would only be able to see things clearly in 2–3 days. He asks you to get in touch with him immediately." Nachmansohn planned to be in London on the 14th for a meeting of the Council for German Jewry and hoped that Krebs and Blaschko would also come. By then he hoped it would become clearer what effect the political situation would have on their trip. He hoped that it would result in nothing more than a postponement for a few days, "but Hitler has played other tricks already (and this is surely not the last)."[74]

During the intervening days the situation grew more threatening. For a time Britain appeared to be ready either to support a French move against Germany or to take measures of its own, and war seemed possible. By the 14th, however, when the Council for German Jewry met in London, Hitler's assurances that he had no aggressive intentions, combined with the lack of will in France or Britain for strong action, were already diminishing the crisis.[75] Krebs came to the meeting, held in the Dorchester Hotel, "where the appeal for raising 3 million pounds in aid of German Jewry was launched," as he recorded in a diary that he began on the next day. Afterward he and Nachmansohn met with Weizmann, where it was "agreed that Nachmansohn and I meet Weizmann in Rehoboth [sic] on the 6 or 7th April."[76]

Returning to Sheffield, Krebs carried out on Tuesday the last of the experiments discussed above, and presumably spent the rest of the week preparing to leave on Saturday, March 21. He was only mildly reassured about the international situation. On the date of his departure he wrote:

> Political situation in Europe highly tense. German papers reject proposals of Locarno Powers ("collective impedance"). No definite prospect for peaceful settlement in sight, although there seems no danger of an immediate war.
>
> Wayne pictures (jocularly) Sheffield bombed before I may return.[77]

The remilitarization of the Rhineland appeared then, as it does today, as a turning point in the political condition of Europe. As Gerhard Weinberg has put it, the result was a "collapse of the post–World War I security system in the face of German action" that was "certain to lead to a complete reorientation of the policies of most of the European powers."[78] On Krebs the event made a deep personal impact. He was among those who had believed up until then that the Allies would not permit Hitler to undo the outcome of the world war, and who thought that if Britain and France had reacted militarily to the occupation, Hitler would have been forced to retreat. When this did not happen, Krebs became markedly more pessimistic about the general future of Europe, and more anxious about his own. If German power were to grow unchecked, then his escape from Nazism could prove to have been temporary.[79]

Leaving Sheffield on Saturday afternoon, Krebs arrived at 6:30 p.m. in London and met Hermann Blaschko's aunt Helene Nauheim, who drove him to her home. With other relatives of Blaschko's family who were concerned about his future, Krebs explained that Blaschko had shown little initiative in the autumn when he refused a job offered him in Palestine, but added that he considered Blaschko's "prospect not hopeless as to Cambridge, and hopeful as to Palestine." On a bright Sunday morning Krebs departed from Victoria Station, after meeting there with a medical investigator who hoped to work in the cancer institute in Jerusalem. On the train he noticed that there were "blooming primroses along the rail, but the trees are still bare." He talked with a German emigrant businessman from Frankfurt "full of Frankfurt local Jewish

Figure 6-2 Georg and Maria Krebs in San Remo.

patriotism." After a quiet channel crossing he arrived in Paris in the evening, where he was met by the Nachmansohns. Dining in a restaurant on the Boulevard St. Michel "opposite to the floodlighted Notre Dame," they discussed "preliminaries for Palestine." Krebs left on a night train from the Gare-de-Lyon, in a sleeping compartment provided with a telephone (which he undoubtedly did not use), and was passing through Provence by morning. At 2:48 in the afternoon his train reached the winter resort seaport of San Remo, on the northwest coast of Italy. "Father and Maria" were waiting for him at the station.[80]

Although Krebs later remembered spending "just one day" with his father and stepmother, his diary indicates that they were probably together for two and a half days. Among the prominent topics of conversation must have been Wolf's situation. Wolf was still waiting to hear whether the General Electric Company in Birmingham, where he had gone for an interview four months earlier, would hire him. They had encouraged him to believe that they were quite interested in him, but told him the decision could not be made at the time because of a coal miners' strike then taking place. Later Wolf read in the newspapers that the

Figure 6-3 Georg and Hans Krebs in San Remo.

strike had ended, but he still received no answer from the company. Near the end of February their father wrote Hans that "Wolf's affairs are not yet decided, but he hopes that they will be settled within a few days." When Hans arrived in San Remo a month later nothing further had been heard.[81]

Hans and his father must also have discussed with anxiety the current political situation. Despite his cheerful recent letters, Georg Krebs appeared to his son very depressed. As Hans put it in 1977, "his whole picture . . . of the world had collapsed. . .as he thought he was a very good German patriot, and did always the right things, was a useful member of the community, and . . . [now he was] one of the proscribed."[82] The early optimism shared by father and son that the Nazis might not last long must have been finally shattered by the events of the preceding three weeks. Now 69 years old, his practice ever more restricted, the elder Krebs must nearly have lost hope that he would live to see the end of this totalitarian regime. Still retaining sufficient financial resources to live comfortably, and with a supportive family, Georg Krebs was far from helplessness or despair. Nevertheless, the only remaining sources of joy for him were his pride in the achievements of his eldest son and his pleasure in the progress of his youngest child. The only useful steps he could take were to do what he could to assist his other son and daughter to leave the Germany to which he felt himself still bound.

Before dawn on the morning of March 26, Krebs took a taxi to the San Remo railway station. On the way the cab ran into a small cart in the darkness, and Krebs was flung into the windshield, breaking it. He sustained three long cuts and several smaller ones on his forehead, but they were not serious enough to delay his departure for Marseilles.   Arriving there at 11:15, he caught another taxi for the harbor, embarked on the steamship *Providence*, and met Nachmansohn on board. The ship sailed at noon under bright sunshine. After three days at sea, land appeared at 9:30 a.m. on March 30, as the ship approached Alexandria. The English Home Fleet was cruising near the harbor, and Krebs was impressed by the "many . . . destroyers, submarines, battleships and sea plane carriers," as well as the "many planes in the air."   After disembarking, Krebs had an experience which suggests that as a world traveler he was still somewhat naive. He paid a "Mr. Schmidt" 18 francs for a visa, "which afterwards appeared to be a complete fraud." That Mr. Schmidt could have taken him in so easily is all the more remarkable, in that Nachmansohn had already informed Krebs before the trip that they would be permitted to leave the ship in Alexandria without visas. Touring Alexandria by taxi, Krebs and Nachmansohn then had dinner in the Cafe Turque and went to a cinema to see *The Merry Widow*. The next morning Krebs walked through the streets of the city, viewed the well-known light-house near the harbor, and had time to post a letter to his parents in San Remo before the ship sailed at 5 p.m. for Port Said.[83]

In the one day they spent in Port Said, Krebs had eyes mainly for the depressing aspects of this Egyptian city at the head of the Suez Canal. During a drive in a horse-drawn cab he noted that "the Arabic quarters are dirty and stinking," and "the 'houses' are partly wooden huts." He saw "scores of 'Händler' on the streets selling bijouterie, stamps, and pornographic photos." Several men tried to solicit him and his traveling companions for prostitutes. Clearly, for a man brought up in a society in which daily life was always orderly even in uncertain circumstances, first impressions of the teeming world of the Middle East were disagreeable. Of a trip on the canal in a motor launch he commented only that it was "rather hot." Krebs was probably quite ready to leave when their ship departed at 5 p.m. for Jaffa.[84]

Twenty-four hours later the coastline of Palestine came into view, but Krebs and Nachmansohn did not leave the ship until 8:30 the next morning, April 2nd, after enduring a tedious passport control process. While waiting, Krebs met a dental surgeon from Berlin named Dr. Krebs, whose parents came from the same region of Silesia in which his own father had been born. In Jaffa they met with several other people who had come to Palestine from Germany. Later in the day they drove through Tel Aviv, and at 4 p.m. set out by car for the northern port of Haifa, taking an inland route through Tulkarm and Jenin. Arriving at the Wollstein House in Haifa by 7 o'clock, they took in the first act of a Hebrew performance of *Othello*, using tickets furnished them by another German settler. On April 3rd they drove from Haifa to an agricultural station in nearby Acre, saw an Arab street bazaar, and traveled east to Tiberias on the Sea of Galilee. They were shown around a Jewish communal village, or *kvutzah*, by a young woman from Vienna. The next day they again toured by car from Haifa, this time to Nazareth and Afula, where they stopped at a hospital and talked with two of its physicians. They were accompanied by three other Germans living in Palestine, and spent the evening with a former director of a shipping company in Königsberg who was now connected with a similar business in Haifa. On Sunday the 5th they went by bus to Jerusalem, where they found the hotel in which they planned to stay fully booked for Passover and were "quartered" instead in a pension.[85]

On Monday Krebs and Nachmansohn turned to the more professional objectives of their journey by visiting various laboratories and scientists in Jerusalem. They spent part of the day at the cancer institute for which Magnes had attempted to recruit Krebs in 1934. The physiological chemistry section that Krebs had originally been offered did not yet exist, but a radiology section headed by Professor Ludwig Halberstädter and a section for the physiology and culture of cells headed by Dr. Leonid Doljansky were in operation. Unfortunately neither Halberstädter nor Doljansky had much time to talk with them. The next morning Krebs walked through the old city and went to a Greek monastery. In the afternoon he and Nachmansohn talked with other officials regarding the cancer institute. On Thursday afternoon they held a meeting with a number of the people they had already met individually, in the office of a Mr. Schocken, who was presumably an administrative officer of the university. This meeting was apparently a setback to whatever hopes Krebs and Nachmansohn may have entertained for an additional section of the institute headed by Krebs. Krebs summarized the outcome by noting that "Schocken points out that he is unable to do anything for further developments, since he has to consolidate the existing parts." The impression with which Krebs left was that facilities could be made available, but that those already present were unwilling to give up any space.[86]

Krebs and Nachmansohn stayed for six more days in Palestine, but Krebs neglected the diary he had kept until then, so that the remainder of their itinerary is lost. During that time they met with Magnes, which probably only confirmed that the University of Jerusalem had little to offer them. They also spent much of the time with Weizmann, both at his Daniel Sieff Institute in Rehovoth, and on their tours of biblical sites. Weizmann advised them not to consider immigrating until they were satisfied that they would have adequate research facilities. It is not clear how much he offered to do himself toward this end. Prior to the trip Nachmansohn had written Krebs that Weizmann intended, after his return to England in May, to undertake a personal fund drive for a science faculty in Jerusalem, the tentative design for which included £20,000 for an Institute for Krebs. The intervening political events may, however, already have upset Weizmann's plans by the time they saw him in Palestine.[87]

While he was in Palestine, Krebs received some discouraging news transmitted by his parents in San Remo. They had received, in the same mail as his "nice" letter from Alexandria, a "less happy" one from Wolf. Just as he and his wife had moved into a temporary apartment in Berlin in preparation for emigrating, they learned that he had been turned down not only in England but for a position for which he had applied in Switzerland. "England," Wolf wrote, "gave as a reason the uncertainty of the political situation. We will not be put off by these failures, but will search immediately for new possibilities." Meanwhile he would follow advice he had received to "remain at AEG until a termination notice is given." It is not evident whether he had reason to believe that such a notice was imminent, or only a general possibility. In forwarding this news to Hans, Georg Krebs suggested again that "perhaps a possibility exists for you to do something for your brother. Whether the setbacks in his efforts up until now are his own fault (as you will be inclined to accept), or due to adverse circumstances, I cannot decide. One must . . . try to help him."[88]

Krebs and Nachmansohn sailed from Palestine, by way of Port Said, on April 15. The visit had been intense for both of them. Most impressive to Krebs was the evidence he saw that the Jews in Palestine were "building up something" in what had been a barren area. Tel Aviv was rising on former sand dunes. At the *Kvutzah* and agricultural station they saw deserts were being converted into fertile lands. The encounters with Germans, some of them friends of his own family or relatives, who had begun new lives here were emotional for him. The prospect that he himself might stake his professional future on Palestine was, however, more remote when he left than when he had arrived. As he sailed westward across the Mediterranean his thoughts must already have been turning toward the work he had interrupted in Sheffield. On Monday, April 20, he and Nachmansohn arrived in Marseilles, took the train to Paris, and spent their last evening together. On the 21st Krebs passed through London and reached Sheffield.[89] Three days later he wrote Norman Edson in Cambridge, "I have now safely returned to this country and I am very anxious about the various branches of research which you and your pupils have carried out during my absence."[90]

IV

On March 12, shortly before Krebs left for Palestine, Wayne had received word that the Medical Research Council would not act until April 1 upon his grant application for an assistant for Krebs. Wayne wrote back immediately to ask if it was possible "to expedite the decision." "We have," he wrote, "a great deal of work for an assistant to carry out and I also feel that it is unfair to keep the prospective employee looking out for an alternative post." The council agreed to consider the application at its meeting of March 20, and was able to inform Wayne on March 24 that his proposal had been accepted. The grant was to take effect on April 1.[91]

It is somewhat mystifying why Wayne pressed the council for so quick a decision on the grounds of the amount of work to be done, considering that Krebs himself was to be away for several weeks. In any case, Krebs must have expressed to Wayne his keen interest in acquiring Johnson as an assistant. Johnson had an honours degree in chemistry and a strong recommendation from the head of the Chemistry Department. He was viewed as the top person in his class, and Krebs thought him to be very bright. His considerable training in organic syntheses could be especially helpful for making compounds needed in metabolic experiments that were not commercially available. Apparently Krebs was confident enough about the council's decision in advance to make arrangements for Johnson to begin while he was gone. Because there was no time then to train Johnson in the biochemical methods used in the laboratory, Krebs assigned him a list of organic compounds to synthesize during his absence. Presumably Johnson started on this task on April 1.[92]

When he returned to the laboratory, Krebs began to teach Johnson how to use the Warburg manometers. As Warburg had done 10 years earlier with him, Krebs first asked Johnson to stand behind him and watch while he himself performed the operations. Johnson, who had been familiar with manometers only through descriptions in the literature, was fascinated to see that they looked exactly as described. He was also impressed with how busy one had to be to take readings every five minutes on each of a row of eight or more manometers. Krebs found Johnson a quiet, humble but composed young man who was very willing to learn and learned rapidly. Soon he gave Johnson responsibility for two Warburg water baths with their full complements of manometers. Johnson quickly perceived Krebs as a determined, incisive man, and also a stern taskmaster. He "would not tolerate looseness in any way," as Johnson later put it. One of the first things Krebs told Johnson was "I expect you to keep well-abreast of the literature, but if I ever see you in the library I'll think you've been wasting your time."[93] What Krebs had in mind with this conundrum was that the "best hours of the day should be spent in the laboratory. That one could read the literature in the evenings and off moments."[94] That is what Krebs did, and it is evident that he did not view the responsibilities of a research assistant as limited to the normal hours of his working day. Johnson soon saw that Krebs neither wasted any of his own time nor expected those who worked with him to waste theirs. Johnson was to carry out a set of experiments each day, to complete the results that night, and to be ready to present them the next morning.[95]

Having grown up in rather straitened circumstances in a steelworking town in northern Yorkshire, Johnson was accustomed to hard work, and did not in general mind Krebs's disciplined regimen. Krebs's rigid expectation for punctuality, however, was sometimes trying. As Johnson began his experimental work, he often had to stay late at night, especially when there were organic syntheses to complete. If he left the laboratory at midnight and appeared a few minutes late the next morning, he would find Krebs looking at the clock or his watch. Johnson felt also that he received little encouragement from Krebs, so that it was hard for him to know how well he was doing.[96]

Part of the difficulty was cultural. Krebs equated punctuality with reliability, as many people did in the German society in which he had been reared. Once he went to Wayne to complain about the fact that his assistant—it may have been either Johnson or Eggleston—was turning up late in the morning. Wayne asked whether the person was putting in a full week's work, and whether

Figure 6-4 William Arthur Johnson in 1937. Photo Courtesy of W.A. Johnson and M. Wainwright.

Krebs was also sometimes keeping him at work at night. When the answer to both questions was positive, Wayne suggested that in this country people were more relaxed about the time, so long as the work was getting done.[97] The sparseness of encouraging words to Johnson was also as inherent in Krebs's background as in his personality. He had grown up in an environment in which he himself had seldom received praise, and in which good work was taken for granted.[98]

Krebs was, on the other hand, remarkably tolerant of mistakes that Johnson made, such as pouring in the fluids from two side bulbs of a manometer flask at once when they should have been done in sequence, or the more costly one of breaking a manometer.[99] If Krebs was supportive in such instances it was in part because he was sensitive to the fact that he himself often made mistakes.[100] He could more easily appreciate that others would also make mistakes than that others could work efficiently under looser time schedules than his own.

We have seen that on several previous occasions when Krebs resumed work after an extended time away from the laboratory, he had taken up a line of investigation different from what he had been following prior to the interruption. This time the opposite occurred. Whatever pondering he did over his work during his journey to the Middle East evidently reinforced his sense of the importance of the investigation of the anaerobic disappearance of pyruvic acid that he had got under way before he left. It is not clear how far along he was in formulating the processes to which pyruvic acid might be connected under these conditions, but we can plausibly surmise that he was persuaded that reactions similar in form to the dismutation reactions he had already proposed to himself held a clue that could unlock the big unsolved problem of the intermediate metabolism of carbohydrates.

Krebs did not continue from the same point at which he had left off. Instead of pursuing the specific connection between pyruvic and succinic acid that Szent-Györgyi's methods had given him occasion to explore, he resumed at a prior stage of the investigation. On Monday, April 27, he extended his earlier effort to determine the ratio of pyruvic disappearance to lactic formation in rat testicle tissue. He operated on a large scale, with duplicates, employing the ceric sulfate method for pyruvate and a method devised by S.W. Clausen to determine small quantities of lactate. The method converted the lactate to acetaldehyde, which was distilled off and titrated with iodine. Following Clausen, Krebs applied a "copper lime treatment" to separate glucose and other interfering substances. "Δpyr." expressed as a volume was 2190 μl; "Δlact." was 963.[101] In distinction from the earlier proportions of pyruvate to lactate, of 1.56:1 and 1.86:1, this one was 2.27:1, suggesting perhaps a theoretical 2:1 ratio. Did it also suggest a specific reaction to Krebs? If so, he did not put it down. His next experiment suggests that at this point Krebs was retreating somewhat from his earlier efforts to test specific hypothetical reactions, in order to establish first some more general features of the type of reaction involved.

The first general question that Krebs wanted to answer was whether the increases in CO_2 production from testis tissue in the presence of pyruvate that he had already observed derived directly from a breakdown of pyruvic acid or from a displacement of bicarbonate from the medium caused by the formation of acids. To distinguish between these possibilities he set up on April 28 an experiment to measure both the immediate change in CO_2 production due to pyruvate during the course of a run and the total change in CO_2 after he had displaced the bicarbonate from the solutions by himself adding an acid. He used four cups, two with tissue and two without, one of each of these with and one without pyruvate. At the end of two hours he tipped sulfuric acid into each medium from a side bulb. By subtracting the difference between the change in CO_2 prior to the addition of the acid in the run with pyruvate and tissue from the run with tissue alone, he calculated that the "extra CO_2 through pyruvate . . . = 109.5 μl." To determine the "increase in total CO_2" due to pyruvate, he had first to subtract from the reading after the addition of H_2SO_4 in the run with pyruvate and tissue the corresponding reading for pyruvate with no tissue, and from this difference subtract the equivalent difference obtained in the two runs without pyruvate. This increase in total CO_2 turned out to be exactly the same as the "extra" CO_2, 109.5 μl. At the end of the experiment Krebs put down the general, but highly positive conclusion:

Thus proved that

pyruvate $\rightarrow$ CO_2 + fixed acid.

He noted further, "pyruvate not determined."[102]

This simple shorthand notation is another example of the immediate response to an experimental result that David Gooding has called "construals" to distinguish them from fully formulated interpretations (see above, p. 117). There lay behind Krebs's conclusion, however, a longer chain of reasoning. The result "proved" that pyruvate gives rise directly to CO_2 because the equality between the "extra CO_2" and the "increase in total CO_2" in the presence of pyruvate indicated (through a subordinate chain of standard reasoning) that the bicarbonate concentration in the medium had not changed through the addition of pyruvate, and therefore that increased acid in the medium was not the source of the CO_2. But how did the same result prove that the pyruvate also gives rise to "fixed acid"? That was how Krebs construed the fact that the bicarbonate in the medium did *not* change. Had the pyruvic acid that broke down not yielded also an equivalent amount of another acid, the medium would have become more alkaline, and there ought to have been an absorption of CO_2. The residue of the pyruvic acid must have included another fixed acid binding the sodium ion initially bound to pyruvate.[103]

We may well ask whether or not Krebs anticipated this outcome. Was it predicted by a particular hypothetical reaction of the form pyruvic acid $\rightarrow$ CO_2 + fixed acid that he had in mind, or was the exact equality between the "extra CO_2" and the "total CO_2" a surprising outcome from which he drew the general inference? No conclusive answer can be given. Up until this point Krebs had not written down in his notebook any chemical equations to express the anaerobic disappearance of pyruvic acid added alone. On the other hand, he had clearly formulated and tested an equation for a reaction between pyruvic and acetoacetic acid (see above, p. 246) that was so similar in form to a reaction fitting the general formula in question that would result from the substitution of a second molecule of pyruvic for the acetoacetic acid that it is unlikely that he had not thought of it:

1) pyruvic a. + acetoacetic a. + H_2O = β-hydroxybutyric a. +
acetic a. + CO_2
2) 2 pyruvic a. + H_2O = lactic a. + acetic a. + CO_2

Whatever he had in mind before the experiment, Krebs could afterward conclude with confidence that he had established two basic features of a reaction, or reactions, whose details he could now explore. He had not yet identified the "fixed acid" (or acids), and because he had not determined pyruvate in the experiment, he did not yet know whether all or only a part of the pyruvate consumed yielded CO_2. He had, however, a solid foundation on which to build. (As on other occasions, the confidence he probably placed in this result at this stage in his investigation can be inferred indirectly from the fact that he later presented this specific experiment to support the same conclusion in the publication to which the investigation eventually led.)[104]

On April 30 Krebs repeated the experiment, this time measuring the pyruvate (by means of the ceric sulfate method.) A null result at the end of the experiment indicated that all of the pyruvate added had been used. From the initial quantity used, he estimated that amount as the equivalent of 448 μl. This time the increase in free CO_2 due to pyruvate (161 μl) did not coincide so exactly with the increase in total CO_2 due to pyruvate (180 μl), although the two figures were probably not far enough apart to disturb the conclusion he had previously drawn. He fixed his attention on the comparison between these quantities and the pyruvate. There was, he commented "less CO_2 than ½ pyruvate."[105] That he put it that way implies that, if he had not done so before the previous experiment, he was by now thinking in terms of a dismutation reaction of the type mentioned above, in which two pyruvic acid molecules yield one CO_2 molecule. This experiment, however, did not fit very well with the expected stoichiometric proportions.

Had Krebs been strongly convinced at this time that he was dealing with a specific single reaction, we might expect him to have pressed forward in the same direction, seeking better agreement between the measured and the theoretical proportions. His next moves suggest instead that the situation appeared more open ended to him, and called for a more general approach. On May 2 he compared the rates of pyruvic disappearance anaerobically and aerobically in three rat tissues; brain, kidney, and testicle. Only in the brain did he obtain full comparative results. They were Q_{pyr} in O_2 = -9.28; Q_{pyr} in N_2 = -7.04. This outcome surprised him. He wrote down, "<u>less</u> anaerobically than aerobically!! Is possible? not accurate, repeat!" When he did so two days later, however, extending the measurements to five tissues, he only confirmed that the anaerobic rate appeared consistently lower than the aerobic rate:[106]

	Q_{pyr} in O_2	Q_{pyr} in N_2
Liver	-8.75	-3.74
Kidney	-16.8	-8.1
Spleen	-5.45	-2.72
Brain	-8.32	-6.98
Duodenal wall	-2.94	-2.09

Since the rapid oxidation of pyruvic acid in tissues in the presence of oxygen was a well-established phenomenon, it appears unclear at first why Krebs doubted that the anaerobic process he was exploring could be slower than the aerobic one. The most plausible explanation is that he viewed the anaerobic reactions of pyruvic acid he was studying as part of the main pathway for the intermediary breakdown of carbohydrates. According to the basic axiom of the field, therefore, these anaerobic intermediate steps ought to proceed at a rate at least as large as the overall reaction.

Krebs now turned to the other element of the general reaction he had identified, the specification of the "fixed acid" formed. Returning to the direction he had taken just before Palestine under the influence of Szent-Györgyi's work, he examined on May 5 the formation of succinic acid in pigeon breast muscle. He tested pyruvate and pyruvate + acetate. Neither added substrate increased the yield of succinate, and acetate *decreased* it in comparison to the control, but in all three a "large amount of succinic formed!!" He remarked, "try washed muscle next time," implying that he suspected that substrates already present in the tissue had masked the effects he was looking for from the substrates added in the medium. On the 6th he switched to the metabolism of yeast, testing the effects of pyruvate anaerobically and aerobically, in comparison to the effects of glucose and a control. He intended to measure the quantity of succinic acid formed, but the acid he added from the side bulb after the run failed to kill the yeast, so the determination was "useless." He did not pursue this digression further. The most likely acid to be formed in a reaction in which pyruvic acid was decarboxylated, whether or not it was the dismutation reaction Krebs was probably entertaining at this time, was acetic acid, and it was to that possibility that he next returned. He had already begun a large-scale control experiment on the 4th, without tissue, presumably to check the spontaneous chemical conversion of pyruvic to acetic acid. After leaving it for two days he measured the "volatile acid" formed, using a steam distillation method. On the same day Johnson carried out an experiment on the formation of acetic acid from pyruvic in rat testis tissue, employing the same analytical method. The volume of volatile acid found corresponded to 34 percent of the pyruvic acid used. Krebs regarded the method as essentially specific for acetic acid, providing that most of the pyruvic acid was used up.[107]

On May 7 Krebs carried out two parallel sets of experiments on rat testis tissue, asking respectively the questions, "acetic from pyruvic?" and "succinic acid?" [from pyruvic]. In the first case the pyruvate was all used, and he obtained volatile acid in "72% of expected quantity." From the data we can see that the "expected" quantity was a ratio of one molecule acetic: two molecules pyruvic, that is, the proportions predicted for the reaction

2 pyruvic acid + H_2O = lactic acid + acetic acid + CO_2

In the experiment concerning succinic acid, he tested for it by adding succinic oxidase after the run. He noted only "considerable pressure with succinoxidase!!" Apparently he did not calculate quantities of succinic acid formed because he wondered, "is reaction specific?"[108]

Where, during this stage of the investigation, did Krebs stand in his theoretical formulation of the problem? The clues contained in these last two experiments are mixed. The paired leading questions suggest that he was still exploring openly what the "fixed acid" was that derived from the anaerobic breakdown of pyruvic acid and that he gave equal attention in this respect to acetic and succinic acid. The reference to the "expected quantity" of acetic acid points, on the other hand, to the more definite hypothesis that the reaction in question was the specific dismutation of pyruvic acid yielding acetic acid, mentioned above. The experimental ratios of pyruvic to lactic acid, acetic acid, and CO_2 so far obtained did not strongly support a conclusion that this reaction alone was occurring. The quantities were consistent, however, with the view that a large part of the pyruvic acid that disappeared could be accounted for by the reaction. There was probably no single experiment that induced Krebs at some point to accept that specific reaction in the same sense that one experiment had "proved" for him the existence of the more general form of the reaction. During this period, however, as he gathered evidence compatible with the reaction, he gradually gained confidence in its occurrence.

What then of the yields of acetic acid and CO_2 in less than the "expected" proportions? Did they indicate that pyruvate also entered other anaerobic reactions? If so, could the formation of succinic acid account for the rest of it? In that case, was the inference to be drawn that these were both "primary" reactions, or did some of the products of an initial reaction disappear in secondary reactions? Or would further experiments improve the above ratios and show that all of the pyruvic acid was accounted for, after all, in a single reaction? At sometime during this stage of the investigation Krebs probably entertained all of these alternatives, but did not have sufficient evidence to choose among them.

The experiment by Johnson mentioned above is the earliest one carried out by him of which there is a record. It indicates that by this time Krebs's new research assistant had already mastered the preliminary exercises that Krebs had given him and was beginning to participate in the ongoing research of the laboratory. Johnson's laboratory notebooks have not survived, so there is not a full record of his daily experiments. The experiments Johnson performed during his first year with Krebs were closely integrated with Krebs's own main lines of investigation. From this point onward, therefore, in following the course of Krebs's research, we cannot assume that all of the experiments relevant to his views and decisions at a given time are recoverable. The general course of Johnson's experiments can, however, be reconstructed from his personal copy of the Ph.D. thesis he completed in 1938, in which he marked in the dates on which each of the experiments he selected to include in it were performed. With this record, and in view of Krebs's recollection that he normally tried out new types of experiments himself before putting his assistants onto them, we can be reasonably certain not to miss significant turning points along the trail.

On the same day that he performed the paired experiments on rat testis just described, Krebs found that in sheep testis "pyruvate leads to increase of succinic acid formation." After a three-day break he tested on May 11 the "influence of washing pigeon muscle on dismutation of pyruvate" and observed that washing finely minced tissue three times "reduces CO_2 production, [but] . . . does not remove it." The title of this experiment implies that by this time he was so confident about the occurrence of the dismutation reaction in question that he could assume, when he added pyruvic acid to a tissue anaerobically, that the increase in the rate of formation of CO_2 was a measure of that reaction. On the same day he tested the effect of pH on CO_2 production from pyruvate in rat testicle tissue and found that the rate "increases with increasing alkalinity!!" Using the same tissue he tested the effect of iodoacetate on CO_2 production and concluded that it "inhibits little!" although the "blank would be inhibited." The experiment itself is a little unclear, since there is no indication that it was anaerobic or that it included pyruvate. The idea that he drew from it, however, that it is "possible to separate this reaction from glucose fermentation!!"[109] is very interesting. For it he reached back to the iodoacetate experiments he had conducted in 1931 under the influence of Einar Lundsgaard's discovery that iodoacetate inhibits lactic fermentation in muscles (see Vol. I, pp. 221–235). Krebs had shown then that adding lactate could restore the respiration, thus separating the subsequent oxidative stages of glucose metabolism from glycolysis. Now he proposed to himself in a similar manner to separate the dismutation reaction from glycolysis.

The next day Krebs returned to the "ratio CO_2/pyruvate" in rat testis. He measured only the "direct" extra CO_2 formed from pyruvate and the decrease in the latter, and in duplicate runs attained ratios of "$\Delta CO_2 : \Delta pyr.$" of 1:2.63 and 1:2.52, compared to "theory, 1:2.50." In the notebook he gave no reason for assuming this theoretical ratio rather than the stoichiometric proportion of 1:2 for the dismutation reaction; but it may have been that he considered that under the given conditions one-fifth of the CO_2 produced remained in solution as bicarbonate. In any case it appears that he had attained a measured CO_2:pyruvate ratio that accorded well with that expected for the reaction in question. Simultaneously he tested the effect of iodoacetate on the reaction and found "no inhibition."[110] Iodoacetate thus did appear able to separate the dismutation reaction from glycolysis.

Now that Krebs had attained CO_2:pyruvate ratios approximating the theoretical ratios for the dismutation reaction, while the lactate:pyruvate ratios measured earlier bracketed the theoretical value, only the yields of acetic acid remained markedly lower than expected for the reaction. It may have been that situation that led Krebs at this point to look for a reaction mechanism that would remove some of the acetic acid produced in the dismutation reaction. Whether or not this was what he had immediately in mind, he did perform on May 13 a set of experiments designed to test a reaction which not only would have that effect, but would connect the process to the formation of succinic acid that he had observed. Under one of these experiments he wrote out the reaction[111]

```
COOH                COOH
 |                   |
CH₃      +    CH₃  → CH₂    +    CH₃
 |            |      |            |
CH₃          CO      CH₂          COH
 |            |      |            |H
COOH         COOH   COOH         COOH
```

That is, two acetic acid + pyruvic acid = succinic acid + lactic acid. This reaction would have several convenient features. It would remove acetic acid without removing lactic acid, and would produce succinic acid. It was, moreover, another anaerobic intermolecular oxidation-reduction reaction, the oxidative condensation of two molecules of acetic acid to form succinic acid being coupled to the reduction of pyruvic to lactic acid. The oxidative condensation reaction also appears to link the reaction with that critical synthetic step in the well-known Thunberg-Wieland-Knoop scheme for oxidative metabolism. If the present reaction could be demonstrated, therefore, it would close a prominent and long-standing gap in intermediary metabolism. It is not evident that Krebs had any independent source for this hypothetical equation, and plausible that he designed it on paper to fit these several functional desiderata.

Although the postulated reaction was anaerobic, Krebs tested for it aerobically in rat testis tissue. He carried out two experiments on this subject, on May 13. The first compared the rate of absorption of oxygen with no additions, with 0.02 M pyruvate, 0.02 M acetate, 0.02 M acetate + 3.3 X 10^{-4} M pyruvate, and 3.3 X 10^{-4} M pyruvate alone. The principal results were "an enormous increase by 0.02 M pyruvate" (Q_{O_2} = 14.1; control, 6.71), "little by 3.3 X 10^{-4} M pyruvate and by acetate," and "no effect of acetate + little pyruvate." In the second experiment, in which he added insulin in three concentrations to acetate + pyruvate, he observed "no effect of insulin."[112] At first sight it appears odd that he should have tried to detect an anaerobic reaction in the presence of oxygen. If, however, such a reaction led to the main pathway of oxidative carbohydrate metabolism, as he apparently supposed it might, then he would expect it to increase the oxygen consumption. In any event, the initial negative result did not dissuade him from pursuing the idea that this reaction, or another reaction of the same general form, connects the initial dismutation reaction with the formation of succinic acid (see Fig. 6-5).

On the same day Krebs began another effort to determine the lactic:pyruvic ratio during the anaerobic disappearance of pyruvate. He measured the pyruvate remaining at the end of the experiment by the ceric sulfate method, and obtained for pyruvic 2520 μl. The lactic acid formed he measured by two methods, that of Clausen and the gasometric method of Avery and Hastings that he had first tried two months earlier in a similar experiment. The first method gave 1145 μl; the second, 1570. The discrepancy led him to conclude that the Avery-Hastings method "gives too high figures!! In copper lime filtrate!! blank as well as experiment. There is a substance which forms CO_2 present, and formed from pyruvate!! See Avery-Hastings. try malic acid." He expressed his sense of the situation forcefully:

ratio lactic :Clausen 1145 Hastings 1570 Wrong!!
     pyruvic 2520   2520

The next day he verified that "malic acid interferes, reacts!!! in pure solution (perhaps precipitated by Ag??)." Moreover, he decided that the pyruvic determination too "gives wrong figures after treatment with Cu-Ca-(OH)$_2$. great

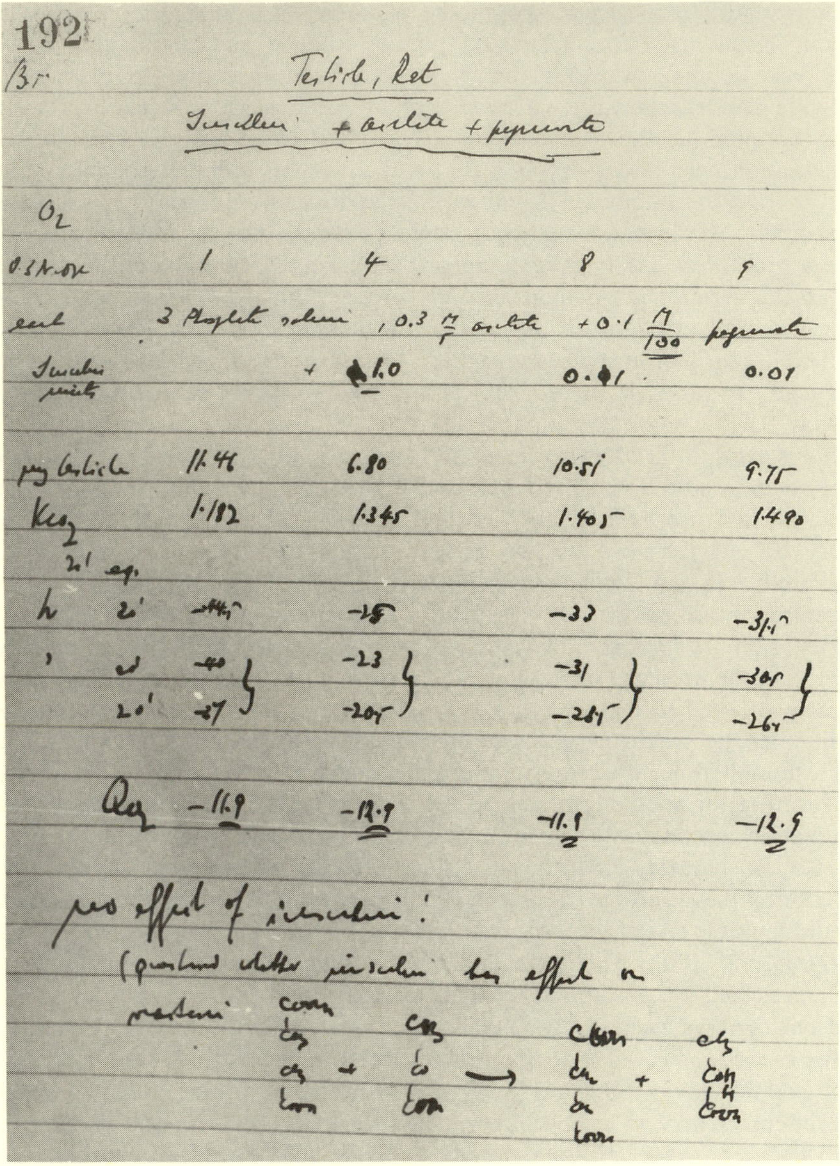
192
O_2
1 4 8 9
0.01
11.46 6.80 10.51 9.75
1.182 1.345 1.405 1.490
Q_{O_2} −11.9 −12.9 −11.9 −12.9
no effect of insulin!

Figure 6-5 Krebs, laboratory notebook, second experiment performed on May 13, 1936.

loss (precipitated!? destroyed by alkali:) .. loss 70%!!!" (This error in determination of the remaining pyruvate would result in a value too large for pyruvate.) Repeating the experiment, he relied on the Clausen method for lactate and modified his procedure for pyruvate so as to avoid adding the copper-lime salts to the portion of the solution used for the determination. He made duplicate determinations of the lactate, which differed somewhat, and yielded respectively 53.3 and 58.0 percent of the pyruvate.[113] Thus, after eliminating a number of methodological pitfalls, he obtained proportions reasonably near to the 50 percent expected from the dismutation reaction. In evaluating and correcting these methodological errors, Krebs was undoubtedly guided in part by the theoretical ratio that he was by now seeking to support.

Thus far Krebs had concentrated his study of the anaerobic disappearance of pyruvic acid on rat testis slices, because that tissue had proved most favorable for the purpose. Now that he had obtained considerable evidence for the dismutation reaction in testis tissue, it was important to broaden the investigation again to other tissues. The reaction could not have the general metabolic significance he thought it might have if it were limited to one or a few special tissues. On May 18 he measured the rate of disappearance in 10 different rat tissues. The anaerobic $Q_{pyruvate}$ values were, in general, similar to those he had obtained for several tissues two weeks earlier when he had compared these with the aerobic rates. Only in liver and brain (and testis) did the rates appear significant: $Q_{pyr}^{N_2}$ in testis, -11.2; in brain, -6.53; in liver, -8.77; in other tissues, -1.73 to -4.17. In an experiment on pigeon red blood cells the same day, he found pyruvate caused pressure changes only slightly greater than the control, and questioned whether there was any effect of pyruvate.[114]

On May 21 Krebs returned to the search for reactions that might produce succinic acid from pyruvic. In rat testis tissue he tested pyruvate + ketogluta-rate, in comparison to pyruvate alone. He did not write down a specific reaction, but it is likely that he had in mind the possibility that α-ketoglutaric acid + pyruvic acid + water = succinic acid + carbon dioxide + lactic acid. The experiment itself did not work out fully, because the "succinic determination [was] not successful"; but from the pyruvate measurements alone he was able to infer that "in testicle ketoglutarate does not effect disappearance of ketonic acids total (not metabolized??)." Perhaps because, as his question implies, rat testis tissue did not metabolize ketoglutaric acid in general, he turned the next day to minced pigeon breast muscle. Now he explored simultaneously the above reaction possibility and the reaction of acetic and pyruvic acid that he had postulated a week earlier. The ceric sulfate test indicated, however, that all of the ketonic acid may have been used up or precipitated with the protein, in the control as well as in the runs with the additions. The effects of the latter were therefore not measurable.[115]

Before he could go further in this latest direction of his investigation, Krebs was diverted to a new factor in the situation brought to his attention by an article in the May 23 issue of *The Lancet*. Entitled "The Biochemical Lesion in Vitamin B$_1$ Deficiency: Application of Modern Biochemical Analysis in its Diagnosis," the article gave the "substance" of a lecture recently delivered by Rudolph Peters, the professor of biochemistry at Oxford. Peters summarized the latest stage of a research effort that he and his associates had been conducting for nearly a decade. In 1927 they had begun with the assumption that acute vitamin B$_1$ deficiency—a condition manifested outwardly in pigeons as "opisthotonus," in which the bird's head nearly touched its back or it displayed cartwheel-like convulsions—is due to abnormalities in the metabolism of the brain. In their early work they found that lactic acid accumulated in parts of the brain of avitaminous pigeons. Switching from postmortem autopsies to tissue slice methods, they showed that in brain tissue from birds with vitamin B_1 deficiency the respiration in the presence of glucose was lowered, and lactate accumulated. The addition of catalytic quantities of vitamin B_1, at first in concentrates and later, when it became available, in pure crystalline form, restored the normal respiration. At this point they concluded "that vitamin B_1 was needed for the oxidative removal of lactic acid."[116]

In order to identify the stage of the oxidative process at which vitamin B_1 acts, Peters and his associates began adding the various known intermediary substrates to the medium of respiring brain tissue slices. With lactate, pyruvate, and glucose, normal tissue and avitaminous tissue responded differently, but not with succinate as the substrate. "Hence we can say definitely," Peters asserted, "that B_1 deficiency is affecting the sugar metabolism at some point related to the 3-carbon stage." The expectation that it was necessary for the removal of lactic acid was refuted when one of Peters's associates found in 1933 that although vitamin B_1 increased the uptake of oxygen in avitaminous brain tissue, it did not cause lactic acid to disappear. Turning to pyruvic acid, another associate then observed, among other things, that pyruvate was never present in normal respiring brain tissue, but was formed from lactate "with avitaminous brain in vitro in absence but not in presence of vitamin"; that "vitamin caused the disappearance of added pyruvate"; and that "the ratio $\frac{\text{extra oxygen taken up}}{\text{pyruvic acid disappearing}}$ with addition of vitamin" indicated that oxygen was consumed in roughly the same amounts in which the extra pyruvate disappeared. "We concluded," Peters stated, "that the vitamin was definitely concerned with the removal of pyruvate, but not directly. The experiments clinched the importance of pyruvic acid as an intermediary metabolite." Peters summarized the metabolic situation with the following scheme:[117]

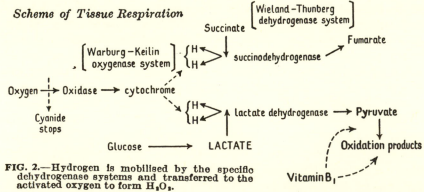

FIG. 2.—Hydrogen is mobilised by the specific dehydrogenase systems and transferred to the activated oxygen to form H_2O_2.

Krebs's reaction to Peters's article can best be gauged from the fact that he carried out his first experiment on the effect of vitamin B_1 on pyruvate in tissues on the very day that it appeared. In his notebook he wrote, "See Peters's *Lancet* of today." He must have read the journal first thing in the morning, abandoned the experimental plan he had for that day, obtained a supply of commercial vitamin B_1 (manufactured by British Drug House), and begun at once. Nor is it difficult to explain his enthusiasm. As the diagram in the *Lancet* paper emphasizes, Peters left it undetermined what step in the oxidation of pyruvate requires vitamin B_1, an inevitable omission, because those steps themselves were part of the "big unsolved problem" of intermediary metabolism. If, as Krebs now supposed, the anaerobic dismutation reactions he was studying were interposed between pyruvate and those aerobic steps, then there was a good chance that it was these reactions on which vitamin B_1 directly acted. Conversely, if vitamin B_1 did prove necessary for the dismutations to occur, that would be strong support for his hypothesis. In the experiment, he measured the formation of CO_2 (whether anaerobically or aerobically is not stated) in rat brain, testicle, and liver slices in the presence of pyruvate with and without vitamin B_1. For liver he did a second run with vitamin B_1 + pyrophosphate, which Peters had found to enhance the effect of the vitamin. Krebs found "no effect of vitamin B_1, or B_1 + pyrophosphate." He concluded only that B_1 was "thus not limiting factor," a very reasonable assumption because he had undoubtedly utilized rat tissues that he had intended to use for a different experiment and that were not likely to be vitamin B_1 deficient. He left as an open question, is "B_1 necessary for dismutation?"[118]

On May 26 Krebs tested the effect on the "pyruvate reaction" of potassium chloride, which caused a "slight inhibition . . . anaerobically!!!!" Evidently surprised, he noted, "try in O_2." He went on instead with vitamin B_1, now confronting more directly the work of Peters by testing it with pigeon brain tissue. The Q_{CO_2} values were control, 1.40; pyruvate, 2.51; and pyruvate + vitamin B_1, 2.68. The main result of the experiment was thus again, "No effect of vitamin." Again he suspected the conditions were unfavorable: "reaction is too slow!!; inhibited by succinic?" His attention was also drawn to an unexpected circumstance within the control aspect of the experiment: "Extra CO_2 with pyruvate only 1–2." That relationship prompted him to speculate about the possible underlying chemical relationships that such proportions might express:[119]

> 1 extra CO_2 corresponds [to] O_2 for oxidation of acetic to ½ pyruvate
> or 1-½ O_2 " " of succinic to pyruvic
> +½ " " of lactic to pyruvic

It appears that he was reasoning on the basis of intermolecular oxidation-reduction reactions that he did not write down, in which acetic, succinic, or lactic acid would be oxidized while pyruvic is reduced. The interest in these notes lies in the glimpse of Krebs scanning even subordinate, contingent relationships in his experiments for leads to further possible theories to test.

The next day Krebs widened his purview, as he characteristically did when a narrowly focused attack did not seem to be yielding positive results. Reverting to rat testicle tissue, he tested the "effect of ions, of succinate, malonate, B_1

[and] insulin" on the "pyruvic acid reaction." Although succinate and B_1 + insulin raised the anaerobic Q_{CO_2} values somewhat, he concluded that there was "no definite effect. Certainly no effect of B_1 + insulin, since same pressure before and after addition." There was also "certainly no inhibition by malonate." He did wonder whether there was evidence of an "increasing effect of succinic???"[120]

After three experiments on, or including, the effects of vitamin B_1 had produced nothing positive, Krebs returned to the line he had been following before the *Lancet* article appeared. He did not abandon his idea that the vitamin might be necessary for the dismutation reaction, but his failure to achieve a quick confirmation evidently lowered his immediate interest in that question below the threshold at which he was willing to prolong the interruption of a promising phase of his main investigation.

Taking up where he had left off four days earlier, Krebs repeated on May 27 the experiment on pigeon breast muscle in which he added pyruvate, pyruvate + ketoglutarate, and pyruvate + acetate and tested for the formation of succinic acid. Because he had suspected in the previous experiment that the ketonic acids were used up, he employed one and a half times as much pyruvate this time. Measuring succinate by the succinoxidase method, he found for the control, for pyruvate, and for pyruvate + acetate quantities ranging from 212 to 236 μl (expressed in microliters CO_2 released in the enzymic reaction), whereas for pyruvate + ketoglutarate it was 1216. "Enormous increase with ketoglutarate!!!!" he commented.[121] These results thus simultaneously appeared to rule out the reaction 2 acetic acid + pyruvic acid = succinic acid + lactic acid, and to indicate that one of the reactions he was looking for could be α-ketoglutaric acid + pyruvic acid + water = succinic acid + carbon dioxide + lactic acid. That outcome gave new direction to the investigation, but it was somewhat less satisfying than if the outcome had been the other way around. As we have seen, the first of these reactions might have linked the initial pyruvic acid dismutation reaction directly to the formation of succinic acid, and connect in addition to the Thunberg-Knoop-Wieland scheme. The second reaction did not lend itself to such a neatly closed system. It might be either an alternative route for the anaerobic disappearance of pyruvic acid or a reaction indirectly connected to the initial dismutation reaction by some further, still unidentified intermediate steps.

Krebs now turned to a basic question underlying the whole of his investigation. His effort to ascertain whether succinic acid can be formed anaerobically from pyruvic acid was framed within the supposition that it can be formed oxidatively from pyruvic acid. This was a widely discussed expectation, but except for the evidence that Toenniessen and Brinkmann had provided from rabbit perfusion experiments in 1927, the reaction had yet to be confirmed.[122] Krebs examined the reaction in an aerobic experiment in rat testis tissue. Measuring the disappearance of pyruvic acid and using Gözsy's malonate method to determine the succinic acid, Krebs calculated that "only 12% of expected quantity found." (The "expected quantity" was based on the assumption that one molecule of succinic acid should be produced for each molecule of pyruvic acid consumed, even though the schemes of Toenniessen and Brinkmann postulated that two molecules of pyruvic acid condensed to give one molecule of succinic acid.) Krebs made no further comment on the minuteness of the observed quantity, but the most obvious explanation was that little succinic acid accumulated because it was further oxidized. On May 29 he started a similar anaerobic experiment but did not complete the succinic acid determination. The next day he tested succinic acid and malonic acid on the respiration of yeast and found that "succinic is not oxidized!!!" and "malonic does not inhibit!!!"[123] With these last three brief forays, the month of May ended. Krebs had made substantial progress in his investigation of the anaerobic disappearance of pyruvic acid, but he still faced major unsolved problems.

V

To his supportive chief Edward Wayne, Hans Krebs was a scientist who charted his own course, whose thinking was largely internal, and who did not pick up his ideas by contact with others.[124] Had Krebs discussed his work more fully with Wayne from day to day, it might have been more evident how much he depended on contemporary developments in his field. He also relied on personal interactions with colleagues more than was obvious to those around him in Sheffield, because he did so mainly by keeping in touch with his former associates in Cambridge. He went back whenever he had a reason to do so.[125] Among these associates, he particularly enjoyed discussions with Norman Edson. During the spring of 1936 he corresponded with Edson over various matters, including the final preparation of their joint paper on uric acid synthesis. Krebs drafted the paper, but he sought and valued Edson's comments on the writing as on the experimental work itself, the last stages of which were mostly carried out by Edson.[126] As Edson neared the end of his time as a student, he became in some senses a junior collaborator for Krebs, and, as Krebs's letter to Edson on his return from Palestine indicates, Edson was himself now directing the research of other students.

As Krebs moved into the central problems of intermediary carbohydrate metabolism, he kept in mind the close connection between this problem and that of the formation of ketone bodies; and as Edson continued the study of "ketogenesis and antiketogenesis" on which Krebs had started him nearly two years earlier, Krebs viewed that work as complementary to his own. Just before his trip Krebs had sent Edson a sample of acetylalanine, for purposes not specified in his accompanying letter, but which must have been connected with his own experiments involving that compound.[127] On April 28 Edson replied to Krebs's request for news on the progress of his little group in Cambridge. Beginning in the characteristic jocular style that so amused Krebs—"Again this miserable worm salutes you. I am glad to know you have returned safely, having escaped the rioters, and I hope you have had an enjoyable time"—Edson reported, among other items, that he had finished his thesis, that the editorial referee had "made a terrible mess of the introduction" to their uric acid paper, and that acetylalanine "was not ketogenic, sad to relate, and a nice theory collapses."[128]

In reply to Edson's "informative, amusing, and delightful" letter, Krebs wrote on May 4 that he hoped to "discuss the experiments soon with you in Cambridge."[129]   Krebs probably came to Cambridge sometime in May.   In addition to discussing the experiments in question, Krebs must have conveyed to Edson a sense of excitement about the new direction his work had taken and his hope that the study of the anaerobic disappearance of pyruvic acid could lead to a major integration of knowledge about the intermediary metabolism of carbohydrates and the ketone bodies. In the mixed language that he often used with Krebs, Edson dubbed Krebs's latest venture "the great Arbeit."   Krebs anglicized Edson's title and referred to it with Edson as "the great work."[130]

Relocations and Dismutations

While Hans Krebs returned from Palestine to an England in which he felt increasingly at home, his brother and sister were still seeking to leave a Germany that felt less and less like home. Wolf's setback with General Electric in Birmingham was particularly distressing to his family, because his prospects there had sounded so favorable and because the company had kept him dangling for four months. Late in April, Lise, worried again about her youngest brother, wondered what would become of him. "Every day," she wrote Hans, "the possibilities for leaving here and settling elsewhere become worse." As usual, however, the family underestimated the advantages that Wolf's engineering successes had brought him. The technical director of AEG had reassured him that he would be allowed to stay on until he could find a suitable position abroad and had armed him with a strong recommendation. It described him as "a very talented engineer, who holds various important patents. Concerning his achievements, diligence, reliability and leadership we express our fullest satisfaction with him." Encouraged by his energetic wife, Lotte, Wolf sent inquiries to every foreign electrical company he could find that might have openings. E.R. and F. Turner Ltd. of Ipswich, England, replied to him that the firm had a vacancy for chief technical engineer of its subsidiary Bull Motors, and invited him to come over for an interview.[1]

Arriving in Ipswich on May 4, Wolf spent the day talking with company officials. Bull Motors made quiet-running electrical machines for use in hospitals and lifts. Wolf had done extensive research on reducing noise in motors and was therefore particularly well qualified for the special needs of Bull Motors. He saw immediately that the company's engineering practices were amateurish by comparison with the exacting standards to which he was accustomed and that he had much to offer them. He came back to Germany pleased and hopeful, and stopped overnight in Düsseldorf, where Lise found him "to my joy in a very good state, physically and mentally." Then he went on to Denmark to follow up another possibility. Returning to Berlin, he received from the managing director of E.R. and F. Turner the news that the company had

Figure 7-1 Wolf Krebs in 1936.

concluded that he possessed "the experience necessary to take over control of our electrical designing department" and had "decided to definitely offer you the position now vacant at a commencing salary of £375 per annum." Wolf cabled his acceptance. The company then made application to the Ministry of Labour, to satisfy the requirements of the Alien Order, that the employment of a foreigner was "reasonable and necessary in the circumstances" and that every effort had been made to find a suitable British subject.[2]

Meanwhile, Lise and her husband were contending with the many obstacles they had to surmount if they were to carry out their decision to emigrate to Palestine. From the time that Hans helped persuade her in England the previous fall that this was the best course for them, they had "pursued our emigration affairs with all energy." They had written many letters, consulted friends, and visited numerous official bureaus, but "everything goes forward slowly." It was necessary first to apply to the Federal Ministry of Finance for authorization to emigrate. By January their application had been approved. There remained great difficulties in obtaining a large financial credit that they were required to deposit in Palestine. They were given a "number" by the Reichsbank, but then had to wait in line until their number was called up—a process that they anticipated might take months or years. There was also the problem of finding means to support themselves after they settled in Palestine. Lise hoped that Adolf could obtain a legal administrative position and that she might be able to resume her career in nursing. Correspondents in Palestine, however, did not paint a bright picture of their prospects. On the advice of friends they decided that they could assess the situation adequately only by making a preliminary trip to Palestine on a tourist visa, for which they applied—and for which they had also to wait, because the monthly quota was limited. Hans helped during his visit to Palestine by making some contacts for them in advance, and he wrote several letters of introduction for them to use when they arrived. There were tight restrictions on the money they could take with them on such an exploratory trip, so that when they learned that they would be able to leave in May, Lise forewarned Hans that she might have to telegraph him to send them money on the way. Meanwhile, the pressure on them grew when the company at which Adolf worked in Düsseldorf was sold and its new owners notified him that he would lose his job in September. All the while Lise was saddened by Georg Krebs's unmovable opposition to Zionism. "Father is not at all in agreement with our plans for the future," she wrote Hans in May, "but he can suggest no other solution for us."[3]

I

In June Krebs pressed onward with the "great work." He moved simultaneously on two subproblems into which the investigation now fell: to solidify his information concerning the primary pyruvic acid dismutation (2 pyruvic acid + $H_2O \rightarrow$ acetic acid + lactic acid + CO_2) and to follow up the most promising lead he had obtained concerning further reactions that might be connected to the basic reaction. On a busy Wednesday, June 3, he conducted five sets of experiments: two concerned with the first of these objectives, two with the second, and one encompassing the larger process under which these two were subsumed.

In the two experiments on the dismutation of pyruvic acid, Krebs tested the effects of several conditions—washing the tissue, adding an extract of liver tissue, adding lactate, and adding NH_4Cl—on the anaerobic formation of CO_2. Only the liver extract exerted an effect. Two experiments employed pyruvic + α-ketoglutaric acid. On one of them he wrote down, "problem is: ketoglutaric + pyruvic = succinic + lactic?? Thus anaerobic oxidation!!" Both experiments, however, measured the aerobic respiration of testis tissue. Apparently he had in mind that if the anaerobic reaction were a "primary step in the breakdown of ketoglutarate," it would increase the subsequent aerobic steps. In the first experiment the results were doubtful. In the second one, using washed tissue, he obtained some distinct effects: an "increase by ketoglutarate" Q_{O_2} = 9.4, compared to control 6.7) and an "additional effect" with pyruvate + ketoglutarate Q_{O_2} = 12.4). He considered the result still not decisive, however, because the blank respiration was large enough to have formed additional pyruvic acid. He thought he should "try other tissues," such as rabbit brain.[4]

In the first of these two experiments Krebs also tested citrate. There was no obvious relationship between this and the rest of the runs. Krebs probably added it mainly to employ a full set of manometers. This parenthetical test was, however, another manifestation of his persistent interest in turning up some new clue to connect the anomalous citric acid with other metabolic reactions. Here, however, there was "no effect of citric acid."[5]

On the same day Krebs tried another strategy to demonstrate that succinic acid formed anaerobically from pyruvic acid in rat testis tissue. Previously he had added malonate at the end of the runs as part of the Gözsy method for determining succinate. Now he added it at the *beginning*, presumably to block any further metabolic reactions which would otherwise remove the succinic acid formed. The "pyruvic acid determination failed," because, as he thought, the "malonate interferes." Estimating from earlier experiments how much pyruvic acid had probably disappeared, he found the measured quantity of succinate to be one-fourth of that expected. Again he based the "expected" quantity on the assumption of a 1:1 ratio between pyruvic and succinic acid, even though the dismutation reaction seems to suggest that the ratio would be 2:1.[6]

The preceding discrepancy as well as some of the features of the ensuing experiments suggest that Krebs was at this time exploring various reaction schemes that he did not write out in his notebook or later, and therefore I have not been able to reconstruct them in detail. In a general way we can surmise that by using malonate to inhibit the removal of succinic acid he felt now in a position to establish the proportions between the quantity of that substance produced and the quantities of its precursors that disappeared, and to check these proportions against those predicted by whatever chemical equations he could devise to connect them theoretically. Possibly in part as a consequence of the preceding result, he began thinking about schemes in which only part of the oxidation went "through" succinic acid. In that case malonic acid could be expected to inhibit only that proportion of the overall rate of respiration.

On June 4 Krebs tested "the influence of malonic acid on [the] pyruvate reaction," measuring the anaerobic CO_2 formation and disappearance of pyruvate in rat testis tissue with and without malonate. Both rates were reduced by about one-third. It is difficult to recover the reasoning behind this experiment. Blocking the removal of succinate should not affect either of the measured factors according to the dismutation reaction in question. He made no comment on the result. In another experiment that day he tested the effects of "malonate on testicle respiration" in the presence of glucose and of lactate. He found inhibitions, increasing over time, of 67.7 and 78.1 percent with glucose and 63.8 and 75.5 percent with lactate, concluding that there was "about equal inhibition with glucose and lactate." It is clear that he attributed these inhibitions to the blocking of that portion of the overall oxidative process that passed through succinic acid and that he tested the measured inhibition against what would be expected according to two different reaction schemes. He wrote down:

According to theory 1/3 or 1/3.25 [of inhibited rate] = 66.6 or 58.5 almost
    theoretical inhibition.
(because according to theory 2/3 or 1.75/3 of oxidation go through succinic
    acid.[7]

If we cannot bring the specific theories that Krebs had in mind just then into sharp focus, we can at least follow his short-term strategy. Malonate inhibition of the removal of succinic acid had provided him with a means to explore theories comprehensive enough to link the anaerobic dismutation reactions into the broader network of pathways comprising the cellular breakdown of carbohydrates.

Krebs pursued this malonate strategy no further at the time. The next day he asked a different question: Do pyruvate and glucose exert additive effects on the anaerobic formation of CO_2 in rat testis tissue? He found "pyr. + glucose more than additive." Adding the rates for pyruvate and glucose tested separately gave (in duplicate runs) $Q_{CO_2}^{N_2} = 18.4$ and 21.5. The rates for the two added together were $Q_{CO_2}^{N_2} = 22.2$ and 23.0.[8] The point of the experiment was to test whether, according to his "summation" criterion (see pp. 153–154), the anaerobic "pyruvic reaction" was, as he thought, separate from glycolysis; the result confirmed his expectation. He had already approached the same question with a different method—the addition of iodoacetate—which had also indicated that the reactions were separable.

On the same day Krebs drew up an experimental "balance sheet" for the products of the anaerobic disappearance of pyruvic acid in sheep testis tissue. Measuring the CO_2, volatile acid, and succinic acid produced with and without pyruvate, he obtained:

$$\begin{array}{lll}
\text{balance sheet: extra } CO_2 & = 9680 & \\
\text{extra volatile acid} & = 5960 & \Sigma\underline{8395} \ [\mu l] \\
\text{extra succinic} & = 2435 &
\end{array}$$

The preformed pyruvic acid (1 mole) was 22,400 μl.[9] He did not measure the pyruvic left at the end. From the fact that he divided the preceding number by two at the bottom of the page, giving 11,200 [μl], we can conjecture that he compared half of the original quantity to the total of the products. What is not evident from the notebook is whether Krebs thought at this point that the balance could be the expression of a definite balanced chemical equation.

Shifting his direction yet again the next day, Krebs asked the question, is "β-hydroxybutyric formed?" in the pyruvate reaction. To find out he applied a method for β-hydroxybutyric acid that Edson had devised by modifying a Van Slyke method. His reason for thinking that it might be formed was probably that Annau and Edson, among others, had observed that pyruvic acid can yield acetoacetic acid in liver tissue aerobically and that acetoacetic and β-hydroxybutyric acid were easily interconvertible. If Krebs had a specific reaction in mind he did not put it down, but from his comment on the results it appears that he anticipated that two molecules of pyruvic acid were consumed to form every molecule of fixed acid and that half of the total fixed acid formed would be β-hydroxybutyric acid. The 0.1 mole of pyruvic acid added to 60 mg of rat testicle tissue was equivalent to 2240 μl. Edson's method gave 52 μl of β-hydroxybutyric acid present at the end. Krebs noted:

$$\begin{array}{ll}
\text{amount of fixed acid expected} & = \sim 600 \ \mu l. \\
& = \sim 300 \ \beta\text{-hydroxybutyric}
\end{array}$$

definite formation of something (but C_4 acids may yield precipitate!!

has to be tried), but yield is only about 1/5 of what would be expected.

His uncertainty over what had been found was probably due to the fact that Edson's method was not entirely specific for β-hydroxybutyric acid and that other C_4 acids present could have interfered with the determination.[10] Krebs did not go on to try to straighten out the situation, but turned at the beginning of the next week to further balance experiments comparing the quantity of pyruvic acid used up to that of the volatile acid (acetic acid) and succinic acid formed. Switching from the rat testicle tissue on which he had relied so heavily during this investigation, he carried out the experiments, on June 8, on guinea pig kidney and testicle slices. The first of these gave an unusual result. The volatile acid formed was equivalent to more than half of the pyruvic acid consumed (Δvolatile acid = 5350 μl, Δpyruvic acid = -8150 μl), but the "Δ"succinic acid, after subtracting the control, was "-375!!"; or, as he remarked, there was "less succinic with pyruvic" than without it. He undoubtedly did *not* conclude that pyruvic acid reacts with succinic acid in a way that removes the latter, but only that the experiment was flawed. The guinea pig testicle experiment fell more in line with expectations:[11]

Δpyruvate	- 6540	
Δvolatile acid	+2842	3787
Δsuccinic equiv.	+ 945	

* * *

The last of these experiments Krebs entered on the first page of a new laboratory notebook. In 1977, when I discussed with him his investigation of the anaerobic reactions of pyruvic acid, we began with this experiment, because the preceding notebook had not yet turned up. He commented on the above figures, "It looked quite good—well, roughly in the right proportions. This [total of Δvolatile acid and Δsuccinic acid] should be half this [Δpyruvic acid]. . . . I believe I assumed that two molecules of . . . pyruvate would give one molecule of succinate. In fact, it was probably only one, because the extra carbon would come from CO_2, so if we divide this by two now, and add 370 to this here, we get almost half." He added, "Well, I was satisfied that it was roughly half."[12] These comments are notable for the intimate way in which Krebs associated elements of recalling what he had thought 40 years earlier, of reconstructing from the figures in front of him what he probably thought then, and of reinterpreting the results on the basis of later knowledge.

In this conversation Krebs did not supply an equation that he might have had in mind to explain his assumption that two molecules of pyruvate would give one of succinate. What he recalled he "had in mind was merely the study of the breakdown point of the carbohydrate such as pyruvate. And I thought any pieces of information would be of interest in elucidating the pathway of oxidation." Knowing something about the hypothetical reactions that Krebs had in mind shortly afterward, I pressed him a little about the "reasoning that you perhaps began with." He commented, "I didn't put down all the mental processes which led to the formulation of a working hypothesis and the design of the experiment. . . . I thought most of the ideas we form prove wrong, and what is important are the experimental data."[13] All of these statements are valid, yet taken at face value they obviously present a one-sided picture of his experimental style. There are more clues about his reasoning embedded in the notebooks than he realized. What we can reconstruct from them shows that the ideas he formed and the experimental data he gathered were inseparable constituents of his investigative trail.

* * *

Now making a much larger change in organisms, Krebs tested the pyruvic acid reaction on June 9 in two types of bacteria, gonococcus and *B. pyocyaneus*. He used cultures grown for 18 hours on agar, then washed in saline solution and suspended in a bicarbonate medium. Measuring the CO_2 that formed anaerobically, he observed that added pyruvate had little effect in the *B. pyocyaneus*, and carried that experiment no further. In gonococcus, pyruvate increased the quantity of CO_2 evolved by 137 μl. There was a decrease of 374 μl of pyruvate, and 187.5 μl more lactate formed with than without pyruvate. Comparing the latter figures, he wrote down:[14]

$$\frac{\Delta \text{ lact.}}{\Delta \text{ pyr.}} = \frac{2.19 - 31.5 + 187.5}{-374} = \frac{1}{1.195}$$

Krebs obviously wrote down this ratio incorrectly (as he noticed immediately when we examined the notebook page in 1977).[15] It should be 1:1.99, and was therefore in very close agreement with the 1:2 ratio expected for the primary pyruvate dismutation reaction. This single experiment was good enough to satisfy him that the reaction occurs quantitatively in gonococcus and he turned back to animal tissue.

On the same day Krebs carried out another set of measurements of products of the pyruvic acid reaction, employing minced sheep testicle as he had done three days earlier, but using different quantities both of the tissue suspension and of pyruvate. This time he measured Δpyruvate as well as Δvolatile acid, CO_2, and Δsuccinic acid. In the results of the previous experiment he had fixed his attention on the sum of the products. This time he picked out the pyruvate:succinate ratio. It was

$$\frac{\Delta \text{ pyruvate}}{\Delta \text{ succinate}} = \frac{17570}{7070} = \frac{2.48^{[16]}}{1}$$

This was a higher proportion of succinic acid than in the earlier experiments, but it still did not approach the 1:1 ratio he had formerly been expecting. At this point he might either have thought that he had still not achieved conditions under which all of the succinate formed was found or else begun to question whether the measured proportions represented the stoichiometry of any definite chemical equation.

During the stage of his research followed so far in this chapter, we have seen Krebs shifting rapidly back and forth between several aspects of his current investigation, pursuing any one subquestion for no more than one or two experiments at a time. We have seen in earlier chapters that digressions in quick succession were characteristic of periods in his investigative pathway when he seemed to be casting around for an opening. A special feature of the present case, however, was that these shifts took place within a circumscribed larger investigative problem from which he did not deviate. He was tacking quickly while seeking a stable overall course. This pattern reflects his special circumstances. In an attack on what he perceived as one of the central unsolved problems of intermediary metabolism, he had reached a reasonably firm base point: the primary anaerobic pyruvic acid reaction. He was convinced that he was on the way to the broader solution, but was unsure how to cross the remaining distance. He saw clearly that to complete the solution he must establish the relation between the reaction he had in hand and the formation of succinic acid. He had been able to link these processes experimentally, but his search for clues to the more specific nature of the connection had not yielded a clear direction to follow. The progress so far was too promising to abandon, but the way ahead was still obscure. Recently Krebs had told Edson that he planned soon to submit a note on his findings to *Nature*—a sure sign that he felt he was onto something so important that he was anxious to publish quickly. Now, however, he had to restrain his impatience. In reply to a question that Edson must have put in a letter to him, Krebs wrote back on June 10, "My preliminary note in 'Nature' must still wait. I do not want it to be a premature note."[17]

At this point Krebs reintroduced into the investigation a substance that had been prominent in its early stage, back in February, before the investigation became oriented around the pyruvic acid reaction. This was acetoacetic acid. On June 11 he tested in rat testicle the "effect of acetoacetic on respiration under different conditions." The conditions in question were acetoacetic acid added alone, added together with acetate, and with α-ketoglutarate (together with controls with acetate and with ketoglutarate alone and a blank). This move may have been a digression from his central problem, a broadening of the investigation to identify other dismutations *analogous* to the pyruvate reaction, but there is some reason to think that he hoped it would lead to the connection he sought between the pyruvate reaction and succinate. The results were "no effect of acetate" (nor were there effects of acetate + acetoacetate, although he did not remark on that), but a "stimulating effect of acetoacetate in the presence of ketoglutarate!!!?" ($Q_{O_2} = 9.1$, compared to 7.4 for ketoglutarate alone and 5.2–5.8 for the other conditions).[18] As we have seen at the end of the last chapter, he had previously found that pyruvate + ketoglutarate increased the formation of succinic acid. This effect of acetoacetate + ketoglutarate on the aerobic respiration gave reason to suspect that this combination too might have such an effect on the formation of succinic acid. Instead of raising that question next, Krebs jumped a step in his reasoning. When he asked, is "succinic acid formed with acetoacetic?" in an experiment with rat testicle tissue on June 12, the combination he tested was not acetoacetate + ketoglutarate, but acetoacetate + pyruvate. Behind these specific moves there was probably a more general consideration. The basic pyruvic acid dismutation reaction could not be connected immediately with the formation of succinic acid, because some intermediate process was necessary to synthesize a C_4 acid from C_3 acids. Acetoacetic and ketoglutaric were already C_4 acids, and ketoglutaric acid could be expected to give rise directly to succinic acid. Acetoacetic acid offered the additional attraction that it was known to be formed in tissues through the condensation of two molecules of acetic acid, providing a connection with one of the products of the pyruvic acid reaction. In the experiment, Krebs compared the quantity of succinic acid formed in rat testicle tissue with pyruvate alone and with pyruvate + acetoacetate. The combination yielded 38 percent more. He concluded, "Increase of succinic acid yield (or ketoglutaric yield) with acetoacetic!!" The parenthetical qualification must have meant that the method used to determine succinic acid was not entirely specific. When he carried out "the same experiment with guinea pig testicle" on the same day, the result was reversed. There was a "decreasing effect of acetoacetic!!!!"[19] For Krebs that result did not contradict the other one, but meant only that the effect he observed in rat testicle did not occur in the same tissue of another animal.

Krebs had now obtained preliminary evidence that three combinations of ketonic acids—pyruvate + acetoacetate, pyruvate + ketoglutarate, and acetoacetate + ketoglutarate—can give rise to succinic acid. By now, if not before, he must have begun to realize that the connection between the pyruvic acid and succinic acid for which he was searching might be not a single pathway, but a family of analogous dismutation sequences.

The growing prominence of α-ketoglutaric acid in his investigation confronted Krebs with the problem that there was no easy way to determine that acid. The two general methods available for α-ketonic acids were the carboxylase and ceric sulfate methods. To see whether either of these methods could be employed "in the presence of pyruvate," a circumstance likely to occur in his experiments, he tested both methods comparatively with the two substances. With carboxylase, he found "ketoglutarate practically *not* attacked!!" and ruled that method out. With ceric sulfate, on the other hand, "both substances react equally," so that the method could be used reliably for ketoglutarate only when pyruvate was not present. With no other choices at hand, Krebs adopted the ceric sulfate method on June 15 to examine the anaerobic disappearance of ketoglutaric acid in guinea pig tissue (the type is not stated). In duplicate runs he obtained for Q_{CO_2} 1.76 and 1.63; for the respective $Q_{ketoglutarate}$, -2.10 and -2.62. He wrote down, "Two ketoglutarate → 1 $\underline{CO_2}$!!!" Clearly he believed he had evidence for yet another ketonic acid dismutation reaction. On the same day he found no evidence for a pyruvic-acetoacetic acid reaction in minced pigeon breast muscle.[20]

On June 16 Krebs pursued the reaction of ketoglutaric + acetoacetic acid, but did not proceed directly to a test of the formation of succinic acid in the presence of this combination. Instead, using rat kidney tissue, he measured the rate of formation of CO_2 anaerobically and the change in the quantity of acetoacetate. The results indicated that there was "no disappearance of acetoacetic," that "Q_{CO_2} increased, but ketoglutaric and acetoacetic [gave] only summation." (That is, the extra CO_2 formed gave $Q_{CO_2}^{N_2}$ for ketoglutaric, 0.85; for acetoacetic, 0.68; for ketoglutaric + acetoacetic, 1.41.) The increases therefore represented only the sum of the separate reactions of the two substances. He wrote, "no indication for a reaction: ketoglutaric + acetoacetic = succinic + β-hydroxybutyric." The negative result is less interesting for us than the fact that he had formulated this particular reaction to test. Its characteristics made it an attractive candidate to connect the pyruvate dismutation reaction to the formation of succinic acid, because the ketoglutarate was so similar in composition to succinic acid, whereas the acetoacetic acid could be linked with the acetic acid arising from the pyruvic reaction. This negative result did not discourage Krebs from pursuing the reaction; it meant to him only that the reaction did not occur under these conditions.[21]

By this time Krebs appears to have sensed that the ketoglutaric-acetoacetic reaction was a key to the solution of his remaining problems, and he oriented his investigation increasingly around it. When he performed another balance sheet experiment for the anaerobic disappearance of pyruvic acid in rat testicle tissue, on June 16, he added the change in the quantity of ketoglutaric acid present to the changes in pyruvic acid, volatile acid, and CO_2 that he had heretofore measured. He handled the problem of determining ketoglutarate by measuring pyruvic acid with the carboxylase method, measuring "total ketonic acid" with the ceric sulfate method, and treating the difference between these results as ketoglutaric acid. Its balance was positive, suggesting that more ketoglutaric acid was formed than was consumed in the formation of succinic acid.[22] Because the pyruvic acid dismutation reaction does not in itself produce ketoglutaric acid, the experiment implied that there must be between these two reactions a third, still unidentified step, of a synthetic character.

On June 17 Krebs deviated slightly from his latest direction to test the effect of vitamin B_1 on the pyruvic acid reaction in rat testicle in liver and muscle extract, obtaining negative results. On the 18th he looked for the pyruvic acid reaction in minced guinea pig brain tissue, finding "considerable CO_2 production from pyruvate," and he tried to determine in a large-scale experiment on the same material whether he could identify ketoglutarate among the products by precipitating it with dinitrophenylhydrazone. He obtained only a cloudy solution. On the same day he began an experiment to test whether succinic acid formed in rat liver tissue in the presence of pyruvate and acetoacetate, but he measured only the disappearance of pyruvate. With rat kidney tissue on that day he did measure the succinic acid formed by ketoglutarate + pyruvate and ketoglutarate + acetoacetate. The result was an "increase in formation of succinic by acetoacetic [+ ketoglutaric], less by pyruvic [+ ketoglutaric]!!" That is, both increased the quantity by comparison with ketoglutaric alone, but the first combination had a greater effect. ($Q_{succinic}$ with ketoglutaric = 0.44; with ketoglutaric + pyruvic, 0.54; with ketoglutaric + acetoacetic, 0.74). This outcome probably encouraged him to think that both reactions occurred, but that the ketoglutaric-acetoacetic acid reaction was more prominent. Next he carried out an experiment on the pyruvic acid reaction in rat testicle tissue to verify that acetic acid was one of the products. The "volatile acid" determination he had been using was quantitative but not specific. Now he applied the lanthanum test, which was qualitative but specific, and obtained an "<u>enormous</u> + + + + reaction."[23]

Over the weekend of the 20th Norman Edson came to see Krebs in Sheffield. Edson had received his Ph.D. from Cambridge a week earlier, and probably wished to consult Krebs in general about his future plans, as well as to go over final changes in their uric acid paper. Krebs met him at the station at 12:41 p.m., undoubtedly immediately after completing the above lanthanum experiment. During Edson's stay they must also have discussed the "great work," and Krebs may have sought Edson's opinion on whether he was far enough along now to send the note to *Nature*. There was probably some time left over for social activities; at least Krebs had written beforehand, "I shall have much pleasure in making arrangements for your adequate reception and entertainment."[24]

Krebs continued during the following week with similar experiments. On Monday he tested the formation of succinic acid anaerobically in rat liver slices in the presence of ketoglutarate and ketoglutarate + acetoacetate. The result, "considerable succinic formation: more with acetoacetate," further consolidated his evidence for the ketoglutarate-acetoacetate reaction that had come to the forefront of his investigation. On the 24th he obtained from anaerobic CO_2 measurements in pigeon brain tissue another "result in favor of ketoglutaric + acetoacetic = succinic + CO_2 + β-hydroxybutyric." The same experiment in guinea pig brain tissue was "not convincing, increase by acetoacetic alone!!" That day he also tried out whether ketoglutaric acid is formed from pyruvic acid in guinea pig kidney tissue in the presence of As_2O_3, presumably on the assumption that the arsenite might block the subsequent disappearance of the ketoglutarate. He found that "265 μl of pyruvic acid are converted into [another] ketonic acid!!"; but this was too little to attempt to isolate the acid and determine whether or not it was ketoglutaric acid. He pursued the same question somewhat differently on rat liver tissue, interpreting again the difference between the change in total ketonic acid and the change in pyruvic acid as the change in ketoglutaric acid. There was a positive "Δ'ketoglutaric'," but less with As_2O_3 than without it. The arsenite was therefore apparently not useful. On June 25 he tested the "effect of [five different] redoxindicators on pyruvic acid dismutation," with uniformly negative results. On the same day he tried out whether α-hydroxyglutaric acid, a compound that he had himself prepared a few days earlier by standard methods, would give rise to acetoacetic acid in rat liver tissues, but it had no effect.[25] He was led to test α-hydroxyglutaric acid because of the possibility that α-ketoglutaric could be reduced to it in a reaction analogous to the reduction of pyruvic to lactic acid. (In this case Krebs clearly remembered in 1977 the hypothesis behind the design of the experiment. After explaining it, he added, "This was an exploration of whether it would be a major metabolite, but the answer was negative.")[26]

On June 27 Krebs tested the pyruvic acid reaction in pigeon liver tissue with muscle extract added. The rates were quite high, and the outcome—that the extra CO_2 produced by pyruvate in the extract (878 μl) about equaled the quantity of pyruvate that disappeared (892 μl)—was clearly anomalous. He wrote down, "Whole of pyruvate → CO_2!!!!?" Clearly he doubted it, adding "probably partly other reactions, such as glycerophosphate + pyruvate = lactic + lactic" (that is, the reactions of the Embden-Meyerhof glycolytic pathway). Even without the additional reactions, the prediction from the dismutation reaction was that half of the pyruvate → CO_2. Even in the late stages of an investigation not all of the experiments fall into line. He tested also for the anaerobic production of succinic acid from pyruvate in pigeon muscle extract, but concluded that "the succinic method [is] not reliable with muscle extract!!" On the 29th he again tested the action of vitamin B_1 on the pyruvic reaction in pigeon brain and liver, in muscle extract. Although he concluded that there was "perhaps effect of B_1," the major effect was that of the extract itself. On July 1 he tried vitamin B_1 on the same tissues of a pigeon that had been kept for 14 days on a rice diet, and was therefore presumably vitamin deficient. Even under these conditions he obtained no effect in the brain tissue and only "perhaps [an] effect on liver in extract." For reasons that are not immediately evident he commented, "There seem to be two reactions with pyruvate."[27]

Krebs performed another experiment on ketoglutarate + acetoacetate on July 3, this time employing rabbit kidney tissue. He measured the rate of formation of CO_2 disappearance of acetoacetic acid, while Johnson measured the succinic acid formed. Krebs expressed the results as follows:

with <u>no</u> acetoacetic
    extra <u>succinic</u> from ketoglut. 1.13 - 0.28 = <u>0.82</u>
    extra CO_2 "   "   2.63 - 1.41 = <u>1.22</u>

    extra succinic from ketoglut. + acetoacetic 2.16 - 0.47 = <u>1.69</u>
    " CO_2 "   "   "   4.46 - 2.35 = <u>2.11</u>
    " acetoacetate disappearing   5.06 - 4.32 = <u>0.74</u>
          (latter not very accurate)
    concentration of ketoglut. and acetoacetate are too high!!
    decrease comparatively small.
    result very much in favor of ketoglut. + acetoacetate → succinic + CO_2
    + β-hydroxybut.[28]

Despite the flaws he noted in this experiment, Krebs saw in it the most convincing evidence yet for the reaction on which he had concentrated much of his attention during the last three weeks. His persuasion undoubtedly derived not from this experiment alone, but was a cumulative effect of these results and earlier ones tending toward the same conclusion.

During these weeks Johnson performed experiments similar to ones Krebs had already tried out. Those that Johnson later picked out to put in his thesis include measurements on May 6, June 6, and June 16 of the extra volatile acid produced from pyruvic acid in several tissues, three on June 24 and 25 on the succinic acid produced from pyruvic acid in rat liver tissue, a similar experiment on pigeon brain tissue on June 29, and one on rat kidney tissue on the 30th. This sampling suggests that Krebs assigned Johnson mainly to gather supporting data for areas of the investigation on which he had established firm footholds, while he himself explored the outer reaches of the investigative territory.[29]

II

Confidence in his evidence for the ketoglutaric-acetoacetic acid reaction appears to have been what Krebs needed in order to finish his delayed note to *Nature*. Scheduled to depart for Oxford on Saturday, July 4, he must have completed nearly all of the writing by the time he performed the experiment, on the 3rd, described above, which seemed to clinch that reaction for him. The note, which arrived at the office of *Nature* on July 9, states Krebs's position so succinctly that it can and should be quoted almost in full. His title, "Intermediate Metabolism of Carbohydrates," was itself a hint of the breadth of the implications he perceived in his findings:

We have found some new chemical reactions in living cells which represent steps in the breakdown of carbohydrates. Pyruvic acid, if added to animal tissues, disappears rapidly not only in the presence, but also in the absence of oxygen. In the presence of oxygen the end-products of the pyruvic acid metabolism are known to be carbon dioxide and water. We find that the primary steps of the oxidation proceed in the absence of molecular oxygen, and as products of the anaerobic oxidation the following substances were identified: (1) acetic acid, (2) carbon dioxide, (3) succinic acid.

The reductive equivalent for the oxidation of pyruvic acid is the conversion of another fraction of pyruvic acid into lactic acid or the homologous reduction of another ketonic acid. The quantitative data suggest that pyruvic acid is metabolized by the following intermolecular oxido-reductions: The first step is a dismutation of pyruvic acid according to the reaction:

(1) 2 pyruvic acid + water → acetic acid + carbon dioxide + lactic acid.

The evidence for the occurrence of this reaction in the tissues which metabolize carbohydrates is conclusive. The subsequent reactions which lead to the formation of succinic acid may be tentatively formulated in the following way:

acetic acid
(2)   +   + 'ketonic acid' → α-ketoglutaric acid + 'hydroxy acid.'
      pyruvic acid
(3) α-ketoglutaric acid + 'ketonic acid' + water → succinic acid +
      carbon dioxide + 'hydroxy acid'.

According to (2) α-ketoglutaric acid is formed by the oxidative condensation of pyruvic and acetic acids, a ketonic acid acting as hydrogen acceptor. According to (3), α-ketoglutaric acid is oxidatively decarboxylated by dismutation. Reaction (3) is analogous to (1).[30]

We may pause here to comment on the foregoing portion of the note. The opening paragraph, read carefully, reveals the broad scope of Krebs's claim. He was not merely describing "some new chemical reactions" that he had found in living cells, but reactions he believed to be the "primary steps of the oxidation" of pyruvic acid. Given the prominent place long accorded pyruvic acid at the transition point between the anaerobic and aerobic phases of the breakdown of carbohydrates, this assertion, stated with such brevity, amounted to claiming (1) that he had found the reaction steps linking these two central metabolic processes and (2) that the assumed transition point was displaced from pyruvic to succinic acid.

The second paragraph stated what these crucial connecting steps were. Krebs appeared to be cautious, differentiating the degrees of certainty he attached to these steps. The entire scheme was "suggested" by the quantitative data. Only for the first step was there "conclusive" evidence, whereas the "subsequent reactions" were "tentatively formulated." Nevertheless the claim as a whole was bolder than the restrained language in which he framed it. From our knowledge of his investigative pathway we can see that Krebs lacked an essential element of proof for his statement that even the "conclusively" established first reaction was the "primary step" in the oxidation of pyruvic acid; that is, his only experiments comparing the rates of the aerobic and anaerobic disappearance of pyruvic acid did not fulfill the fundamental requirement that the partial reaction take place at least as rapidly as the overall reaction. (Possibly Johnson had performed experiments, of which there is no record, that appeared to meet this criterion, but subsequent developments make it unlikely that he could have attained that outcome consistently.)

The "subsequent reactions" that Krebs formulated "tentatively" were, in fact unequally tentative. Reaction (3) was a generic formulation based mainly on experimental evidence for two reactions of this type—that for α-ketoglutaric + acetoacetic acid and for α-ketoglutaric + pyruvic acid. Reaction (2) appears to have a quite different status. I have identified no experiments in the record of Krebs's investigation designed to test this particular form of reaction. I may well have missed them; but it is plausible that Krebs had not yet found evidence for the intermediary step necessary to account for the *formation* of α-ketoglutaric acid and that he constructed reaction (2) on paper to fill the gap. His willingness to do so suggests that the two reactions for which he did have evidence left a well-bounded space that did not appear to him too large to bridge through such an inference.

The note continued:

> The experiments suggest that different 'ketonic acids', such as pyruvic acid, acetoacetic acid, oxaloacetic acid,[1] [the footnote referred to Szent-Györgyi's recent paper described in Chapter 4], or their homologues may be concerned in reactions (2) and (3), and may possibly take the place of pyruvic acid in (1). It seems that acetoacetic acid reacts preferentially in (3), and it is therefore of great interest that we find in tissues which metabolize carbohydrates a specific system which catalyzes the oxidation of β-hydroxybutyric to acetoacetic acid by molecular oxygen[2] [this footnote referred to a recent paper by Quastel giving evidence for such a system]; β-hydroxybutyric acid may thus act as a carrier for molecular oxygen according to the scheme (4).

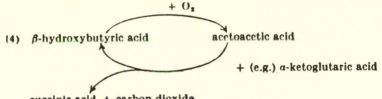

As is indicated in (4), α-ketoglutaric acid is not directly oxidized by molecular oxygen, but through the intermediation of another ketonic acid. It has long been known that there are links between carbohydrate breakdown and 'ketone bodies', and it is now possible to describe this link, or at least one of the links, in chemical terms.[31]

The basis for Krebs's statement "It seems that acetoacetic acid reacts preferentially in reaction (3)" may have been no more than the experiment of June 18 in which ketoglutarate + acetoacetate produced succinic acid in rat kidney tissue at a rate modestly greater than ketoglutarate + pyruvate did. His "preference" for that version of the generic reaction appears to have been influenced by the larger scheme to which acetoacetic acid lent itself. The boldness of the scheme depicted is self-evident. The mechanism that Krebs proposed would not only connect the anaerobic and aerobic phases of carbohydrate metabolism, but provide the link that had eluded biochemists for decades between those processes and the formation of the ketone bodies.

In an astute discussion of the development of Krebs's views based on his published papers, Steven Benner noted in 1976 that this scheme was cyclical, and added that " 'cycle' was a magic word, ever since the ornithine cycle."[32] Krebs did not use the word cycle here, however, and the process depicted was less analogous to the ornithine cycle than to Szent-Györgyi's recently published fumarate-succinate scheme. Just as Szent-Györgyi's closed loop would function as a carrier for hydrogen, Krebs's loop would serve "as a carrier for molecular oxygen." While discussing this paper in 1977, Krebs himself recognized Szent-Györgyi's scheme as the inspiration for his own.[33] There is little indication that in 1936 Krebs hoped to model his solution to the problem of carbohydrate metabolism after his discovery of the ornithine cycle. The predominant pattern in his mind was the coupled oxidation-reduction dismutation reaction; if he was consciously following any model in addition to that of Szent-Györgyi, it was the current Embden-Meyerhof pathway for the glycolytic phase of carbohydrate metabolism.

The penultimate paragraph of the paper further broadened its implications:

> The oxidative formation of succinic acid from pyruvic acid has been discussed by previous workers. The new feature is the demonstration that this oxidation is brought about by anaerobic oxido-reductions. The reactions described seem to occur in all animal tissues which metabolize carbohydrates. They occur also in bacteria. Reaction (1), for example, is quantitatively realized in gonococci.[34]

Here too, the reach of Krebs's assertion can only be appreciated by elaborating on his condensed statement. He was not merely posing a pathway that might occur alongside the one that previous workers had "discussed," but claiming that he had found the pathway for which they had been looking. When we recall that the formation of succinic acid from pyruvic acid had been widely discussed as the crucial uncertain stage within the Thunberg-Knoop-Wieland scheme, and particularly the variant proposed by Toenniessen and Brinkmann, then Krebs can be seen as aiming to resolve that long-standing problem. From succinic acid the pathway through the C_4 dicarboxylic acids and back to pyruvic acid was already clear. Consequently Krebs was putting forth a reaction scheme that could potentially connect existing pathways together into one comprehensive scheme leading from carbohydrate foodstuffs to their complete oxidative breakdown and link up with fatty acid metabolism as well.

If this interpretation relied on the *Nature* article alone, it would clearly be reading too much between the lines of Krebs's tightly crafted statements. The record of his investigative trail leaves no doubt, however, about his persistent interest in the Thunberg-Knoop-Wieland and Toenniessen-Brinkmann schemes of oxidative metabolism, and we shall soon see from Krebs's subsequent moves corroborating evidence for what he had in mind here. As should now be obvious, there lurked behind the circumspect tone of this paper a theoretical framework of breathtaking potential sweep. It is no wonder that Krebs perceived himself as engaged in a "great work," and understandable that he was anxious to publish a sketch of the structure that he hoped to substantiate; but was this not still a premature publication? Why could he not wait until he had firmer and more extensive evidence for these conclusions; until, for example, he had more than a single experiment on gonococcus to support his view that the reactions "seem to occur in all animal tissues which metabolize carbohydrates." Why not procure direct evidence for reaction (2) before formulating it even tentatively in a published paper? Krebs was, in some respects, an extraordinarily patient man;

but under some circumstances he was in a great hurry, and this appears to have been such a situation. The confidence with which he presented his scheme approached insouciance. Even more than in previous instances we see here that his conviction that most of the ideas he formed would turn out to be wrong did not inhibit him from exposing them to the view of his peers.

The character of Krebs's scheme also reveals, more powerfully than any of his earlier publications, the strength of his commitment to the goal of linking up metabolic reaction sequences into complete pathways. It suggests also the prominence that we should give, among his intellectual mentors, to Franz Knoop. No one had articulated more forcefully than Knoop that the "final goal" of physiological chemistry was to "present a scheme that puts together an unbroken series of equations . . . from the foodstuffs . . . all the way to . . . the final oxidative products."[35] No one had worked harder to realize that goal than Hans Krebs.

If the ideas underlying the note to *Nature* were daring, the writing itself was masterly. Aside from the fact that Krebs did not waste a single word, he stated his position so prudently, qualified his explicit claims so carefully, and implied his broader claims so inconspicuously that he protected himself from any appearance of grandiose aims. He had laid a foundation on which he could build a large superstructure if future developments justified it, but which he could otherwise leave behind him with little embarrassment.

III

On his way to Oxford on the weekend of July 4, Krebs passed through Cambridge to talk with Norman Edson. They had just completed the uric acid paper, and it is a measure of the value Krebs placed on Edson's opinions that he had written, on June 30, "I have just posted the paper and I am sending you a carbon copy. I have followed all your suggestions, and if some more points turn up we might consider them when we correct the proof."[36] It is reasonable to surmise that in Cambridge he sought Edson's views concerning the next phase of the "great work." A letter that Krebs wrote shortly afterward indicates that he must have queried Edson on information from Edson's ketogenesis experiments bearing on the role of acetoacetic acid and that Edson made the point that NH_4Cl "does not seem to inhibit acetoacetic acid breakdown."[37]

The main object of Krebs's trip to Oxford was to meet David Nachmansohn, who was there to present a paper about his current research on the rate of decomposition of acetylcholine. Nachmansohn had written that he had "all sorts of things, both professional and personal, to discuss."[38] Their time together in Palestine had deepened their friendship (as evidenced by the fact that Nachmansohn addressed Krebs now in his German letters in the "du" form), and he probably sought Krebs's advice about his experimental work as well as a plan to spend several weeks in August doing research at a laboratory in Plymouth. Undoubtedly they also spent much of their time together discussing the discouraging direction that events had taken in Palestine since their visit.

On the very day that Krebs and Nachmansohn sailed from Palestine, several Arabs had held up an automobile on the road from Tel Aviv to Haifa—a route that Krebs and Nachmansohn had traveled 12 days earlier—robbed the passengers, and killed three Jews. This event marked the outbreak of a series of "disorders," including riots in Jaffa on the 19th, which led the newly formed Arab National Committee to declare a general strike on the 20th. The cause of the crisis was the collision between the growing momentum of the movement for the Jewish national home in Palestine and the emergence of Arab nationalism. Developments in Europe since the Nazis took power were propelling German and central European Jews to immigrate to Palestine in swelling numbers, limited only by the quotas set by the British Mandate. With a large influx of capital, some of which Jewish settlers brought with them, some raised by donations in Europe and the United States, the Jewish community was spurring rapid urban development and seeking land for agricultural expansion.   The Zionist movement foresaw the day when Jews would constitute the majority in Palestine.   To Palestinian Arabs that prospect was threatening.   The Arab National Committee demanded a prohibition on further Jewish immigration and on the transfer of Arab land to Jews.[39]

Krebs's sister Lise and her husband arrived in Palestine in mid-May, in the middle of the crisis, which cast a pall over their visit and their prospects. Because of the unsettled conditions they saw little of the land, spending most of it with two families, one of them related to Adolf, who had previously immigrated.   Their hosts were not encouraging, and none of the contacts that Hans and others had provided for them turned up hopeful leads for employment. They sailed away on June 24, chastened by their experience, but still convinced they had no choice except to carry on with their plan to emigrate in September.[40]

For Hans the situation in Palestine further dimmed the possibility that his future lay there. Weizmann was now too preoccupied by the political tensions to devote much effort to the science faculty at the University of Jerusalem.[41] During June a radiologist, probably a member of the existing section of the cancer institute there, came to Sheffield to speak with Krebs. The conversation only made Krebs more pessimistic. He wrote Nachmansohn a letter (that has not survived) to which Nachmansohn replied

> Your opinion of Boris Levin [the radiologist] is unfortunately—as always—correct.   One must let him go his own way.   When we departed from Palestine we had no idea what a difficult time the people there faced. It has become a true showdown struggle. How fortunate it is that England is supporting our interests so firmly.[42]

Nachmansohn referred probably to the refusal of the British government to halt Jewish immigration to Palestine. Even he, however, was coming to acknowledge that the circumstances in Palestine at present were unfavorable for their venture.

While Hans Krebs was meeting Nachmansohn in Oxford, Wolf Krebs arrived in England to stay. Less than two weeks after his interview in Ipswich, E.R. and F. Turner sent word that it had received the necessary permission to hire a foreigner and that he could come immediately to take up his new job. Wolf quickly gave notice at AEG, received his final salary and payments for his patents, and was asked to sign a statement that he had no further claims on the company. He declined to do so. He made arrangements to ship his furniture and prepared to leave. Permitted to leave Germany with only £5 each, he and Lotte carried with them also some valuable items that his father had arranged for him through a cousin and that they could afterward sell. When they were ready to leave Berlin, about 20 people, including one of the senior designers for AEG, came to the train station to see them off: a gesture of support that touched Wolf deeply. Their departure from Germany was uneventful. Arriving by ship at Norwich on July 4, they took the train to Ipswich, where the managing director greeted them at the station.[43]

Wolf was taken immediately to the company office to be introduced to his new responsibilities as head of the Design and Development Department of Bull Motors. His particular charge was to design industrial and "super silent" AC motors. His wife was taken separately to the home of one of the officials to be introduced to the practical ways of English life. Notwithstanding the apprehensions that his family had had for Wolf's future, their concern that he was not aggressive enough, and their admonitions to Hans to do something for him, Wolf had, in the end, made his own way out of Germany on the strength of his record as a highly talented, inventive, and productive engineer. He too had proved, in his own profession, to have "international value." Georg Krebs must have found some consolation in the fact that both of his sons were now safely relocated.[44]

IV

On July 8 Krebs was back at work in Sheffield. One of the first things he did was to look up Edson's first paper on ketogenesis, where he found that, contrary to what Edson had told him in Cambridge, he had reported an experiment which "suggests that ammonium chloride actually does inhibit the acetoacetic acid break-down." Writing back immediately, Krebs asked, "Do you think it worthwhile going into this matter?" Another matter related to his current investigation that Krebs must have discussed with Edson was the possibility that mesoxalic acid (ketomalonic acid) would influence the pyruvic dismutation reaction. With his letter he sent a sample of mesoxalic acid that he had prepared that day, and himself performed an experiment with it the next day. Neither mesoxalic nor tartronic acid exerted an effect. On the 10th he added pyruvic acid to yeast anaerobically, and found that after one hour none of it had disappeared. With Jensen rat sarcoma tissue, pyruvate decreased the anaerobic formation of CO_2.[45] None of these experiments therefore opened up a new lead.

On July 11 Krebs embarked on a series of tests of vitamin B_1 on the pyruvate reaction in the tissues of avitaminous animals, including pigeons, chickens, and rats. He had obviously planned for these experiments well in advance, for the animals had been fed rice diets for several weeks, until they showed symptoms of vitamin deficiency. During the next two weeks he performed 13 such experiments. In the majority of them, vitamin B_1 had little or no effect, but by the time he finished the series on July 23, he had attained marked increases in the rate at which pyruvate disappeared in avitaminous rat liver and kidney tissue.[46] The final paragraph in his *Nature* article stated,

> The work of Peters and the findings of Simola suggest that vitamin B_1 is a co-enzyme for dismutations of the type of the reactions (1) and (3). Experiments on tissues of vitamin B_1 deficient rats and chickens show that such is the case.[47]

Krebs submitted the article just before *beginning* the above experiments. It is not unlikely that he added or revised this last paragraph sometime before the paper appeared on August 15.

During the same period that he concentrated his main effort on the vitamin B_1 effect, Krebs also pursued the question of what other ketonic acids might participate in reaction (3) of his anaerobic oxidation scheme. As his note in *Nature* indicated, an obvious candidate was oxaloacetic acid, suggested in part by Szent-Györgyi's work showing that oxaloacetic acid is both formed and reduced in animal tissue. On July 9, Johnson performed an experiment on the dismutation of α-ketoglutaric and oxaloacetic acid in rat kidney cortex slices, measuring the formation of succinic acid. The results were:

	$Q_{succinate}$
Control	0.31
+ Ketoglutarate	1.47
+ Oxaloacetate	1.64
+ Both	2.04

The experiment thus supported the view that these two ketonic acids underwent a reaction of the type shown in Krebs's scheme. Krebs would have formulated the specific reaction as α-ketoglutaric acid + oxaloacetic acid + $H_2O \rightarrow$ succinic acid + CO_2 + malic acid.[48] An interesting implication of this reaction is that it contained an alternative to Szent-Györgyi's view that oxaloacetic acid is "overreduced" to succinic acid. In Krebs's reaction the succinic acid arose from α-ketoglutaric acid through a known oxidative decarboxylation reaction, while the oxaloacetic acid underwent a single-step reduction to malic acid. Given his skepticism about the concept of "overreduction," Krebs may have been particularly pleased to obtain evidence for a possible way around this leapfrogging feature of Szent-Györgyi's scheme.

Turning to another possible dismutation involving oxaloacetic acid, Krebs included on July 13 a test of pyruvate + oxaloacetate in an experiment on chicken brain slices, the rest of which was part of his investigation of vitamin

B_1. Measuring the quantities of oxaloacetate consumed, he found that "oxaloace-tate disappears rapidly"; but since CO_2 formed with oxaloacetate + pyruvate not much more rapidly than with pyruvate alone, he wondered if "Q_{CO_2} [is] due to pyruvate?" In heart muscle slices, on the other hand, he noted with emphasis that more CO_2 formed with oxaloacetate than with pyruvate. On the 15th, in another experiment appended to a set on vitamin B_1 in rat testicle tissue, oxaloacetate formed CO_2 at the same rate that pyruvate did; but on the 19th in a similar experiment on rat brain tissue, "oxaloacetate + pyruvic [yielded] more than pyruvic" (Q_{CO_2} = 4.98, compared to 3.15).[49]

These results indicated that oxaloacetic acid entered an anaerobic reaction, but did not point unequivocally to a pyruvic-oxaloacetic dismutation as the reaction being measured. That is probably why on the 20th Krebs tested the influence of oxaloacetic acid alone in rat testicle tissue. Recrystallized oxaloacetate raised Q_{CO_2} from 1.49 to 5.61. "Considerable CO_2 production!!" he noted. The next day he compared the effects of pyruvate, oxaloacetate, and pyruvate + oxaloacetate, with less positive results: Q_{CO_2} for control = 1.62; for pyruvate, 5.92; for oxaloacetate, 4.46; for pyruvate + oxaloacetate, 6.00. There was, he commented, "no additive or cumulative effect on testicle."[50] The status of a pyruvic-oxaloacetate reaction therefore remained in doubt. There is no indication of whether or not Krebs entertained the possibility of a dismutation involving two molecules of oxaloacetic acid.

On July 23 Krebs wrote Edson:

> I shall be passing through Cambridge on Saturday next, July 25th, at about one o'clock. I think it may be handy to discuss the great work on this occasion, so I shall call at the laboratory at about one o'clock and shall be very pleased to find you there.
> The proof of the uric acid paper arrived yesterday.[51]

Clearly Krebs did not anticipate that Edson would need much time to prepare for his visit. He made this stop on his way to attend a conference on microbiology in London.[52] He went to the meeting, outside his normal professional orbit, probably because he intended to expand his investigation of the reactions in bacteria, and hoped to profit from the expertise of those accustomed to working with these organisms. He was away from his laboratory until August 5. There is no record of what he might have done in the time left over from the conference, but it is evident that this summer he did not take an extended holiday. Perhaps he had no incentive comparable to the previous two years to do so.

While Krebs was gone, Leonard Eggleston and William Johnson carried on by themselves with experiments that Krebs had set out for them to do. He could leave them on their own with confidence, for both of them were proving to be exceptionally competent and reliable. Eggleston was scientifically untrained and served as a technician performing operations for the experiments that Krebs recorded and interpreted in his own notebook. When Krebs left for London, however, Eggleston began a separate notebook in which to enter the experiments he did in the absence of his chief. His conscientious, meticulous nature is evident in the careful way he wrote out the full details of the procedures and data in a clear, even hand. He ventured no conclusions. In Krebs's absence Eggleston performed five experiments on the effects of vitamin B_1 on the disappearance of pyruvate: on liver tissue from an avitaminous rat and chicken, a normal chicken, and two normal rats. In most cases vitamin B_1 increased the rate at which pyruvic acid was consumed.[53]

Johnson was already working more independently than Eggleston. Although Krebs normally instructed Johnson each morning on what experiments he wanted done, he left it to Johnson to plan how to go about them, and Johnson kept his own laboratory records. Occasionally Krebs would give Johnson a "lovely" summary of where he was expecting the investigation to go, but in between these times, he did not talk a lot with him. Krebs soon saw that Johnson was developing into an excellent experimenter; but because Krebs rarely showed enthusiasm, Johnson did not realize how highly Krebs thought of his work. To Johnson, Krebs remained a hard person to work for, but one for whom he nevertheless quickly came to have great respect. The shy young man was awed by a chief who seemed to him always to know where the research would end, even before he had the evidence.[54]

While Krebs was away, Johnson probably continued to do experiments on the pyruvic acid reaction and the formation of succinic acid in animal tissues. The only experiment from this period incorporated later into his thesis, however, was a measurement of the formation of succinic acid from fumaric acid in rat kidney. This experiment and a similar one that he had already performed in June along with a larger number with pyruvic acid[55] suggest that while Krebs concentrated on the dismutation of ketonic acids, he was keeping in mind a wider range of anaerobic oxidation-reduction reactions, and also that he continued to be broadly influenced by the work of Szent-Györgyi.

Sometime between July 23 and August 5, Krebs wrote down in his laboratory notebook the following scheme for the formation of citric acid (see Figure 7-2, following page):

He thus entertained two alternative reactions:

malic acid + acetic acid → citric acid - 2H or,
malic acid + pyruvic acid = (C_7 tricarboxylic acid) - 2H → citric acid + $2CO_2$.[56]

What prompted him to write down these two hypotheses to account for the formation of citric acid in tissues?

* * *

When I came across this page in Krebs's notebook in 1977, I naturally thought that the notation at the bottom, "with testicle? see Thunberg," pointed to a likely source for these equations. There was no obvious connection with the experiments preceding it, and I wondered if something Krebs had just read by

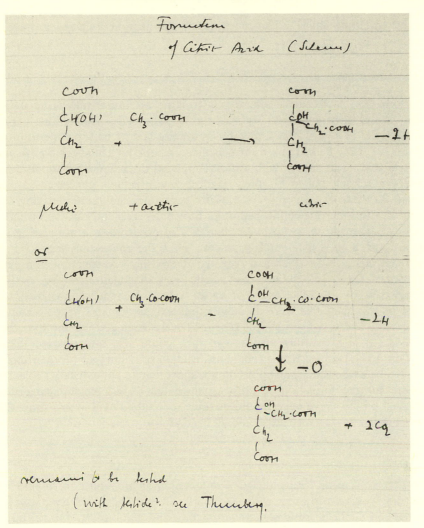

Figure 7-2 Krebs laboratory notebook No "12," p. 89.

Thunberg stimulated him to think again about this subject. When I pointed the note out to Krebs he recommended that I look up the reprints from Thunberg in his file. I returned the next day convinced that I had found the answer. There was a paper published by a student of Thunberg not long before the time at which Krebs put down this scheme, in which the author discussed the dehydrogenation of malic and citric acid and also identified citric acid in testicle tissue. Krebs considered the possibility that this work had influenced him, but after a few minutes of discussion he recognized no direct connection. "I can't really throw much light on the circumstances which gave rise to formulating this page," he said, "but it was certainly Thunberg and his school who kept an interest in citrate alive." The reactions on the page he believed to have been based on "paper chemistry."[57]

* * *

Returning to this page after reconstructing his investigations up until that point, I feel clearly that Krebs was right. The reference to Thunberg was probably not to a particular article by Thunberg that he had just read, but only a reminder, after he had posed the schemes, to look up Thunberg on the subject of citric acid. The query "with testicle?" related the scheme not to work by the Thunberg school, but to his own ongoing investigations concentrated on that tissue. As we have seen on several earlier occasions, the question of how to link citric acid to other metabolic pathways hovered persistently in the background of his mind, ready to surface whenever a way occurred to him to bring it into contact with a current research problem. Here the connection appears to have been that malic acid, which had recently entered his investigation as a putative product of the α-ketoglutaric–oxaloacetic acid dismutation, could readily be imagined to give rise to citric acid by an oxidative condensation reaction with either of two acids that were already prominent participants in the dismutation reactions. The two hypothetical reactions were analogous to the hypothetical reaction (2) in his *Nature* article, and they could also be supposed, like that reaction, to be coupled with a reduction reaction providing a hydrogen acceptor.

Krebs's laboratory notebooks do not show that he immediately subjected his scheme for the formation of citric acid to the experimental testing that he indicated on the above page should be done, but subsequent events show that he kept in mind having such experiments carried out in his laboratory.

V

The first experiment Krebs himself conducted upon his return on August 5 dealt with the "course of pyruvic acid reaction in N_2 and O_2" in rat liver tissue. He carried out duplicate runs both anaerobically and aerobically, for 30, 60, 120, and 180 minutes each. $Q_{pyruvic}$ ranged between 5.7 and 7.5 for the anaerobic runs and between 9.6 and 12.0 for the aerobic ones. The values over time were, he noted, "unusually constant!!" and the rate in "O_2 about double N_2!!"[58] This strong reinforcement of earlier experimental indications that pyruvic acid was consumed anaerobically at a considerably lower rate than aerobically must have raised serious questions in Krebs's mind about the view he had stated in his *Nature* article, that the anaerobic dismutation reaction was the primary step in the oxidative breakdown of pyruvic acid.

The next day Krebs measured the effect of malic acid on the respiration of rat testicle tissue. This was probably a general preliminary test of the substance that appeared on theoretical grounds to be involved in more than one way in the anaerobic reactions he had under consideration. Malic acid substantially increased the rate during the first 30 minutes, but over the next 60 minutes the difference between it and the control gradually disappeared. For the time being

Krebs did not go further in this direction. On the same day he measured the influence of narcotics on the anaerobic metabolism of pyruvic acid in rat testicle tissue and the aerobic and anaerobic disappearance of pyruvic acid in chorion and placenta tissue.[59] Then he was ready to begin the extended investigation of bacterial metabolism that he had evidently contemplated for some time.

As we have seen, Krebs had already performed experiments with bacteria from time to time, particularly while working with Marjory Stephenson and her student Donald Woods during his last year at Cambridge. Most of his experience, however, had hitherto been with animal tissues. He now turned more extensively to microorganisms for several concurrent reasons. He wanted to examine more fully whether the pyruvic acid reaction he had established occurred also in bacteria, as his single experiment on gonococcus earlier in the summer indicated. The reaction might be especially convenient to study in bacteria because they were easy to keep alive for a long time and gave large manometer readings. He remembered that the success Warburg had had in studying the absorption spectrum of the *Atmungsferment* in yeast was due partly to the ease with which the organism could be handled. Finally, he was interested in bacteria because of the differences between their metabolic reactions and those of higher organisms. As he came to see that the anaerobic oxidations he was studying consisted of a family of specifically different reactions following a common general pattern, he wished to explore the range of reactions that might occur in different classes of organisms.[60]

Krebs's move in this direction was eased by the cooperation of the Sheffield Bacteriology Department, located almost next door to his own laboratory. Professor J.W. Edington and Dr. J. MacA. Croll of that department supplied him with various strains of bacteria, showed him how to grow the bacteria in large quantities, and made their sterilization facilities available to him.[61]

Despite the overall simplicity of bacteria, their metabolic processes appeared in some respects more complicated than the corresponding processes in higher organisms. It had been long known, for example, that bacteria fermenting glucose produced multiple products, including lactic, acetic, fumaric, and succinic acid, alcohol, carbon dioxide, and hydrogen, and in some cases more esoteric compounds such as acetylmethylcarbinol and butyleneglycol. Moreover, different types of bacteria formed different products, and under varied conditions the same bacteria produced different substances. At the time Krebs entered the area of bacterial metabolism, there was special interest in the question of whether the Embden-Meyerhof pathway that had just been established for muscle glycolysis and yeast fermentation would apply to bacterial metabolism; and if so, what modifications were necessary to account for the variable products of bacterial fermentations. Ever since Neuberg had shown in 1911 that yeast can ferment pyruvic acid, there was also interest in pyruvic acid metabolism in bacteria. *Bacterium coli* (later renamed *Escherichia coli*, or *E. coli*), the common intestinal bacterium employed most often in metabolic studies, was known to break pyruvic acid down anaerobically, yielding about the same end products as glucose did. The intermediate steps involved remained uncertain, however, and were a current subject of investigation.[62]

Krebs began his bacterial experiments not with *B. coli*, but with *Staphylococcus* aureus. In 1977, when he reexamined the first such experiment in his notebook, he "could not reconstruct why I used this particular organism."[63] Such a reconstruction does seem possible, however. In the back of his notebook he wrote down a reference to a recently published paper by M.G. Sevag and N. Neuenschwander-Lemmer, from the Robert Koch Institute in Berlin, "On the Dehydrogenation of Lactic Acid by Staphylococcus." Employing two species of *Staphylococcus*—aureus and albus—Sevag and Neuenschwander-Lemmer gave as their reasons for choosing these organisms that their metabolism was "the simplest possible." They contained a minimum of enzyme systems, and were therefore highly favorable for the investigation of single-step breakdown reactions. They demonstrated that staphylococcus forms pyruvic acid by dehydrogenating lactic acid, but found that pyruvic acid itself was not further attacked.[64]

If Krebs took his cue that the simplicity of the metabolism of staphylococcus made it suitable for his purpose from Sevag and Neuenschwander-Lemmer, he evidently did not accept their conclusion that these bacteria cannot break down pyruvic acid. The first thing he did with the organism, in fact, was to test whether the pyruvic acid reaction he had been studying in animal tissues took place there too. On August 7 he measured the anaerobic formation of CO_2 and the aerobic respiration in the presence of glucose and of pyruvic acid. Both rates increased more with pyruvate than with glucose, and the quantities of extra CO_2 produced (227 μl) relative to that of pyruvate consumed (503 μl) satisfied almost exactly the 1:2 ratio expected for the dismutation reaction. He found that the lactic acid formed (313 μl) was "too much." Nevertheless his first experiment on this organism gave strong evidence that it carries out the same reaction he had found in animal tissue and in another bacterium, gonococcus. Moving quickly, he tested the effect of vitamin B_1 on the reaction. The results were negative, but this time when he measured Δlactic acid and ΔCO_2 resulting from pyruvic acid, the ratio of the two products came out exactly "1:1 as expected." The next day he tried vitamin B_1 again, this time in three concentrations, and the anaerobic formation of CO_2 now increased nearly in proportion to the concentration of the vitamin.[65] *Staphylococcus aureus* thus appeared to be a cooperative organism for the extension of the "great work" to bacterial metabolism.

Seeking to expand the range of organisms still further, Krebs picked up from a recent paper by W.H. Schopfer in the *Archiv für Mikrobiologie* that the mold *phycomyces nitens* might be suitable for the study because it was sensitive to vitamin B_1. Obtaining a 12-day-old culture of phycomyces from the Sheffield Botany Department, he cut off its sporangia, suspended it in saline, and measured its gaseous exchange anaerobically with vitamin B_1 and pyruvate. Little happened. "No decent pressure change," he commented. "Unsuitable for metabolic work, since plant grows in air normally!"[66]

Krebs now interrupted his investigation of anaerobic oxidations to perform two sets of experiments on the inhibition of respiration in rat tissues. The inspiration for them may have come from Thunberg. At a time which cannot be dated precisely, but could well have been during this period, Krebs reread the classic 1920 paper by Thunberg that had so strongly shaped subsequent thought and investigation of intermediary oxidative metabolism (see Vol. I, pp. 19–21). On the "references" page that he customarily kept in the back of his laboratory notebook, Krebs wrote down the following points from Thunberg's article that attracted his attention:

> T. Thunberg. Skand. Arch. 40.1 1920. See also B.Z. 258.
> p. 34.   acetic → succinic discussed.
>   41.   Oxal. and malonic acid inhibit! Oxalic see also Skand. Arch. 22, 430–1908.
>   Oxalic inhibits CO_2 production <u>more</u> than O_2 absorption!
>   53.   Inhibition of respiration by tartronic acid
>   67.   Inhibition of CO_2 by maleic acid 10^{-4} molar!

Immediately following these notes Krebs copied out some rates of respiration and fermentation of *Staphylococcus aureus* from a recent paper by Akiji Fujita and Takeshi Kodama. Next he wrote down the following:

<u>citric acid</u>. from malic + acetic or pyruvic

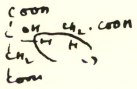

try anaerobic pressure change   citric acid + methylene blue
                 Is acid formed?
                 <u>how much MB reduced?</u>
                 <u>pyruvic</u> formed?[67]

Because the meaning we can attach to these notes obviously depends on the circumstances under which Krebs wrote them, the uncertainty about their date leaves any interpretation tentative. At first glance it appears that, contrary to my previous view, his reading of Thunberg must have been the source for Krebs's "scheme for the formation of citric acid." That relationship is made unlikely, however, by the fact that the reactions Krebs put down here, which partially duplicate that scheme, were not in Thunberg's article (nor in the other two references to Thunberg). It is more likely, therefore, that in accordance with his note "see Thunberg," he read these articles as a consequence of drawing up his scheme. In his reading he probably saw that Thunberg had not discussed the formation of citric acid (although Thunberg included that acid in his experimental results), but in the course of looking for that topic he found other views that interested him. Among them was Thunberg's argument for the synthesis of succinic acid from two molecules of acetic acid that had played so central a role in the schemes of oxidative intermediary metabolism that he, Knoop, and Wieland afterward proposed.

The notes that Krebs put down on citric acid did not refer directly to discussions in Thunberg's papers, but may well have been reflections stimulated by reading them. The first line appears to be a reinforcement in the light of Thunberg's views on dehydrogenation, of the reactions Krebs had already been considering for the formation of citric acid. The rest of these telegraphic notes suggest, however, that he contemplated also employing Thunberg's methylene blue method to study the breakdown of citric acid. The final query, "pyruvic formed?" raises the intriguing possibility—since pyruvic acid also entered into one of his alternative schemes for the formation of citric acid—that Krebs entertained here the idea that citric acid is involved in a cyclic process. If so, however, there is no evidence that this was more than a passing thought.

Krebs took no direct action on these ideas about citric acid. The views and observations of Thunberg that apparently did lead to Krebs's next experiments were those about inhibitions referred to in the above note on the lines beginning with page references 41, 53, and 67. The statement "Oxalic and malonic inhibit" summarized Thunberg's observation that in the presence of either of these acids tissue extracts decolorized methylene blue more slowly than the controls did. That result led Krebs back to Thunberg's paper of 1908, in which he had measured the effects of potential substrates on the respiration of isolated tissue and made the striking observation summarized in Krebs's statement that "Oxalic inhibits CO_2 production more than O_2 absorption." Krebs's notes to pages 53 and 67 refer to experiments on the respiration of isolated tissues described in Thunberg's 1920 paper. The note "See also B.Z. 258" referred to a more recent article by Thunberg on the specificity of dehydrogenases. There Thunberg had examined the specificity of succino-dehydrogenase by testing the activity of a number of derivatives of succinic acid. He explained the inhibition exhibited by some of them by distinguishing the "fixation specificity" of the dehydrogenase from its hydrogen activation specificity. Compounds whose molecules were similar in structure to the natural substrate of the enzyme could be fixed to it without being activated. The inhibitory action of malonate on the oxidation of succinate, discovered by Quastel, could be explained also in this way as a competition between malonate and the similar succinate molecule for the dehydrogenase molecule. Malonate could be fixed on the enzyme, but did not react further, making the enzyme inaccessible to succinate.[68]

Thunberg's observations and discussion probably suggested to Krebs a strategy for searching systematically for inhibitors that could block the specific reactions of intermediary oxidative metabolism that interested him. On August 9 Krebs tested the "inhibition for respiration" in rat kidney tissue of maleic, oxalic, and mesaconic (cis-methyl-fumaric) acid. Mesaconic had little effect, whereas the other two substantially reduced the rate of oxygen uptake. Maleic and oxalic acid he also tested in the presence of succinic, fumaric, malic, and lactic acid. The result was that "maleic seems to inhibit fumaric and malic, not succinic oxidation. Oxalic seems to inhibit succinic mainly!"[69] Since oxalic acid was similar in structure to succinic acid, and maleic was the *cis*-isomer of fumaric acid, these effects could be fitted into the explanatory framework given by Thunberg.

The next day Krebs extended this approach, examining in guinea pig tissue the "inhibition of brain respiration" by aconitic, tricarballylic, citraconic, and tartronic acid. The choice of the first two of these compounds is especially interesting because these two molecules are similar enough to citric acid to imply that he was looking for an inhibitor that might block its oxidation. These substances, however, exerted "no certain effect." He tested oxalic and aconitic acid on the anaerobic pyruvate reaction in testicle tissue, again without finding any major effects.[70]

We can readily appreciate how attractive this approach must have seemed to Krebs when he began the preceding experiments. Methods to block specific metabolic reactions so that the intermediate products would accumulate had long been a major desideratum in the field. Those few agents that did act in this way, such as malonate and arsenite, were of great value; but, for most of the substrates already implicated in oxidative metabolism, nothing comparable was available. If he could identify additional inhibitors of the oxidation of some of these specific substrates, the task of establishing the pathways that connect them would be greatly eased. Despite the indication from the first experiment that maleic acid acted in such a manner on the oxidation of fumaric acid, Krebs must have decided after this brief foray that the overall results were not promising enough to divert him further from his current investigation.

Through the rest of August, Krebs continued his study of the dismutation reactions along the same lines as before, dividing his time about equally between animal tissues and the newer extension to staphylococcus. On the 11th, he performed an experiment on rat liver slices similar to the first one he had conducted in February to test the dismutation reaction acetoacetic acid + pyruvic acid + H_2O = acetic acid + β-hydroxybutyric acid + CO_2. The result was that "pyruvic causes increase of β-hydroxybutyric formation and acetoacetic acid disappearance, and CO_2 production; but not equivalent."

Extra β-hydroxybutyric	+ <u>1.10</u>
Extra acetoacetic	- 1.52
Extra CO_2	+ 2.27[71]

"Equivalent" quantities would have been the 1:-1:1 ratio predicted by the dismutation equation.

In the next experiments, on "staphylococcus + B_1," Krebs added the vitamin in four concentrations and observed, in contrast to the previous experiment with rat tissue, an "enormous effect of B_1" on the anaerobic formation of CO_2 from pyruvate. Repeating the experiment with small quantities of B_1, he found a substantial increase, and on August 13 with even smaller quantities he still attained a "definite effect." Because he had found in the literature little data concerning the metabolism of the bacterium, he next measured the "oxidations of *Staphylococcus aureus*" in the presence of four substrates: glucose, *l*-lactate, *d*-lactate, and pyruvate. He graphed the curves of oxygen absorption over time. The total quantities of oxygen absorbed, he noted, were equivalent to $2\frac{1}{2}O_2$ per molecule of glucose, "almost $2O_2$," per molecule of lactate, and "almost $1O_2$," per molecule of pyruvate. He inferred that "glucose [is] probably not broken down through lactic acid (since glycolysis is too slow!) and rate is too slow, and final O_2 figure is too slow!"[72] The metabolism of this simple organism thus appeared to diverge from the pattern found in animal tissue.

Up until this point Krebs had carried out all of his experiments involving vitamin B_1 with commercial ampoules from Hoffmann-La Roche or Glaxo. Peters had used crystalline B_1 in the most recent of the experiments on the consumption of pyruvate in chickens, but this pure form of the vitamin was still hard to procure. When Krebs did receive some from Hoffmann-La Roche, he tried it out first on the disappearance of pyruvate in rat liver tissue, on August 14, and observed to his surprise, "no effect of crystalline B_1!!" In contrast, a yeast extract that he tested simultaneously exerted an "enormous effect."[73]

Now performing almost every day at least one experiment on animal tissue and one on staphylococcus, Krebs observed another peculiarity of the bacterial metabolism on August 17, when he found "no measurable decomposition of ketoglutarate and acetoacetate!!" when either was added alone to a suspension, whereas pyruvate alone greatly increased the anaerobic formation of CO_2. In a second experiment he found that although ketoglutarate "does not react alone," it gives in combination with pyruvate an "increase of CO_2" over that caused by pyruvate alone. Oxaloacetate added alone gave "quickly about theoretical figure!!" for its decarboxylation, and also increased the "rate in the presence of pyruvic!" This behavior made him wonder if the oxaloacetate, whose decarboxylation normally yields pyruvic acid, might undergo a reaction in which the other carboxyl group was removed, giving rise to malonic acid. He could test for that possibility, he thought, by trying whether succinoxidase was inhibited afterward in the suspension. On rat kidney tissue that day he repeated the experiment with acetoacetic + pyruvic acid that he had performed six days earlier on liver tissue. As before, he observed an increase in the disappearance both of pyruvate and of acetoacetate when pyruvate and acetoacetate were added together, but the two increases were not in "equivalent" amounts (he did not measure the β-hydroxybutyrate this time). A similar experiment in kidney tissue was incomplete because the pyruvate determination "failed to work!!"[74]

On August 18 Krebs carried out a single experiment on staphylococcus, finding again that ketoglutaric acid increased the rate of CO_2 formation from pyruvic acid, but acetoacetic acid did not exert an analogous effect. The next day he found that crystalline vitamin B_1 failed to increase the pyruvic acid reaction in staphylococcus, whether or not it was added in yeast extract. This was a far more dramatic difference from the behavior of commercial vitamin B_1 than the similar result in animal tissue, because the commercial vitamin had exerted large effects in the bacterium. In rat tissue that day he again tried to attain equivalent proportions between the products of the reaction of acetoacetic and pyruvic acid. This time he succeeded, at least for two elements of the equation. The extra β-hydroxybutyric acid formed when the two substances were added together was "exactly equivalent to [the] acetoacetic" that disappeared. If that result confirmed an expectation, there was nevertheless a surprise in the fact that in one of the controls "pyruvic [added alone also] gives β-hydroxybutyric!!!!" ($\Delta\beta$-hydroxybutyric for pyruvic was 0.75, for pyruvic + acetoacetic, 4.08).[75] These mixed results must have left Krebs uncertain about the underlying situation. The stoichiometric proportion that previously had eluded him pointed to the dismutation reaction as the sole source of the β-hydroxybutyric acid, whereas the result for pyruvic alone indicated that another reaction was also yielding β-hydroxybutyric acid from pyruvic acid.

Matters went on similarly for the rest of the month. On the 20th Krebs obtained an increase in the anaerobic CO_2 from pyruvic + oxaloacetic acid in guinea pig brain and kidney tissue. On the 21st and 22nd two experiments on staphylococci using various oxidation-reduction indicators that he thought might catalyze the pyruvate reaction inhibited it instead. In several further experiments yeast extract continued to activate the reaction, from which he drew an interesting query: "yeast extract is general activator for metabolism!! (not for growth): or is increase due to growth!!!!! Not probably since no nitrogen available!! Try amino acids?" On the 24th in pigeon brain tissue a mixture of ketoglutaric + acetoacetic acid yielded a smaller increase in anaerobic CO_2 than did either substance separately, an outcome that he noted laconically "is unexpected." In rat kidney on the 25th, on the other hand, the same two substances together gave "more than summation." From the 26th to the 28th he examined the effects of various conditions, including glucose and pyruvate as substrates, increasing quantities of yeast and bacteria concentration, on the pyruvate reaction in staphylococci, and carried out a large-scale experiment. After one of these experiments he observed that the "organisms grow in N_2 on account of pyruvic acid dismutation!! energy giving process!! try again!"[76]

Despite such points of interest, and a consolidation of the data supporting some aspects of his view of the anaerobic dismutation reactions, August as a whole marked a lull in the progress of the "great work." Further experiments revealed complexities and divergent results under different conditions or with different organisms and tissues. The vitamin B_1 effect that Krebs thought he had confirmed when he submitted his *Nature* article disappeared with the purification of the vitamin. It remained doubtful whether the reactions he had formulated in his paper as the potential primary steps along the main route of oxidative carbohydrate metabolism would prove to be so, or whether other types of reactions accounted for some of the pyruvic acid that disappeared. What appeared at the end of June as a rapid thrust to the heart of the "big unsolved problem" must have seemed by the end of August to be settling down into a long campaign for small further gains along a more diffuse front.

While Krebs fixed his own experimental work on the dismutation of the ketonic acids in the summer of 1936, he directed some of Johnson's experiments to another, well-known anaerobic reaction, the reduction of fumaric acid to succinic acid. In June and July Johnson performed at least two such experiments, measuring the effect of adding fumarate on the formation of succinic acid in rat liver and kidney tissue. In both cases the effects were positive.

$Q_{succinate}$

	Liver	Kidney
Control	0.36	0.18
+ Fumarate	0.72	1.50

By late August Johnson had switched to minced pigeon breast muscle. After measuring the formation of CO_2 from pyruvate on the 26th, he measured both CO_2 and succinate from fumarate on the 26th and 28th. Again the increases were marked.

	26th		28th	
	$Q_{succinate}$	Q_{CO_2}	$Q_{succinate}$	Q_{CO_2}
Control	0.64	0.36	0.43	1.1
+ Fumarate	3.15	1.46	1.57	1.9[77]

Because there is a record of only a sampling of Johnson's experiments, we cannot follow the course of his investigation closely, and consequently cannot reconstruct in fine detail the stages in the thought of his chief that guided this work. Whatever Krebs had in mind when Johnson began the experiments with fumarate early in the summer, the measurements of CO_2 and succinate together that Johnson was carrying out by the end of August indicate that by then, if not sooner, Krebs had formulated the hypothesis that the anaerobic reduction of fumaric to succinic acid was a component of a coupled oxidation-reduction system. The other component was a substrate that undergoes an oxidative decarboxylation analogous to those occurring in the dismutation reactions of the ketonic acids.

If all went according to plan, Krebs traveled southwest after completing a set of experiments on Friday, August 28, in order to spend the weekend with David Nachmansohn in Plymouth. They had planned this reunion for some time, but Nachmansohn still had to press Krebs to come. Having heard nothing, he had written on the 22nd,

> What has happened about your coming here? I am expecting you definitely this weekend. <u>I am sure you will spend here a marvelous and useful time.</u> The landscape is beautiful, we can hope for passable weather, and there are some nice people here. . . . I hope that your research is continuing to be just as exciting [as in July], but a 3 day break is good and important for any investigation. You lose 10% in time and gain 30% in productivity. (I will show you the appropriate curves here.)[78]

Nachmansohn knew the habits of his friend well. The fact that Eggleston recorded experiments in his separate notebook on August 28 and 29[79] supports the assumption that Krebs did tear himself away from his research for the trip.

In Plymouth the talk was probably more about science than about Palestine. Nachmansohn had come to believe by then that Weizmann was too overburdened, and perhaps too old, for them to expect much from him in the near future. He thought that Weizmann should turn over to someone else the building of the science faculty in Jerusalem. Although Nachmansohn was hopeful that the worst of the political crisis in Palestine was over and that economic growth would quickly resume, he had already written Krebs in July that "we must put our plans aside for the immediate future. Perhaps in a few months things will change."[80]

Krebs was probably more eager than usual to share his own recent scientific progress with Nachmansohn. Earlier in the summer he had written about his forthcoming *Nature* article in such a way that Nachmansohn had written back, "I am completely inspired by all the new things you have found, and am very happy about it. It is truly another great step forward in knowledge of metabolism, and I am especially joyful that it is you who has achieved it. You are really ripe for a Nobel prize, if it were up to me you would receive it this year."[81] Krebs would have taken this judgment as a sign of his warm friend's enthusiasm, rather than as a realistic prediction, but the tone of Nachmansohn's response implies that Krebs must have expressed his own sense of the significance of his results in expansive terms. In Plymouth, Nachmansohn probably also sought Krebs's advice about his own promising research on acetylcholine, and showed him the experiments he was doing during his month in Plymouth.

VI

By 1936 considerable progress had been made in the identification of certain dialyzable coferments (or coenzymes), whose presence was essential for the catalysis of oxidative cellular processes. Otto Warburg's discovery of the "yellow enzyme" and identification of this pigment—a small molecule in the class of flavin compounds—led the way. In 1933 Richard Kuhn, Paul György, and Theodor Wagner-Jaurreg isolated from milk another such substance which they named lactoflavin. Although Krebs took no part in such investigations, he followed them closely. In the late summer of 1936 he became increasingly interested in testing the effects of these and other "cofactors" on the anaerobic oxidation reactions he was studying. In August he wrote to Warburg asking for some of the yellow enzyme. On August 22 he tested lactoflavin on staphylococcus, but observed no clear effect. On September 1 he tested some of the yellow enzyme he had received from Warburg's laboratory on the anaerobic formation of CO_2 in staphylococcus. It doubled the rate, whereas a new sample of crystalline vitamin B_1 obtained from Adolf Windaus had no effect.[82]

On the same day Krebs measured the rate of respiration of staphylococcus with lactate and with glucose added, and found this time that "lactate gives same as glucose."[83] This result contradicted that of the similar experiment he had carried out in early August, and must have thrown into question the conclusion he had then drawn, that "glucose is probably not broken down through lactic acid."

On September 2 Krebs branched out in three ways at once on the bacterial side of his investigation. In one experiment he tested "staphylococci + aldehyde." He had in mind the possibility that in this organism the original type of Cannizzaro dismutation discovered by Parnas and by Battelli and Stern in 1911 might take place, yielding an acid and alcohol corresponding to the aldehyde. He found, however, "no pressure change with 5 mg acetaldehyde" (or, as he put it in 1977, "nothing happened").[84] The second new dimension was to pursue with bacteria the anaerobic reactions of fumaric acid that Johnson had been up until then examining in rat tissues and pigeon breast muscle. The third new move was to test fumarate both in staphylococcus and in *B. coli*. It is probable that he had made a more general decision to pursue his study of anaerobic oxidations comparatively in both types of bacteria.

When he made these moves Krebs had formulated the hypothesis that one of the "partners" in the coupled anaerobic reaction in which fumaric acid is reduced to succinic acid was lactic acid. The lactic acid would be oxidized to pyruvic acid, which would in turn give rise by decarboxylation to acetic acid and CO_2. This was an obvious choice in the context of recent experiments by others showing that staphylococcus and *B. coli* both oxidize lactic acid anaerobically to pyruvic acid and that *B. coli* decarboxylates pyruvic acid.[85] The conversion of lactic to pyruvic acid was a dehydrogenation, so Krebs could envision the reaction between fumaric and lactic acid as a dismutation in which lactic acted as the hydrogen donor, fumaric as the acceptor. In the experiment with staphylococci, fumaric + lactic did cause a "definite pressure"—that is, it increased the formation of CO_2 anaerobically—but the control containing fumaric alone caused "about the same" change. The experiment as a whole was inconclusive because the "total absolute figures [were] small." The results with *B. coli* were more striking. Here the combination yielded by far the largest increase, but the fact that fumaric acid without lactic also exerted a substantial anaerobic effect equally impressed him. He wrote down:

> fumaric alone reacts!!
> much more fumaric + lactic
> fumaric + lactic = succinic + pyruvic
> pyruvic → acetic CO_2, H_2.[86]

The experiment thus appeared to verify his hypothesis. In his notation about the subsequent reactions of pyruvic acid he included H_2 among the products because of the special property of *B. coli*, studied particularly by Marjory Stephenson, to produce hydrogen gas (see p. 190).

Krebs had included "fumaric alone" undoubtedly as a control, and the fact that it also "reacts" anaerobically evidently came as a surprise to him. He did not put down an equation to explain this reaction, but it undoubtedly occurred to him that two molecules of fumaric acid might dismutate, one being reduced to succinic, the other oxidized to oxaloacetic, which would in turn be decarboxy-lated.

On September 3 Krebs found that both crystalline vitamin B_1 and the yellow enzyme inhibited the pyruvic acid reaction in liver tissue, a result that he questioned. Yeast extract doubled the reaction in *B. coli*. Repeating the experiment of the previous day with fumarate + lactate in *B. coli*, he obtained a similar result: "again considerable pressure with fumaric: more with fumaric + lactic!" Exploring the possibility that the reduction of fumaric acid might be coupled with oxidative deamination, he tried fumaric acid + *l*-aspartic acid in rat kidney tissue. There was a "definite increase [of Q_{CO_2}] by fumarate" alone, as there was in *B. coli*, but "fumarate + aspartic gives less CO_2."[87] He did not determine NH_3, presumably because the CO_2 measurement was in itself enough to discourage further exploration in this direction.

The next day Krebs did not himself perform any experiments, but Johnson measured the changes in CO_2 and lactate in a pigeon breast muscle suspension in the presence of fumarate. His result ($\Delta CO_2 = 510 \ \mu l$, Δlactate = -594 μl) fit well with the view that the reaction Krebs was examining in bacteria occurred also in animal tissues.[88] (The difference between the procedure that Johnson used—adding fumarate alone and measuring the decrease in lactate present—and the one that Krebs used—comparing the change in CO_2 with fumarate and with fumarate + lactate—is interesting. Since there is no way to tell whether or not this was the first experiment by Johnson directed at the possibility of a fumaric-lactic reaction in pigeon breast muscle, however, it is difficult to comment further on this difference.)

According to the dates in his notebook, Krebs may have been away from his laboratory on September 4 and 5, but put in a busy work day on Sunday the 6th. One of the four sets of experiments he did that day was to test the yellow enzyme, yeast extract, and a cozymase preparation just received from Karl Myrbäck in Stockholm on the pyruvate reaction in staphylococci. Krebs had written Myrbäck two weeks earlier to request the substance, in order to test "whether cozymase perhaps plays a role" in the "strong accelerating effect" of yeast extract on the decomposition of pyruvic acid in animal tissues that he had recently observed. In staphylococci the yellow enzyme and cozymase + yellow enzyme were both "active," but the readings were low, and Krebs suspected that the staphylococci were too old. In a second experiment that day he tested "aldehyde + CO_2" on *B. coli*. He was asking the "question whether" the reaction

$$\begin{array}{ccc} CH_3 & & CH_3 \\ | \quad\quad O & & | \\ HCO \ + & \rightarrow & COOH \\ \quad\quad H_2 & & \\ + \ CO_2 & & + \ HCOOH \end{array}$$

occurred. This equation—another variation on the anaerobic dismutation theme, which he apparently constructed on paper—would account for two of the

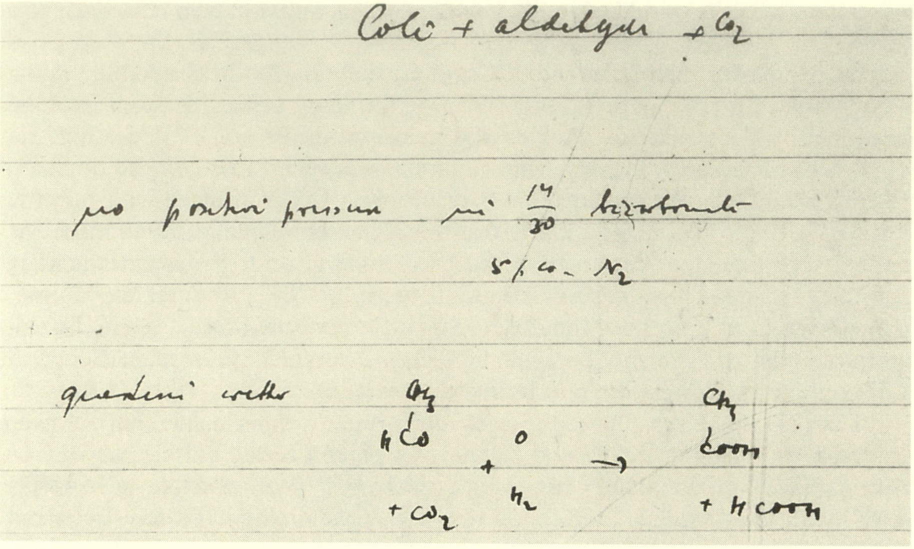

Figure 7-3 Hypothetical reaction posed by Krebs in laboratory notebook No "12," p. 159.

prominent products of these bacteria, acetic and formic acid. There was, however, "no positive pressure."[89] It was another of his prolific ideas that turned out quickly to be wrong.

On the same Sunday Krebs began an examination of the effects of malonic acid and other possible inhibitors on the reactions of fumaric and succinic acid in pigeon breast muscle. His immediate objective was to check some of the effects of malonate that Szent-Györgyi and his coworkers had described in their publications of the preceding year. Beyond that, however, Krebs undoubtedly believed that these effects were pertinent to the anaerobic reactions of fumaric acid that he and Johnson had been studying and on which he now wished to center his attention.

Szent-Györgyi's group had reported several observations related to these substances. Ilona Banga had shown that "malonate strongly inhibits" the respiration of pigeon breast muscle, but that "fumarate + malonate respires again like fumarate alone." Using a micromethod that he devised for determining fumarate, Bruno Straub found that when he added fumarate + malonate aerobically to a muscle suspension, a considerable proportion of the fumarate disappeared, but that when he did it anaerobically the fumarate did not disappear. Szent-Györgyi explained these results in terms of his current succinate-fumarate-oxaloacetate hydrogen transport scheme. The fumarate that gradually disappeared aerobically was reduced to succinate. Normally the succinate was reoxidized, but malonate blocked that process by inhibiting succinoxidase. Respiration therefore ceased when all of the fumarate was removed. Malonate stopped respiration in the absence of added fumarate within two minutes, the length of time necessary to reduce the small quantity of fumaric acid normally contained in the tissue. It was the observation that the fumarate was not removed anaerobically that led Szent-Györgyi to believe that fumaric acid was not directly reduced, but had to be oxidized to oxaloacetate, which was "overreduced" to succinate;[90] it was Krebs's dissatisfaction with Szent-Györgyi's explanation that now led him to do "some experiments to see for myself what the phenomenon looked like."[91]

As in other situations where he followed up the results of his contemporaries, Krebs did not repeat any of the exact experiments reported in the papers from Szent-Györgyi's laboratory, but explored "the phenomenon" as it related to his own investigative concerns. He did employ pigeon breast muscle coarsely minced as they did, but in the first experiment, in which he added fumarate by itself and in combination with malonic, maleic, tartronic, and malic acid, including besides a run with malic acid alone, he measured the CO_2 production, as he had been doing previously in his studies of anaerobic dismutation reactions. The results on which he remarked were that there was "CO_2 production from fumaric! inhibited by malonic? and tartronic. Not influenced by maleic, malic. Malic leads to CO_2 production too!!" The doubtful inhibiting effects that had probably been the main subject thus impressed him less than the observation that fumaric acid alone forms CO_2 anaerobically in pigeon breast muscle, as he had already found earlier that it did in bacteria. It was apparently more surprising to him that malic acid acted similarly.[92]

The aerobic experiment that Krebs conducted on the same day was not strictly comparable to the anaerobic one. He tested the effects of malonic and maleic acid on the respiration of the same tissue suspension in the presence of fumaric acid, and the effects of malonic only in the presence of succinic acid. There occurred in the controls an "enormous absorption of O_2." Despite there being "too much pressure," he concluded that "it is clear that malonic inhibits fumaric acid oxidation!!! So does maleic."[93] Malonic acid also inhibited succinic acid oxidation, but since that was a well-known effect, Krebs did not comment on it. The inhibition of fumaric acid oxidation was contrary to what Szent-Györgyi's group had reported, and therefore invited further examination.

In his next experiment, on the "effect of malonate on oxidation of succinic and fumaric in muscle," Krebs minced the pigeon breast muscle finely, by keeping the second disk on the Latapie grinder, and he diluted the suspension in 10 parts of phosphate saline instead of the two parts used in the preceding experiments. He added malonate in two concentrations. He found that, under these conditions, fumarate added without malonate brought "no increase" in the respiration and that there was "increase by fumarate in presence of malonate, but less than by succinate." None of this fit the view that fumaric acid restored the respiration inhibited by the effect of malonate on the oxidation of succinic acid.[94]

Switching to guinea pig brain tissue on Tuesday the 8th, Krebs tested the effects of fumarate, malate, and fumarate + malate on the oxygen consumption. There was a (small) "overall increase by fumaric and malic" acid, with no further increase due to the two together. The rate of respiration in the control with no added substrates was "very high," and the "falling off [during the second hour was] not inhibited by malic and fumaric." The "conserving" effect of fumarate on the respiration of pigeon breast muscle thus did not appear in guinea pig brain tissue. On the same day Krebs tested the effect of malonate on succinic and fumaric acid in kidney slices from the same animal and found that malonate alone inhibited respiration by 64 percent during the first 20 minutes, malonate + fumarate by 22 percent, and malonate + succinate by 70 percent. "Less inhibition by malonate ($^M/_{100}$ with fumarate!!"[95] Here, finally, was a result in accord with those of Szent-Györgyi.

On the same day Johnson performed an experiment with pigeon muscle slices immersed in muscle extract which confirmed the older observations of Thunberg and of Quastel that malonate inhibited the anaerobic reduction of fumaric acid to succinic acid. Johnson also carried out another experiment showing that lactic acid disappeared when fumaric acid was added anaerobically to muscle tissue.[96]

Having seen for himself only a small part of the phenomena that Szent-Györgyi and his associates had reported, and some phenomena hard to fit with their observations, Krebs returned to his preoccupation with the reaction of fumaric and lactic acid. In four rat tissues in a medium containing lactate he measured the effects of fumarate on the anaerobic formation of CO_2. He found "definite increases in testis and kidney; in liver first inhibition, later increase." In brain tissue the pressures were too small to yield meaningful results. In the tissues in which he obtained increases he noted, however, that the "rates are too small to explain oxidation of lactate in intact tissues." That is, the anaerobic oxidation reaction he had in mind (see above, p. 314) evidently occurred in these animal tissues, but if this result were to hold generally, the reaction could not constitute the main pathway for the further decomposition of the lactic acid produced by glycolysis. On the same day a combination of the yellow enzyme + cozymase and of the yellow enzyme + crystalline vitamin B_1 each increased the rate of the anaerobic pyruvic acid reaction in an old suspension of staphylococcus.[97] Continuing his short-term oscillations between experiments on animal tissues and on bacteria, Krebs next pursued the fumaric-lactic acid reaction in *B. coli*. In two parallel sets of experiments, one using a fresh suspension in bicarbonate saline, the other with *Kochsaft* added, he measured the CO_2 formed and, assuming all the fumarate added was used up, computed the fumarate:CO_2 ratio. The experimental ratios were 3.10 and 3.05. The expected ratio, he noted, was fumarate:CO_2 = 2. The reaction sequence that this proportion implies he had in mind was:

2 fumaric acid + 2 lactic acid = 2 succinic acid + 2 pyruvic acid
2 pyruvic acid + H_2O = lactic acid + acetic acid + CO_2

The measured ratios of 3:1 clearly indicated that something more was taking place. Perhaps in order to ascertain what else might be consuming fumarate, he tried the next day the same suspension with fumarate + lactate and that combination + three other substrates: succinate, acetate, and pyruvate. In the first case the ratio was 3.25, whereas with each of the additional substrates it was somewhat higher (3.59, 3.34, and 3.72 respectively).[98]

Shortly before this time Sydney Elsden, a student of Marjory Stephenson, had come to Sheffield to spend a few weeks working under Krebs's supervision "on problems relating to the formation of succinic acid by various bacteria." Elsden had hoped to bring his manometers with him, but when the Cambridge Biochemistry Department was unable to supply them for him in time, Krebs assured him that "we have a few manometers to spare and it will be perfectly all right if you make a start without your own manometers." On September 11 Krebs recorded in his own notebook the results of an experiment Elsden had carried out on "Coli + fumaric + lactic": 680 μl of fumaric acid added yielded 227.5 μl CO_2 and 642 μl of succinic acid, proportions roughly consistent with the reactions described above in the preceding paragraph. With "fumarate alone (no lactate)" Elsden had found 23 μl CO_2 and 83.8 μl succinic acid. Krebs wrote down a reaction that he thought might fit this situation:

equation: 3 fumaric = 2 succinic + 1 acetic + CO_2 + Na H CO_3

The next day he himself tested both fumaric and ketoglutaric acid with lactic acid, in the same *B. coli* suspension he had used the day before, measuring the CO_2 formed anaerobically. Prior to the experiment the suspension had been aerated, and he thought that part of the lactate had been oxidized "as long as air was available!!" before the fumarate was added. The ketoglutaric acid was "not metabolized!!!" As in several previous experiments he noted that "fumarate alone reacts," and this time he wrote down the reaction that he thought occurred in that case:[99]

2 fumaric → oxaloac[etic] + succinic

Krebs had now formulated three different dismutation reactions involving fumaric acid. Clearly his conception that there existed a family of analogous metabolically significant anaerobic oxidation reactions was permitting him to exercise his scientific imagination rather freely. On the same day he repeated the experiments with fumarate + lactate on *B. coli*, but measured this time, in addition to the CO_2 directly formed, the bicarbonate formed in the medium. He found a "huge increase in bicarbonate!! and free acid from fumaric + lactic." Although he did not say so, this result must have cast doubt on the validity of the 3:1 fumarate:CO_2 ratios obtained in the experiments of two days earlier, when he had not taken into account the CO_2 liberated in the form of bicarbonate.[100]

Perhaps recognizing that he had not dealt adequately with the observations made by Szent-Györgyi's group, Krebs took up again on September 13 the question, "Does fumarate abolish the inhibition by malonate?" Using cat muscle tissue, he obtained a "strong inhibition [of respiration] by 0.01 M malonate," on which 0.02 M fumarate "has no effect." With minced pigeon breast muscle the result was more positive:

Fumarate abolishes partly malonate inhibition, not completely.

Although not in full agreement with Banga's finding that fumarate did restore the respiration completely, Krebs's result was apparently enough to satisfy him, and he went on with a new phase in his own line of investigation, the search for additional anaerobic reactions. On the next day he tested in *B. coli* the effects of a combination of acetate + formate on the anaerobic pyruvic acid reaction. His purpose was to see whether the reaction long known in *B. coli*, "pyruvic → formic + acetic" as he wrote it down, was reversible. The absence of any marked effect by adding the two products together as substrates convinced him that it was not. On the same day he tested malic + fumaric acid anaerobically on *B. coli* and obtained more pressure than from fumarate alone, a preliminary indication that he was on the track of yet another oxidation-reduction reaction involving fumaric acid. Finally he tested two ketonic acid dismutation possibilities: oxaloacetic + pyruvic acid, which he had already tried with indecisive results on animal tissue during the summer, and a new combination, oxaloacetic acid + acetic acid. In *B. coli* the first of these combinations raised the quantity of succinic acid and CO_2 formed more than oxaloacetic acid alone did. The second combination exerted no effect.[101]

In his experiments with fumaric acid up until now Krebs had been concerned primarily with its role as the compound reduced in anaerobic oxidation reactions, particularly the reaction in which the compound oxidized is lactic acid. Now he turned to the question, How is fumaric acid (and malic acid) oxidized? One answer that he had already entertained for several years was that it was through the Thunberg-Knoop-Wieland scheme, or a variant of it. Now, however, he approached the question from the orientation of Szent-Györgyi's work. Szent-Györgyi had shown that malic or fumaric acid can be oxidized directly to oxaloacetic acid, but he had also postulated that molecular oxygen reacts with these substances through some undefined carrier system interposed between the succinic-fumaric system and the cytochrome system. The recent work of David Green, an American research fellow in the Cambridge Biochemistry Department whom Krebs had known well while he was there, gave Krebs an idea about what this system might be. The anaerobic oxidation of α-glycerophosphate to glyceraldehyde phosphate was a key step in the Embden-Meyerhof glycolytic pathway; but Green had identified in rabbit muscle tissue a "glycerophosphate oxidase" that catalyzed the same reaction aerobically, and gave evidence that this enzyme was different from the enzyme responsible for the glycolytic reaction. Working with Flora Jane Ogston, Green showed that glycerophosphate was the only compound apart from succinic acid that reacted with molecular oxygen directly through the cytochrome system.[102] It occurred to Krebs that glycerophosphate–glyceraldehyde phosphate might constitute the intermediate carrier that Szent-Györgyi had placed between the cytochromes and the succinic-fumaric system. If so, these substances should interact according to the reaction

fumaric

glyceraldehyde phosphate +   [or] → glycerophosphate + oxaloacetic

malic

Taking up this question on September 15, Krebs first examined the "specificity of succinic oxidase," asking "is glycerophosphate attacked?" That is, was Green's glycerophosphate oxidase really a distinct enzyme, or were glycerophosphate and succinic acid oxidized by the same enzyme? Using washed pigeon breast muscle (presumably the washing removed oxidases other than succinic oxidase), Krebs compared the rates of respiration in the presence of glycerophosphate and succinate and concluded that the former was oxidized at only 4.8 percent of the rate the latter was. Moreover, malonate inhibited the "glycerophosphate oxidation only slightly (18%)," in contrast to its known effect on succinate oxidation. Methylene blue increased the rate of oxidation, and the oxygen absorbed by glycerophosphate was "more than the theoretical yield," that is, more than the calculated quantity required to convert it to glyceraldehyde phosphate. These results, though not unambiguous, suggested that glycerophosphate was not oxidized by succinic oxidase. To test the glyceraldehyde-fumaric (or malic) acid reaction he had in mind he added glycerophosphate and glycerophosphate + fumarate aerobically to finely minced pigeon breast muscle in the presence   of malonate.   Presumably he expected that under these conditions the malonate would prevent the oxidation of succinic acid present or formed by the reduction of fumarate, some of the glycerophosphate would be oxidized to glyceraldehyde, and the latter would oxidize fumaric acid.   He commented on the results, "α-glycerophosphate increases respiration in the presence of malonate, more than fumarate. Effect < starts? > larger, but falls, probably because the reaction glycerin aldehyde + fumaric stops owing to decomposition of coenzyme." With pigeon brain tissue he tested fumarate + glycerophosphate aerobically, without malonate, and obtained "no decent effects (though small increases) perhaps respiration without substrate too high." Thus, although these experiments appeared inconclusive, he construed the results as though the reaction in question probably took place.[103]

The next day Krebs tested the effect of As_2O_3 on "succinoxidase and glycerophosphate oxidase," that is, on the rates of oxidation of α-glycerophosphate and succinate in washed pigeon breast muscle. The result, that the second reaction was largely suppressed, whereas the former was unaffected, must have removed any remaining doubt in his mind that these were separate enzymes (though he repeated the experiment on the 21st to make sure, and obtained the "same result [with a] new suspension").[104]

On the same day (September 16), Krebs returned to a more familiar aspect of his ongoing investigation, the anaerobic reaction of oxaloacetic and pyruvic acid as an example of the generic ketonic acid dismutations formulated in his *Nature* paper. Testing that combination in two rat tissues, testicle and brain, he found in the former that oxaloacetate + pyruvate caused more CO_2 to form than either added alone ($Q_{CO_2}^{N_2} =$ for the two $= 6.72$; for pyruvate, 5.75; for oxaloacetate, 5.28). With brain tissue the effect was more striking ($Q_{CO_2}^{N_2} =$ for pyruvate + oxaloacetate $= 6.00$; for pyruvate, 3.87; oxaloacetate not tested). He wrote down "+ oxaloacetate + pyruvate, big increase." Later that day, using tissue from another rat, he obtained only a "small effect of pyr. + oxaloacetate" in spleen tissue, but an even higher rate in testicle tissue ($Q_{CO_2}^{N_2}$ for oxaloacetate + pyruvate $= 8.31$). In each of the experiments except for that on the spleen he tested the effects of As_2O_3 and observed "complete inhibition."[105] It is probable that the rates of anaerobic CO_2 formation he obtained here were higher than any he had seen with analogous reactions and that these results led him to believe that the reaction oxaloacetic acid + pyruvic acid + H_2O = malic acid + acetic acid + CO_2 held a "special interest" in the metabolism of animal tissues.

On September 17 Krebs moved along three different salients of his research front. The first set of experiments, on "oxidation of acetate by *Coli*," he conducted probably to explore the question what becomes of the acetic acid formed as a consequence of the anaerobic reaction between fumaric and lactic acid. The metabolic fate of acetic acid in animal tissues was, as we have seen, one of the long-standing mysteries of the field. It was quite likely, however, that bacteria dealt with this prominent substrate in a different manner. In the experiment he found that acetate added aerobically was rapidly metabolized. The amounts of extra O_2 absorbed and the total extra CO_2 produced in the form of gas and bicarbonate accounted, however, for only 69 and 70 percent respectively of the quantities theoretically required to oxidize acetate to CO_2 and H_2O. The acetic acid, he inferred, was incompletely oxidized. The sodium added with the acetate was almost completely liberated (90 percent). "Therefore," he concluded, a "neutral substance or weak acid must be formed from acetic." He wondered if that substance might be the known product of bacterial metabolism, "acetyl methyl carbinol?"[106]

The second experiment Krebs performed on the 17th, opening yet another extension of his current investigative program, was based on earlier work by his Cambridge colleague Marjory Stephenson. In 1931 Stephenson and Leonard Stickland had identified in a number of bacteria, including *B. coli*, an enzyme that activated molecular hydrogen. The basic reaction they expressed as $H_2 \rightleftarrows$ 2H. The activated hydrogen could be transferred to various hydrogen acceptors, which were reduced in the process. Among the compounds they showed could accept the hydrogen were standard indicators such as methylene blue, inorganic salts such as nitrate, molecular oxygen, and a metabolite such as fumarate, which was reduced to succinate. Because the action of hydrogenase on fumaric acid was analogous to the anaerobic reactions of fumaric acid that Krebs had been studying, the reaction "fumaric + $H_2 \rightarrow$ reduced" attracted his interest. He tested it by adding fumarate to fresh *B. coli* suspensions in atmospheres respectively of 5 percent CO_2 and H_2, and 5 percent CO_2 and N_2. He then measured the quantities of succinic acid formed. In hydrogen gas the yield was 97 percent of that calculated for the reduction of the fumarate present; but since "about 26%" of the fumaric acid in nitrogen gas also formed succinic acid, only 71 percent of the fumaric acid consumed in H_2 could be attributed to the "hydrogenase" reaction.[107]

This experiment was probably part of a joint research project that Krebs had arranged to carry out with Sydney Elsden during the five weeks that Elsden worked with him. Although Krebs's experiment has the appearance of an initial foray into the domain of hydrogenase reactions, Elsden must have begun related experiments earlier. Elsden's experiments were, in fact, either prompted by, or led to, doubts in Krebs's mind about the existence of hydrogenase. Elsden found some evidence in *B. coli* for the existence of another variant of the anaerobic oxidation reactions involving fumaric acid:

Fumaric acid + formic acid = succinic acid + CO_2

Consequently Krebs inferred "that hydrogen may react with fumaric acid through the intermediation of formic acid and since all the substrates which can be reduced by hydrogen seem to react with formic acid, it appeared to be unnecessary to assume a hydrogenase."[108] The reactions of formic acid with hydrogen were associated with two other enzymes, formic dehydrogenase and hydrogenlyase, that Stephenson had distinguished from hydrogenase (see above, pp. 190–192).

In the last of the three sets of experiments he conducted on September 17, Krebs again added glycerophosphate + fumaric acid to a pigeon breast muscle suspension, using this time coarsely minced tissue, and As_2O_3, which his previous experiment on the subject had shown to suppress the oxidation of succinate, in place of malonate. He tested afterward for oxaloacetate, using his aniline citrate method. The main results, he noted, were that "glycerophosphate + fumaric are oxidized!!!" and that "some ketonic acid formed!!" Presumably he interpreted this result, as he had the similar previous one, as an oxidation of fumaric acid consequent to a prior oxidation of the glycerophosphate to glyceraldehyde phosphate. The experiment was weakened, however, by the fact that "glycerophosphate has inhibiting action" (an effect not evident to me in the data, since oxygen consumption with glycerophosphate alone was greater than in the control and glycerophosphate + fumarate was greater than with fumarate alone). Wondering if he had used "impure stuff?" because "according to [Frank L.] Pyman" the coauthor of the method he had looked up for its preparation, it "contains enzyme poisons," he prepared fresh glycerophosphate the next day following the directions of Pyman and Herbert A. Stevenson. That same day he carried out an indecisive experiment on pyruvate + acetate in *B. coli* which indicated that the two together produced "about the same" amount of CO_2 anaerobically as pyruvate alone, or even "(rather less!!)." Then, after having performed multiple sets of experiments for 13 consecutive days, he took a weekend off.[109]

Krebs began the week of September 21 with another experiment related to his interest in the reactions of succinic acid, glycerophosphate, and fumaric acid. The oxidations of all three of these substrates in pigeon muscle he now found to be inhibited by As_2O_3, but in "different degree" (fumarate > succinate > glycerophosphate). Shifting back to the ketonic acid dismutation phase of his multifaceted investigation, he tested ketoglutarate + pyruvate in *B. coli*, and obtained no overall increase in the yield of CO_2, although there was an initial increase in the rate of its formation. "Removes lag," he put down. Elsden measured the succinic acid formed and also found no increase. On the 23rd Krebs carried out a comparable experiment with oxaloacetate + pyruvate and obtained, as he had recently also in animal tissue, a large increase in the yields of both CO_2 and succinate[110]—a result that probably reinforced his feeling that, among the several analogous ketonic acid dismutations he had studied, this version was somehow particularly important.

On the same day Krebs examined the metabolism of a suspension of propionic acid bacteria that had been growing very well in broth agar for the past 18 hours. Harlan G. Wood and Chester H. Werkman had shown during the past year that these bacteria, named for their characteristic product, utilize CO_2 while fermenting glycerol through a series of reactions leading to propaldehyde and propionic acid. They provided also "indirect evidence" that the same organisms form succinic acid by the condensation of two-carbon compounds. Krebs was attracted to the prospect that propionic acid bacteria carried out further variations on the dismutation reactions he had been studying. When he tried out a concentrated suspension of the culture he had grown, however, he found "no <u>pressure</u> anaerobically with this organism."[111]   Even though the bacteria were known to convert glycerol to propionic acid anaerobically, Krebs did not spend further time coaxing his organisms to work without air for him.

On September 24 Krebs tested "acetoacetic + pyruvic anaerobically" in *B. coli*. The combination, he wrote down, "gives CO_2!!", thus confirming that the reaction whose occurrence in rat tissues had launched him seven months before onto his long preoccupation with dismutations took place also in this organism. On the next day he tested in the same organism a combination he had not previously tried, the anaerobic "oxidation of fumaric [and] of glucose." When added together, these substances produced more bicarbonate and more free CO_2 than fumarate did by itself. What reaction did Krebs have in mind? The fact that he wrote down "fumarate more quickly broken down in the presence of glucose" indicates that he did not suppose that these two substances reacted directly with one another; nor was it plausible that they would. Most likely he expected that glucose would give rise through glycolysis to glyceraldehyde phosphate, which would then react with fumaric acid through the reaction he had previously formulated for that case.[112]

In the last experiment Krebs also measured the aerobic rate of oxidation of glucose in the same bacterial suspension. Apparently he wished to compare the rate of the anaerobic and aerobic reactions. The exact connection he wished to explore is not clear, but when he repeated the overall experiment the next day (on a "suspension aerated for about 1 hour" to consume oxidizable intercellular substrates) he sharpened the comparison by measuring the aerobic rates of oxidation for both glucose and fumaric acid (separately). Anaerobically he compared the free CO_2 + bicarbonate and the succinate produced by glucose, fumarate, glucose + fumarate, and a blank. The combination of glucose + fumarate produced more both of total CO_2 and succinate than did fumarate alone, whereas glucose alone and the blank produced no succinate. His interpretation of this aspect of the experiment seems somewhat at odds with the results themselves. He concluded, "Glucose accelerates oxidation of fumaric," even though the increased succinate would appear to derive from a reaction in which fumaric acid is reduced. Probably his view of the situation was influenced by the reaction he had previously formulated, that is, the oxidation of fumaric acid to oxaloacetic acid by glyceraldehyde phosphate. The comparison between the aerobic and anaerobic reactions of fumarate added by itself led him to infer:[113]

CO$_2$ formation from fumaric acid too small to explain rate of oxidation in O_2.
1 fumaric and $3O_2$ for complete oxidation, therefore fumaric $Q_{O_2}^{fumaric}$ should not [be] more than 3 x $Q_{O_2}^{fumaric}$, actually $Q_{O_2}^{fum} = 97$; $Q_{O_2}^{fum} = 5.1 = 1/19!!$ [sic]

That is, the rate of oxidation of fumaric acid aerobically was 19 times the anaerobic rate. The question that this experiment seemed to answer negatively was whether the anaerobic reaction (presumably a dismutation in which an intracellular substrate was reduced) was a primary step in a pathway leading to the complete oxidation of fumaric acid. The posing of the question is further evidence of Krebs's persistent quest for steps along a "main" pathway of oxidative metabolism.

On September 28 Krebs tested "*Coli* + Fumaric + Glycerol," measuring the "CO_2 and bicarbonate exchange." It was well known that *B. coli* and other bacteria fermented glycerol slowly, yielding succinic acid. Krebs carried out parallel sets of experiments combining glycerol with fumarate and glycerophosphate with fumarate (with controls for each). On the next page of his notebook he wrote out an unusually extended "interpretation" of the results of this experiment.

2 Reactions: (1) fumaric + glycerol → succinic + acetic + ~~CO$_2$~~ formic?
$$= - 2NaHCO_3 + 2CO_2$$
(2) fumaric + malic → succinic + acetic + $2NaHCO_3$
$$→ + NaHCO_3 + 1CO_2$$
Now if there is a decrease in $NaHCO_3$ the same quantity accounts for reaction 1.

For reaction 2 Krebs assumed that malic acid was present when he added fumaric acid, because fumaric and malic acid are rapidly interconverted in *B. coli*. He next began calculating, from the measured changes in bicarbonate and yields of free CO_2 in the main experiment and its controls, the change in bicarbonate and CO_2 due to the addition of fumarate and glycerol. Δbic. was -147.5 μl, the net production of CO_2 was 56 μl. Then he reasoned:

If (1) and (2) form at the same rate, <u>no</u> bicarbonate change, but $3CO_2$ formed, therefore 1/3 of 56 [μl. CO_2] due to fumaric decomposition.
         2/3   "   "   "   glycerol.

(Because in reaction 2, in which the source of CO_2 is the oxidation of fumaric acid, $1CO_2$ is produced; whereas in reaction 1, in which the source is glycerol, $2CO_2$ are produced. The bicarbonate produced in reaction 1 is canceled by the decrease in reaction 2.) Reasoning further, he wrote:

Δ CO_2 from fumarate in the presence of glycerol 56/3 = 19. fumarate alone ~ 57
Glycerol has <u>no</u> effect on fumarate oxidation!!

(Because the total CO_2 resulting from the combination of fumarate + glycerol was about equal to that in the control with fumarate alone, glycerol did not change the overall rate of oxidation of fumarate.) "Glycerophosphate calculation," he added, is "difficult since $NaHCO_3$ is formed in dephosphorylation." He looked up the pK values for α-glycerophosphate and phosphate to use for such a calculation, but did not carry it out.[114]

This page can be viewed as another example of what David Gooding calls experimental "construals" to distinguish such provisional, immediate interpretative responses from fully worked out theoretical conclusions (see p. 117). Krebs appears to have been writing as he thought, and the tentative nature of his inferences is self-evident. He was trying out a possible interpretation, checking to see whether it would fit his data. The next day he recognized a flaw that he had not noticed when he wrote out the page. At the bottom he added "29.9. Theory incomplete, since fate of formate is unknown."[115] This is probably typical of many more examples of the way in which he reflected on the work of the day. It happened to survive because he put it in his notebook instead of on scrap paper.

On the same day that he carried out the preceding experiment, Krebs returned to the reactions of fumaric acid with hydrogen gas in *B. coli*. This time he added fumarate in an atmosphere of 2 percent CO_2, 60 percent H_2, and 40 percent air. The pressure change he regarded as a measure of the absorption of H_2, since he assumed that in the respiration of the organisms the quantity of O_2 absorbed equaled that of the CO_2 released. He found that the "H_2(?) uptake stops after about two H_2??" That is, the quantity of gas absorbed corresponded to twice the quantity of fumaric acid consumed. The question marks indicate, however, that he was too uncertain about his assumptions to draw conclusions from this relationship. Later on the same day, he measured the anaerobic CO_2 formation in guinea pig brain tissue in the presence of fumarate + glucose and observed only a "doubtful!!" increase. Carrying out the "same experiment with rat testicle tissue" the following day, he found that "CO_2 formed but no succinic formed!!" Below that he put "queer." His puzzlement implies that he was now viewing this reaction as one expected to reduce the fumaric acid to succinic acid. As we have seen, he had in similar previous experiments with *B. coli* referred to fumaric acid not as reduced, but as oxidized. Perhaps the "interpretation" of the experiment of September 28 reflects a transition in his point of view. Even in reactions in which some of the fumaric acid may be oxidized, some of it would also be expected to be reduced in its role as a hydrogen acceptor.[116]

Two parts of an experiment the same day on "*Coli* + fumarate + α-glycerophosphate + formate" yielded contrasting results. Glycerophosphate + fumarate again gave large increases in CO_2, and in succinic acid as well. Formate + fumarate gave, on the other hand, little more of either than did fumarate by itself. This result, either by itself or in conjunction with similar experiments that Elsden may have performed, led Krebs to conclude that "formic acid does not react with fumarate." The theory by which he had believed he could dispense with hydrogenase thereby "collapsed."[117]

Krebs attained a more positive outcome that day in a test of "*Coli* + oxaloacetate + various substances." As we have seen, the dismutation reaction of oxaloacetic and pyruvic acid in animal tissues already appeared important to him. In *B. coli*, however, he tried oxaloacetate with three other substances that were prominent products of bacterial metabolism—formate, lactate, and acetate. The most striking results came from the control using oxaloacetate alone:

About <u>half</u> of oxaloacetic → succinic
half of   "   → CO_2!!

This outcome supported Szent-Györgyi's view that oxaloacetic acid acts as a hydrogen acceptor oxidizing intercellular substrates, although he would not have interpreted this part of the reaction following Szent-Györgyi, as a direct "overreduction" of oxaloacetic to succinic acid. The rest of the experimental results were mixed. "lactate [+ oxaloacetate] increase CO_2 (owing to reaction fumaric + lactic → CO_2)", but did not increase succinic acid. "Acetate increase <u>succinic</u>!!" he noted with emphasis, but did not increase CO_2. Formate + oxaloacetate had no effect, just as formate + fumarate had had no effect.[118] The experiment did not, therefore, lead to an identification of any specific oxidation-reduction reactions in which oxaloacetic acid took part.

On the last day of September Krebs filled the last pages of his current laboratory notebook with another experiment on fumarate + glycerol in *B. coli*, as well as fumarate + pyruvate. He measured CO_2 and succinic acid. He included also measurements of the respiration of the *B. coli* in air with glycerol alone and fumarate alone, in order to estimate the magnitude of the anaerobic glycerol + fumarate reaction in comparison to the overall rates of oxidation of the two substrates. Elsden contributed to the experiment measurements under the same conditions of the effects of acetate alone and of lactate alone. Krebs calculated that the acetate in Elsden's experiment yielded succinic acid and CO_2 in the ratio 2:1, while lactate produced the ratio 1.65:1.[119]

Through most of September Johnson apparently continued to carry out measurements of the succinic acid and CO_2 produced anaerobically in pigeon breast muscle from fumaric acid and from pyruvic acid. During the last week of the month, however, he began to conduct more varied experiments. On the

28th he included among them a measurement of the formation of citric acid anaerobically from oxaloacetic acid + pyruvic acid in rat testis tissue. The two substances together yielded 459 mm³ of citric acid, compared to 93 mm³ for pyruvic acid alone—a remarkable increase. A similar experiment on kidney tissue gave a much smaller increase. On the 30th he investigated for the first time (recorded in his thesis) the "formation of citric acid from malic acid." This result was not as large as the previous one, but nevertheless positive. Pyruvate + dl-malate yielded anaerobically 77 mm³ of citric acid, compared to 20 mm³ from pyruvate and 16 for the control.[120] It is evident that this experiment constituted a test of one of the two schemes for the formation of citric acid that Krebs had written down in his notebook during the summer. The outcome was encouraging enough that he decided to have Johnson continue in this direction.

On this same last day in September, one day after he had found that formate did not react with fumarate in B. coli, Krebs wrote to Marjory Stephenson:

> I hasten to report that I have completely surrendered: Hydrogenase, formic hydrogenlyase and formic dehydrogenase are three separate things. My capitulation is complete and without reserve and I only hope that the Spanish loyalists will not suffer a similarly complete defeat. There is only one positive result of my fight on this front, a renewed admiration for your excellent work.

"Elsden's time here," Krebs added, "is now drawing to its close. I enjoyed his presence in the lab very much indeed," even though Elsden had not "learned as much as is good for him." Krebs was probably not set back in any serious way by his "defeat" in a contest with Stephenson that had undoubtedly been playful from the beginning. For every idea he lost, he had others to put forward. Even the five short weeks of "joint work" with Elsden had brought forth "two main results":

> The first is the demonstration that fumarate oxidizes glucose, glycerol and lactic acid with the same rate as does molecular oxygen. This, in conjunction with the fact that the oxidation by molecular oxygen is inhibited by malonate to the same extent as is the oxidation of succinic acid, makes it almost certain that molecular oxygen reacts in B. coli through the intermediation of succinic acid. The second result is a definite proof that pyruvic acid is converted into succinic acid. We find under anaerobic conditions that 20% of the metabolised pyruvic acid is converted into succinic acid. The mechanism of this reaction is, however, obscure. There is no equivalent formation of lactic acid and in this way B. coli differs from the animal tissues.
>
> I do not want to bother you with further details in this letter, but I should be very glad if I may discuss matters with you in Cambridge during the meeting of the Biochemical Society.[121]

One of Krebs's other close connections at Cambridge was also "drawing to its close," as Norman Edson finished up his work there and prepared to return to New Zealand to take up a teaching position. On September 17 he sent Krebs a draft of his "final paper," in collaboration with Luis Leloir, on ketogenesis. "The writing-up has presented many difficulties," he acknowledged, "because the work is so incomplete." He worried also that "much of this work is a repetition of Quastel et al. or filling in of a few facts that they missed." Krebs agreed with the first part of this assessment. "The paper contains a great number of very interesting data," he wrote back on the 25th, "but I do not feel very pleased with the present shape of the paper. I have no definite suggestion to make as to how it can be improved." After pointing out which points in the paper "seem to be fully investigated" and which seemed not "to be completed," he added, "I suppose you do not want to postpone publication, but would you perhaps agree to publish those parts only which you have finished." Krebs was a rigorous critic of the writing of all of his students, and the fact that he enjoyed the wit and valued the advice of Edson did not dilute his judgments. Edson, like most of his students, respected that judgment for its clarity, honesty, and validity, and he hastened to make the suggested changes before his imminent departure.[122]

In September still another relocation touched Krebs's life, as his sister Lise wrote from Düsseldorf that she was preparing to leave permanently for Palestine at the end of the month. Her husband had already left, and Hans had wired him £5 in Marseilles to help him cover his immediate expenses. Lise stayed behind to finish packing their belongings and to make other last-minute arrangements. Early in September rumors that immigration to Palestine would be halted caused at least 2000 German Jews to leave so quickly from Marseilles that three extra steamships were placed in service. Georg Krebs regarded this haste as a "mass psychosis," but Lise did not regret having taken "the step." After completing her preparations she spent the rest of her remaining time in Germany in Hildesheim, even though she wished that both she and her father could be "spared these days of parting." Presumably she sailed, as planned, on September 29.[123]

VII

Krebs's research effort during September 1936 can be viewed at a general level as a straightforward extension of the study of anaerobic dismutations that had occupied him through the summer. To the group of ketonic acid reactions with which he had begun he added a second class of dismutations oriented around fumaric acid. To the two types of animal tissue—pigeon breast muscle and various rat tissues—and staphylococcus in which he had been studying these phenomena, he added a second type of bacteria, B. coli, as a major research material. Within each category of dismutations he added further specific reactions to the list of those whose physiological occurrence he was testing. Within these broad outlines, however, the short-term order of his experimental progression defies logical analysis. Seemingly more easily diverted than ever, he seldom performed more than one or two experiments concerning any one of these reactions before taking up a new one or reverting to one previously studied. He shifted nearly as frequently between his two main subproblems, the ketonic reactions and the reactions related to fumaric acid. Although Johnson seemed to specialize in pigeon muscle experiments, Krebs went back and forth repeatedly between bacteria and animal tissues during the month.

The fragmented character of Krebs's investigative pathway at the daily level may appear as merely another example of what his later associates called his "grasshopper style," but it reflects also some of the more particular circumstances of his current situation. To some degree the "great work" was losing the sharp focus that it appeared to have at the time he wrote his *Nature* article, and he was not confining himself anymore to the conceptual structure set out there. He still aimed at the ambitious goal of putting together a comprehensive scheme for the reactions of oxidative carbohydrate metabolism, but he was less certain what its shape might turn out to be.

The investigative trail that Krebs was following in this period was, however, probably more purposefully organized than the day-by-day pattern of experiments reveals. Mentally he was, in a sense, investigating all of these reactions at once, so that the order in which he actually performed experiments on them depended more upon contingent circumstances than on the overall state of the problem or of the individual subproblems. Another scientist in a similar position might have worked more systematically, have kept at each particular reaction until he or she had clinched the evidence for it or rejected it, before proceeding to the next one. It can be argued, however, that Krebs's scattered approach was particularly well adapted to the situation he faced. Surveying multiple possibilities, but lacking any deep theoretical grounds for choosing from among them the reactions either most likely to occur or to be metabolically important, he could forage the experimental terrain surrounding him more efficiently by "scouting" it than by investing a great deal of time in any one corner of the territory at hand.

In ways that we cannot trace from day to day as we can trace his experimental pathway, Krebs was probably searching also during this time for the more comprehensive conceptual architecture within which to situate the reactions that would prove to be physiologically significant. His "interpretation" of the experiment of September 28 on glycerol and fumarate affords us perhaps a small glimpse of the way in which he was thinking about these problems. Here too, he was probably very flexible, trying out not only different individual reactions on paper, but linking them in different manners. As the September 28 example shows, the experimental evidence was indeterminate enough to lend itself to various hypothetical reactions. At a broader level, his picture of the patterns into which the individual reactions might fit was probably also diffuse, in part because the "models" that influenced him provided two divergent ways to view anaerobic oxidation-reduction reactions. The Embden-Meyerhof glycolytic pathway connected such reactions into an extended sequence in which the compounds oxidized and those reduced were each regarded as substrates—that is, as intermediates in the reaction chains connecting foodstuffs with final decomposition products. The Szent-Györgyi scheme treated the oxidizing molecule as a component in a carrier system operating upon a variety of substrates and did not focus attention on the connections between the substrates themselves. I believe that at this point in his own investigative pathway, Krebs viewed the same reactions sometimes from the perspective of one of these models, and sometimes in the light of the other. Sometimes he saw fumaric acid entering as a substrate, with a particular second compound such as lactic acid, into a dismutation reaction that he thought of as a potential step in a metabolic main line. At other times he saw not only fumaric acid, but oxaloacetic acid and glycerophosphate as examples of a class of carrier systems "promoting" the oxidation of a series of substrates. Sometimes he must have anticipated that a single, or at most several specific pathways for the oxidation of carbohydrates would emerge, at other times he may have envisioned a more diffuse network of analogous dismutation reactions. Similarly, in his parallel experiments on animal tissues and bacteria, he sometimes appeared to be employing widely separated organisms in order to identify reactions common to all organisms; but sometimes he appeared to be engaged in two parallel but distinct projects, seeking to differentiate bacterial and animal metabolism as well as to find the points of analogy between them.

Main Routes and Carriers

During 1936 Hans Krebs was only one of a number of biochemists who directed their attention at problems closely connected with the intermediate stages of oxidative carbohydrate metabolism. The work of Szent-Györgyi's school was stimulating a great deal of interest, particularly in the role of the C_4 dicarboxylic acids in tissue respiration. Early in the year, for example, Guy Greville, a former associate of Frank Dickens now at the Courtauld Institute of Biochemistry in London, undertook a "critical study of Szent-Györgyi's theory." To acquaint himself with the methods behind it, Greville worked in Szent-Györgyi's laboratory in Szeged. On his return Greville confirmed some of the experimental results of the Szeged group, but raised questions about the conditions under which some of them were obtained, and concluded that "the establishing of Szent-Györgyi's theory will be very difficult." At Cambridge, Frederick J. Stare and Carl A. Baumann in the Molteno Institute confirmed that catalytic quantities of fumaric acid "preserve" the respiration of pigeon breast and pig heart muscle. In the Biochemistry Department, Jean M. Innes also began experiments intended to test Szent-Györgyi's conclusions. Krebs knew Stare, and must have kept in touch with these developments during his visits to Cambridge. His own growing interest in Szent-Györgyi's views and methods conformed, therefore, to the trend in his intellectual milieu.[1]

In Cardiff, Juda Quastel and his associates continued to investigate problems in fatty acid metabolism closely linked to carbohydrate metabolism. In 1936 Quastel and Maurice Jowett took up problems in the oxidative metabolism of glucose in the brain, including the anaerobic breakdown of pyruvic acid, that again intersected with aspects of Krebs's current work. A native of Sheffield, Quastel was one of the few people in Krebs's immediate research field who occasionally visited him in his laboratory. They did not, however, develop a personal rapport. Quastel perceived Krebs as friendly and polite, but not someone with whom he could discuss scientific problems. Krebs thought Quastel an able, imaginative scientist, but a difficult colleague who was generally dissatisfied.[2]

The scientist whose investigative pathway most closely paralleled that of Krebs at this time was his former student Hans Weil-Malherbe (who had now added his wife's maiden name to his surname). By the beginning of 1936, Weil-Malherbe had assembled the necessary manometric apparatus and the time to pursue energetically the study of glutamic acid metabolism in brain tissue that he had begun under Krebs's supervision a year earlier. In February he published the first of a series of papers concerning his independent research on that subject. Along the way he formed the hypothesis that α-ketoglutaric acid was not only the product of the deamination of glutamic acid, but an intermediate in the oxidation of pyruvic acid. This view led him to undertake, during the spring and summer of 1936, experiments on the aerobic and anaerobic breakdown of pyruvic acid in brain tissue that were similar to the experiments Krebs took up at almost the same time in other tissues. Like Krebs, Weil-Malherbe measured the succinic acid formed when pyruvic acid and other substrates were added to tissue slices or minces. Weil-Malherbe, however, devised a better method for determining succinic acid with a purified preparation of the enzyme succinic oxidase.[3]

Although Weil-Malherbe had written Krebs at the end of February urging him to spend a weekend with him and his wife, Krebs apparently did not do so, and there seems to have been no contact between them during the months when their research interests converged. When Krebs's note in *Nature* appeared in mid-August, Weil-Malherbe was caught by surprise, and quickly submitted a note of his own, on September 2, beginning, "Experiments which have been in progress in this laboratory during the past six months have had results very similar to those described by Krebs in a recent letter."[4]

His own recent experiments demonstrated, Weil-Malherbe wrote, "the formation of succinic acid from pyruvic acid, acetic acid and α-ketoglutaric acid . . . both anaerobically in minced brain and aerobically in the minced brain poisoned with malonic acid." Weil-Malherbe had thus adopted the same experimental strategy that Krebs did to prevent with malonic acid the succinic acid that formed aerobically from breaking down. He employed anaerobic experiments not for the same reason as Krebs did—that is, not to make the anaerobic dismutation reactions of the ketonic acids the immediate subject of study—but as another means to obviate the oxidative breakdown of the succinic acid produced. For these results and his earlier finding that α-ketoglutaric acid was probably formed in the course of pyruvic acid oxidation, Weil-Malherbe proposed two possible explanations, the first of them being itself a double process:

> 1) there are two alternative paths of succinic acid formation from pyruvic acid, one leading to α-ketoglutaric acid by the condensation of two molecules of pyruvic acid and subsequent decarboxylation, the other starting with decarboxylation of pyruvic acid to acetic acid and subsequent condensation of two molecules of acetic acid.

The first of these paths was Weil-Malherbe's own proposal; the second, as he indicated by a footnote reference, was Thunberg's well-known condensation reaction. The second overall "possible explanation" was that "α-ketoglutaric acid is formed by condensation of one molecule of pyruvic acid and one molecule of acetic acid. It has not yet been possible to decide which is correct."[5]

The rest of Weil-Malherbe's note dealt with his study of the decarboxylation reaction of α-ketoglutaric acid, using the dye brilliant cresyl blue as a hydrogen acceptor, from which he inferred that a "dehydrogenation precedes the decarboxylation, thus excluding the possibility of an aldehyde as intermediate." He suggested a decarboxylation mechanism. Finally, he noted,

> Pyruvic acid liberates carbon dioxide and forms succinic acid anaerobically even in absence of glucose or other hydrogen acceptors, no doubt because it acts itself as hydrogen acceptor, being partly reduced to lactic acid. The system β-hydroxybutyric acid $\rightleftarrows$ acetoacetic acid, quoted by Krebs, is only one of many reversible oxidation-reduction systems in the cell, which act as hydrogen acceptors guaranteeing the progress of vital oxidations independently of the varying oxygen supply.[6]

By the time Weil-Malherbe's note appeared in late September, Krebs was himself exploring hydrogen acceptor systems other than the β-hydroxybutyric–acetoacetic acid pair that he had featured in his *Nature* article two months before. Moreover, he did not conceive of these as "reversible oxidation-reduction systems," but as carrier systems analogous to Szent-Györgyi's cyclic fumarate-oxaloacetate-succinate scheme. His thinking was, therefore, probably not modified by Weil-Malherbe's commentary on a scheme that he was already partially leaving behind him. Nevertheless, Weil-Malherbe was showing himself able to make independent contributions to the broader problem of oxidative carbohydrate metabolism. Like Krebs, he aimed to resolve the long-standing question of whether, and if so how, pyruvic acid gave rise to succinic acid; also like Krebs, he sought to connect new reactions with those incorporated into schemes of oxidative carbohydrate metabolism. Weil-Malherbe focused his attention particularly on the compound—α-ketoglutaric acid—to which his special area of study of glutamic acid metabolism in brain had led him. In keeping with his broader range of experience, Krebs explored a broader list of possibilities.

Among the prominent substrates of oxidative metabolism whose links with a coherent reaction scheme remained most elusive was, as we have seen, citric acid. In August 1936 there appeared in *Hoppe-Seyler's Zeitschrift* a brief paper addressed to the question of "the formation of citric acid." The authors were Franz Knoop and Carl Martius. Trained as a chemist in Breslau, Martius had gone to Munich in 1928 for a year to study with Heinrich Wieland, where he began to turn to biochemical problems. Having no position for Martius there, Wieland sent him to Stockholm to work with Hans von Euler. When Martius returned to Germany, Wieland still had nothing to offer, and sent him this time to Knoop in Tübingen. To Martius, Knoop seemed autocratic, difficult to get along with, and unhappy in Tübingen. Knoop drove regularly back to Freiburg on weekends. After Martius had completed some work begun in Wieland's laboratory, on cholic acid, and asked Knoop what he should do next, Knoop said, "You can work on the synthesis of citric acid. If you take oxaloacetic and acetic acid and put them together, you will get citric acid." Martius objected that that would not work, because acetic acid is a very stable substance. Knoop suggested that he could put some enzymes in the solution. Martius still thought the reaction chemically impossible, and preferred to try oxaloacetic acid with the much more reactive pyruvic acid. When he mixed the two substances in an alkaline soda solution at low temperature, and oxidized the mixture with hydrogen peroxide for 20–30 hours, he isolated the calcium salt of citric acid in 35 percent yield.[7]

In the paper in which Knoop and Martius described the reaction:

$$\cdot HOOC\cdot CH_2\cdot \underset{\underset{COOH}{|}}{CO} + H\cdot CH_2\cdot CO\cdot COOH = HOOC\cdot CH_2\cdot \underset{\underset{COOH}{|}}{COH}\cdot CH_2\cdot CO\ COOH$$

$$+ H_2O_2 = HOOC\cdot CH_2\cdot \underset{\underset{COOH}{|}}{COH}\cdot CH_2\cdot COOH$$

they added that "statements in the literature make it probable that the physiological synthesis proceeds similarly. Investigations based on the reaction herein discovered are in progress." Reflecting Knoop's long-time interest in reversible metabolic reactions, they also raised the question of whether a "splitting in the reverse direction is possible in the organism."[8]

Knoop and Martius's paper appeared a few days after Krebs had written down in his laboratory notebook two different hypothetical reactions for the formation of citric acid, involving malic acid rather than oxaloacetic acid. Although there is no direct reference to their article in his records of the time, he must have read this paper with great interest.

I

During the first few days of October 1936, Krebs concentrated his attention on the reactions connected with fumaric acid in bacteria. In three successive sets of reactions all completed on Friday the 2nd, he tested the effect of malonate on the Q_{O_2} in *B. coli* of a number of substrates he had been studying, including succinate, formate, acetate, glycolate, fumarate, glucose, lactate, glycerol, α-ketoglutarate, pyruvate, malate, and glycerophosphate. Probably his objective was to ascertain which of them were oxidized through succinic acid, for their oxidation would be expected to be inhibited by the agent known to block succinoxidase. The results were "uneven" for several of the substrates. With succinate itself, acetate, glucose, and pyruvate, there were substantial reductions; with fumarate and lactate, minor reductions, with α-ketoglutarate, none. In the third set of experiments, the data from which he did not fully reduce, he found some reduction in the O_2 absorption with fumarate, but not with malate. This difference raised a question. He had regarded the addition of malate and of fumarate as equivalent, because he supposed them to be rapidly interconverted in the bacterial cells. "Is malate oxidation fumarate oxidation?" he asked himself. "No in the presence of malonate," he answered. "Malonate inhibits perhaps fumarase!!" (that is, the enzyme responsible for the conversion of fumarate to malate). Despite this observation, he did not follow up the general approach represented in these experiments, probably because the overall pattern of results discouraged the idea that this method could separate in a clear-cut way the substrates oxidized through succinic acid from those oxidized by some other route.[9]

The experiments that Johnson had carried out in early September on fumaric and lactic acid in pigeon breast muscle had yielded approximately 1:1 ratios of CO_2 formed:lactic acid consumed, in conformity with the view that the lactic acid was oxidized to acetic acid + CO_2. In a similar experiment that Krebs had performed on September 12 with *B. coli*, however, he had noted that there was "more than 1 CO_2 for 1 *dl*-lactate; [i.e.,] 136:112!!!." He did not comment at the time on the possible significance of the difference. Sometime during the next two weeks, however, he drew from this result (or a subsequent result from an experiment I have not identified in which the ratio may have approached 2:1) the conclusion that in *B. coli* fumarate must oxidize lactate beyond the stage of acetic acid. That inference prompted him to test, on October 3, the anaerobic "oxidation of acetate by fumarate in *coli*." The acetic acid was oxidized rapidly, and he calculated that 69 percent of it was oxidized, "as obtained before in O_2." That is, as much acetate was oxidized anaerobically by fumarate as had been oxidized in the earlier aerobic experiment of September 17. Moreover,

$$\frac{\Delta \text{ succinic}}{\Delta CO_2} = \frac{778}{417} \quad \text{almost equivalent!}$$

That is, he apparently expected that the reaction in question would produce two molecules of succinic acid for each molecule of CO_2—an expectation that is hard to interpret, since the reaction he apparently had in mind was fumaric acid + acetic acid → succinic acid + CO_2. At any rate, he had now found evidence for yet another anaerobic oxidation reaction, and one that further differentiated the metabolism of *B. coli* from that of animal tissues.[10]

The equality in the quantities of acetic acid oxidized anaerobically with fumarate in *B. coli*, and oxidized aerobically in an earlier experiment supported the view that this anaerobic reaction lay on the main oxidative pathway for acetic acid. On October 5 Krebs arranged a single set of experiments to test whether the same relation held for the "oxidation of lactic acid by *coli*." Aerobically he measured the increase in O_2 consumption caused by lactate in a phosphate buffer solution and the increase in CO_2 formation in a bicarbonate medium. Anaerobically he measured the total CO_2 formation and the succinic acid formed. The results were: aerobic, $\Delta O_2 = 448 \ \mu l$, $\Delta CO_2 = 474 \ \mu l$; anaerobic, $\Delta CO_2 = 439 \ \mu l$, Δsuccinic acid = 806. The succinic acid was slightly less than the 810 μl calculated for complete oxidation, but on the whole these were very satisfying figures. They showed that the anaerobic and aerobic oxidations of lactic acid were essentially quantitatively equal.[11] Krebs construed these equalities to mean that the overall oxidation of the respective substrates acetic acid and lactic acid proceeded via their anaerobic oxidation by fumarate, the succinate produced in these reactions then being oxidized aerobically. By the time he completed these experiments he had probably consolidated in his mind the idea that succinic acid is one of very few substrates that are oxidized by molecular oxygen, and that most cellular oxidations are molecular oxidation-reductions of the type he had been studying, which lead to those few substrates.

On the same day Krebs followed up on the experiments on oxaloacetic acid in *B. coli* that he had last performed on September 28. This time he paired oxaloacetate with formate and with acetate. The results were:

	CO_2		Succinic acid	
Oxaloacetate (504 μl added)	251 μl		264 μl	
Oxaloacetate + formate		206 μl		254 μl
Oxaloacetate + acetate	301 μl		301 μl	

Seeing an unusual pattern in this data, Krebs wrote down an extended interpretation of the results:

> result: half of added oxaloacetate is reduced to succinic; half is split, broken down. [He apparently drew this inference from the fact that 264 μl of succinic acid formed from the 504 μl oxaloacetate consumed.]
> reactions:
> oxaloacetic + formic → fumaric + CO_2 (since formic does not reduce fumaric, but is oxidized!) [This reasoning he based on the experiment of September 29, in which formate did not react with fumarate, and which had led him the next day to "capitulate" to Marjory Stephenson.]
> fumaric + acetic → succinic → CO_2
> formic inhibits CO_2 output, since it reduces oxaloacetate to fumarate before it is broken down, but no decrease in succinic since fumarate reacts with other substrates. [He was giving here an explanation for the above result that the addition of formate reduced the CO_2 formed from 251 to 206 μl, while leaving the quantity of succinic acid essentially unchanged.] Acetate increases yield in CO_2 and succinic, since fumaric + acetate → succinic + CO_2!! [Here he was interpreting the result of the oxaloacetate + acetate portion of this experiment in terms of the conclusion drawn from the experiment of two days earlier on the "oxidation of acetate by fumarate in *coli*."]

Krebs appears here to have been writing as he reasoned rather than stating conclusions already thought through. As soon as he had put these ideas down, he realized that they did not hold up. He added:

not conclusive:

animal: oxaloacetic + pyruvic → CO_2 + acetic + malic bacteria:
pyruvic → formic + acetic
oxaloacetic + formic → CO_2 + acetic!!![12]

His recognition that it was easy to construe these results in various ways illustrates how fluid Krebs's theoretical position still was as he strove to construct a coherent picture of cellular oxidations.

About the same time that he wrote down these reflections, Krebs received an answer to his letter to Marjory Stephenson:

> Dear Dr Krebs,
>   Many thanks for your letter: I am left wondering whether your "conversion" is the result of experiment or a defence mechanism against a flow of letters!

She added that she had been exploring the possibility of extending Elsden's stay with him, but that the terms of his grant precluded it, and also invited Krebs to come to supper, lunch, or tea in Cambridge on October 17.[13] Krebs responded quickly, "My conversion was genuine and the result of experiments." After summarizing the reasoning that had earlier led him to think that hydrogenase was an "unnecessary" assumption (see p. 323), he wrote:

> Now we have found that formic acid does not react with fumarate and the theory collapsed. We have just found the real reaction by which fumaric acid is oxidised in *B. coli*: formic acid + oxaloacetic acid = malic acid + CO_2. I do not in the least propose to construct a "defence mechanism against flow of letters." On the contrary, I am anxious to provoke a flow of letters.[14]

The reaction in question was a variant on the reaction "oxaloacetate + formic → fumaric + CO_2" included in the above reflections in his laboratory notebook. Given that malic and fumaric acid were rapidly interconverted in *B. coli*, these reactions were essentially equivalent. Thus, in the wake of a theory that had just "collapsed" and of reflections on several reactions that he had realized were, on the whole "inconclusive," he wrote concerning one of the latter, "We have just found the real reaction." These juxtapositions show how irrepressible a theorist he had become.

On the same day that he wrote Stephenson (October 6), Krebs performed an experiment on "*coli* + pyruvic HCO3/CO_2, in acid solution," the results of which again confronted him with an indeterminate situation. Pyruvic acid added anaerobically caused "more CO_2 (or H_2?); than bicarbonate!!!!" When he repeated the experiment under the same conditions with formate (whether added alone or with pyruvate he did not state), he found "No CO_2 formation!!!" and "No H_2"; that is, there was no change in pressure to attribute to the formation

of either gas. "Therefore," he wrote, "in acid solution perhaps <u>lactic</u> formation."

pyruvate = lactic + CO_2 + acetic?
or pyruvate + formate = lactate + CO_2[15]

The first of these alternatives was the pyruvic dismutation reaction that he had
long had in consideration, the second a new variant on the general dismutation
scheme. It is not evident why Krebs thought that the absence of CO_2 or H_2
could be explained by either of two reactions in which CO_2 was a product. At
any rate, as in his reflections of the previous day, he saw here that the same
experimental situation could be explained in more than one way and that he was
not in a position to choose between them.

Yet another interpretative ambiguity arose on the same day in an experiment
on "*coli* + glyceraldehyde." Three weeks earlier (see p. 321), Krebs had
postulated the reaction "glyceraldehyde phosphate + fumaric acid → glycerophosphate + oxaloacetic acid." Now he compared the anaerobic formation of
CO_2 from glyceraldehyde + fumarate with the aerobic formation of CO_2 from
glyceraldehyde alone.   Aerobically there was "oxidation +, increasing."
Anaerobically there was "quick CO_2 production with fumaric acid." Now,
however, Krebs questioned the reaction he had earlier assumed to be occurring:
"Remains to be analysed!! Is fumaric or glyceraldehyde oxidized? Determine
bicarbonate!!"[16]

Instead of pursuing the question posed here, Krebs broke off this series of
experiments on *B. coli* to return to animal tissue. On October 7 he tested the
anaerobic reaction of "glucose + fumarate" in rat testis tissue. There was a
"formation of CO_2 in both" the control and with fumarate, but to his great
surprise there was "<u>no</u> effect of fumarate!!!!!!"[17]

* * *

When I drew this emphatic statement to his attention in 1977, Krebs
responded, "So it is entirely different from bacteria. That is what it means.
That is due to the fact that fumarate cannot be reduced to succinate." When I
asked whether this was one of the first indications that animal tissues did not
react that way, he was less certain: "I think there were already earlier observations, but I don't know whether that was the first."[18]

Despite the six exclamation marks, Krebs did not at the time draw the
conclusion that he gave in retrospect as the meaning of the experiment. Ten
days later, as we shall see, he stated that fumaric acid is rapidly reduced to
succinic acid in muscle tissue. His immediate reaction to the experiment must
therefore only have been surprise that the result was anomalous. Nevertheless,
Krebs was not *mistaken* in what he said in 1977. His statement was not a
memory, but an interpretation of the experiment in the notebook that lay before
him. In 1977 the experiment meant to him just what he said it did. In October
1936 it had not yet acquired that meaning.

* * *

On the same day that he carried out this and one other experiment on rat tissues, Krebs returned to the organism with which he had begun his intensive study of bacterial metabolism in August, apparently to compare in *Staphylococcus* the reactions he had more recently studied in *B. coli*. In the first new experiment with *Staphylococcus* he verified that "lactate + fumarate [have a] distinct effect" on the anaerobic formation of CO_2. In an experiment on fumarate + glucose, on the other hand, he obtained results that were "difficult to interpret." There was a "1. definite reduction of fumaric," but "2. no increased CO_2 formation from glucose + fumaric." In another experiment the same day on aerobic oxidations, he encountered further anomalies. There was "no oxidation of fumarate" and only "slight oxidation of succinate!!!" in contrast to a "quick oxidation of glucose." Glycerol, he found in an experiment the next day, was oxidized "more rapidly than glucose," *dl*-lactate less quickly than glycerol, and this time there was "no oxidation of succinate."[19] Having made no clear-cut advance in two days with *Staphylococcus*, Krebs turned his full attention to another of his current preoccupations.

During the first week of October, Johnson had performed several further experiments on the anaerobic formation of citric acid. To estimate citric acid he employed a colorimetric method to determine the pentabromoacetone recently published by George W. Pucker, Caroline C. Sherman, and Hubert B. Vickery, modifying the procedure to decompose any oxaloacetic acid present that would otherwise interfere with the determination. On October 1, Johnson tested in rat testis tissue both of the hypothetical mechanisms that Krebs had formulated in August: from malic acid + acetic acid and malic acid + pyruvic acid. The former did not raise the quantity significantly above the controls, whereas the latter increased the yield by four times. Two days later, Johnson compared the effect of pyruvate + malate with that of pyruvate + oxaloacetate. The former again increased the quantity of citric acid substantially, but the latter gave one and one-half times as much as the former did. On the 5th he turned to minced pigeon breast muscle, finding that pyruvic acid alone increased the anaerobic $Q_{citrate}$ from 0 to 0.12, whereas pyruvate + oxaloacetate raised the rate to 0.68.[20]

The succession of substrate pairs that Johnson tested during the last week of September and the first week of October featured the two reaction possibilities for the formation of citric acid that Krebs had sketched out in his notebook in August and the combination pyruvate + oxaloacetate from which Carl Martius had just synthesized citric acid chemically. The record does not directly reveal to what extent Krebs was stimulated to pursue this line of experimentation by Knoop and Martius's paper and to what extent he was following the internal logic of his own intermittent prior efforts to account for the physiological origin of citric acid. There is no explicit trace of an impact of their paper on him during the period in which it ought to have occurred; but it would have been unusual for him to have missed a paper so pertinent to his interests, and his retrospective discussion in 1977 of the general development of his views implies that their publication helped to direct his thinking.[21]   The most plausible interpretation is that seeing their paper reinforced the interest in the question that his notebook reaction schemes indicate that he was pondering just before it appeared and that the experiments Johnson then took up represent a combined test of his own ideas and the reaction Martius and Knoop had demonstrated chemically.   It turned out that both their reaction and one of the two that he himself had formulated (that of pyruvate + malate) appeared promising. By the time that Johnson completed these experiments in early October, Krebs considered that he had discovered the general reaction through which citric acid is formed in animal tissue.   Experimentally they had established that "very considerable quantities" of citric acid formed from pyruvic acid together with either malic or oxaloacetic acid. He represented the summary reaction involved, however, as "malic acid + pyruvic acid → citric acid."[22] Thus neither the greater quantity of citric acid formed from pyruvic acid + oxaloacetic acid in animal tissues in his own laboratory nor Martius's chemical synthesis induced him to prefer that alternative to his own scheme.

Believing that he was finally on the track that might integrate the elusive citric acid into known metabolic paths, Krebs took up on October 8 the "oxidation of citric acid by muscle." He measured the effect of citrate alone and in the presence of malonate on the rate of oxygen consumption in coarsely minced pigeon breast muscle. Given that malonate was regarded as a specific inhibitor of the oxidation of succinic acid, what does the design of the experiment imply about Krebs's expectations? If he had included measurements of succinic acid, we might infer he anticipated that citric acid could give rise to succinic acid, which would accumulate in the presence of malonate. That he actually measured only the effects on the respiration suggests that he was not thinking in such a direction, but looking for an effect analogous to the restorative effect that Szent-Györgyi had found with fumarate, and that is what Krebs in fact found. The results were:

$$Q_{O_2}$$

	1st period	2nd period
Control	240	205
Citrate	235	202
Malonate	10	20.5
Citrate + malonate	155	102

He wrote down

no effect of citrate in the absence of malonate, but enormous effect with malonate. Citrate oxidation does restore as much as fumarate. Is citrate intermediate in the oxidation of fumarate?[23]

The effect was at least as much as he could have hoped for. Citrate not only restored oxidation in the same manner as fumarate, but the quantitative similarity of their effects raised the possibility that citrate lay along the main pathway through which fumarate is oxidized. If so, Krebs had already at hand, in the reaction of malic and pyruvic acid, the connection between fumarate and citrate. This was clearly a hot trail.

To answer the question of whether citric acid was an intermediate, Krebs turned to the fundamental criterion of whether it was metabolized as rapidly as the overall reaction sequence of which it was supposed to be a part. In this case the criterion was not applied to the ordinary rate of oxidation of the substances involved—citric acid was long known to raise the cellular respiration as much as any substrate did—but their oxidation in the presence of malonate, which would presumably isolate that fraction of the cellular oxidations which they shared. The results—O_2 absorbed with fumarate = 184; with dl-malate, 179; with citrate in low concentration, 42; with fumarate in low concentration, 110— were not supportive. He summarized the results:

1) citrate is less quickly oxidized than fumarate and malate.
2) dl-malate and l-malate same rate (not in citric acid formation!!)
3) small quantity of citric less effect than small quantity of fumaric.

The outcome left him in doubt about the question at hand: "citric not intermediate?" he wondered, but decided only that it was "not definite!!!"[24] Although he did not treat the result as a refutation of his hypothesis, it had markedly cooled the trail, and for now he followed no further along it.

Krebs's motive for testing both dl-malate and l-malate in the preceding experiment was that the natural isomer was very expensive. If dl-malate could serve his purpose—that is, if the l-isomer were metabolized without interference from the d-isomer—then he could use the racemic mixture instead. That being apparently the case, he looked up the literature on its preparation and wrote out what appeared to him the best method to employ.[25] Despite the support he was receiving, cost was still a factor with which he had to reckon in his research.

At the same time Krebs looked up in *Beilsteins Handbuch der Organischen Chemie*, the standard reference work on methods, a procedure for preparing citramalic acid (see Figure 8-1, following page): His reason for doing so was that he thought "if pyruvic (or lactic) reacts with pyruvic as does oxaloacetic or malic in the citric acid synthesis: citramalic acid would be formed."[26] Instead of pursuing this analogy, however, he returned to the unsettled questions that had arisen in his experiments with *B. coli* just before his latest venture with citric acid.

In the experiment of October 6 on pyruvic acid in *B. coli*, Krebs had probably been aiming to ascertain whether the basic pyruvic acid dismutation reaction that he had been studying in animal tissues since the previous spring, and that he had been able to demonstrate easily in staphylococcus in August (see p. 306) applied also in *B. coli*. This organism proved, however, to be more capricious. On the 9th he measured the "CO_2 formation and bicarbonate

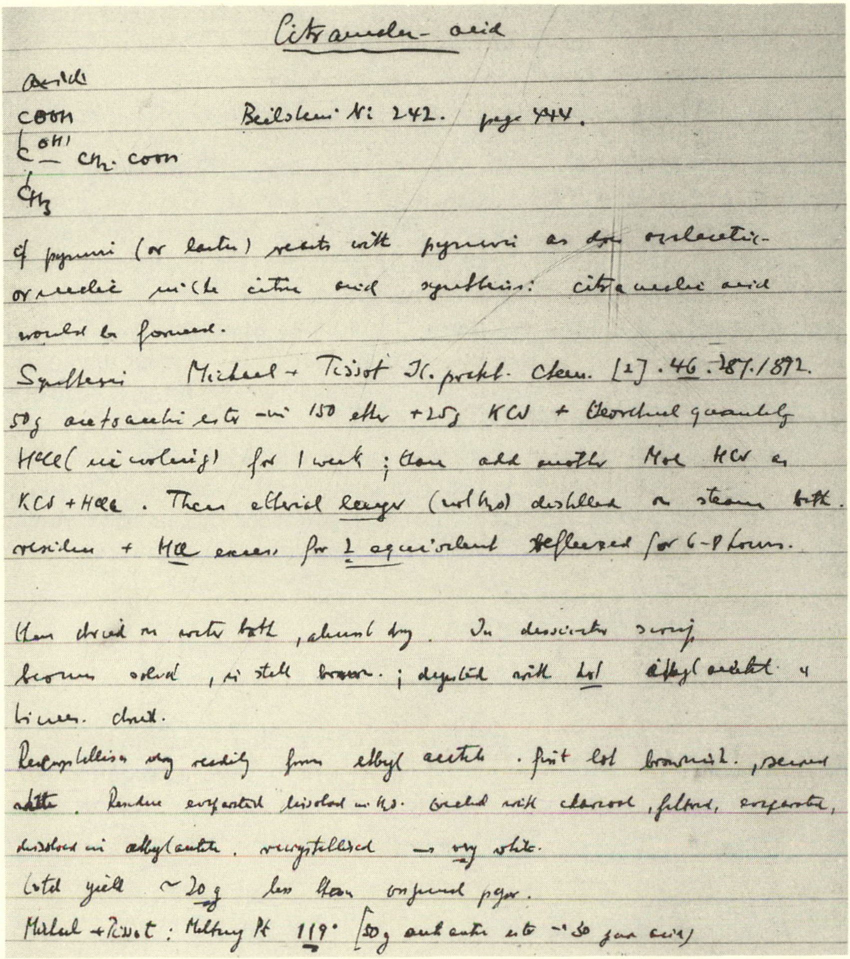
Citramalic acid

COOH
C OH — CH_2·COOH
CH_3

Beilstein No 242. page 444.

If pyruvic (or lactic) reacts with pyruvic as does oxaloacetic acid in the citric acid synthesis: citramalic acid would be formed.

Synthesis Michael + Tissot J. prakt. Chem. [2]. 46. 287. 1892.

50g acetoacetic ester in 150 ether + 25g KCN + theoretical quantity HCl (in cooling) for 1 week; then add another Mol HCN as KCN + HCl. Then etherial layer distilled on steam bath. residue + HCl excess for 1 equivalent refluxed for 6-8 hours.

then dried on water bath, almost dry. In desiccator syrup, brown solid, is still brown.; digested with hot ethyl acetate

Recrystallises very readily from ethyl acetate. first lot brownish, second white. Residue evaporated dissolved in H_2O. treated with charcoal, filtered, evaporated, dissolved in ethyl acetate, recrystallised → very white.

total yield ~20g less than original paper.

Michael + Tissot: Melting Pt 119° [50g acetoacetic ester → 30 g acid]

Figure 8-1 Krebs laboratory notebook No "14," p. 23.

decrease" with pyruvic acid in a fresh *B. coli* suspension and found that the pyruvate gave "almost 1 CO_2" per molecule and that there was an equal *decrease* in the bicarbonate. These simple proportions were not enlightening, since the pyruvic acid reaction was expected to yield one molecule of CO_2 for two molecules of pyruvate. Having used "old pyruvate," he repeated the experiment with fresh, and obtained this time a "yield of CO_2" corresponding to 86.5 percent of the pyruvate consumed. Still later the same day he repeated the experiment with a "low bicarbonate" concentration in the medium, and found this time "much more gas than bicarbonate decomposed!!" Because of the special properties of *B. coli*, he had to ask himself whether "H_2 formed?" There was "more than 1 gas for 1 pyruvic!!"[27] He did not attempt further measurements of the CO_2:pyruvic ratio; for the time being the variability of the proportions and the complication due to the formation of hydrogen apparently defeated his effort to extend to *B. coli* the reaction about whose occurrence in other tissues he was now confident.

It may have been the possibility that H_2 formed in the last of these experiments that prompted Krebs next to take a look at the process known to be the source of that gas, the hydrogenlyase reaction established by Marjory Stephenson: $H_2COOH \rightarrow 2H_2 + CO_2$. On October 10 Krebs examined the "splitting of formate by *coli*." In fact, he compared the effects on the change in gas pressure due to added formate and to pyruvate. He observed a "definite gas evolution from formate, [but] much more from pyruvate," an outcome that led him to wonder if there might be a " 'pyruvate hydrogen lyase'?"[28] Not pausing longer to find out, he moved on to another of his current uncertainties about *B. coli*, whether in the reaction of glyceraldehyde + fumaric acid it was the former or the latter that was oxidized.

When he had left that question open on the 6th, Krebs had noted that he should "determine bicarbonate" in order to complete the analysis. On the 10th he did so, in addition to measuring the free CO_2 produced by fumarate, fumarate + glyceraldehyde, and glyceraldehyde. The measurements of the CO_2 showed:

Δ CO_2 in fumarate 211

Δ 1034 [μl]

in fumarate + glyceraldehyde 1245

The difference on the right he treated as the CO_2 arising from the oxidation of glyceraldehyde by fumarate, comparing the figure to 1492 μl required for the complete oxidation of the glyceraldehyde present. From the bicarbonate measurements he found

Δ bicarbonate in fumarate   =   + 240.5
in fumarate + glyceraldehyde   + 395.5

From this increased yield of bicarbonate he inferred that there was a "definite oxidation of fumaric acid in the presence of glyceraldehyde." Because he was interpreting the increased formation of free CO_2 as due to the oxidation of glyceraldehyde, the answer to his question about whether fumarate or glyceraldehyde was oxidized appeared to be that both substances are. Two days later, in a similar experiment using a small quantity of glyceraldehyde, he reaffirmed this conclusion. Of the 105 μl of extra CO_2 formed by glyceraldehyde + fumarate, he attributed 41 to fumaric acid, that quantity corresponding to the "extra HCO_3^-" yielded by the combination. That left 64 μl, or "54% of total," due to the glyceraldehyde.[29] This outcome meant that two reactions were occurring at once; fumarate oxidized glyceraldehyde anaerobically, while glyceraldehyde also "promoted" the anaerobic breakdown of fumaric acid.

II

The preceding experiment was the last one that Krebs performed prior to the Biochemical Society meeting scheduled for five days later in Cambridge. During the intervening days Leonard Eggleston recorded two experiments on the "effect of vitamin B_1 on the output of CO_2," in various tissues of an avitaminous rat and two on the effect of yeast extract, pyruvate, and the yellow enzyme on the CO_2 output of *Staphylococcus*.[30] That procedure may indicate that Krebs had already left Sheffield. In any case he was probably busy finishing the paper he planned to present at the meeting. The title of his talk, "The Oxidative Breakdown of Carbohydrates," had been set well in advance, but for its substance he drew on the experiments he had carried out down to the time he interrupted them to prepare for the meeting. He must therefore have had to put together the best overview of the subject he could in the midst of an investigation that had not reached a natural resolution. The character of the manuscript of his talk suggests that he may have written it in something of a hurry.[31]

Krebs opened with a broad assessment of the general state of his subject:

> During the last few years considerable progress has been made in the analysis of the <u>anaerobic</u> breakdown of carbohydrates in living cells, and we know now, at least for certain types of cells such as muscle and yeast, many details of the chemical mechanisms by which carbohydrate is converted into lactic acid or alcohol respectively. Very little, however, is known about the mechanism which brings about the <u>oxidative</u> breakdown of carbohydrates. We know a few partial reactions and some enzymes which appear to be concerned with the oxidative breakdown of carbohydrates, for instance the conversion of lactic into pyruvic acid under the influence of lactic dehydrogenase, but the facts available so far are too scanty for drawing a more detailed scheme of carbohydrate breakdown or to define the exact role of the enzymes concerned.[32]

There was nothing novel in this point of view. Franz Knoop had expressed the situation similarly in 1931 (see Vol. I, p. 9), and the emergence of the Embden-Meyerhof pathway since then had only heightened the contrast between what was known about the anaerobic and aerobic phases of carbohydrate metabolism. For Krebs, however, this statement was the public surface of a personal quest that he had first taken up in 1933, abandoned after nine months of unsuccessful effort, and renewed again early in 1936. To draw "a more detailed scheme of carbohydrate breakdown" had by now become his driving scientific goal.

"In the course of some work which was carried out in surviving tissues and bacteria," Krebs continued,

> we have found a few new reactions which, in my view, play a role as intermediate processes in the oxidative breakdown of carbohydrates. I propose to describe in this paper the new reactions first and to discuss afterwards their place in the physiological breakdown of carbohydrates.[33]

If we compare these sentences with the opening of his *Nature* paper four months earlier: "We have found some new chemical reactions in living cells which represent steps in the breakdown of carbohydrates," among which he identified the pyruvic acid dismutation reaction as the "first step" in the oxidative metabolism of carbohydrates (see p. 293), we can quickly see that the claims Krebs now wished to make were more qualified than those he had then put forth. He had already retreated from his earlier view that he had found the "primary steps" in that process. The vicissitudes in his experimental pathway during those intervening months reveal clearly how he had been pressed toward a more cautious position.

The first reaction Krebs described was the same anaerobic dismutation of pyruvic acid that had figured most prominently in his *Nature* paper: "2 pyruvic acid + water = lactic acid + acetic acid + CO_2." He summarized the evidence for the reaction: the rapid anaerobic disappearance of pyruvate added to tissues, the measured formation of lactic acid, CO_2, and acetic acid. "The agreement between theory and facts is not quantitative," he acknowledged, "in that the amount of carbon dioxide and acetic acid is about 10% low whereas lactic acid figures tend to be too high. These deviations are due to side reactions which I am to discuss later." The dismutation reaction was, he held, "common to all tissues," even though he had not been able to demonstrate in all of them the appearance of acetic acid. "We find the same dismutation of pyruvic acid in certain bacteria," he added, "in staphylococci and gonococci." Inconspicuously omitted from his discussion of the reaction was *B. coli*. As we have seen, in the week before giving the paper he had still not been able to obtain satisfactory data for the yield of CO_2 in the reaction in that organism. Briefly he enumerated the other dismutations of the same type that "seem to occur generally if ketonic acids are present." These included the reactions of pyruvic + acetoacetic acid, oxaloacetic + pyruvic acid, and ketoglutaric + acetoacetic acid—essentially the same list that he mentioned in the *Nature* paper. He did not have time to present the evidence for these reactions, but did comment that the interaction between oxaloacetic acid and pyruvic acid (yielding malic acid + acetic acid + CO_2) was "of special interest . . . because the rate of this reaction is higher than that of the other dismutations."[34]

Turning to "another type of reaction which we have found in animal tissues and in bacteria," Krebs mentioned the "long-known" rapid reduction of fumaric acid to succinic acid in muscle under anaerobic conditions, and asked what substance "acts as hydrogen donator?" From an analysis of the carbon dioxide production curve presented on a slide (which has not been preserved), he argued "that the reducing agent in muscle is lactic acid which undergoes oxidation to pyruvic acid." The pyruvic acid subsequently dismutates, yielding succinic acid, acetic acid, and CO_2. Presenting these equations and the experimental evidence for them also on slides, he noted that the reaction was "first verified in pigeon's breast muscle, but we find it also in *B. coli* and in staphylococcus aureus." After summarizing these results he pointed out a "most remarkable difference between *B. coli* and muscle in respect to the carbon dioxide formation per molecule of lactic acid." In muscle the proportions were 1:1, but in *B. coli* "about two molecules of CO_2 are obtained from one lactic acid," a difference that led him to the conclusion that acetic acid may be further oxidized in *B. coli* and to the experimental results showing that "such is the case."[35]

Krebs next enumerated several other substances "which are oxidized by fumaric acid in *B. coli*: glucose, glycerol, glycerophosphate, glyceraldehyde, and glyceric acid." He discussed only the case in which glucose was the "substrate." The quantitative agreement between the oxygen consumption in air and the anaerobic CO_2 production in the presence of fumarate "comply well," he asserted, "with the view, first expressed by Szent-Györgyi for muscle respiration but abandoned later by him, that molecular oxygen reacts through the intermediation of succinic acid."[36] Apparently Krebs had in mind a mechanism that Szent-Györgyi had proposed in 1934, that succinic acid directly catalyzes the transport of hydrogen from substrates to molecular oxygen. One year later Szent-Györgyi had replaced that theory with his fumaric-succinic acid carrier mechanism.[37]

In muscle the situation was quite different: "fumaric acid oxidizes lactic acid to acetic acid only and so far we were unable to find other oxidations by fumaric acid in muscle." "What then," Krebs added, "is the fate of acetic acid in muscle?" That was, of course, a question that biochemists had been asking, not only for muscle but for animal tissues in general, for more than a quarter of a century. At this point he had only "some suggestions" to offer. The quantities of succinic acid that appeared in the anaerobic dismutation experiments on pyruvic acid were too large to explain as due to the reduction of a close precursor such as fumaric acid. Recalling the old demonstration by Toenniessen of the formation of succinic acid from pyruvic acid, and some related results of Dorothy Needham, Krebs claimed that there was "conclusive evidence" for "the synthesis of succinic acid from . . . compounds" containing three or two carbon atoms. "The mechanism by which succinic acid is formed is not yet clear in every detail," he acknowledged, "but certain points are clear." The process is "an oxidative condensation of two molecules," and the open questions were "(1) which substance acts as hydrogen acceptor . . . and (2) which molecules actually condense to form succinic acid?" Although confident that the answer to the first point was a second molecule of pyruvic acid reduced to lactic acid, on the second point Krebs ducked: "It is not possible to decide whether two molecules of acetic or whether one acetic and one pyruvic acid molecule react in the primary process. I am not going to discuss this point any further, since it has already been dealt with by H. Weil."[38] Here Krebs was undoubtedly referring to the recent paper in *Nature* in which Hans Weil-Malherbe too had written that it was not possible to decide between alternative reaction possibilities. Thus on this critical question Krebs once again ran up against the same obstacle that had defied removal ever since the formulation of the Thunberg-Knoop-Wieland and Toenniessen-Brinkmann schemes during the 1920s. Here too Krebs was in retreat from his position of the previous July (see pp. 293–294), when he had presented a "tentative formulation" of a set of reactions connecting pyruvic acid to succinic acid.

In the next paragraph Krebs withdrew from the implication in his *Nature* article that the conversion of pyruvic acid to succinic acid is the sole pathway for pyruvic acid oxidation.

> There is no doubt that pyruvic acid can be converted into succinic acid in animal tissues as well as in *B. coli*, and since succinic acid may undergo an oxidation to fumaric, oxaloacetic and pyruvic acid we are justified in concluding that succinic acid can be a stage in the oxidative breakdown of pyruvic acid. But is this path the only course of pyruvic acid oxidation? This assumption would be justified if it could be shown that the rate of succinic acid formation is equivalent to the rate of pyruvic acid oxidation.[39]

Parallel aerobic and anaerobic experiments showed, however, that the rate of succinic acid formation can account for only about 20 percent of the pyruvic acid oxidized in the presence of air in *B. coli* or in testis tissue. Although it might be argued in the latter case that this result was due to "some artificial experimental conditions," such could not be the case with *B. coli*, "and we may therefore safely conclude that for *B. coli* the path through succinic acid is not the only path of the oxidative breakdown of pyruvic acid." A second path was available in the more direct degradation of acetic acid arising through the intramolecular dismutation that he had already discussed. (He inserted afterward in his draft, "pyruvic being split into formic + acetic acids.") "The way through succinic acid," in this organism at least, was only a "side reaction" providing "the catalyst, succinic acid, which is necessary for the oxidation of a number of substrates."[40]

In animal tissues too, Krebs believed there was a second path.

> Experiments on animal tissue which were undertaken with a view of examining the question whether there is another mechanism of pyruvic acid oxidation led to the discovery of a reaction which appears to be of general biochemical interest; if pyruvic acid is added to tissues under anaerobic conditions together with malic acid or oxaloacetic acid very considerable quantities of citric acid are formed. We found this reaction so far in pigeons' muscle in rat kidney and in rat testis, that is in all the tissues which we have studied.[41]

The incompleteness of our record of Johnson's experiments precludes certainty about when the first of these results was obtained. All of those summarized in this paragraph, however, are included in the series that Johnson carried out between September 28 and October 9, and these are the earliest data on the formation of citric acid contained in his thesis. It is, therefore, most plausible that the "discovery" that Krebs presented here was less than three weeks old. The interpretation of the reaction found, however, was not entirely certain:

> The synthesis of citric acid represents an oxidation of malic acid and pyruvic acid. Malic and acetic acids if added to tissues do not yield

citric acid; pyruvic acid must be present. It is so far not possible to decide as to whether pyruvic acid is necessary as hydrogen acceptor only or whether it is also the condensing reactant, in other words it must be left open whether the decarboxylation occurs before or after the condensation.

Sometime after he had typed his manuscript Krebs reconsidered his position and revised the first two sentences to read:

Citric acid may be conceived as arising from malic and acetic acid by oxidative condensation. Malic and acetic acids, however, if added to tissues do not yield citric acid; pyruvic acid must be present.[42]

The uncertainty that Krebs expressed here, more prominently in the revised paragraph than in the original, over the question of whether pyruvic acid enters a condensation reaction directly or only after decarboxylation to acetic acid, corresponded to the same question that had long remained "open" concerning the condensation reaction within the Thunberg-Knoop-Wieland scheme. It is interesting that Krebs did not apparently feel the same uncertainty about whether malic acid entered the reaction directly or only after oxidation to oxaloacetic acid. In this regard he seemed unmoved by Martius's chemical synthesis of citric acid from oxaloacetic and pyruvic acid.

Moving on to the question, what becomes of the citric acid formed in this reaction, Krebs wrote:

Citric acid is rapidly oxidised in muscle and other animal tissues if air is admitted. Citric acid may therefore be considered as an intermediate in the breakdown of pyruvic and malic acid, but I wish to emphasise at once that it appears to be a side reaction only. [Afterward he wrote in here: "since not more than 20% of the pyruvic acid oxidized reacts to form citric acid under our conditions."] The physiological role of this side reaction is not clear nor do we know anything about the products of oxidation of citric acid.[43]

Thus Krebs had been able to connect the formation of citric acid to his satisfaction with other metabolic reactions, but its decomposition remained as obscure as it had been ever since Thunberg had identified it as one of the few substances rapidly oxidized in animal tissue. Krebs had, as we have seen (see pp. 11-13), been thinking about this problem off and on since at least the summer of 1933; but the scheme he had tried out then led nowhere, and he still had no further ideas on the subject. We should emphasize here that his conception of the oxidation of citric acid as a "side reaction" did not direct him to look for reactions that might reconnect the oxidation of citric acid with other paths of carbohydrate decomposition. The problem was to identify *any* plausible intermediate products of the decomposition of citric acid that were further metabolized in animal tissue.

Leaving citric acid, Krebs next discussed the "undefined" intermediate system through which, as Szent-Györgyi had shown, molecular oxygen reacts on substrates. "Although I am not in a position to make a conclusive statement about the mechanism through which malic acid is oxidised," he conceded, "I should like to say that experiments favour the view that malic acid is oxidised by dismutation with glyceraldehyde or glyceraldehyde phosphate." Mentioning that Flora Ogston and David Green had shown that "glycerophosphate is the only substance apart from succinic acid which is known to react with oxygen through the cytochrome system," Krebs suggested that "the system glycerol-glyceralde-hyde seems to act as an oxygen carrier for malic acid in a way which is analogous to the role of the succinic-fumaric acid system in the oxidation of lactic acid." He adduced as "direct evidence for this reaction in *B. coli*," an experiment showing that "the amount of fumaric or malic acid oxidized anaerobically is considerably increased if glyceraldehyde is added."[44] Although Krebs had carried out a number of experiments on fumaric acid and glyceralde-hyde, none of those recorded in his notebook appears to provide the clear-cut evidence for this reaction invoked in his paper, and it may well have been an experiment by Elsden, mentioned in his letter to Marjory Stephenson, that Krebs presented. It is also not evident to me, given that malic and fumaric acid added to the medium were readily interchanged in the organism, why Krebs chose to represent the reaction unambiguously as an oxidation of malic rather than fumaric acid.

Krebs finished his paper with "a few words correlating the reactions" he had discussed "to a scheme which of course is meant to be provisional only."

The essential feature of the scheme is the idea that the oxidations are partly brought about by anaerobic dismutation. There are three oxidizing agents known in muscle (and in *B. coli*): (1) fumaric acid, (2) ketonic acids (pyruvic, oxaloacetic acids), (3) glyceraldehyde.

Molecular oxygen reacts only to restore fumaric acid and glyceral-dehyde after their dismutations. The first step in oxidative carbohy-drate breakdown, we assume, is that chain of dismutations which is known as fermentation or glycolysis leading to three carbon com-pounds, in animal tissues mainly lactic acid. Lactic acid is oxidised by dismutation with fumaric acid. The pyruvic acid formed is oxidised by another molecule of ketonic acid, pyruvic or oxaloacetic acid. The ketonic acids are also the agent for the oxidative condensa-tion of succinic acid. Glyceraldehyde is the oxidising agent for malic acid.

The reactions given account for the complete oxidation of carbohy-drates since two molecules of pyruvic acid are eventually oxidised, through the stages of succinic, fumaric and oxaloacetic acid to one molecule of pyruvic acid and three CO_2.

The evidence for part of the scheme is conclusive. Not conclusive-ly shown is the dismutation between glyceraldehyde and malic acid, and it remains doubtful whether the whole oxidation of pyruvic acid passes through succinic acid. The scheme is incomplete in that it does not contain the citric acid synthesis. More work has to be done first to elucidate the role of this side reaction before it can be incorporated into a scheme.[45]

Measured by his own best standards, neither Krebs's paper itself nor the conceptual scheme it presented were tightly constructed. In style it lacked the economy of words characteristic of his more finished writing. In substance it sketched a scheme for the oxidative breakdown of carbohydrates that was not only "provisional," but also ambiguous. In the body of his paper he presented two classes of generic dismutation reactions that he believed "played a role as intermediate processes." That multiplicity would seem to imply a network of parallel, interlocking pathways. When he attempted to "place" such reactions into a larger scheme, however, he did not give free rein to the implication that there were analogous routes. Instead he linked certain specific examples of these reaction types into one sequence intended to account for "the complete oxidation of carbohydrates." The path led through succinic acid, despite the fact that earlier in the paper he had viewed the "path through succinic acid," at least in *B. coli* and perhaps in animal tissues as well, as a "side reaction." In both *B. coli* and animal tissues he had defined a "second path": in the former case through the more direct degradation of acetic acid, in the latter through the formation of citric acid. Similarly the glycerophosphate-glyceraldehyde mechanism that he proposed as analogous to the fumaric-succinic system of Szent-Györgyi pointed not toward a single pathway, but to a fork leading toward different end points. There thus appears in Krebs's approach an unresolved tension between an experimental identification of families of analogous reactions associated in diffuse patterns and a conceptual drive to delineate a scheme restricted to one, or at most two, pathways.

There was a corresponding ambiguity in Krebs's treatment of reactions observed in different organisms. On the one hand he stressed the "remarkable" difference between the fumaric-lactic acid reaction in animal tissue and in *B. coli*, from which he had inferred that the latter further oxidize acetic acid in a way that the former do not. On the other hand, he utilized "direct" evidence for a malic acid-glyceraldehyde reaction that he had been able to attain only with *B. coli*, as support for the general occurrence of this reaction. The general scheme with which he concluded his paper did not differentiate reactions that occurred in different organisms. Here too there was an unresolved tension between his recognition that metabolic processes varied in different tissues and organisms and his desire to find a common scheme.

The scheme that Krebs presented in this paper began with the same dismutation reaction that he had proposed in July as the "first step" in the oxidative breakdown of carbohydrates. From there on, however, it diverged from the scheme presented in his *Nature* article. The tentative but sharply delineated sequence with which he had then connected pyruvic acid to the formation of succinic acid had now disappeared, replaced by an acknowledgment that the "mechanism by which succinic acid is formed is not yet clear in every detail." In compensation the new scheme incorporated dismutation reactions of fumaric acid and a glycerophosphate-glyceraldehyde mechanism based on lines of investigation that Krebs had taken up only after he wrote the *Nature* article.

In four months the details of Krebs's picture of the oxidative breakdown of carbohydrates had shifted dramatically. There remained, however, a persistent, dominating theme, expressed most saliently in his statement that "the essential feature of the scheme is the idea that the oxidations are partly brought about by anaerobic dismutation." Equally interesting in retrospect is what his scheme did <u>not</u> feature. We may notice that in the paragraph above claiming that the reactions he had given account for the "complete oxidation of carbohydrates," he described an oxidation of pyruvic acid, through succinic, fumaric, and oxaloacetic acid to pyruvic acid and CO_2. The sequence was an implied skeleton version of the familiar Thunberg-Wieland-Knoop scheme. It was also a reaction circuit returning to the initial substance. The fact that Krebs mentioned this sequence twice in his paper, without pausing to draw attention to any general pattern exemplified in it, is strong evidence that a "search for cycles" was far from his mind.

III

The meeting of the Biochemical Society on Saturday, October 17, was held in the familiar surroundings of the Cambridge Biochemical Laboratory. Krebs was not the only speaker prepared to "draw" a scheme concerned with oxidative carbohydrate metabolism. In the first paper on the program Hans Weil-Malherbe spoke on "pyruvic acid oxidation in brain." After describing his new enzymic method for determining succinic acid, Weil-Malherbe said that he had found with its aid "a large increase of succinic acid after incubation of minced brain with pyruvic or α-ketoglutaric acid." With acetic acid there was a smaller and inconstant increase in the succinic acid formed. He had found that acetic acid can be formed from pyruvic acid, and "in presence of certain poisons, especially pyocyanin, formation of acetoacetic acid from pyruvic acid is found." The latter point caught Krebs's interest sufficiently so that he wrote down across the top of his program, "evidence for acetoacetic from pyruvic."[46]

Weil-Malherbe then advanced "a scheme of reactions in brain":

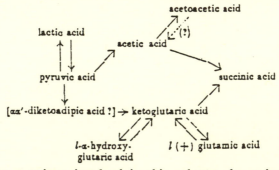

Interpreting the reactions involved in this scheme, he pointed out that the decarboxylation of ketoacids begins with a dehydrogenation and that ketonic acids can be their own hydrogen acceptors: "one molecule of ketoacid is decarboxylated at the expense of another molecule which is reduced to the hydroxyacid."[47]

We should note that, despite the superficial appearance of his diagram, Weil-Malherbe was not proposing a cyclic scheme, but two pathways leading from pyruvic to succinic acid, with associated side reactions. Like Krebs, however, Weil-Malherbe probably assumed a further succession of the reactions of the Thunberg-Knoop-Wieland theory leading from succinic acid back to pyruvic acid.[48] Weil-Malherbe was obviously thinking along similar lines to Krebs, although within a somewhat more restricted scope. Their respective schemes were overlapping, though not necessarily convergent attempts to order the similar though not identical reactions that they were uncovering through similar experimental methods. Despite the complementary nature of their investigations, Krebs and his former assistant were apparently not during this period systematically exchanging views.

Krebs read his own paper later in the day—he was ninth on the program—and was followed immediately by Jean Innes, whom Dorothy Needham introduced to speak on "the role of the 4-carbon dicarboxylic acids in muscle respiration." Attempting "to confirm Szent-Györgyi's theory that fumarate acts as a catalyst for transference of oxygen to the substrates in muscle," Innes found that at the concentrations of fumarate she used, "the oxygen uptake was in no case greater than that necessary for complete oxidation of the fumarate disappearing, which was not to be found as succinic, oxaloacetic, pyruvic, or malic acids." She concluded that "the added fumarate was being used as a substrate for oxidation rather than as a catalyst."[49] Krebs did not write anything on his program about this talk, and probably did not regard the result as remarkable. Even though he too was influenced by Szent-Györgyi's conception, Krebs already tended to view fumarate sometimes as a catalyst and sometimes as a substrate. In his own talk he had just commented:

> Szent-Györgyi conceptions appear in a somewhat modified form in the new scheme and although Szent-Györgyi's theories cannot be considered correct in the form in which they were represented it must be said that his conception contained a considerable amount of truth.[50]

The talks by Weil-Malherbe, Krebs, and Innes must have conveyed a cumulative impression at the meeting that biochemists in Britain were concentrating considerable effort on the problem of oxidative carbohydrate metabolism. The high point of the meeting, however, was not this subject, but a demonstration by Frederick Bawden, J.D. Bernal, and Norman Pirie of liquid crystalline plant viruses.[51] The participants at this session were being shown one of the landmark achievements in the biochemistry of their time.

As usual, Krebs took advantage of his return to Cambridge to keep up his scientific ties there. His conversations with Norman Edson and Luis Leloir were also farewells with his two former students on the eve of their departures for faraway homelands. Assuming that he had accepted her invitation, Krebs probably also had a long talk with Marjory Stephenson. He had an especially important reason to meet with Francis Roughton, for they hoped to spend the rest of the weekend, and perhaps the following days, finally "doing" their paper on the role of carbamino ornithine as an intermediate in the ornithine cycle (see pp. 215–217).[52] Apparently neither of them had done anything further about their joint project since Krebs included a brief description of the hypothesis and its supporting evidence in his review article on amino acid metabolism nearly a year earlier. From subsequent developments we can infer that when they talked the matter over they decided that it would be best to gather further experimental data before writing the paper. At any rate, when Krebs returned to Sheffield he began, on October 21, a series of several experiments on urea synthesis pertinent to their theory. In a preliminary measurement using *dl*-ornithine, he obtained unusually low rates of formation. Asking himself, "Is low Q_{urea} due to *dl*-ornithine[?]" he next compared the rate with that substance and with some *d*-ornithine he had obtained from Ernest Baldwin (Krebs must have brought it back with him from Cambridge). He noted that the sample was "not quite safe!!" probably because he thought that Baldwin's skills as a bench chemist did not equal his talents as a teacher. Nevertheless the *d*-ornithine yielded urea at higher rates than did the *dl*-ornithine. Using Baldwin's material again, Krebs tested the "effect of ions in urea synthesis," omitting respectively from the normal saline solution in four runs: K and Ca; Na, Mg, and PO_4; Na; and K, Ca, Mg, and PO_4. The latter two combinations lowered the rate somewhat, but not enough to undermine their view that the concentration of CO_2 was the crucial factor influencing the rate. Turning now to that factor, Krebs examined "the effect of CO_2 pressure" by comparing the rate of urea synthesis in ordinary bicarbonate saline in an atmosphere of 5 percent CO_2 in O_2, with a special phosphate saline in pure oxygen. The Q_{urea} in the former was 20.9, in the latter 10.3. These were, he noted "enormous figures," and he was evidently surprised that there was "still considerable rate (half)" in the supposedly bicarbonate free medium. One explanation that occurred to him was that glucose in the medium might have given rise, through lactic acid, to bicarbonate.[53] There were apparently still a few complications to iron out before the investigation was completed. For now, however, he put the question aside and returned to the main line of his research on oxidative carbohydrate metabolism.

From the record of his subsequent experiments it appears that after giving his paper at the Biochemical Society on October 17, Krebs recognized that he had a great deal more data to gather before he could make further progress toward the broad scheme of oxidative carbohydrate metabolism he had outlined there. The very number of the reactions for which he believed he had evidence meant that he had an enormous amount of work to do in order to establish definitively their chemical balance sheets and their rates relative to the overall rates of respiration in different organisms and tissues under varied conditions. For the rest of October and into November he focused his effort on a closer analysis of the inhibition of succinoxidase by malonate and by fumarate, in part for the "technical reason" that he needed to identify factors that might interfere with the determination of succinic acid. Both he and Johnson multiplied experiments on measurements of the products of the reaction of fumaric acid in muscle tissue.[54] Krebs was evidently now placing more priority on the consolidation of the results thus far attained, and less on their interpretation, than he had during the headier phase of the "great work." Still he ventured from time to time new variations on his earlier efforts to link the anaerobic reactions into sequential schemes. On November 11, for example, he designed an experiment on "β-hydroxy-butyrate + pyruvate in pigeon brain" to explore the idea that

> it is conceivable that 1) 2 pyruvic + $H_2O \rightarrow$ lactic + acetic + CO_2 is in fact 2) pyruvic + β-hydroxybutyric = lactic + acetoacetic [plus] 3) pyruvic + acetoacetic + H_2O = acetic + CO_2 + β-hydroxybutyric and the increasing effect of acetoacetic on CO_2 production could then be explained (2) + (3) = (1)!! If this is true, acetoacetic can be replaced by β-hydroxybutyric (or <u>malic</u>)!! Therefore . .

He measured the effect of pyruvate + dl-β-hydroxybutyric acid on the anaerobic formation of CO_2 in pigeon brain tissue. The results (Q_{CO_2} control = 2.68, + pyruvate, 3.79; + pyruvate and β-hydroxybutyrate, 4.26) did not, however, impress him. "No definite effect," he remarked. "Try liver, and add insulin,"[55] but he did not do so at once.

On the same day Krebs also pursued the carbamino ornithine question, seeking further data on the dependence of the rate of urea formation on the CO_2 pressure. As he had already done 15 months earlier (see pp. 198–199), he prepared saline solutions with graded concentrations of bicarbonate, in equilibrium with corresponding proportions of gaseous CO_2. He carried out the experiment on rat liver slices, at 40°C, probably because Roughton wished to have figures at various temperatures. Krebs obtained a modest rise in Q_{urea} (2.03, 2.66, and 3.50), between 1.25, 2.5, and 5 percent CO_2,[56] but not the "rapid increase" that took place according to his *Annual Reviews* article.

Beginning the next day a series of experiments with *Staphylococcus aureus*, the bacterium that had appeared in preliminary experiments in August favorable for the study of dismutations, Krebs applied to that organism experiments similar to those he had in the meantime carried out in animal tissues and *B. coli*. On November 14 he observed that 22.41 μl of fumarate added with pyruvate increased the quantity of CO_2 formed anaerobically by 31 and 55 μl, that is, by more than the amount required to oxidize the fumaric acid. There was, therefore, a "catalytic effect of fumarate" in this organism.[57]

On November 18 Krebs tried again to establish the "effect of CO_2 pressure" on the synthesis of urea, employing the same solutions and gas mixtures as he had used a week earlier, this time with liver slices from a well-fed rat. The d-ornithine came this time from Hoffmann-La Roche, and the results were satisfying:

$\%CO_2$	10	5	2.5	1.25	0
Q_{urea}	10.72	8.70	7.89	5.35	2.78

For comparison he ran a similar experiment with no added ornithine. He graphed the results showing the desired curves,[58] and went on the next day with his investigation of *Staphylococcus*.

On November 19 Krebs gave a lecture to the Medico-Chirurgical Society. Since he worked in the laboratory on this and the following day, it must have been a local Sheffield society. Speaking to physicians and surgeons, he took the opportunity not only to situate his own work within a larger picture of carbohydrate metabolism, but to connect it with medical problems. He began:

> The work which I am going to discuss today deals with certain aspects of the carbohydrate metabolism in tissues. Work on these lines, although of no immediate practical clinical importance, will be of importance, (I think), for those who are interested in the analysis of the disorders of carbohydrate metabolism.
>
> Of all disorders of carbohydrate metabolism, diabetes mellitus is the most important one from the practical point of view. The discovery of insulin has solved, to some extent, the practical clinical problem of diabetes . . . .
>
> The work which I am going to report was undertaken with the view of shedding light on these problems . . . it will be obvious that the normal <u>physiological</u> mechanisms must be clear before we can tackle the <u>pathological</u> mechanisms, but unfortunately our knowledge of the <u>normal</u> carbohydrate metabolism is still very scanty indeed. The object of research on these lines must be, at the present time, in the first place the analysis of the physiological processes. There is no doubt that it will be comparatively easy to explain the pathological processes once the normal mechanisms are fully known.[59]

Framing his problem in this way was obviously in part a way to make contact with the interests of his audience, but it was not merely a tactic devised to suit a local occasion. To contribute knowledge that would ultimately advance medicine was, I believe, one of Krebs's deep underlying motivations. In the three years since his departure from Germany he had cut off almost all of his direct ties with clinical practice. Nevertheless, he retained his medical roots and the sense of responsibility that he had for a time manifested as a concerned hospital physician. The frequency with which he included insulin in his experiments despite nearly uniform negative results attests to more than an intellectual interest in this remarkable substance. The discovery of the miraculous ability of insulin to effect a "cure" from the suffering of diabetes had taken place while he was a medical student. The goal of finding out how insulin works could well have symbolized for Krebs one way to fulfill his personal obligation to the medical world he had left.

After giving an argument for the "working hypothesis" that insulin acts as a chemical catalyst, Krebs asserted in his talk that to analyze "carbohydrate metabolism and its disorders means to establish the nature of these partial steps [in the breakdown of carbohydrates]. There is no doubt that in this series of reactions into which the oxidation of carbohydrates has to be resolved, we shall come across a reaction in which insulin plays a role and another reaction where the ketone bodies come in." Explaining why experiments on isolated tissues had proved the most successful method for studying these reactions, and claiming that the "metabolism of the excised tissue is essentially the same as that in the body," he then turned to the analysis of the partial steps. Passing quickly over the digestive processes in the alimentary canal, he next described the general pattern of the reactions of the anaerobic phase in the breakdown of hexoses. The 10 partial reactions involved (those of the Embden-Meyerhof pathway, although he did not identify them as such) he presented on a chart but did not discuss in detail, summarizing the process as "an intramolecular rearrangement." Weaving in his medical theme, he pointed out that the fact that diabetic animals produce normal quantities of lactic acid demonstrated "that insulin is not required for the splitting of sugar into lactic acid."[60]

Moving on to the aerobic phase of carbohydrate breakdown, Krebs made a remark that reveals clearly the degree to which his own approach was shaped by the features of the recently established anaerobic pathway: "Obviously," he stated, the mechanism of the breakdown of lactic acid in the presence of oxygen "will be a series of several reactions and what is wanted on [sic] this field is a scheme analogous to the scheme . . . for the formation of lactic acid from glucose." It was in this field, he said, "where my own work comes in. I am not yet in a position to offer a complete scheme, but some advance has been made . . . ." He had found that "under certain conditions lactic acid may be broken down although oxygen is absent," but only if fumaric acid is added to the tissue. Presenting the same reaction that he had featured in his Biochemical Society paper—"fumaric acid + lactic acid = succinic acid + pyruvic acid"—he stressed a point that he had not made then, that the succinic acid formed in the reaction "is readily reconverted to fumaric acid if oxygen is available. Fumaric acid acts thus like a catalyst."[61]

"The product of the oxidation" being pyruvic acid, he continued, "we have now to investigate the question of how pyruvic acid is broken down." Again he invoked reactions that take place in the absence of oxygen:

pyruvic + pyruvic → CO_2 + acetic + lactic
pyruvic + acetoacetic → CO_2 + acetic + β-hydroxybutyric

The second of these two reactions, he pointed out, represents a link between carbohydrate metabolism and the ketone bodies. As he had done in his talk to the Biochemical Society, he now asked, "What is the fate of the acetic acid formed by oxidation of pyruvic acid?" The answer he gave was, however, different from what he had said then:

There is evidence suggesting that two molecules of acetic acid are condensed to form succinic acid, according to the following scheme:

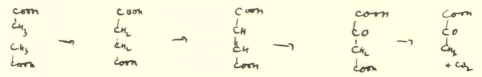

Succinic acid thus yields eventually by further oxidation one molecule of pyruvic acid, and as a result, two pyruvic acid molecules are oxidized in this way to one molecule of pyruvic acid. I do not propose to go into particulars, but I may say that there is no doubt that this series of reactions occurs in tissues, although it remains to be shown whether there are alternative paths.[62]

Drawing attention to the implications of this scheme for ketogenesis, Krebs pointed out that the reduction of acetoacetic acid to ß-hydroxybutyric acid "is brought about by an intermediate in carbohydrate metabolism." This "is probably not the only link between carbohydrate and fat metabolism, but we may say, at least, that we visualize the line on which explanations will be found." Digressing from his analysis of the intermediate steps themselves, Krebs then discussed the implications of Warburg's discovery of the "yellow enzyme" and of the discovery of lactoflavin, for the eventual understanding of the enzymes concerned with these steps. Returning finally to his initial theme, he acknowledged that

> None of the partial reactions known so far is affect[ed] by insulin and we are therefore not able to say anything definite about the role of insulin. The scheme of lactic acid breakdown is still incomplete, and insulin must be concerned with those parts which are obscure. But I think the hope is justified that work along these lines will reveal the function of insulin in not too distant times.[63]

His talk to the Medico-Chirurgical Society revealed, as openly as anything that Krebs wrote during the early years of his scientific career, the links between his current work, his vision of the general direction of his enterprise, and some of his hopes for where it might lead. A comparison of the scheme of aerobic carbohydrate breakdown as he depicted it in this talk with the version he had presented just one month earlier to his biochemical colleagues is also revealing. The general pattern and core reactions remained the same. The fact that he omitted here some of the reactions he had included there—especially the formation of citric acid and the reaction of malic acid with a glycerophosphate-glyceraldehyde system—can be attributed to the simplified picture that he wished to convey to a less specialized audience. The new emphasis on the catalytic role of fumaric acid suggests, on the other hand, a development of his thought giving increased prominence to the point of view of Szent-Györgyi. The discussion of the link between carbohydrate breakdown and ketogenesis retrieved a feature of the scheme presented in his *Nature* article that he had passed by in his Biochemical Society paper, and was included probably to bring out the implications of his work for the problem of diabetes.

The most striking difference between the two papers lies in Krebs's treatment of his question about the fate of acetic acid. Where he had previously left it deliberately uncertain how succinic acid is formed, he now invoked the original Thunberg condensation reaction of two molecules of acetic acid and presented "a scheme" that was the familiar Thunberg-Knoop-Wieland scheme. Did he do so again for simplicity, or had he changed his mind on this question during the short interval between his two lectures? Because he chose not to enter details concerning the evidence for these reactions, we cannot be certain. This shift is, however, a further sign that his theoretical framework continued to be malleable.

We can also notice that Krebs drew the scheme as a linear reaction sequence, obscuring the feature that these reactions formed a closed circuit. Except for the omission of some details, this was exactly the same form in which he had depicted the Thunberg-Knoop-Wieland scheme three and a half years earlier, before he had himself entered this field, in his review article, "The Decomposition of Fatty Acids" (see Vol. I, p. 419). Shortly after Krebs wrote his latest talk, Hans Weil-Malherbe submitted to the *Biochemical Journal* a paper detailing his work on the formation of succinic acid. Weil-Malherbe also situated his investigation within a framework provided by the "Thunberg-Wieland theory." As in his preliminary papers, he supported the existence, in animal tissues, of alternative routes from pyruvic acid to succinic acid. "It is probable," he wrote in his conclusion, "that the further oxidation of succinic acid passes through the stages of fumaric, malic and oxaloacetic acids; pyruvic acid is formed by decarboxylation of the latter and the oxidative cycle starts again."[64] In his ongoing investigations of the Toenniessen and Brinkmann variation of the scheme, K.A.C. Elliott too had recently described the removal of pyruvic acid as "a cycle of reactions," and pictured them in such a form:[65]

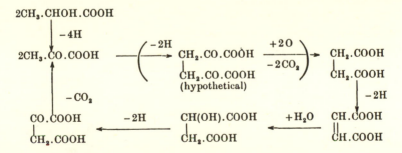

The fact that what others perceived explicitly as a cyclic process Krebs described in such a way that this property was virtually submerged permits a stronger inference than just that he was not fixing his attention on cycles. His mental vision of the problem seemed to be preempted by other patterns. What these patterns were are quite clear: anaerobic dismutations, Szent-Györgyi's carrier mechanism, and the exemplar set by the Embden-Meyerhof pathway. In more general terms he maintained what was by now a traditional perspective, that the breakdown of carbohydrates was an extended sequence of partial reactions, divided into an anaerobic and an aerobic phase, and leading from the foodstuffs at one end to the final products of oxidation at the other.

IV

On Saturday, November 21, Krebs made another of his frequent trips to Cambridge, this time to participate in the oral examination for Donald D. Woods. Woods had completed his thesis (which Krebs had read), and his examination represented the last step in completing his degree requirements. Turning down an offer by Hopkins and his wife of "hospitality for the night," Krebs left Cambridge on the same day to see his brother in Ipswich. It was their first meeting since Wolf and Lotte had arrived in England. Hans visited them in a new house into which they had recently moved after spending two months in a boarding house. When Wolf told Hans what his salary as chief engineer was, Hans pointed out that he himself had started at Cambridge at a much lower level, and said that Wolf should be very satisfied with what he got—as, indeed, Wolf was.

If Hans asked Wolf about his work at Bull Motors, he learned that Wolf had already made substantial contributions to their product line. Wolf had quickly seen that their "super-silent" DC motors attained that quality merely by being built very massively and solidly, and that with the experience he had brought with him from AEG he could easily redesign them to achieve noise reductions more efficiently. He began, however, by designing for them a new range of AC motors that would be more broadly applicable in industry than were the expensive special-purpose motors on which the reputation of the company rested. By the time Hans visited him, Wolf had probably already been able to provide new designs for both open-casing and enclosed fan-cooled motors in sizes ranging from one to 100 horsepower.[66]

Hans stayed in Ipswich through Sunday and probably returned to Sheffield on Monday. He had time to carry out two sets of experiments on Tuesday and Wednesday—on pyruvate + glycerol in staphylococci and on the effect of ions on urea synthesis for the carbamino ornithine project[67]—before taking the train for Leeds on Thursday to give another general lecture.

Krebs came to the University of Leeds at the invitation of Frederick Challenger to lecture to the Chemistry Department and the Leeds district of the Chemical Society. Challenger had asked him to do "something general rather than too specialized," because "the audience will contain a large number of students." Krebs responded by preparing a talk entitled "The Biological Breakdown of Carbohydrates" that again situated his own work within a broader context. The perspective in which he placed it this time, however, was not the implications of the work for pathology, but the general character and functional significance of the fermentation and oxidation of carbohydrates.[68] Substantively he went over much of the same ground that he had covered in his Medico-Chirurgical Society lecture, but at a level of detail intermediate between the succinct account he had given there and what he had presented earlier to the Biochemical Society. A comparison of these three lectures, each oriented around his own current investigations, but quite different in character, shows that Krebs had acquired a surprising capacity to adapt the same material with sensitivity to the interests and backgrounds of varied audiences.

In the portion of the lecture describing his own field of activity, Krebs presented a picture of the oxidative breakdown of carbohydrates similar enough to that in his Medico-Chirurgical Society lecture so that it is unnecessary to summarize it. He did, however, reintroduce some of the qualifications present in his Biochemical Society paper that he had eliminated from his presentation to a clinical audience. The most pertinent of these relates to the topic of the fate of acetic acid in animal tissues, in which he maintained, as he had there, that it was impossible to decide whether pyruvic acid is necessary to the condensation reaction forming succinic acid from acetic acid only "as the acceptor for hydrogen or whether acetic and pyruvic acid condense" together in the reaction.[69] It is difficult to tell whether Krebs was wavering back and forth, whether he had simply decided not to bring up such uncertainties when he had addressed himself to clinicians, or whether he had written the last lecture earlier than the Medico-Chirurgical lecture, even though he delivered it a few days later. In any case it is clear that on this crucial point he was, as he liked to describe himself, "feeling his way into the unknown" and that there was some groping along the way.

For us, the most compelling aspect of this lecture was not Krebs's recapitulation of the current state of his investigation, but a very general discussion, with which he opened his talk, on "carbohydrate breakdown as an energy-giving process." These two paragraphs provide the clearest contemporary statement of the broad biological outlook within which he pursued his study of carbohydrate metabolism, and an insight into the particular intellectual attraction that these problems held for him:

> Among the numerous chemical processes which occur in living cells the breakdown of carbohydrate appears to be of special interest, because it is one of those reactions which are closely linked up with the vital activities of the living organisms. These close relations between carbohydrate breakdown and life manifest themselves by the fact that a continuous decomposition of carbohydrates is necessary for the maintenance of the life of the majority of organisms, animals, moulds, yeasts, bacteria.
>
> Carbohydrate is needed by living organisms because all living organisms require a constant supply of energy, and the breakdown of carbohydrate is the chief source of energy. Energy may also be obtained by living cells from the breakdown of fat, or protein, and a few other substances, but carbohydrates are the preferential source of energy in many organisms. The energy obtained is required for doing work (movement) or for chemical synthesis (especially in the growing organism), and sometimes for producing light or electrical energy; it

is, however, not always clear why living cells require energy, as there are some cells such as the nervous cells in brain, which do not seem to utilize their energy—except for the production of heat. But we may take it—although we do not always see the reason—that consumption of energy is an inherent property of living matter and it will be obvious that the study of the energy-giving reactions is of greatest, I may say fundamental, interest for those who wish to study the nature of life.[70]

His references to "life" and to reactions "closely linked up with the vital activities" might appear on the surface to indicate that Krebs still maintained, as he had in 1932, that such chemical processes are "bound to the life of the cells" (see Vol. I, pp. 355–358). Passages later in his lecture make clear, however, that this was no longer his view. After tracing the history of fermentations to the work of Louis Pasteur, Krebs noted that Pasteur had identified biological oxidations with life itself. In spite of Buchner's success in "separating fermentation from life," the impression had persisted that energy-giving processes are inseparable from cells. During the last 20 years, however, "many oxidative reactions have been separated from cells . . . and the view is now widely accepted that oxidations and fermentations are due to chemical mechanisms which, on principle are the same as those with which the organic chemist deals. No room is left in the field for the so-called vital forces."[71]

There was nothing notably original about Krebs's introductory discussion of carbohydrate breakdown as an energy-giving process. It reflected what had become, over the course of nearly a century, a biological commonplace. What was important was that this viewpoint had become the driving force behind Hans Krebs's investigative enterprise—one that increasingly oriented and organized his daily research effort. We might go further and surmise that the forceful nature of this perspective, counteracting his propensity to scatter his effort, was providing him with the sense of high priority that increasingly channeled his investigative pathway and pressed him forward where he might earlier have been tempted to digress to less complicated problems. Warburg had taught him that a scientist should take on the large central problems of his time, but to Krebs problems were always as numerous as "pebbles on a beach." His growing conviction that for *him* the energy-giving reactions of carbohydrates provided *the* central problem of his time seemed to be imparting to his hit-and-miss research style a steadiness of purpose that was another sign of his scientific maturity.

Krebs arrived in Leeds in the afternoon. Challenger met him at the train and took him to an early dinner. Krebs delivered his lecture in the Lecture Theatre of the new Chemistry Department building at 7 p.m. In introducing him, Challenger remarked that although Krebs was unfortunately not a Yorkshireman, he was the next best thing; someone who had settled in Yorkshire. Krebs stayed overnight in the Great Northern Hotel and wasted no time getting back to Sheffield on Friday morning. He arrived in his laboratory in time to carry out two sets of experiments on *Staphylococcus*, testing the effects respectively of lactoflavin and of yeast on the pyruvate reaction.[72]

The next week Krebs was again on the lecture circuit. After performing on Monday another experiment on *Staphylococcus* and one on the effects of pyruvate + β-hydroxybutyrate and acetate on pigeon muscle, he left for Cambridge on Tuesday, December 1, to read a paper at the Biochemical Tea Club. His talk bore the same title as the one he had given a month earlier in the same place, but there is no record of whether he treated the subject in the same way. The next evening he was in Birmingham to repeat his lecture on the "biological breakdown of carbohydrates" at the Birmingham and Midlands Section of the Institute of Chemistry. The announcement for his lecture stated that "the Lecture will be an account of research in which the Lecturer, using a technique of considerable interest to Chemists, has made many valuable contributions."[73] Once again he was back at work in his laboratory on the following day.

Ironically, even while Krebs was delivering these lectures, ongoing experiments in his laboratory were weakening one of the pillars in the scheme of oxidative metabolism that he was presenting. Through October and November, Johnson performed an extended series of measurements of the changes in the quantities of CO_2, succinate, and lactate due to the presence of fumarate added anaerobically to pigeon breast muscle. The aim was to ascertain whether the increases of the first two and the decreases of the latter were in proportions consistent with the net reaction fumaric acid + lactic acid → succinic acid + acetic acid + CO_2. In the early experiments the changes were in the expected sense; but as Johnson continued he obtained irregular results for lactic acid, and in some cases its quantity actually increased during the course of the experiment. By the time Johnson had completed this series, on November 23, Krebs must have begun to doubt whether lactic acid was in fact the preferred oxidative equivalent for the reduction of fumaric to succinic acid in animal tissues.[74] It was just that reaction that provided the centerpiece in the sequence of anaerobic reactions constituting the scheme of carbohydrate breakdown that Krebs was discussing in his lectures. As will be seen below, he continued to believe that the reaction between lactic and fumaric acid occurred; but gradually his latest reaction scheme must have begun to lose the sharp outlines he had attempted to give to it.

Early in December Krebs undertook, along with his ongoing experiments on *Staphylococcus* and several other types of bacteria, a series of experiments on animal "tissues, pyruvic → lactic." These included rat liver, kidney, and intestine on the 7th and guinea pig brain and pigeon liver on the 9th. He measured ΔCO_2, Δpyruvate, and Δlactate.[75] This was the basic reaction he had been studying for over six months, and these experiments appear to have been done mainly to gather more definitive data concerning a known phenomenon, rather than to be new ventures.

* * *

The above comment on the character of these experiments is no doubt influenced by the fact that the data derived from them actually did appear in a paper Krebs published three months later. When I drew that to his attention in 1977, he commented:

HAK: Yes. It often happens that when I felt there was material for a paper I repeated the crucial experiments in the light of the information [I then had]. One could organize the experiments in such a way that they were more clear cut.
FLH: So this must have been that kind.
HAK: That must have been that stage that they became suitable for a conclusion.[76]

When we held this conversation, however, neither Krebs nor I were conscious that some of the conclusions he had recently drawn were already looking less conclusive. In that light I believe that these experiments can be viewed as a stage in another retreat from a comprehensive theory of oxidative carbohydrate breakdown to a more rigorous but more limited description of some of the anaerobic oxidation reactions occurring in animal tissues.

* * *

At about the same time Albert Szent-Györgyi sent Krebs his last copy of a reprint of a new set of articles that he and his group had published entitled "On the Meaning of Fumaric Acid for Animal Tissue Respiration." The papers had appeared in *Hoppe-Seyler's Zeitschrift* in September, and Krebs had probably seen them in the Sheffield library shortly before he received the reprint. Perhaps because Szent-Györgyi's work was now so pertinent to his own, he reread the papers in the reprint and marked several passages that particularly interested him.[77]

Szent-Györgyi and his associates reported substantial experimental progress in their study of the catalytic role of fumaric acid in tissue respiration. Concentrating their attention on the relationship between oxaloacetic and fumaric acid, they had realized that their previous determinations of oxaloacetic acid were unreliable, because the analytical method they had used had not clearly differentiated oxaloacetic from pyruvic acid, and they had devised new methods specific for each substance. Recognizing also that "in the field of intermediary metabolism qualitative methods can be misleading," they set a rigorous standard for measuring by quantitative balance methods "the chemical changes in small quantities of tissue within short time periods (2–10 minutes)." In this way they established that oxaloacetic acid added to pigeon breast muscle disappeared at a rate more than high enough to account for the respiration of the muscle during the time intervals chosen.[78] They thus presented their case in accordance with the stringent criterion of the field for the required quantitative relationship between a partial and an overall reaction.

Addressing themselves to the "numerous theories of intermediary metabolism based on the unproven rapid catalytic decarboxylation of oxaloacetic acid" (yielding pyruvic acid, as in the Thunberg-Knoop-Wieland theory), which would, if correct, "make the catalytic reduction of oxaloacetic acid accepted by us impossible," Szent-Györgyi and his group demonstrated to their satisfaction through balance measurements that oxaloacetic acid was reduced more rapidly than it was decarboxylated and that the source of the pyruvic acid which did appear in their experiments was not oxaloacetic acid, but some unknown three-carbon molecule, a "triose."[79]

Beyond buttressing their basic theory with these further experiments on pigeon breast muscle, Szent-Györgyi's associates extended their study to other tissues, finding that the system applied to liver tissue, but not to malignant or embryonic tissue. Szent-Györgyi also proposed a new explanation for the Pasteur reaction, that is, for the suppression of glycolysis under aerobic conditions. Szent-Györgyi interpreted Meyerhof's work as establishing that muscle glycolysis is "actually an oxidation of triose through pyruvic acid. Respiration is, according to our investigation an oxidation of triose through oxaloacetic acid." Oxaloacetic and pyruvic acid therefore compete as acceptors for the activated hydrogen of carbohydrate substrates, so that glycolysis occurs only when oxygen is not present to produce oxaloacetic acid. This interpretation especially caught Krebs's attention, and he put a large encircled exclamation point beside it in his copy of the paper.[80]

Although Szent-Györgyi stated that "the objective of the present work is to deepen further the theory of fumaric acid catalysis" that he had presented in his earlier papers, the theory itself appears to have changed in the process. He now discussed the oxidation of fumaric acid and the reduction of oxaloacetic acid, while barely referring to their connection with succinic acid. There was no further mention of the concept of "overreduction." Had Szent-Györgyi abandoned this explanation of the conversion of oxaloacetic to succinic acid that he had put forth a year before and replaced the oxaloacetic-fumaric-succinic acid loop that he had described then (see p. 254) with an oxaloacetic-fumaric system? Or had he simply fixed his attention entirely on one portion of the larger system? On that point his discussion was silent. Szent-Györgyi's theoretical position appears to have been more malleable even than that of Hans Krebs had been during the past six months.

One of the passages that Krebs marked in Szent-Györgyi's reprint was in the article by Bruno Straub on "decarboxylation of oxaloacetic acid through muscle tissue," which supplied the experimental basis for Szent-Györgyi's assertion that that acid is reduced so much more rapidly than it is decarboxylated in muscle tissue that the latter reaction can be neglected. Krebs put a large exclamation point next to the statement that in unwashed muscle tissue 7 percent of the oxaloacetic acid present is spontaneously decarboxylated in 10 minutes.[81] Krebs may have been interested in this information largely for technical reasons; but it could not have been lost on him that if Szent-Györgyi's interpretation of these results was correct, then the Thunberg-Knoop-Wieland theory was called into question more broadly than over the controversial step connecting pyruvic or acetic acid to succinic acid. The step oxaloacetic acid → pyruvic + CO_2 was common to all of the variants of the scheme, including those that Krebs had been incorporating into his own schemes of oxidative carbohydrate breakdown.

Two further points in the Szent-Györgyi reprint stimulated Krebs to carry out experiments.   On December 12 he examined with "liver, fumarate + pyruvate summation. CO_2?" noting "see A. Szent-Györgyi."   There was, however, "no effect."[82]   I have not been able to identify what it was in the papers of Szent-Györgyi or his associates that prompted Krebs in this case.   The source of a second experiment (even though he did not explicitly refer here to Szent-Györgyi) is, however, very clear.   To judge from the density of his exclamation marks and the fact that he also copied the page number on the cover of his reprint, Krebs must have reacted with some excitement to a footnote in the contribution by Ilona Banga, "On the Oxidation of Fumaric and the Reduction of Oxaloacetic Acid in Minced Muscle Tissue."   Appended to a statement in the text that "from the [above] results the important methodological conclusion can be drawn that the disappearance of oxaloacetic acid can be viewed as a measure of its reduction," the footnote read:

> Under artificial conditions the oxaloacetic acid can, to be sure, disappear in other ways.   For example, oxaloacetic acid disappears with great rapidity when glutamic acid is added.   In this case the oxaloacetic acid reacts with the glutamic acid, a reaction catalyzed by the muscle.   Other amino acids are also able to react with less intensity with oxaloacetic acid.   The carbonyl group probably condenses with the amino group.   It is not impossible that this process is significant for oxidative deamination.[83]

Why this suggestion would be enticing to Krebs is not hard to understand.   It offered a potential new link between one of his current preoccupations, anaerobic reactions of the C_4 dicarboxylic acids, and the problem of deamination that he had pursued intensively in former years.   We may recall that, in spite of his discovery of the enzyme d-amino acid deaminase, he had been disappointed not to have made greater progress on the mechanism of deamination for the naturally occurring amino acids (see pp. 162–163).   Banga's footnote raised the possibility that one reason might have been that deamination is not the straightforward oxidative process that he and others had supposed it to be, but a coupled oxidation-reduction reaction.   Accordingly, on December 14 he tested in rat kidney tissue (the material with which he had carried out much of his earlier work on deamination) the "anaerobic desamination [sic] of amino acids," comparing the effects of added glutamate, glutamate + oxaloacetate, and glutamate + fumarate.   Afterward he measured with the Parnas method the NH_3 formed.   The results—"No deamination by fumaric or oxaloacetic acid"—must have dampened his enthusiasm for the new idea, but he tried the next day a more general approach by testing the effect of malonate aerobically on the deamination of glutamic acid.   Perhaps what he had in mind was that by inhibiting succinic oxidase, malonate might reduce the amounts of fumaric and succinic acid formed within the tissue and, if the reaction suggested by Banga did exist, indirectly affect its rate.   There was, however, "no inhibition of deamination."[84]   This second negative outcome apparently dissuaded him from following Banga's lead any further, and he resumed his current lines of investigation.

On December 15 Krebs wrote Szent-Györgyi a long letter thanking him for the reprints. "The papers are extremely useful to me," he wrote. "I have to look them up almost daily and as we have no departmental library, I find it very useful to have the reprints." Even taking into account that these remarks were intended to be courteous, they indicate that Szent-Györgyi's work was at this time exerting a strong impact on Krebs. He did not, however, accept Szent-Györgyi's positions passively:

I found your recent work most inspiring and although I do not agree with every detail of your theories, I am certain, on account of your work and of my own experience that your fundamental ideas are correct. In confirmation of your theory I have found that oxaloacetic acid oxidises pyruvic acid according to the scheme: oxaloacetic acid + pyruvic acid = malic acid + acetic acid + CO_2. This seems to be a special case in which fumaric acid acts as a carrier for molecular oxygen in the way which you suggested. I find, however, some difficulty in assuming that the whole of the respiration is brought about by oxaloacetic acid. You state in your paper that respiration is an oxidation by oxaloacetic acid. If this be so an oxidation of carbohydrates would be observed under anaerobic conditions if an excess of oxaloacetic acid is available. In actual fact there is some respiration if tissues are incubated with oxaloacetic acid under anaerobic conditions, but the carbon dioxide evolved is very much less than that formed in the presence of oxygen, and so far the oxidation of pyruvic or homologous ketonic acids are the only oxidations which can be brought about by oxaloacetic acid.[85]

The next paragraph in Krebs's letter shows that he was aware that the theoretical structure presented in Szent-Györgyi's latest paper diverged from the position Szent-Györgyi had taken in his previous paper on the subject in 1935. Instead of confronting Szent-Györgyi with an inconsistency, however, Krebs astutely brought the older view back into consideration as a complement to the newer one:

I believe that your view expressed some time ago that the system succinic acid-fumaric acid is a carrier for molecular oxygen holds to some extent. We have found that fumaric acid oxidises lactic acid to pyruvic acid and malic acid to oxaloacetic acid if added to muscle under anaerobic conditions, at the same time fumaric acid is reduced to succinic acid. More clear cut are experiments carried out with *Bacterium coli*. In this organism molecular oxygen can be completely replaced by fumaric acid in the case of the oxidation of glucose, lactic acid, glycerol, acetic acid. It is possible to oxidise glucose with fumaric acid under anaerobic conditions in *B. coli* and the amount of carbon dioxide formed and the rate of the carbon dioxide formation is identical with that in the presence of air. The papers [sic] are in agreement with the assumption that in *B. coli* succinic acid is the only substance which reacts with molecular oxygen.[86]

The portion of this paragraph dealing with *B. coli* echoes the same point that Krebs had made in his Biochemical Society talk in October (except that I doubt that the view that Krebs then described Szent-Györgyi as having "abandoned" is the same one that he referred to in the present letter as one Szent-Györgyi had "expressed some time ago." It is just possible that Krebs had already seen Szent-Györgyi's latest paper then. I have assumed, however, that Krebs was alluding there to an earlier shift in Szent-Györgyi's position.). The first portion of the above paragraph shows that, in spite of the irregular results of Johnson's recent experiments on the fumaric acid–lactic acid reaction in pigeon breast muscle, Krebs was still maintaining that the reaction takes place there. He had, however, given up the position, taken in his Biochemical Society paper, that lactic acid is the only substance oxidized by fumaric acid that he had been able to find in muscle. Where he had previously invoked another carrier system, glyceraldehyde-glycerophosphate, for the oxidation of malic acid, he now attributed the oxidation of both malic and lactic acid to the succinic acid–fumaric acid system. There is no record of experiments performed by either Krebs or Johnson in the intervening period on the fumaric acid–malic acid reaction in animal tissues. Apparently he had changed his mind about the interpretation of his earlier data.

The final substantive paragraph in Krebs's letter indicates that Krebs's view of the situation in animal tissues was evolving more broadly away from what he had formulated in his lectures in October and November:

> In animal tissues the position seems to be more complicated in that fumaric acid cannot completely replace molecular oxygen, nor can oxaloacetic acid replace it and I find that the experimental replacement of molecular oxygen by the supposed carrier is the crucial experiment by which the role of the carrier can be demonstrated since the carbon dioxide output which occurs in the presence of fumaric or oxaloacetic acids in muscle under anaerobic conditions, is less than the output in the presence of oxygen, I assume that there are respiratory mechanisms in which fumaric and oxaloacetic acid are not involved.
>
> There are many details which I would like to discuss with you but I feel that I should not trouble you with too long letters.[87]

Tacitly Krebs was confronting Szent-Györgyi here over divergent experimental criteria applied in the two laboratories to answer the same question, for it had been precisely the central claim of Szent-Györgyi's latest papers that his group had proven quantitatively that oxaloacetic acid is reduced more than rapidly enough to account for the whole respiration of the tissue. His basis of comparison was the rate of disappearance of the oxaloacetate and the rate of oxygen consumption. Krebs chose, for his version of the "crucial experiment," to compare these two rates through the common measure of the carbon dioxide produced, and reached the opposite conclusion.

Since the purpose of Krebs's letter was to *discuss* certain aspects of Szent-Györgyi's theories with him, we should be cautious about inferring that the above paragraphs directly reflect Krebs's overall opinion about these theories. Nevertheless the statement "I am certain . . . that your fundamental theories are correct" is too strong to dismiss as a mere compliment. The letter is consistent with other indications that during this period Krebs was approaching oxidative carbohydrate metabolism increasingly through Szent-Györgyi's perspective. In an essay he prepared for a book dedicated to Frederick Gowland Hopkins, entitled *Perspectives in Biochemistry*, Krebs wrote that Szent-Györgyi

> suggests that certain simple metabolites, such as oxaloacetic and fumaric acids, act as intermediary hydrogen carriers in cellular respiration.
>
> The work which is reported in this article developed along lines similar to those which Szent-Györgyi discussed in his stimulating papers.[88]

The deadline originally set for submission of the essays for this volume had been January 1, 1937.[89] Some features of the essay suggest that Krebs finished it somewhat later, and discussion of its contents will therefore be deferred. Nevertheless this passage reinforces the view that by the end of 1936, Szent-Györgyi's theoretical framework was becoming a dominant influence on Krebs's investigative orientation. The passage also amplifies the indications in his letter and his talks that he viewed that framework differently than Szent-Györgyi himself did. Where Szent-Györgyi sought to define a unique carrier system, Krebs saw the idea of a carrier composed of simple metabolites as generalizable. Where Szent-Györgyi made a sharp distinction between substrates and carrier molecules, Krebs perceived oxaloacetic and fumaric acid as *metabolites* that *acted* as carriers.

Krebs's letter to Szent-Györgyi contains also clues about how his own scheme of oxidative carbohydrate breakdown was faring by mid-December. The inclusion of both malic and lactic acid as substrates oxidized by fumaric acid must have diminished the special place he had given to the fumaric acid–lactic acid reaction as the source of the pyruvic acid in his scheme, and his "assumption that there are respiratory mechanisms in which fumaric and oxaloacetic acid are not involved" makes us wonder if he was any longer attempting to fix on any specific sequence of reactions to account for the "complete oxidation" of carbohydrates. Perhaps he envisioned parallel or intersecting pathways made up of analogous dismutations, the sequence that he had earlier described representing one of several alternative routes through the system, but I suspect that the picture of an unbroken reaction sequence that had been his goal had for now simply grown diffuse, and that Szent-Györgyi's theoretical framework provided him with an alternative conceptual space within which to pursue the problem. The emphasis on a number of carrier systems interacting with various substrates diverted attention from his unsolved problem of how these various substrates were linked into metabolic pathways.

Szent-Györgyi replied quickly to Krebs's letter:

Many thanks for your friendly lines, which interest me very much and have given me much pleasure. Certainly this group of C_4 dicarboxylic acids is interesting. It will require much work, however, before one can really state anything quantitative and definitive about them. The circumstances are very complicated and, at least in animal tissues, make the interpretation of results extremely difficult. Oxaloacetic acid is very unstable. When it is in excess it is also attacked decarboxylatively. Then pyruvic acid is formed. Pyruvic acid also originates through the oxidation of trioses. Pyruvic acid, however, is not an indifferent substance, it interferes strongly in the equilibrium of the cell and inhibits the respiration. Similarly the C_4 dicarboxylic acids are not indifferent. If one adds a little more [of them] one obtains an unspecific toxic effect of dibasic acids. This is especially disturbing in experiments with oxaloacetic acid, which is quickly reduced, and of which one must therefore add more if one wishes to replace the oxygen for a longer time, as is often necessary in respiration experiments. All in all, the matter is damned difficult. Therefore I am not persuaded that the insufficient anaerobic CO_2 production in the presence of oxaloacetic acid is proof of a restricted role for this system. Naturally the exclusive dominance of this C_4 dicarboxylic acid is not proven either and my own conclusions will certainly need to be modified on many points. That, however, makes no difference. The important thing is that the core be correct, and that the investigation is advanced through our experiments. Your lines assure me that this is in fact the case.[90]

Szent-Györgyi's letter to Krebs reflects some of the hallmarks of his scientific style. A highly productive experimentalist with a fertile imagination, Szent-Györgyi was well aware of unresolved complexities in the problems with which he dealt, yet proposed theories freely, uninhibited by the frequency with which he had later to modify or discard them. He even prided himself on his "wild" theories.[91] In a letter to Sir Henry Dale in 1931 announcing that he was launching himself into research on cancer, Szent-Györgyi wrote, "Now I am the happy father of 800 rats and innumerable theories of which I hope to disprove half a dozen this year."[92] His ideas and his work were, however, not taken lightly; he was as likely to discover something very important as to announce a theory that would soon collapse, and he was in 1936 on the verge of receiving a Nobel Prize for the isolation and identification of vitamin C four years earlier. It is not accidental that the word "stimulating" became associated with Szent-Györgyi's current theory of respiration. The theory changed each time he discussed it, and contemporaries did not accept it as a proven theory, but it also offered badly needed fresh insight into a set of problems that had long resisted solution. That Hans Krebs in particular should have been attracted so strongly to a theory, critical aspects of which did not satisfy him, illuminates the scientific styles of both men. Although Krebs had none of the outward flamboyance of the extraordinarily charming Hungarian scientist, there was underneath his more reticent manner a kindred venturescme spirit. We have repeatedly seen Krebs also publicly putting forth suggestions and theories that he might soon afterward have to modify or withdraw. Moreover, he reached out regularly for ideas and results of his contemporaries that he could test in his own way. It is revealing of Krebs's attitude toward scientific theories that he could find one with which he did not fully agree nevertheless "most inspiring." Krebs's own scientific creativity rested, in fact, to a considerable extent on his capacity to assimilate ideas from the work of others while utilizing the flaws that his critical scrutiny readily uncovered to give direction to his own effort.

V

On December 11, 1936, a controversy that had rocked England reached a climax, when King Edward VIII gave up his throne rather than his intention to marry a divorced woman. Krebs heard the news while he was walking in Sheffield with Nan Wayne. They went together to a cinema and watched the abdication ceremony on a news film.[93]

This public event nearly coincided in time with a significant personal event in Krebs's life. While he was eating lunch at the staff table in the university refectory, a botanist and friend, John Lund, sat down next to him with a young woman. Lund took a stamp out of his wallet to give her for her collection. Krebs, who had once collected stamps himself, took the opportunity to strike up a conversation about them.[94]

The young woman was Margaret Fieldhouse, a teacher in a nearby convent school, who drove the short distance to the refectory about twice a week to have lunch with friends. She and Lund had sat at the staff table this time only because the rest of the tables were full. Even though Krebs did not say very much, she found him "totally absorbing." After she had sat next to him a second time, she knew already that she was going to break her current engagement to a young man who was at the time away in Trinidad in the foreign service. Krebs was also immediately attracted to Margaret. She was young and very pretty, shy but spontaneous and forthright, an altogether warm and engaging woman.[95]

The daughter of a Yorkshire family, Margaret grew up in Wickersley, a village near Sheffield. Her father owned several shops. After attending a Catholic primary school where her aunt was headmistress, she was sent at the age of eight to a boarding school near Birmingham at which her mother had served as a supply teacher. It was run by a French order of nuns. Margaret stayed there until she had attained her school certificate at 16, then came back to Sheffield to study chemistry, botany, and geography at the Notre Dame Convent, a day school. By then she thought that she would like to be a teacher. Although she had never learned to cook, she decided to train for domestic science, and went to the University of Manchester to study that subject. Dietetics was beginning to make an impact on domestic science, so she learned something about recent advances in the practical knowledge of nutrition.[96]

In the spring of 1936, Margaret was hired to begin teaching in the fall at a newly built convent school in Sheffield. Besides instructing older girls in domestic science, she would have a class of small children to teach. To prepare for that work she spent July and August at Lady Margaret Hall, Oxford, in a Montessori training program. Maria Montessori herself taught some of the classes, and Margaret found her inspiring.[97]

Margaret began teaching at the convent school in September. In addition to her Montessori pupils and domestic science classes for both 11 and 16 year olds, she was asked to teach ancient Greek history to 12 year olds. She had never studied ancient history herself, but admired the "dauntless" attitude of the nuns who were willing to tackle any job that had to be done, and she enjoyed having to work up the new subject in a hurry. The half hour that she had for lunch each day therefore came in the midst of a demanding schedule. Fortunately she could drive in two or three minutes from the convent school to the university refectory, arriving in a jaunty red MG sports car of which she was quite proud.[98]

Soon after Hans and Margaret had met by chance they began to see each other by design. Sometimes he called on her at her family's house, sometimes she came to his boarding house, where his landlady served tea and biscuits in his neatly kept sitting room. She learned quickly how preoccupied he was with his science. She had never before encountered anyone who went about his work with such a sense of urgency. Since she was hardly idle herself, however, his relentlessly disciplined work schedule did not leave her unoccupied. They quickly found a rapport that bridged their differences in age, temperament, and cultural background. She saw that his willful personality was softened by an appealing simplicity that revealed itself most directly in his sense of humor. The same things struck both of them as funny. They often told each other about amusing phrases they had heard, and roared with laughter each time they were repeated.[99]

Full Circle

On December 27 or 28, 1936, Hans Krebs made another trip to Cambridge, in order to meet with Francis Roughton so that they could finally finish their work on carbamino ornithine.[1] There is no record of what Krebs and Roughton actually worked out together. By this time or shortly afterward Roughton himself wrote an outline and draft of their proposed paper, together with numerous calculations that he had carried out, all in his elegant, meticulous handwriting. The draft was entitled "Carbornithine as an Intermediate in the Synthesis of Urea by the Liver," by Hans Adolf Krebs and Francis John Worsley Roughton. It began:

In previous work on the synthesis of urea by the liver (summarised by Krebs, 1934) the first stage in the process has been formulated thus: -

R CH_2 NH_2 + CO_2 + NH_3 → R CH NH CO NH_2 + H_2O
   (ornithine)   (citrulline)
where R = COOH CH (NH_2) CH_2 · CH_2

It is possible, however, that this complex reaction may actually take place in two stages, each consisting of a more well-known type of reaction, viz:

(i) R CH_2 NH_2 + CO_2 → R CH NH COO^- + H^+
                       (carbornithine)
(ii) R CH NH COO^- + NH_3 → R CH NH CO NH_2 + OH^-
                       (acid amide of carbornithine)
carbornithine thus being considered to be an intermediate compound between ornithine and citrulline.[2]

After pointing out that the general physicochemical mechanism of carbamino reactions, including the equilibrium and velocity constants, had been completely worked out by Carl Faurholt in the 1920s and that more recent writers, including himself, had only confirmed this "masterly work" and shown its physiological importance in the transport of carbon dioxide, Roughton wrote that "the purpose of this paper is to see whether the reaction is also important in urea synthesis."

A priori CO_2 should combine appreciably, at physiological pH, with the δ-NH_2 group of ornithine, but not with the α-NH_2 group, for the pK of the ionisation of the δ group (considered as a cationic acid, i.e., R $CH_2NH_3^+$ → R CH_2 + H^+ R CH_2 NH_3 → R CH_2 NH_2 + H^+) is near to 8.0, and hence at pH 7.4 a considerable proportion should be present in the form R CH_2 NH_2, which is capable of combining with CO_2: On the other hand the α-NH_2 group of ornithine has a pK value near 10.0, and hence like glycine, is only present to the extent of 1% or less in the uncharged -NH_2 form at physiological pH. It may be remembered that CO_2 cannot combine with -NH_3^+ groups.

For quantitative argument, however, the equilibrium constant of reaction (1) must be known for ornithine: the first task of this paper was to measure this constant, and to calculate, therefrom, the amount of carbornithine which can exist at physiological CO_2 pressure, pH and temperature. The amount so calculated proves to be appreciable.

Further evidence, bearing on the two-stage theory, was sought by comparing the effects of CO_2 concentration and bicarbonate concentration on the rate of urea synthesis by liver slices, with their effects upon the amount of CO_2 combined with the δ-NH_2 group of ornithine. Under suitable conditions the two effects are found to run parallel, and hence the theory that carbornithine may be the first intermediate in urea synthesis by the liver is rendered not only plausible, but also probable.[3]

There followed several pages detailing the calculation of the equilibrium constant and the experimental procedure, employing Roughton and James Ferguson's three syringe method, that they had used to measure the necessary quantities (see pp. 215–216).

Krebs must have left Cambridge with the feeling that the reaction step he had been seeking ever since 1932 to complete the ornithine cycle was now filled in. His satisfaction, however, was short lived. Less than two weeks later, on January 9, 1937, he received from Roughton the news that

I was in London yesterday and took the opportunity of consulting [Albert] Neuberger in regard to the active coefficient of ornithine ions of various kinds. He told me something very important, which, if true, makes our work on carbornithine as an intermediate between ornithine and citrulline wrong. His point was that the pK of about 8.5, which we had attributed to the δ-NH_2 of ornithine, must on chemical grounds really be attributed to the α-NH_2 of ornithine, and that in our experimental work we have therefore been measuring the carbamino combination at the α-position, which is obviously of no interest in urea synthesis. The pK of the other amino group, according to [Carl L.A.] Schmidt is about 10.7, and therefore according to our formula, the amount of CO_2 bound there would be only about 1/100th of the amount of CO_2 bound at the NH_2 group with the pK of 8.5. What do you think of this? It looks to me as if we have been saved from dropping a brick.[4]

After mailing this letter Roughton thought of an alternative to the reaction steps that he now believed to have been ruled out, and quickly sent off an additional note on a postcard:

Further to my letter of Jan. 9, is it possible that the two stages of the reaction are

$$CO_2 + NH_3 \rightarrow NH_2\,COOH \rightarrow NH_2\,COO^- + H^+$$
$$NH_2\,COO^- + \text{ornithine} \rightarrow \text{citrulline?}$$

At pH 7.4 and 37°, as in your tissue slice experiments, an appreciable proportion of the ammonia should be in the NH_3 form.[5]

Krebs was not ready to concede that the carbamino ornithine theory was wrong. On January 12 he wrote back:

I have been thinking hard about Neuberger's criticism and I have just consulted Glasstone. At first thought he considered our previous assumption correct, but when I showed him your letter he began to doubt. The problem is obviously to decide which of the two amino groups forms the zwitterion and Glasstone does not know of any evidence which would decide between the two possibilities. He was very much interested in the problem, but he said it is a tricky one and he does not see a possibility of deciding. He would like to know Neuberger's arguments.

I looked up various papers and in [Edwin J.] Cohn's paper (Erg. Physiol. Vol. 33) I found the weak basic group of lycine attributed to the ϵ-amino group. This seems to agree with Neuberger's view. The idea occurred to me that perhaps two zwitterions may occur, one formed by the carboxyl group + α-amino group and one by the carboxyl group + δ-amino group. I don't see any reason why there should only be one type.

Are the dissociation constants of the various amino valeric acids (α and δ-amino valeric acids) of any use in this problem? Data are available in the literature (Cohn p. 844).

Your suggestion that $NH_2\,COO^-$ may react with ornithine would

save a part of the original conception, but I am by no means convinced
yet that Neuberger's criticism is correct.

. . .

I am going to write Neuberger to ask him about his evidence and
I should like to write to you again when I have discussed the matter
further with Glasstone.[6]

Samuel Glasstone, with whom Krebs consulted, was a physical chemist at
Sheffield with a particular interest in the rates of chemical reactions.[7]
   The contrast between the responses of Roughton and of Krebs to Neuberger's views is illuminating. Roughton may have been more easily persuaded in
part because he had talked the matter over with Neuberger, whereas Krebs had
at this point only seen a brief summary of Neuberger's argument. A more basic
reason is that Roughton was deeply familiar with arguments based on equilibrium
constants, and found this one immediately compelling. The only surprising thing
is that he needed Neuberger to bring the criticism to his attention. Krebs
undoubtedly understood the argument, but he did not habitually rely on such
considerations, and his letter suggests that he thought about the problem more
qualitatively than Roughton did. Moreover, he clearly felt a greater stake than
Roughton did in "saving" the "original conception."

I

When Krebs resumed his research in the new year of 1937, on Monday, January
4, he first took up a question based on the report by Hans Weil-Malherbe that
he had noted down at the Biochemical Society meeting in October (see pp.
352–353), that acetoacetic acid is formed from pyruvic acid in the presence of
pyocyanin. Using liver tissue, Krebs tested the effect of pyocyanin on the
aerobic formation of acetoacetic acid from added β-hydroxybutyric acid and on
the change in the quantity of acetoacetic acid added anaerobically. In the first
case there was an increase in respiration, but the "oxidation of β-hydroxy[butyric
is] not influenced by pyocyanin." In the second case the quantity of acetoacetic
acid diminished more rapidly with pyocyanin than without it. "Increased
reduction?!!" he asked, then wondered, "Why should pyocyanin cause
appearance of acetoacetate in Weil's experiment[?]"[8] Having used liver instead
of brain tissue, and β-hydroxybutyric in place of pyruvic acid, Krebs had not
directly tested Weil-Malherbe's conclusion. Perhaps he had assumed that β-
hydroxybutyric acid would be an intermediate in the reaction his former
associate had observed. Although he did not immediately carry the experimental
investigation further, he pursued in his mind the question of how pyruvic acid
is converted to the two ketone bodies. It was an old question, but one in which
Weil-Malherbe's result had perhaps renewed Krebs's interest.
   Through the rest of the first week in January Krebs conducted experiments
on the anaerobic and aerobic reactions of acetate, fumarate, and succinate in *B.
coli*. On Saturday he tried out propionic acid bacteria again after a six-month

lapse, but went back on Monday the 11th to *B. coli*. During the next two weeks he concentrated most of his attention on this organism, although he returned once to *Staphylococcus* and tested one more bacterium, *B. proteus*. Johnson continued to gather data on the anaerobic pyruvic acid reaction in animal tissues.[9]

Sometime during the two weeks after he performed the experiment related to Weil-Malherbe's result, Krebs formulated the hypothesis that an intermediate stage in the conversion of pyruvic acid to either of the two ketone bodies is the condensation of pyruvic and acetic acid to give acetopyruvic acid, according to the scheme:[10]

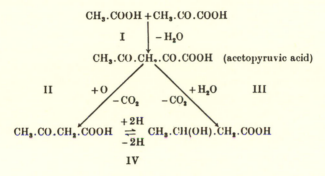

Johnson synthesized sodium ethyl acetopyruvate from sodium ethylate and hydrolyzed the ester to give sodium acetopyruvate. With this preparation he performed a preliminary test of Krebs's hypothesis on January 20, measuring the effect of acetopyruvate on the anaerobic formation of CO_2 in rat liver and kidney tissue. In both cases he obtained a large increase. The next day he followed up this encouraging start by comparing the rate of formation of β-hydroxybutyric acid anaerobically from acetopyruvic and from pyruvic acid. The scheme predicted that in the absence of oxygen acetopyruvic acid should form β-hydroxybutyric acid through a dismutation reaction and that pyruvate should have a similar effect only if the tissue produced sufficient acetic acid to support the preceding condensation reaction. The result (Q_{CO_2} control = 0.63; + pyruvate, 0.52; + acetopyruvate, 1.17) was in accord with the hypothesis.[11]

Further progress depended on devising a method to determine acetopyruvic acid. As usual, Krebs favored a manometric approach. While Johnson carried out the above experiments, Krebs tried out a method based on the "splitting of acetopyruvic acid in yeast." When he added yeast carboxylase to acetopyruvate in an acetate buffer solution, 195 μl of CO_2 evolved (corrected for controls), compared to 224 μl calculated for the complete decarboxylation of the compound. "Acetopyruvate reacts!!" he put down with apparent enthusiasm: "roughly correct yield!!" Johnson proceeded on the same day to employ the new method in order to measure the anaerobic disappearance of acetopyruvate in rat liver tissue. The result, Q = 2.34, was again an encouraging indication that acetopyruvic acid was metabolized under these conditions. The scheme predicted that aerobically acetopyruvic acid should yield acetoacetic acid by oxidative decarboxylation. On the 22nd Johnson found that acetopyruvate raised the rate of formation of acetoacetic acid in liver tissue in the presence of oxygen from Q = 1.4 to 1.9. Two days later he compared the aerobic effects of acetopyruvate and pyruvate and obtained the supportive figures: $Q_{acetoacetic\ acid}$ control = 0.37; + pyruvate, 1.20; + acetopyruvate, 1.63. He consolidated the situation on the 26th by obtaining similar results in two further experiments. Between January 27 and February 2, he measured substantial increases in the anaerobic formation of CO_2 with acetopyruvate in pigeon breast muscle and liver tissue, sheep liver, and sheep kidney.[12]

Johnson's experiments appeared to be verifying Krebs's latest theory with remarkable ease. The scheme offered an alternative to the various mechanisms for the formation of the ketone bodies that had been at issue for more than two decades. It also suggested a quite different link between carbohydrate metabolism and the ketone bodies than Krebs himself had proposed the previous summer in his *Nature* article (see p. 293). What the new scheme did not address was his central concern with the route from pyruvic acid through succinic acid to the complete oxidative breakdown of carbohydrates.

While Johnson was directed onto the new experimental track opened up by the acetopyruvic acid scheme, Krebs stuck with his ongoing line of investigation into anaerobic reactions in bacteria. Increasingly he focused his study of the bacterial dismutation of pyruvic acid on *Staphylococcus*, where that reaction appeared to comprise the primary route for the breakdown of pyruvic acid, and to employ *B. coli* mainly for the reactions of fumaric acid, which seemed in that organism to play a particularly important role as a hydrogen carrier in the sense defined by Szent-Györgyi. Around the beginning of February he came to the conclusion that the measurements he had made up until then of the reactions of fumarate with various substrates in *B. coli* sufficed only to decide the "preliminary question" of which substrates may enter into such reactions. To attain "complete" quantitative data on the rates of the reactions would require more complicated experiments in which he measured "in commensurable units" the rates of oxidation of the substrates by fumarate and by molecular oxygen. Choosing the formation of CO_2 as the common unit, he recognized that since carbon dioxide may be formed from bicarbonate in the medium if the substrate gives rise to an acid, or be absorbed if it combines with a product of the reaction, he must in each case measure both the "free CO_2" produced and the change in bicarbonate. Additional controls were necessary to account for the possibility that the substrate may also give rise to CO_2 by fermentation and that another gas in addition to CO_2 may be produced. The "complete method" that Krebs worked out required him to employ up to 17 manometers in a given experiment. On February 2 or 3 he performed the first such mammoth experiment, using "*old coli*" and choosing glucose as the substrate to be paired with fumarate. He recorded 16 simultaneous manometer measurements at 20-minute intervals for 140 minutes. From his results he inferred that "fumarate does not inhibit fermentation," and suspected from a "pause" in the pressure change in one of the manometers that CO_2 alone formed at first but that later on the bacteria might also have produced H_2. He carried out another such experiment with a fresh culture and obtained essentially the "same result" over the first 100 minutes, but "later [there was] too little CO_2."[13]

During January Juda Quastel had apparently made one of his occasional visits to Krebs in his laboratory and, finding him engaged in bacterial metabolism, wrote a letter afterward which may have made suggestions about reactions in *B. coli* based on his own pioneering investigations in that field during the 1920s. Krebs replied on February 1,

> Thank you very much for your letter and for the references. Some of your previous suggestions are not confirmed by my present work. For instance, the Cannizzaro which you suggested to occur in *B. coli* if pyruvic acid is added does not take place in this organism. It occurs in a number of cocci and in animal tissues. In *B. coli* the fermentation of pyruvic acid is a most complicated process in which among other products, succinic acid is formed.
>
> The reaction: formate + fumarate = CO_2 + succinate which you suggest on the basis of some growth experiments does not seem to occur in *B. coli* and I am unable to explain the effect of formate which you discovered, on the grounds of any metabolic processes.
>
> However, these are minor points and I should like to say that I find a great deal of useful information and inspiration in your work.
>
> With best wishes for your trip to America.[14]

Krebs's comments on the complicated processes in which pyruvic acid appeared to be involved suggest one of the reasons that his investigation of *B. coli* was itself turning out to be a long, complicated process. This exchange is typical of the negotiations that investigators in different laboratories carry on over small discrepancies in their respective results. The final paragraph was not an empty compliment, for Krebs's laboratory notebooks and papers both made repeated references to Quastel's publications. Quastel may well have overlooked the general praise, however, and fixed on the specific disagreements as further evidence that Krebs failed to appreciate his contributions.[15]

A warmer exchange took place between Krebs and his old colleague at Freiburg, Rudolph Schoenheimer. Schoenheimer wrote on February 4 to thank Krebs for sending him some reprints: "I scarcely need to tell you with what great interest I read your work, and I would like to wish you also from my heart the best of luck." Now at the College of Physicians and Surgeons of Columbia University, Schoenheimer reported that

> I have had very good fortune with respect to my working conditions here. They are much better than in Freiburg and we have begun investigations that would have been impossible to undertake in Freiburg. The application of deuterium is providing much amusement, unfortunately, however, in view of the tedious methods of measurement things do not go very quickly. Fortunately there is much interest

in it here, so that I have an ample number of people working with me.[16]

The investigations to which Schoenheimer alluded were, of course, his epoch-making applications of the isotope deuterium to the study of intermediary metabolism.[17]

For more than half a year the question of a future in Palestine for Krebs had lain dormant. Early in February that possibility came once again to the fore. On the 5th Walter S. Cohen, a trustee of the Cancer Research Trust at Hebrew University, wrote to inform him that "the question of establishing a Department of Physiological Chemistry," as the third department in the cancer research laboratories, "is now under consideration, and the Trustees would very much like to have a talk with you on the subject if it is possible for you to come to London." Notwithstanding his repeated earlier disappointments over this subject and the skepticism that they had engendered, Krebs agreed to come on Saturday, February 13. Assuming that this date was kept, Krebs met in London with another of the trustees, Dr. V. Idelson, Cohen being unable to be there. At the meeting Idelson apparently raised the prospect that Krebs could be offered a two- or three-year contract as head of the Department of Physiological Chemistry.[18]

On the same trip to London, Krebs probably got together with his two old friends, Hermann Blaschko and David Nachmansohn. For both of them Krebs was coming to assume the role of expert scientific critic, as well as trusted counselor. Repeatedly Krebs tried to bolster Blaschko's self-confidence, encourage him in his research, dissuade him from the belief that others were against him, and urge him to be more responsive to job possibilities that came up.[19] With Nachmansohn Krebs sometimes had the opposite problem, to counteract an impulsive tendency to act hastily. In January Nachmansohn had sent him a paper that he had written, and asked for criticisms. As Krebs related the situation to Blaschko, who had suggested the idea of doing the paper to Nachmansohn, Krebs had found Nachmansohn's discussion "very unsatisfactory and wrote back that in my opinion it must be fundamentally changed," but Nachmansohn had replied that Krebs should "correct it as much as possible, and send it as quickly as possible to the publisher. He considers it better on the whole to publish something than to publish nothing at all, even if the matter is not first class. I therefore sent in the corrected paper, even though I did not consider it a good idea."[20] Much as Nachmansohn and Blaschko relied on his advice, Krebs found them difficult to persuade to accept what he advised. They must have tested his sometimes limited patience, but he remained a supportive colleague and loyal friend. He also judged rightly that beyond their present difficulties both of his friends were talented scientists with good future prospects.

In the laboratory, Johnson worked on the acetopyruvic acid investigation throughout the second and third week of February, but Krebs recorded no experiments in his own notebook from February 10 to 16, and only four between February 17 and 24.[21] It is not clear what caused his experimental pace to slacken during this period. Perhaps the day he spent in London was the beginning of a longer trip. He may also have used part of the time to write two scientific papers that he finished shortly afterward. For the first portion of this interruption in his main lines of research he was probably also preoccupied with the carbamino ornithine problem.

Late in January Albert Neuberger had put his "opinions about the dissociation of ornithine into writing" and sent copies of his carefully structured two-page discussion to both Roughton and Krebs. Neuberger pointed out that the ionization of ornithine proceeds in three stages, the first being the dissociation of the carboxyl group. The second stage, however, might be the dissociation of either the α-NH_3^+ or the δ-NH_3^+ group, the third stage being the dissociation of the other one. Because experimental measurements can give only the overall constant for the total dissociation of both groups, the question of the order in which the two dissociate must be approached through considerations about the influence of other substituents, their charges and distances from the dissociating group, and comparisons with simpler compounds. "The δ-NH_3^+ group," he wrote, "can be regarded firstly as an aliphatic amine (pK $\sim$ 10.6) which is under the influence of two opposing charges." Taking into account the distances of these charges, he estimated the pK as 10.2. "The α-NH_3^+ group can be compared with valine and leucine (pK about 9.8); the introduction of a positively charged group in γ position will cause the pK to decrease to about 9.4." From these and certain additional arguments Neuberger concluded that the second step in the dissociation of ornithine was

$$NH_3^+ \ (CH_2)_3 \ CH \ (NH_3^+) \ COO^- \rightleftarrows NH_3^+ \ (CH_2)_3 \ CH \ (NH_2) \ COO^- + H^+$$

The third step was the equivalent dissociation of the δ-NH_3^+ group.[22]

The implications of this conclusion for Roughton and Krebs's theory were self-evident. The form of ornithine in which the α-amino group was in the undissociated state that permitted CO_2 to combine with it could be present in "appreciable" quantities. The form required by their theory, with the δ-amino group in that state, could not be.

The receipt of Neuberger's letter apparently prompted Roughton to ask Krebs to gather some additional data on the production of urea in liver slices at varying CO_2 pressure, but this communication has not survived. To comply with the request was one of the things Krebs was trying to do in early February, perhaps particularly on the 10th and 11th. Something went wrong, however, and on the 12th he wrote Roughton:

> I am sorry that I cannot give you any fresh data . . . . I have carried out a number of experiments, but they were not satisfactory. I enclose one of the previous experiments in which all the data for which you asked are recorded.

After answering a question about the relation between the thickness of the tissue slices and the excess CO_2 concentration in them, Krebs apologized a second time:

I should have written before, but I have been trying to get some results during the last few days. Unfortunately various accidents interfered.

Nevertheless, Krebs still held his ground:

I do not yet feel convinced that Neuberger is correct. I take it from his resumé that the α-amino group of ornithine is expected to behave like the α-amino group of any α-amino acid, whereas the δ-amino group should react like any aliphatic amine. Am I right, now, in assuming that ordinary α-amino acids such as glycine or leucine do not form significant quantities of a carbamino compound under physiological conditions, whereas methylamine does? If this is true the carbamino compound of ornithine would preferentially be that of the δ-amino group.[23]

Whatever further conversations on the subject took place between Roughton and Krebs have been lost. Krebs was obviously at a disadvantage in debating about arguments with which he was less thoroughly familiar than either Neuberger or Roughton; yet he was not irrational to resist a conclusion which may have been based on strong indirect inferences, but was at the time impossible to prove with decisive data. There is no record of how long Krebs held out, or whether he ever explicitly capitulated. It is plausible that he persisted in his belief that he might turn out to be right after all, but that, unable to think of a way to support his view experimentally, he gradually gave up hope of solving the problem in the near future.

* * *

Krebs discussed his work with Roughton on carbamino ornithine several times with me. In July 1977, I had seen only the two letters from Roughton of January 1937. Krebs's comments suggested that Neuberger's views immediately persuaded them to give up an idea that had, anyway not been of major importance:

—We didn't follow it up any further, no.
—It was very much of a side line, because my main interest . . . is clear from my published papers. It was concerned with finding the answer to a specific question and . . . it turned out . . . it was the wrong question.[24]

We went back over the subject in May 1979, after I had examined the Roughton papers. Krebs still recalled Neuberger's opinion as compelling:
—Neuberger, who was very experienced in this, put me right.
—It is a clear-cut story. We both agreed and both accepted this story.

I then showed him the two letters in which he had written that he was not convinced that Neuberger was correct. He seemed mildly surprised, but did not comment directly, except to say "all I know [was that] we abandoned the idea." He began, however, to discuss Neuberger's argument in a different tone. It was "not straightforward, but we respected Neuberger very much." A little later he added, "I wonder what physical chemists now would say to how straightforward this is. I don't know whether Neuberger was actually himself completely convinced, or whether he only suggested this."[25]

The shift in Krebs's attitude that occurred during this conversation was not a change in his historical account of how he had reacted to Neuberger in 1936, but a new reaction in 1977, in which he again raised a question about Neuberger's opinion after many years in which he had regarded the matter as closed. He was not, of course, reopening the question of whether carbamino ornithine might actually be an intermediate in the urea cycle, for that question had long since been settled in a different direction.

A few days later, while discussing other subjects, I read to Krebs a passage from a letter he wrote in 1946 to a colleague who had sent him a copy of a review article he had written on glutamine:

Is your sympathy with a glutamine mechanism of urea formation not based on certain psychological circumstances rather than experimental evidence? I can't help feeling that you are desperately in love with glutamine (which of course is quite proper for you; I have been guilty of similar love affairs). When in love it is difficult to keep one's judgement cool; the object of one's passion appears all important and one tends to credit it with virtues which it perhaps does not possess.[26]

When I asked him if he could think of "any ideas that you were particularly fond of," he answered,

Well, I can quote what you showed me yesterday, that I said carbamino groups being the precursor of citrulline. And I think I said something to the effect of rescuing something of the concept. Now in my choice of words there was definitely some expression of affection for the idea.[27]

It would be tempting, but misleading, to claim that in confronting Krebs with two letters that he had forgotten I succeeded in removing a facade of objectivity with which he had retrospectively surrounded his attitude toward the carbamino ornithine theory. His resistance to Neuberger's conclusion may have been a short-term effect that had not only faded from his memory, but was not in conflict with the summary statements that "we didn't follow it up further," "we both accepted this story," and "we abandoned the idea." On a longer time scale they did all of these things. On the other hand, the fact that he did forget about his initial resistance may have been conditioned by his belief that a scientist ought always to be ready to abandon a theory in the face of compelling contrary evidence. His spontaneous identification of the views expressed in the two letters with his acknowledgment in another letter of "love affairs" with scientific ideas reveals his recognition that he could not always conform to that ideal behavior. Every scientist whose investigative pathway I have followed in close detail experienced at some time or other tensions between the objective imperative to abandon ideas in the face of contradicting "facts" and a subjective attachment to some of the ideas being pursued. Nor are such tensions necessarily obstacles to scientific success, for, as the foregoing episode illustrates, there are no wholly objective indicators to mark the point at which a scientific theory has been confronted with decisive contradictory evidence.

II

By early 1937 Krebs was discovering a pleasant way to spend even more of his day than usual in the laboratory. Margaret Fieldhouse now came there instead of to the refectory for lunch. When she arrived Hans, regularly sent her across the street to Mrs. Edlington's shop to buy a tea cake—a soft roll with butter and ham or bacon on it—for each of them, while he boiled some water in a beaker for their tea. They ate together sitting on laboratory stools, but he interrupted their conversation at frequent intervals with "just a minute Margaret" in order to make a set of manometer readings. Soon he taught her to read the manometers for him, while he attended to something else. Among the beneficiaries of this new arrangement was Leonard Eggleston, who was relieved for a time from one of his normal responsibilities.[28]

Hans explained to Margaret in a general way how the manometer experiments were conducted, but she was not particularly interested in learning about the details. She thought that it was quite a privilege to be allowed to read the manometers. Sometimes when she arrived in the laboratory, however, she found him concentrating so intently on his work that he scarcely took notice of her.[29]

Looking back on this daily routine from the distance of half a century, Margaret described it as a way that Hans could carry on his courtship without diverting himself. "There was no romance to it at all," and the situation reminded her "how quickly I had to learn that I was to receive no full-time attention." On weekends, however, Hans was more relaxed. Often they went for a walk into the Derbyshire countryside or the cinema—the "pictures," as they called it—two activities that both of them enjoyed very much.[30]

During February Hans asked Margaret out to a dance. She drove to the occasion in her own car, however, and met him at the door. Hans greatly enjoyed dancing, even though his rather jerky, old-fashioned waltz steps were not up to the standard to which Margaret had been accustomed. Afterward he showed how unaccommodating he could sometimes be. Leaving the dance, he simply said good-bye at the steps and cycled off. When she had found her own way to her car it would not start, so she had to go back to the dance and ask someone else to give her a ride home.[31]

Hans also introduced Margaret to the Waynes, who invited both of them to dinner. She too found them very warm and helpful. One of the diversions that

Hans allowed himself at this time was to participate occasionally in a play put on by the university Dramatic Society. Soon after he had begun seeing Margaret regularly, Hans asked her to come to a play—perhaps it was Thornton Wilder's *The Happy Journey*—in which he had a small part. Margaret was less impressed by the few lines Hans delivered than by what she thought was a striking performance by Nan Wayne.[32]

The first of the two research articles that Krebs completed in February 1937 was entitled "Metabolism of Ketonic Acids in Animal Tissues." Johnson was listed as a second author. This paper can be viewed as the full version of the preliminary paper that Krebs had published in *Nature* nine months earlier. He presented detailed evidence for most of the reactions that he had then claimed to "have found" in cells. His conclusions were, however, more qualified than in the *Nature* paper, and his claims were narrower. Some of the similarities and the differences are readily apparent from a comparison of the short *Nature* article quoted in Chapter 7 (pp. 294–297), with the introductory paragraphs of the new article:

> In this paper experiments are described which show that ketonic acids can react in animal tissues according to the general scheme:

$$R.CO.COOH + R'.CO.COOH + H_2O \rightarrow R.COOH + CO_2 + R'.CH(OH).COOH \quad ......(1)$$

α-ketonic   α-ketonic   carboxylic   α-hydroxy-acid
acid I   acid II   acid

or

$$R.CO.COOH + R'.CO.CH_2.COOH + H_2O \rightarrow R.COOH + CO_2 + R'.CH(OH).CH_2.COOH$$
$$......(2).$$

α-ketonic   β-ketonic acid   carboxylic   β-hydroxy-acid
acid   acid

> Examples are given in which α-ketonic acid I as well as α-ketonic acid II in (1) are represented by pyruvic acid. In other cases the α-ketonic acid in (2) is pyruvic acid or α-ketoglutaric acid and the β-ketonic acid in (2) acetoacetic acid or oxaloacetic acid.

> The reactions 1 and 2 elucidate a mechanism by which α-ketonic acids are broken down in the animal body. Although it has long been known, from the work of Embden, that α-ketonic acids undergo oxidation to the fatty acids which are shorter by one carbon atom, the question of the mechanism of this oxidation remained open. According to (1) and (2) the oxidation of α-ketonic acids is not brought about by molecular oxygen, but by a dismutation, that is to say by an intermolecular oxido-reduction. The oxidizing agent for the ketonic acid is a second molecule of ketonic acid which is reduced to the corresponding hydroxy-acid.

> The reactions (1) and (2) appear to play a role in the course of the normal oxidative breakdown of carbohydrates, of fats and of the carbon skeleton of amino-acids. This will be discussed in full in subsequent papers.[33]

His overall theme thus remained the dismutation reactions that had preoccupied

Krebs ever since the days when he had bantered with Edson about the "great work." We can see that the "general scheme" he presented here was a stronger, more general statement of the same view that he had set forth in his *Nature* article as (1) a specific reaction of two molecules of pyruvic acid, (2) a reaction of α-ketoglutaric + "ketonic acid," and a subsequent comment that "different" ketonic acids may take the place of those depicted in the examples. In contrast to the *Nature* article, however, he no longer linked individual reactions of this type into a sequence; and where he had formerly asserted that certain of these reactions constituted "primary steps" in the oxidation of pyruvic acid, he now made only the indefinite claim that reactions of this general class "appear to play a role in the course of the normal oxidative breakdown of carbohydrates, of fats and of the carbon skeleton of amino-acids." Although he thus hinted that these reactions might have a very broad significance, he now avoided committing himself about their location within the larger picture.

In the body of the paper Krebs was also cautious, placing more emphasis on the data and drawing less extensive inferences from them than in his previous treatments of the subject. He presented in most detail the evidence for the basic reaction:

$$2 \text{ pyruvic acid } + H_2O \rightarrow \text{acetic acid } + CO_2 + \text{lactic acid}$$

that he had been studying for nearly a year. He discussed separately the quantitative data for the consumption of pyruvic acid and for the production of each of the three products. In each case he listed numerous results, some that he had obtained himself, but most of them derived from the many experiments Johnson had performed on this reaction. In testis and brain tissue, Krebs noted, these products "account for the bulk of the metabolized pyruvic acid. In other tissues, however, the low yield of acetic acid leaves a considerable fraction of pyruvic acid the fate of which remains to be explained." The "search" for other possible products had yielded two substances: "succinic acid is found in small amounts, β-hydroxybutyric acid in considerable amounts in some tissues such as muscle and in smaller quantities in other tissues such as testis."[34]

In another sharp contrast with his treatment of the same question in his *Nature* article, as well as in his lectures during the fall, Krebs now wrote, "We do not propose to discuss the mechanism of the succinic acid formation in this paper, but confine ourselves to presenting a few data which show the magnitude of the succinic acid production in various tissues in the presence of pyruvic acid." His choice of the word "confine" hints that Krebs may have had to restrain himself from further discussion of a question that he thought he had solved at the time he wrote his *Nature* article and that he had discussed freely in the fall lectures. He did include a brief discussion of the significance of the formation of the other anaerobic product of pyruvic acid, β-hydroxybutyric acid, in the light of the results of other investigations, from Embden in 1912 until Hans Weil-Malherbe's latest publication, but he proposed no specific chemical links. In this case he apparently withheld the mechanism that he had currently in mind, pending the completion of Johnson's experiments on acetopyruvic acid.[35]

The change that had taken place in Krebs's approach is most obvious in the final portion of this section of his paper, entitled "Balance Sheet of the Anaerobic Metabolism of Pyruvic Acid"

> Five substances, lactic, acetic, succinic, β-hydroxybutyric acids and CO_2 have now been identified as products of the anaerobic metabolism of pyruvic acid. The quantities formed indicate clearly that there are several reactions concerned with the removal of pyruvic acid. It is not possible to express the anaerobic fate of pyruvic acid by one formula.

In some tissues the basic reaction shown above accounted for up to 80 percent of the pyruvic acid removed. In others, in which the yield of acetic acid was low, he could "explain the data" through equations such as

14 pyruvic acid + $5H_2O$ → 5 lactic acid + 4 β-hydroxybutyric acid + 9 CO_2 + acetic acid

He viewed these not as single reactions, but as net reactions, the sum of a main reaction and one or more secondary reactions.[36]

In his Biochemical Society lecture Krebs had presented the same basic situation in a different way. Acknowledging then too that the measured products of the basic pyruvic acid dismutation reaction did not always fit the predicted proportions, he described the discrepancies as "deviations" due to "side reactions," while maintaining that the primary reaction was "common to all tissues." At one level he seemed five months later merely to be presenting the same conclusion with a different emphasis, combining the effects of the main and secondary reactions instead of separating them, but this shift signified an underlying retrenchment. We have seen that through much of the year he had been seeking to sort out through these balance methods reactions that represented steps in metabolic pathways. Lumping such reactions together into "net reactions" that represented the same data in various ways and did not pretend to "express the anaerobic fate of pyruvic acid by one formula" was a tacit admission that for this crucial case he had not succeeded in this endeavor.

Summarizing the evidence that other ketonic acids "may react similarly to pyruvic acid," in accordance with his generic reaction scheme, he presented three reactions—that of pyruvic + acetoacetic acid, ketoglutaric + acetoacetic acid, and pyruvic + oxaloacetic acid—that he had discussed in his previous paper and lectures. Here he did not go into the question of whether the "balance sheets" accorded quantitatively with the assumption that these were single reactions. Regarding the reaction

pyruvic acid + oxaloacetic acid + H_2O → acetic acid + CO_2 + malic acid he made the only statement that was bolder than his previous discussions of the equivalent reactions. Where he had said in his Biochemical Society lecture that this reaction was "of special interest" because of its high rate, he now said that "the rate is higher than that of the analogous reactions . . . in most tissues (brain, testis, kidney, muscle) and in view of the fact that under physiological conditions oxaloacetic acid is available in tissues, it is likely that [this reaction] is the preferential way by which pyruvic acid is oxidized."[37] If that were true, then, in spite of the central attention he had long given to the reaction of two molecules of pyruvic acid as the basic type dismutation reaction, one of the homologous reactions that he had treated as a variant on the basic reaction might turn out to be physiologically of more fundamental importance.

In his "Discussion" section Krebs did not develop a single broad conclusion, but made six partially separable points. The most general claim was that the two reactions represented in his introduction "indicate the manner in which animal tissues decarboxylate α-ketonic acids. Animal tissues do not possess a 'carboxylase' of the type occurring in yeast" (which there converts pyruvic acid to acetaldehyde + CO_2). His second point raised the question of "whether anaerobic reactions are the only reactions by which pyruvic acid is metabolized in tissues" and answered that since the aerobic rate of metabolism of pyruvic acid is in most cases considerably higher than the anaerobic rate, "the anaerobic dismutations cannot account for the whole pyruvic acid breakdown." This conclusion represented a tacit refutation of the position he had taken in the *Nature* article that these anaerobic reactions formed the "primary steps" in the oxidation of pyruvic acid. Point three was that the amount of CO_2 produced by the anaerobic processes accounted for from 20 up to 80 percent of the total respiratory CO_2. "It is not yet clear," he added, "by what process the rest of the respiratory CO_2 is formed. The decarboxylation of oxaloacetic acid is another possibility, but its quantitative significance cannot yet be assessed." Here Krebs tacitly disagreed with Szent-Györgyi's recent assertion that oxaloacetic acid is not decarboxylated in animal tissues. In his fourth point, however, Krebs expressed his qualified acceptance of Szent-Györgyi's theory that oxaloacetic acid is an oxygen carrier in animal tissues. "The work reported in this paper shows that Szent-Györgyi's view holds for the special case of the oxidation of α-ketonic acids." In point five he recapitulated his discussion of the relation between pyruvic acid and the ketone bodies; in point six he referred to the oxidation-reduction reactions described in his paper as examples of intermediate carriers for hydrogen transport of the types suggested by Quastel and by David Green; and finally he reported that crystalline vitamin B_1 had failed to exert the effect on the dismutation of pyruvic acid earlier shown "in preliminary experiments" by crude commercial preparations.[38]

Krebs submitted the paper on ketonic acid metabolism in animal tissues to the *Biochemical Journal* on March 1 along with a separate paper, of which he was sole author, entitled "Dismutation of Pyruvic Acid in *Gonococcus* and *Staphylococcus*." As we saw in Chapter 7 (p. 307), when he moved into the field of bacterial metabolism extensively, in August 1936, *Staphylococcus aureus* was the organism that he began to use to study the anaerobic reaction of pyruvic acid. The first experiments indicated that the organism was suitable for that purpose. Nevertheless, during September and the following months he experimented mainly with *B. coli*. Though metabolically more complicated, *B. coli* could be grown more easily, whereas the metabolically simpler *Staphylococcus* required a more exacting medium. In late November and mid-December he again concentrated his attention for a while on *Staphylococcus aureus*. When he compared the rates at which pyruvic acid disappeared aerobically and anaerobically, he found that the latter could actually exceed the former, in contrast to animal tissues in which—as he noted in his conclusions to the paper described above—the anaerobic rate was generally lower. It appeared, therefore, that the reaction

$$2 \text{ pyruvic acid } + H_2O = \text{lactic acid } + \text{acetic acid } + CO_2$$

played a more prominent role in the metabolism of *Staphylococcus* than in animal tissues. In this organism Krebs studied almost exclusively this particular reaction, and employed it particularly to examine the effects on the reaction of activators such as yeast extract, the yellow enzyme, and oxidation-reduction indicators, the latter turning out instead to be inhibitors.[39] During January and February he performed a few more experiments on *Staphylococcus aureus* and also tried out other forms of cocci.

As the title of his paper indicates, whereas he had discussed a family of analogous dismutations in animal tissues, Krebs discussed only the specific dismutation reaction of two molecules of pyruvic acid in *Staphylococcus*. Although he found evidence for other side reactions, yielding especially succinic acid when large quantities of pyruvic acid were present, he concluded that "the data indicate that this 'dismutation' is the preferential reaction by which pyruvate is broken down in these organisms."[40]

In one experiment that appears to have been an exception to the above, Krebs had, in late December, tested ketoglutaric acid with pyruvic acid in *Staphylococcus*, and observed that "ketoglutarate not split alone, but accelerates splitting of pyruvate." In other words, he construed the result not as evidence for a dismutation reaction of ketoglutaric acid with pyruvic acid, but as an effect on the standard pyruvic acid dismutation. The next day, in an experiment on activators of the pyruvic acid dismutation, he included ketoglutarate and fumarate, along with several coenzymes. In a subsequent experiment he included also lactate. All three of these substances increased the rate.[41] Reporting these results in his paper, he wrote,

> Further work is also necessary before the effects of lactic acid, fumaric acid and α-ketoglutaric acid can be explained. The last two substances increase not only the rate, but also to a small extent the yield of CO_2. They act in quantities which are small compared with the amount of pyruvic acid that they may cause to react; their action appears thus to be catalytic.[42]

Viewed from our perspective there appears to be an obvious parallel between this situation and the one that Krebs had encountered five years previously with ornithine. A substance regarded as a metabolic substrate acted catalytically on another metabolic reaction. Had Krebs been modeling his approach to the reactions of pyruvic acid on his experience with urea synthesis, we might have expected him to have looked for an explanation patterned after the explanation of the ornithine effect. That he drew no such inferences reinforces the view that he was not looking backward to his early success for answers to his current questions.

The character of the two papers that Krebs sent off for publication at the end of February suggests another way in which his approach to the problem of oxidative carbohydrate metabolism was changing during the winter of 1936–37. In his lectures during the fall he was not only attempting to link individual reactions into paths that would account for the complete oxidation, but was employing various tissues and organisms to find evidence for reactions belonging to a scheme that would be common to all tissues. In these two papers, on the other hand, he was treating the metabolism of animal tissues and of a class of bacteria as distinct problems. In *Staphylococcus* he examined a single dominant dismutation reaction. In animal tissues he examined a family of ketonic acid dismutations. In neither case did he now consider the reactions of fumaric acid that he was at the time studying extensively in another class of bacteria, *B. coli.*

One explanation for these differences would be that they are only artifacts due to the different formats within which he presented his views in the fall and in the spring. The earlier general schemes appeared in lectures, two of which were broad overviews that he developed for more general audiences; the later, more restricted discussions appeared in specialized research papers. Perhaps as he gathered more and more data in an investigation that had now gone on for nearly a year, he was merely subdividing his subject, covering the domains within the broad problem one by one in separate publications. His paper on animal tissues did promise subsequent papers with a "full" discussion of implications only touched on in this first paper.

There is reason, however, to think that a real shift in his viewpoint was occurring. He had begun his investigation, Krebs related in 1977, suspecting "that the basic processes would be similar [in all tissues] on account of the fact that anaerobic glycolysis was already known to be the same basically in muscle and in yeast cells."[43] Although retrospective, this comment rings true, because of the powerful impact that we have seen the Embden-Meyerhof pathway exerted on his thinking. That did not mean that he expected the reactions to be identical in all organisms; but the differences would be restricted, just as a single step near the end of the Embden-Meyerhof pathway accounted for the production of lactic acid in muscles instead of the acetaldehyde + CO_2 that appeared in yeast. In his Biochemical Society lecture he treated as similarly restricted divergences the way that *B. coli* and animal tissues handled acetic acid.

By early 1937, however, I think that Krebs was finding the differences in the metabolism of the several organisms he was studying too large to be reduced to minor variations on a common theme. Thus the pyruvic acid dismutation reaction was so dominant in *Staphylococcus* that he came to regard it as the main—or, as he called it, the "preferential"—pathway for the oxidation of pyruvic acid, whereas he was beginning to perceive the oxaloacetic-pyruvic acid reaction in that role in animal tissues. Although it may be that he restricted his discussion to the ketonic acid dismutations in animal tissues because that was enough to fill one paper, the reason may also have been that he had by then concluded that in animal tissues fumaric acid is not reduced. That realization would strongly differentiate respiration in animal tissues from respiration in *B. coli*, where he was finding a dominant role for the reduction of fumarate. Correspondingly, he may have barely mentioned the dismutation of pyruvic acid in *B. coli* in his article on that process in *Gonococcus* and *Staphylococcus* because he recognized that the reactions of pyruvic acid in the former type of bacteria were so much more complex than in the latter as to require separate treatment.

The global picture of oxidative carbohydrate metabolism that Krebs had been striving to construct through most of 1936 thus appeared by March 1937 to be dissolving into partial schemes separated both by reaction types and by types of organism. We may wonder whether Krebs had generally retracted his theoretical ambitions as well. Did he now foresee a long, pragmatic research program intended mainly to gather more rigorous quantitative data concerning the various reactions he had identified, or was he still aiming to synthesize these data into a larger architectural framework? Did the experience of giving up repeatedly the broad theoretical positions he had temporarily occupied chasten him or reinforce his determination to try again? We cannot penetrate to the private state of his mind during the spring of 1937, but from his general temperament we can surmise that he simply kept on going, in the optimistic assumption that something new might turn up.

III

Through most of March, Johnson went on with experiments with acetopyruvic acid, confirming the earlier results favorable to the theory that the substance was an intermediate in the formation of the ketone bodies. Krebs pursued anaerobic oxidations in *B. coli*. During the first week he examined the hydrogenlyase reaction, in which formic acid is broken down to yield H_2, and the hydrogenase reaction in which H_2 acts as the hydrogen donor, reducing substrates. After a few scattered experiments on other dismutations, the aerobic oxidation of formic acid, and two experiments to see whether alcohol underwent a coupled oxidation in *B. coli*[44]—it did not—he turned on March 20 to the general question of how acetic acid is broken down in *B. coli*. In his Biochemical Society lecture he had stated that acetic acid is rapidly oxidized anaerobically in the presence of fumaric acid in this organism, unlike the situation in muscle. Now, however, he explored other possible mechanisms. The first question he formulated was, "Is H_2 formed from acetic acid?" through the reaction sequence

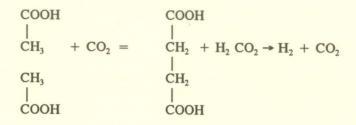

$$\begin{array}{c} \text{COOH} \\ | \\ \text{CH}_3 \\ | \\ \text{CH}_3 \\ | \\ \text{COOH} \end{array} + CO_2 = \begin{array}{c} \text{COOH} \\ | \\ \text{CH}_2 \\ | \\ \text{CH}_2 \\ | \\ \text{COOH} \end{array} + H_2\,CO_2 \rightarrow H_2 + CO_2$$

as he wrote it down in his notebook. ("It's all based still on the Thunberg idea," Krebs commented in 1977 when he looked back at this experiment, "but in bacteria the situation would be possibly different from higher organisms, I mean there could be additional reactions and this was one of them.") What he had in mind in 1937 was not the Thunberg-Knoop-Wieland scheme, but a dismutation reaction incorporating Thunberg's acetic acid condensation and in which CO_2 acted as the hydrogen acceptor. The fact that CO_2 was essential for the growth of *B. coli* prompted Krebs here and in several other ways to ask whether it could play such a metabolic role. In this one ingenious equation, therefore, Krebs managed to combine three ideas that particularly interested him. To test the theory he added acetate anaerobically to *coli* suspensions grown anaerobically on formate and measured the gas formed. In contrast to formate, which he tested comparatively, acetate did not increase the quantity above that formed by the control. "<u>No</u> H_2 from acetic!!" he put down, and one more attractive hypothesis had to be dismissed.[45]

In place of this unconfirmed idea Krebs now posed another idea, that formaldehyde might be an intermediate in the oxidation of acetic acid. To test the possibility he compared the effects of acetate and formaldehyde on the rate of oxygen consumption in a *B. coli* suspension. Both substances increased it considerably. He graphed the results, which showed that acetate was oxidized more rapidly during the first hour, but that afterward the gap narrowed as its rate fell while that of formaldehyde continued steadily.[46] The outcome therefore neither strongly confirmed nor ruled out formaldehyde as an intermediate, and Krebs did not comment on whether he thought it a favorable or unfavorable result. The experiment appeared at least to indicate that *B. coli* can metabolize formaldehyde in some fashion.

On the 24th Krebs examined whether there is a "dismutation of formalde-hyde." Measuring the anaerobic rate of formation of gas in the presence of formaldehyde in *B. coli grown* aerobically, he observed a "quick, constant rate," suggestive of a dismutation. If formaldehyde were dismutated, the reduced product would be methyl alcohol, and the question next arose, can the methyl alcohol be reoxidized in the bacteria? When he tested the "oxidation of methyl alcohol" in a fresh culture at pH 6.8, he observed "<u>no</u> effect." A second problem with the previous experiment was that the increase in gas pressure might have been due to H_2 rather than CO_2. Accordingly he next asked, "Is H_2 formed from formaldehyde?" To find out, he repeated this experiment on the same day with alkaline hydrosulfite to absorb the CO_2 and observed no pressure change, in contrast to a comparative experiment with formate. "No hydrogen from formaldehyde!!" he wrote, "whilst formate reacts rapidly." This outcome increased the likelihood that in the earlier experiment the gas pressure did signify a dismutation reaction. The next day he pursued that question further, asking "Does anaerobically grown *coli* dismute formaldehyde?" and finding a large increase in gas pressure, he answered affirmatively: "readily split."[47]

If formaldehyde were dismutated, the oxidative product would be formic acid, which should show itself by a general acid formation detectable as a decrease in the bicarbonate in the medium. When he tried the experiment on *B. coli* grown aerobically, comparing the effects of formaldehyde with two other aldehydes, he observed "definite acid formation from formaldehyde, very little acid formed from acetaldehyde and iso-valeric aldehyde!" The fact that no hydrogen had formed from formaldehyde now led him to wonder whether that substance would prevent it being formed from formic acid, that is, "Does formaldehyde inhibit formic hydrogenlyase?" He found an "almost complete inhibition!!!" Finally on this same very busy day he arranged a set of experiments that enabled him to measure both the H_2 and CO_2 formed anaerobically, with a culture grown anaerobically, and again tested formaldehyde separately and with formate. From the results he concluded (1) that "formaldehyde $\rightarrow$ H_2 (about ½ equivalent) (or 1/3 ?) rate increasing due to disappearing inhibition," (2) that formaldehyde again inhibited hydrogenlyase, and (3) that the "rate of dismutation [is] very slow!!" "Therefore," he concluded, "dismutation is probably only [a] side reaction."[48]

This sequence of experiments must have left Krebs puzzled about the role of formaldehyde in *B. coli*. It appeared to be metabolized, but not in a way that fit into any scheme he could devise. That was the way the matter had to stand while he interrupted his laboratory research for two weeks.

By this date the acetopyruvic acid investigation was completed, and Krebs was ready to send off his third paper of the month to the *Biochemical Journal*. Johnson, who had performed nearly all of the experiments, was again the second author. "Previous attempts to explain the mechanism of ketogenesis from pyruvic and acetic acids were," Krebs asserted, "merely speculative. Our own experiments do not support any of the previous assumptions." Presenting the scheme that is here reproduced above on page (379), he wrote:

> It is certain, as will be shown in this paper, that both ketone bodies are formed from acetopyruvic acid, but it is not yet possible to decide whether acetopyruvic acid reacts according to (II), or to (III) or in both ways. Since acetoacetic and β-hydroxybutyric acids are readily interconvertible, as suggested by reaction (IV), the net effect of (III) can also be brought about by (II) + (IV), and that of (II) by (III) + (IV).[49]

The data they presented showed not only that acetopyruvic yielded acetoacetic acid aerobically and β-hydroxybutyric acid anaerobically in a number of animal tissues, but that it consistently did so more rapidly than did pyruvate, acetate, or pyruvate + acetate. By the most exacting standard available for demonstrating an intermediate metabolic reaction, therefore, they had built for this mechanism a convincing case.[50]

While he pursued his research interests in March, Krebs also worried about his latest Palestine dilemma. He was more reluctant than ever to leave England, or to give up his permanent position in Sheffield for a short-term appointment, and apprehensive that his research in a cancer laboratory would be restricted to problems deemed pertinent to that disease. The appeal of working in Palestine was nevertheless still too strong for him to allow him easily to decline. As he had done before, he sought the advice of Weizmann. Nachmansohn had already written Krebs late in February that, although Weizmann "would very much like to have you in Palestine, he also sees the difficulties very accurately. In the first place you cannot exchange a life-long post for a 2–3 year contract. He says . . . that the whole matter is only possible if the University is brought in and you receive a permanent position." Nachmansohn suspected that Magnes, who was still chancellor of the university, had had a hand in creating these unsatisfactory conditions. On the other hand, Nachmansohn thought that the requirement to work on cancer was no serious problem; it could be satisfied by "making 3 experiments a year" in that field.[51] Krebs, who disliked expedient compromises, was not content to handle the matter that way. On Sunday, February 28, he made the trip to London to see Weizmann, but Weizmann was detained by some other obligation and missed their appointment. Apologetically he wrote on March 2 that "I will be very glad to get together with you on your next visit to London, and to discuss everything with you."[52] There is no direct record of whether this meeting actually took place.

When Idelson asked him for his views on the Cancer Research Institute, Krebs took a strong, straightforward position. On March 24, he replied:

> 1. The most essential point, in my view, is the freedom of the scientific research in the institute. A narrow interpretation of the term "cancer research" is bound to hamper the success of the work. The history of cancer research shows that essential progress . . . can only be made together with the general progress on related fields . . . . I am confident that it is not against the spirit of the Deed if the workers in the institute are given complete freedom in the choice of the subject of their research . . . . I do not feel in a position to consider the further points in detail before this fundamental question has been satisfactorily settled.

He nevertheless briefly outlined the sums he thought would be required for equipment, running expenses, and the salary for the head of the proposed department. Taking a cue from Weizmann's opinion, he stated that "the position of the head of the department should be a permanent position as it is usual in the case of senior academic posts in universities and in research institutes."[53]

When he sent off this letter Krebs turned his attention to a more immediate and happy prospect. The Easter holiday had arrived, and he planned to spend it with Margaret. The Waynes had recommended to them a little hotel at

Portscatho, on the coast of Cornwall. Hans and Margaret drove off from Sheffield with the top on his old Austin roadster up, because there was still snow on the ground. They made slow progress through rainy weather along the winding roads that led to the southwest, and stopped for the night in Taunton. As they descended toward the coast the next day they found that there were already primroses and other spring flowers in the hedgerows. They felt that in two days they had journeyed from one climate to another.[54]

Along the way Margaret encountered a setback that taught her something about Hans's personality. With one of her earlier boyfriends she had been taking a German class, and felt by this time that she was making good progress. After carefully thinking through a sentence in German and waiting for an appropriate moment, she tried it out on Hans. He pretended not to understand what she was saying. When she said, "That was German," he denied that it was. She never again attempted to speak German with him.[55]

It may well have been that Hans refused to encourage Margaret's German in part because he was himself determined to put German behind him as he assimilated himself to England. He showed great interest in her English upbringing, wanting to know in particular what she had read as a child. Altogether their holiday at Portscatho was a very good one. They went hiking and shared their strong mutual interest in wildflowers. Unlike Hans, Margaret also knew a lot about birds, and Hans expressed amazement that she could identify them by their songs. They returned from the beauty of spring in Cornwall with a growing attachment to one another.[56]

Willful and unromantic as Hans was, and undivertable as he was from his working habits, he did not dominate all areas of his relationship with Margaret. Both were strongly independent, and both made their own decisions without encroaching on the interests of the other. In areas outside his science, Margaret exerted as much influence on Hans as he did on her. He accepted gracefully her corrections of his spoken English, and when she let him know that she could not bear the way he combed his hair straight back, he quickly began to part it from the side.[57]

Back in his laboratory on April 6, Krebs took up the investigation of formaldehyde in *B. coli* where he had left it before his holiday. Testing formaldehyde (in two concentrations) both anaerobically and aerobically, the latter again in comparison with acetate, he now obtained low readings under both conditions. He remarked that the "oxidation is slow this time just <u>doubles</u> the blank: suggests coupled oxidation!!" and that "anaerobic dismutation with 0.01 M [formaldehyde] very slow, cannot explain oxidation." Reasoning from these views that "if oxidation of formaldehyde is coupled, perhaps more rapid oxidation if 2 substrates are present," he tried formaldehyde in combination with acetate. The "coupled oxidations" he had in mind here were not dismutations, but reactions of the type that David Keilin had especially studied, in which the oxidations of two substances are coupled together (see p. 67). Krebs obtained, however, "no marked effect." The next day he tested combinations of formaldehyde with lactate, glucose, and formate, and found "strong almost complete inhibition of lactate and glucose oxidation by <u>formaldehyde</u>!!" There was "no inhibition of blank and formate oxidation," but (although he made no note of it) neither were there any increases indicative of a coupled oxidation. Having found no mechanism through which the observed oxidations of formaldehyde could be explained, he pursued the question no further. (After reviewing these experiments in 1977, he characterized them as "all exploratory tests which didn't lead anywhere.")[58]

After this exploration ended without success, on April 7 Krebs returned to the role of fumaric acid in the anaerobic oxidation of various substrates in *B. coli* and spent most of the next four weeks gathering further data regarding these reactions. During the same time he gave Johnson the task of extending the experiments he had begun in the fall on the formation of citric acid in animal tissues from pyruvate + malate. Between April 13 and April 22 Johnson measured the quantities of citric acid produced from these two substances in rat testis and liver, pigeon brain, and minced sheep brain.[59]

IV

In Tübingen Carl Martius spent part of the winter of 1936-37 thinking about how citric acid may be broken down in plant and animal tissues. The mechanism, suggested by Thunberg and others, that the first step was the formation of acetonedicarboxylic acid by decarboxylation, was based solely on chemical considerations, and made implausible in Martius's opinion by the fact that acetonedicarboxylic acid could not be found in yeast, bacteria, or animal tissues. Martius also approached the problem initially from the point of view of the chemical properties of the potential intermediates. Citric acid was a stable substance. Reflecting on how it might be broken down, he had the idea that if the tertiary OH group could be displaced to a neighboring carbon atom, the resulting compound, isocitric acid, would be readily dehydrated. From a biochemistry lecture he had once heard in Stockholm he remembered the way malic acid is degraded by the removal of water to fumaric acid, and he thought that citric acid ought to be dehydrated in an analogous way to yield aconitic acid. The product of the ordinary chemical dehydration of citric acid, *trans*-aconitic acid, was unreactive; but he had recently read a paper indicating that the other isomer, *cis*-aconitic acid, was unstable, and he thought that this compound might be the active intermediate. The reactions he posed would thus be (see following page, top formula):

Martius then worked out theoretically what the expected products of the dehydration of isocitric acid would be. The first step would yield oxalosuccinic acid which, he assumed, would then spontaneously lose CO_2 to give α-ketoglutaric acid (see following page, 2nd formula). From that point the further reactions were well known. By decarboxylation, α-ketoglutaric acid would lead to succinic acid. Familiar with Szent-Györgyi's work, Martius expected the sequence to continue from succinic acid through fumaric, malic, and oxaloacetic acid.[60]

To test his theory, Martius again drew on the analogy to the malic acid–fumaric acid reaction. That reaction had been demonstrated enzymatically

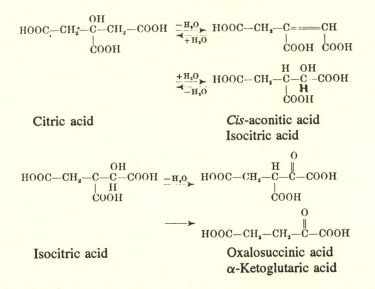

Citric acid   *Cis*-aconitic acid
                             Isocitric acid

Isocitric acid   Oxalosuccinic acid
                          α-Ketoglutaric acid

by means of a "liver enzyme" preparation, and Martius adopted the same method. Preparing a solution of dried acetone extract of beef liver in a phosphate buffer, he added sodium citrate and methylene blue. The dye was rapidly decolorized, and after six hours he was able to isolate from the filtrate a dinitrophenylhydrazone of α-ketoglutaric acid. Using the same enzyme preparation he showed that *cis*-aconitic acid can be converted to citric acid and that *cis*-aconitic and isocitric acid reduce methylene blue more rapidly than does citric acid itself.[61]

Martius thus did not test his theory by identifying and measuring the rates of the postulated reactions in surviving tissues. His approach was an extension of the way in which Thunberg had worked two decades earlier: to combine the criterion that the substances in question reduce methylene blue anaerobically, with theoretical considerations about the manner in which they can be linked into reaction chains (see Vol. I, pp. 18–21). The only product whose enzymatic formation he demonstrated directly was α-ketoglutaric acid, four steps removed from citric acid. The experiments were simple, and he carried them out quickly. He succeeded, where others had failed, in offering a persuasive theory for the breakdown of citric acid, mainly by the careful manner in which he thought the problem through. He drew on his extensive background knowledge of organic reactions, but he recognized that enzymic reactions do not necessarily follow the course predicted by spontaneous chemical processes and gave priority to reactions analogous to known physiological reactions. The "key" to his solution was to see that the shortest route may not be the best route. He had sufficient confidence in his reasoning to incorporate two "extra" anaerobic steps in order to introduce an isomer of citric acid more likely on chemical grounds to undergo the subsequent oxidative steps. He then deduced theoretically a product requiring two such steps, verified experimentally that this product occurred, and had sufficient confidence in his reasoning to conclude that the complex mechanism he had devised was the only plausible pathway that could be imagined to connect its beginning with its end point. His theoretical structure was strong enough so that he required only minimal experimental support to corroborate that it was a highly probable physiological process.

According to his own recollection, Martius did this work without the knowledge of his chief, Franz Knoop. When Martius afterward told him about it, Knoop immediately recognized its significance. They wrote up a short preliminary paper which Knoop, as editor of *Hoppe-Seyler's Zeitschrift*, was in a position to get rapidly into print.[62]

Martius and Knoop's paper entitled "The Physiological Decomposition of Citric Acid" began:

> Inasmuch as work has obviously been done repeatedly on the decomposition of citric acid, we would like to announce briefly that we believe that we have in the Tübingen Institute discovered the basic reaction mechanism.

They discussed first the chemical arguments against a mechanism producing acetonedicarboxylic acid and concluded that it could not be supported "through physiological analogies." "If one wishes to base [the decomposition] on known oxidation mechanisms," they went on,

> it is therefore better to regard citric acid as an acetate-substituted malic acid. Malic acid is readily converted to fumaric acid, the latter to oxaloacetic acid. These are familiar biological processes, the enzymes for which have moreover often been investigated. If similar reactions occur in this situation there will arise, by way of the unsaturated aconitic acid a β-carboxylated α-ketoglutaric acid. [Theodor] Wagner-Jauregg has already identified the presumed intermediate product isocitric acid, and we have found that it disappears more rapidly than citric acid itself does. The ketonic acid formed is both a β-ketonic acid and a substituted oxaloacetic acid that, according to Wieland, easily splits off the carboxyl group standing in a β-position with respect to a carbonyl group. Thereby arises α-ketoglutaric acid, which is then easily broken down through succinic, fumaric, and oxaloacetic acid to pyruvic acid and is thereby connected to the oxidation products of the carbohydrates and several amino acids. Moreover, succinic and pyruvic acids are often found as natural decomposition products, and their origins can in this way be explained.

Briefly mentioning the experimental confirmation of this theory by the isolation of α-ketoglutaric acid as described above, they asserted that:

> The mechanism developed above gives an explanation for the formation of this substance. An alternative derivation for the acid

isolated can scarcely be found. But it gives the derivation developed here firm support, in fact, it seems to us to prove it.

. . .

We have repeatedly emphasized the great significance of a physiological reversibility of biochemical reactions. Therefore it appears to us an interesting fact, that as intermediate decomposition products both of the acids appear that can according to our previous publication condense so readily again to citric acid: oxaloacetic and pyruvic acid.[63]

This final paragraph was due mainly to Knoop, who had argued for many years that the metabolic breakdown reactions must be reversible in order to account for the synthesis of biochemical compounds. Martius himself was somewhat skeptical of this idea. He was more interested in the possibility that citric acid "plays an important role as an intermediate in the synthesis of protein." For him, α-ketoglutaric acid lay on a pathway connecting citric acid with glutamic acid, one of the building blocks of proteins.[64]

Martius and Knoop's two-page paper was "submitted" to the editor of *Hoppe-Seyler's Zeitschrift* on February 12, 1937, and rushed into the issue that appeared on February 24. The numbering of its pages, "I–II," suggests that it may have been tacked onto an issue that was already nearly in press. Sometime during the third week in April this issue reached the Sheffield University library, where Hans Krebs ran across it during one of his routine checks of the latest literature in his field.[65]

The impact that this brief preliminary announcement made on Krebs is easy to imagine. For several years he had "hovered" over the question of how citric acid may be decomposed in animal tissues, without coming upon anything promising. During the previous fall he had satisfied himself that he had evidence for a dismutation reaction through which citric acid is formed, and at this very time Johnson was performing further experiments to solidify that evidence. Krebs had presented his views on the formation of citric acid prominently in his Biochemical Society lecture in October, where he had described it as a "side reaction" and emphasized that he did not "know anything about the products of oxidation of citric acid" (see p. 349). In the newest issue of *Hoppe-Seyler's Zeitschrift* he now had before him what might well turn out to be the solution to that long-standing mystery.

Martius's reaction mechanism did far more than to fill a void concerning the products of the oxidation. Krebs must have perceived at once that he had been mistaken to believe that the oxidation was a "side reaction only." The new mechanism reconnected the products with the main path through which the C_4 dicarboxylic acids—long the center of attention in schemes of oxidative metabolism—were supposed to be decomposed. The final paragraph of the paper even suggested that the process returned to its starting point by reconstituting citric acid. Krebs may also have noticed that the classification of this sequence of reactions as an example of "physiological reversibility" was inappropriate. A reversible reaction was understood to be one in which the process followed the same pathway in both directions. The suggested return from α-ketoglutaric to citric acid lay, however, along a different route from the proposed mechanism for the formation of α-ketoglutaric acid from citric acid. A process that returned in such a way to its starting point was defined as a cycle.

Either immediately, or after a very short period of subsequent reflection, Krebs made up his mind to test Martius and Knoop's mechanism through his own metabolic methods. The initial questions that he probably posed to himself were, Do the reactions occur in tissues other than liver, and are the rates sufficient to regard this as a "preferential" metabolic path rather than a mere sideline?[66]

Undemonstrative though he was, Krebs must have returned from the library with more than a trace of excitement. He did not, however, tell Johnson what he had in mind.[67]

On April 26, Johnson measured the effect of 0.02 M citrate on the respiration of minced heart muscle. The oxygen uptake in 30 minutes increased from 1280 to 1440 μl. This experiment did no more than fit what had been known since the early experiments of Battelli and Stern and of Thunberg. The next experiment that Johnson performed, on April 28, was more significant. It was a measurement of the "formation of succinate from citrate aerobically in presence of malonate" in finely minced pigeon breast muscle. A more direct test of Martius's mechanism would have been an experiment on the formation of α-ketoglutaric acid from citrate. From his early work on deamination Krebs had available the arsenite method for blocking the further decomposition of α-ketonic acids (see Vol. I, pp. 342–343), and he had since 1933 used the dinitrophenylhydrazone method for determining α-ketoglutaric acid. For succinate, however, he had both a more convenient manometric method and Quastel's malonate method for inhibiting the oxidation of the succinate formed. In recent months he had employed these methods frequently and examined the methods themselves extensively. Assuming, therefore, the validity of Martius's conclusion that the α-ketoglutaric acid formed from citrate would be further broken down to succinic acid, he began with the succinate test in place of the more direct one. The results were striking:

Substrate added (μl)	Succinate found
None	20.5
0.02 M citrate	64.5
0.1 M malonate	54.5
0.1 M malonate + 0.02 citrate	387.0
0.033 M malonate	77.0 and 69.5
0.033 M malonate + 0.02 citrate	488.0 and 336.0[68]

This must have been one of the most dramatic single experiments ever performed in Krebs's laboratory. Citrate not only increased the quantity of succinic acid that accumulated, but the effect was large enough to be unmistakable. Although much still remained to be done to work out and demonstrate the overall process in detail, Krebs must from this point on have had little doubt about its reality. With the connection between citric and succinic acid made, all the work he had already done on pyruvic acid, the dicarboxylic acids, and the formation of citric acid would have begun to fit into the outlines of a coherent picture.

* * *

No documents directly record that Krebs saw such a picture at this point, nor did he ever recall a discrete point of recognition during our extensive conversations on this subject. I have nevertheless a strong sense of certainty about it, a certainty based on my own personal experience upon reaching this juncture in the narrative. Although I have been gathering information and thinking about this matter for more than 10 years, I have not previously tried to identify precisely how or when Krebs came to perceive this cyclic pattern, because I was persuaded that that could not be done until I had reconstructed all of the investigative events preceding it. As I wrote the above paragraphs I came to a point at which it became clear and obvious to me that the outlines of a cycle had emerged. I cannot see how it could not have been equally clear and obvious to Krebs, nor does anything in the contemporary record suggest that it could have been clear to him before this point. That does not mean that he necessarily experienced an instantaneous "flash of insight," only that the general picture must have come into focus within a short period bounded at the beginning by his encounter with Martius and Knoop's paper and at the end by the above experiment. One might object that hindsight provides me with a picture that was not available to Krebs on April 28, 1937. I cannot dispel such a claim, only reply that it feels compelling to me that it was the vivid experience of reconstructing the prior events, rather than my background knowledge of subsequent events, that led me to recognize this as the time during which Krebs must have recognized the crucial pattern.

* * *

On April 29 Johnson repeated the experiment on the formation of succinic acid from citric acid, using one-third the concentration of citrate employed in the previous one, and still obtained a substantial effect. On May 1 he tested four concentrations, (0.04, 0.02, 0.01, and 0.0013 M) and observed positive effects for all but the smallest of these.[69] During this time Krebs continued his investigation of fumarate in *B. coli*. Not until May 4 did he himself perform an experiment on "citric acid in muscle (pigeon breast)." The success of Johnson's preliminary experiments now made it important to examine in greater detail the formation of both α-ketoglutaric and succinic acid. The situation was complicated by the fact that Hans Weil-Malherbe had recently shown that the presence of α-ketoglutarate or malonate could interfere with the succinoxidase test for succinic acid. Krebs drew upon Weil-Malherbe's methods to cope with these problems. In the experiment, Krebs first tested the effect on the formation of succinic acid of citrate added anaerobically. In this case it would not be necessary to use malonate, because succinic acid would not be oxidized under anaerobic conditions. Martius had, of course, described the second stage of the reaction sequence leading from citric to succinic acid as oxidative. Perhaps it was Krebs's prior preoccupation with anaerobic oxidations that led him to check whether in this case too the oxidations might be brought about by dismutations. The results, however, were inconclusive.[70]

At the same time Krebs carried out an aerobic experiment designed to test for the formation of both α-ketoglutaric and succinic acid. For succinic acid he used malonate to inhibit further decomposition, and removed it afterward, in separate runs, by two methods devised by Weil-Malherbe. In one of them he extracted succinic and malonic acid in ether and destroyed the malonate with permanganate. In the other he destroyed the malonate by autoclaving. The experiment using the latter method yielded a negative result, but with the former, citric acid increased the quantity of succinic acid found from 78.6 to 322 μl. For α-ketoglutaric acid he used arsenite to prevent further decomposition. After the experiment was over he oxidized the α-ketoglutarate to succinate and determined it with the succinate method. Citric acid increased the yield from 13.5 to 74 μl. Finally, to identify α-ketoglutaric acid more positively, he carried out a large-scale experiment with arsenite and determined it, with the method he had developed in 1933, by precipitating the dinitrophenylhydrazone.[71] In the space of seven days Krebs had attained strong evidence that citric acid did give rise, in accord with Martius's reaction mechanism, to substantial quantities of α-ketoglutaric and succinic acid in pigeon breast muscle.

After carrying out an experiment on the "coupled oxidation of formate" in *B. coli* on Thursday, May 6, Krebs departed for a long weekend to attend a meeting of the Physiological Society in Cambridge. He had two main reasons, other than the meeting itself, to go. One was to support Hermann Blaschko, who felt that David Keilin did not have a high regard for his work and might give a negative evaluation of the Ph.D. thesis that he was completing. Krebs was certain that Blaschko was mistaken, but wished to talk the matter over with Keilin. For his own research interests he wanted to discuss bacterial metabolism with Donald Woods and Marjory Stephenson. He had "compiled a great deal of new material," he had written Woods ahead of time, "on which I should be glad to have your and M.S.'s opinion."[72]

* * *

In 1977, when we examined the first experiments recorded in Krebs's laboratory notebook on the formation of α-ketoglutaric and succinic acid from citric acid (I did not then know about Johnson's earlier experiments), I asked him whether at the time he did this initial experiment he was "in outline at least thinking of the cycle?" He replied, "Maybe there was the hypothesis, yes. I

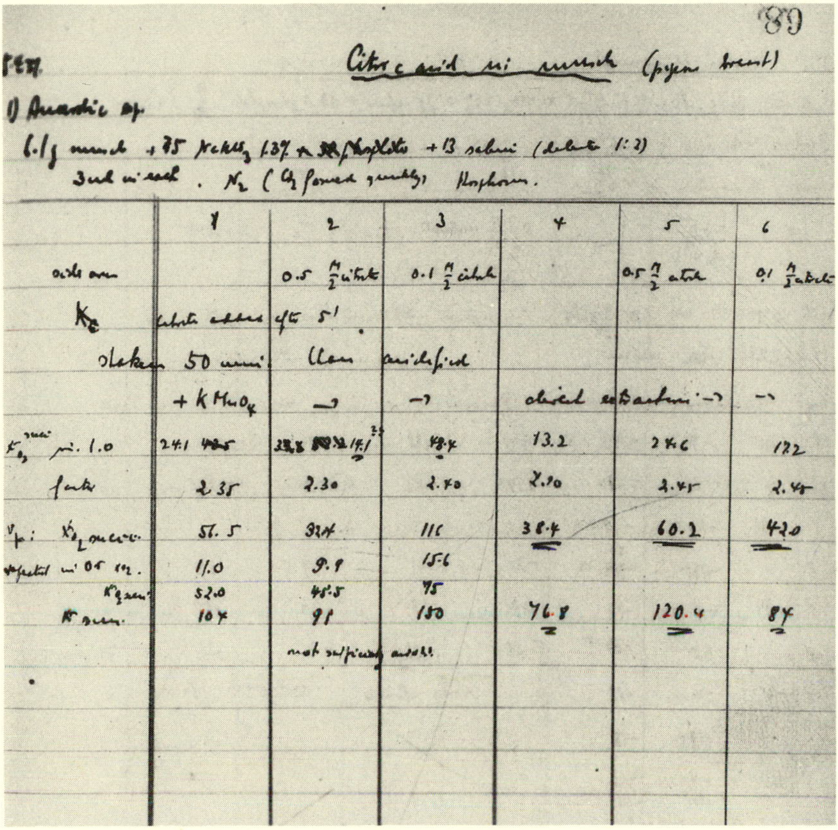

Figure 9-1 Anaerobic and aerobic experiments on citric acid in pigeon breast muscle, performed by Hans Krebs on April 5, 1937.

think it is likely that the subsequent experiments were designed to test whether there was evidence for the cycle."[73] He was not so much remembering that he had had a hypothesis in mind then as inferring from the logic of the situation that he must have had one.

* * *

That he had formed the hypothesis that there was a cycle does not necessarily mean either that Krebs had formulated a tightly connected cycle of reactions or that he perceived the hypothesis as so significant that it should become his first order of business. Still deeply immersed in his long investigation of fumaric acid in *B. coli*, and perhaps stimulated by his discussion of bacterial metabolism with Woods and Stephenson, he returned from Cambridge with his highest priority still on that investigation. Perhaps he expected that he could soon bring it to a close and then turn his full attention to the new investigative vista that he had just opened. On his first day back, May 11, he

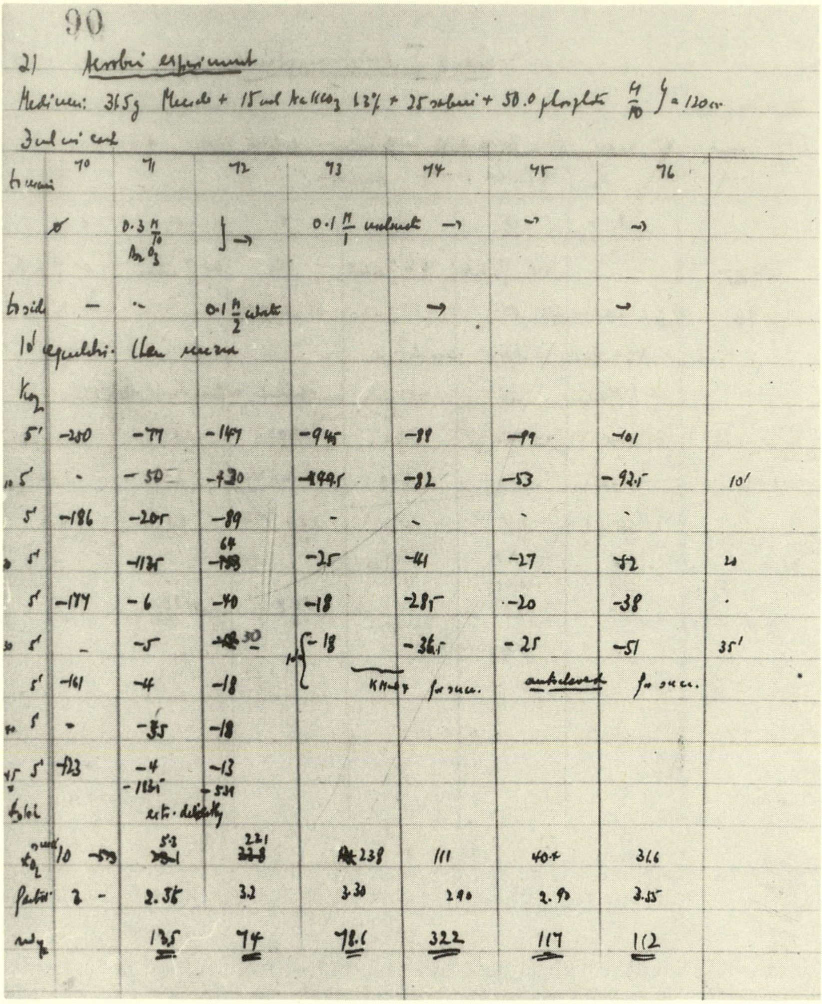

Figure 9-2 Aerobic experiments on citric acid performed on April 5, 1937.

performed three more bacterial experiments and one experiment with pigeon breast muscle on the formation of α-ketoglutaric acid from citric acid.

In the latter experiment Krebs was simultaneously measuring the quantity of α-ketoglutaric acid formed and settling on a procedure for performing such determinations in future experiments. In parallel experiments with citrate added to finely minced pigeon breast muscle in the presence respectively of arsenite and malonate, he added trichloroacetic acid after one hour, filtered, and tested the filtrates with dinitrophenylhydrazone. In the first solution he found a "precipitate," in the second "less precipitate," but there was "no great difference." The method of forming the dinitrophenylhydrazone of α-ketoglutaric acid thus appeared to be, under these experimental conditions, ineffective. He then proceeded to a "determination of succinic." The idea was first to convert α-ketoglutarate to succinate and then estimate succinate as a measure of the α-ketoglutarate. The experiment with malonate, which would yield succinate in the first place, served here as a control. He tried two methods for oxidizing the α-ketoglutarate. First, he autoclaved the filtrate, which he noted "splits ketoglutarate" according to Weil-Malherbe. The subsequent succinic acid determination yielded 171 μl O_2, compared to 99 μl with the malonate control. When he "oxidised with $KMnO_4$" instead, the succinate determinations yielded respectively 324 and 280 μl O_2. A "direct" determination with no prior oxidation gave " ~ 0" O_2, but when "repeated" with $NaHSO_3$ gave 16.5 μl, representing succinic acid present at the end of the experiment. "Thus in As_2O_2," he put down, "16.5 μl O_2 = succinic; 324 μl O_2 = ketoglutaric."[74] The experiment served both to confirm that citric acid gives rise to large quantities of α-ketoglutaric and to persuade him that the method of oxidizing the ketoglutarate with permanganate and estimating the resultant succinate was the most suitable method to measure quantitatively the formation of α-ketoglutaric acid.

During the next two weeks Krebs experimented entirely on *B. coli*. None of Johnson's experiments between May 11 and 18 are recorded. On Wednesday, May 19, Johnson performed an experiment on the effects of malonate on the anaerobic reduction of fumaric acid to succinic acid in pigeon breast muscle. There was little effect on the rate, which was very low already in the control. He repeated the experiment on the 21st and obtained this time a clear-cut partial inhibition ($Q_{succinic\ acid}$ decreased from 4.35 to 2.35). These experiments showed that malonate blocked the succinic-fumaric acid reaction in either direction. Johnson had, however, performed the same type of experiment during the previous fall, and it is not evident how their repetition here was connected to Krebs's new point of view. A second type of experiment that Johnson performed during the same period was, on the other hand, clearly derived from Martius and Knoop's paper. As described above, Martius had tested his theory that citric acid gives rise to *cis*-aconitic acid somewhat indirectly, by showing experimentally that *cis*-aconitic acid can be converted to citric acid in the liver enzyme preparation. He did not mention this experiment in the preliminary paper, so it must have been coincidental that Johnson too tested for the formation of citric acid from *cis*-aconitic acid in minced pigeon muscle. Johnson prepared *cis*-aconitic acid by boiling citric acid for three hours with 50 percent sulfuric acid. He then added *cis*-aconitic acid anaerobically to the muscle suspension and after 80 minutes determined the citrate. The rate of formation was substantial ($Q_{citric\ acid}$ = 3.15, compared to 0 in the control). On the 21st Johnson added *cis*-aconitic acid to an extract formed by grinding the muscle tissue in quartz sand, and found that a "large amount" of citric acid formed.[75] These experiments led Krebs to think that *cis*-aconitic acid was an intermediate not in the decomposition of citric acid, as Martius had stated, but in its synthesis.[76] This difference of view reflects the underlying difference in the scientific background and approach of these two men. Martius reasoned in terms of reaction mechanisms. *Cis*-aconitic acid fit between citric and isocitric acid on the path leading to the oxidative decomposition steps, because that sequence provided a rational chemical process. Less influenced by such arguments, Krebs was more inclined to accept the changes observed in intact isolated tissue as the best guide to the order of reactions in a normal metabolic pathway.

Not long after he had seen it himself, Krebs told Johnson to read Martius and Knoop's paper, and Johnson consequently understood himself in these experiments to be gathering evidence about whether the postulated reactions occur in muscle tissue and at a quantitatively significant rate. One day a visitor to the laboratory asked Krebs what they were working on, and, as Johnson later related it, "Krebs just sketched out on a piece of paper what his ideas were, and this came around in a beautiful circle." The circle was probably not perfectly drawn, and may have looked like this:

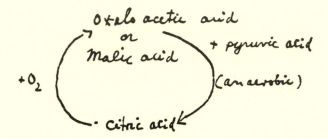

The visitor was impressed. It was the first time that Johnson saw what Krebs had in mind.[77]

Johnson's experience was not unique among students of Krebs. Although he sometimes gave lucid explanations of the objectives of the experiments he asked them to perform, Krebs often left them in the dark about where he was heading.

Krebs's sketch as I have reconstructed it assumes that citric acid is synthesized in the way he had already described in his Biochemical Society lecture in the fall, by means of an anaerobic dismutation of pyruvic acid with either malic or oxaloacetic acid. As we have seen, Johnson was working further on these reactions just prior to the appearance of Martius and Knoop's paper. That Krebs was now attempting to integrate his older application of the dismutation approach to the formation of citric acid with his new view of a cycle incorporating citric acid is supported by the fact that Johnson repeated these experiments again on May 24 and 27, measuring the rates of formation of citric acid in the presence of pyruvate, pyruvate + malate, and pyruvate + oxaloacetate.[78]

On May 27 Krebs himself finally undertook two more experiments clearly connected with the metabolism of citric acid in animal tissues. On that day he tested the effects of malonate, citrate, and their combination on the rate of respiration of rat kidney slices. He probably chose this tissue to extend the range of the investigation because it was one of the few in which the capacity of citric acid itself to raise the rate of respiration had been demonstrated. In the experiment malonate alone inhibited the respiration strongly (Q_{O_2} in control =

17.2; with malonate, 4.70). Malonate with citrate inhibited only slightly (Q_{O_2} with citrate = 23.2; with citrate + malonate, 13.8). In the light of the theory he had then in mind, he may have construed this result to mean that the oxidative reactions inserted between citric acid and succinic acid could still continue in the presence of the agent that blocked the overall respiratory process. Krebs did not pursue further experiments on this tissue, probably because he felt that at this stage of his investigation it was preferable to stay with the tissue already employed most extensively for experiments of the type it involved.[79]

On the same day, using the familiar finely minced pigeon muscle suspension, Krebs tested the effects of malonate on the respiration in the presence respectively of citrate, α-ketoglutarate, and fumarate. In each case malonate substantially reduced the rate. When Krebs looked over this experiment in 1977, he commented, "I don't know that I could have drawn any conclusions from this."[80] That retrospective opinion appears to be an accurate assessment. Nor is it clear what question he had in mind in performing it. After this set of experiments he again concentrated entirely on *B. coli* (except for one experiment on a different type of bacteria) for two more weeks. Between May 27 and June 3 the only experiment by Johnson that is recorded was another one on the formation of citric from *cis*-aconitic acid.

Given his intense and by now long-standing desire to establish the paths of the oxidative breakdown of carbohydrates, and given the conviction he must by now have acquired that the cycle that he was developing promised to solve that problem, it is puzzling that Krebs put so little of his own experimental effort at this time into that investigation. Prone as he was to shift easily from one problem to another when an opening occurred, it is difficult to see why he would continue to give his priority to another investigation from which he must no longer have expected much more than to gather further supporting data for conclusions already reached. The most plausible surmise is that although he had established enough about the characteristics of the cycle to see it clearly in outline, and although he had enough evidence to suggest that it constituted an important metabolic pathway, he did not, for the moment, see how to proceed to clinch his case.

On June 4 Johnson measured the formation of citric acid anaerobically in minced pigeon muscle in the presence of oxaloacetic acid alone, and obtained both an absolute quantity (459 μl in 40 minutes) and a rate ($Q_{citrate}$ = 4.6) distinctly higher than those attained in all of the earlier experiments with combinations of oxaloacetate or malate with pyruvate.[81] That result, together possibly with the results of similar unrecorded experiments, may well have induced Krebs to doubt the validity of the reactions he had previously formulated for the formation of citric acid. It was obvious that oxaloacetic acid must condense with some other molecule in order to form the larger citric acid molecule, but no longer so evident that the other molecule was pyruvic acid. At the same time, the experiment confirmed one of the basic requirements for the validity of the cycle itself, that citric acid is synthesized from one of the C_4 dicarboxylic acids at a rapid rate.

On the same day Johnson performed experiments on citric acid added to pigeon muscle in the presence of arsenite and of malonate.   Instead of determining respectively the α-ketoglutarate and succinate expected to form, he measured the disappearance of the citric acid. The results were in the expected sense. Malonate raised the rate of disappearance from Q = -3.5 and -7.4 to Q = -9.8 and -10.3. Arsenite raised the same rates from Q = -7.4 to values ranging from -10.9 to -16.9. These results could be interpreted as showing that the inhibitors "check the synthesis of citric acid but leave the breakdown unaffected." At the same time he tested the effect of arsenite on the anaerobic formation of citric from oxaloacetic acid. The quantity of citric acid found decreased from 375 to 49 μl.[82] This result too could be interpreted as checking the synthesis, not by blocking the reaction of α-ketoglutaric acid as in the previous case, but by blocking the reaction of oxaloacetic acid, itself an α-ketonic acid.

On June 7 Johnson performed two experiments—one on the anaerobic formation of citric acid from oxaloacetic acid and one on the effect of citrate on the respiration of pigeon muscle—that were similar to previous ones. Another experiment he performed on the same day, however, included a significant additional feature. He employed a relatively small quantity of citrate (0.15 μl of 0.02 M solution) and compared the "extra" O_2 absorbed in the experiment with the amount required to oxidize completely the added citrate. The former exceeded the latter by ratios of 2.95 and 2.64:1.[83] That result signified that citric acid was not merely being oxidized, but was exerting a catalytic effect on other oxidations. Krebs interpreted the result to mean that the citric acid oxidized was being regenerated in a cyclic process, just as Szent-Györgyi had interpreted the catalytic effect of fumaric acid.

In the next two days Johnson carried out two similar experiments in which he added citrate alone and in combination with glycogen, glycerophosphate, hexose diphosphate, and glucose. The overall increases in oxygen consumption were smaller than in the preceding experiment, but in each case the quantities were larger in the presence of the second substance than with citrate alone.[84] The design of these experiments indicates that Krebs had in mind that the catalytic effect of citric acid is exerted somehow on the oxidation of substrates involved in carbohydrate metabolism.

The order and precise timing of developments in the investigation of the role of citric acid that Krebs pursued with Johnson's assistance between late April and early June must remain more tentative than in the accounts of earlier episodes in Krebs's research path, because this account relies in part on the incomplete record of Johnson's experiments. I have treated as the first of their type several experiments that may instead represent refined versions of prior experiments that he left out of his thesis. The overall time period in question is so compressed, however, that I believe a full record would produce only small perturbations in the pattern of events as I have depicted them.

* * *

When in reviewing the experiments recorded in Krebs's laboratory notebook we had reached the end of the first week in June, I asked whether he remembered "a very sharp point in which it seemed that obviously the case was strong enough." He answered, "Sharp not in the sense of a day . . . one day or another, but the evidence gradually took . . . convincing shape. In this sort of work it is very unusual that one experiment settles matters once and for all."[85] The foregoing reconstruction corroborates his recollection. With a few exceptions, each of the experimental results on the subject that he and Johnson obtained fit consistently into the cyclic picture he was constructing, but no single experiment decisively proved that the reactions in question were connected in the manner in which he placed them. Moreover he had not attempted to fulfill the stringent requirement that each of the postulated intermediates react at a rate at least equal to the rate of a common measure for the overall process.

* * *

By June 7 or 8, if not sooner, the evidence had taken convincing shape for Krebs, and he was once again impatient to publish his conclusions as quickly as possible. He planned another short paper for *Nature*. Although he may, in his customary way, have begun drafting pieces of a paper earlier in his investigation, he could not have completed it until after Johnson carried out the above experiments on the catalytic effect of citric acid. By June 10—only two days later if the above experimental chronology is valid—the paper was complete and ready to send off.

The rapidity with which Krebs completed his latest contribution to *Nature* is all the more impressive, considering that on the day before he typed out the finished version, he had gone to Cambridge for a mid-week sherry party at the Roughtons. Alice Roughton was famous in Cambridge for elegant parties, and Krebs commented in 1977 about this trip, "One didn't mind in those days to travel 120 miles to go to an evening sherry party. Stay overnight and come back first thing in the morning."[86] In this case, however, his motives were not only to enjoy the sociability of the evening and to see friends. He used the occasion also to talk again with David Keilin about the concerns of Hermann Blaschko, of whose potential as a scientist he had a high opinion, despite Blaschko's diffidence about himself. Afterward he was able to encourage his pessimistic friend with the news that "Keilin has written a splendid referees report" on his thesis, and then "he spoke of the 'great calamity' that a man of your capability should be dependent upon small annual grants." Krebs also found time during what for him must have been an exceptionally busy period to read Blaschko's thesis for himself. He thought very well of it.[87]

The paper that Krebs sent to *Nature* on June 10 was entitled "The Role of Citric Acid in Intermediate Metabolism in Animal Tissues." Krebs had again achieved a masterpiece of succinct, lucid presentation:

> Citric acid is readily oxidised in animal tissues. One gramme of
> pigeon breast muscle (dry weight) for instance can oxidise about 100
> mg. citric acid per hour. Martius and Knoop working with "citrico-

dehydrogenase" from liver discovered recently that the oxidation of citric acid yields α-ketoglutaric acid. We have been able to confirm Martius and Knoop's results with other tissues and with O_2 as oxidising agent. We find no significant quantities of α-ketoglutaric acid if citric acid is oxidised in unpoisoned tissue, CO_2 being the chief end-product, but if specific inhibitors such as malonate (M/30), or arsenite (M/100) are present, the oxidation of citric acid is incomplete; large quantities of α-ketoglutaric acid and also small amounts of succinic acid appear.

Citric acid, however is not only oxidised in the tissues, it can also be formed at a rapid rate. It is synthesized from oxaloacetic acid and an unknown constituent of the tissue, probably a triose. The synthesis of citric acid does not require molecular oxygen and can therefore best be demonstrated if the tissue is kept anaerobically and if oxaloacetic acid is added. Under these conditions pigeon breast muscle forms up to 100 mg. citric acid per gramme per hour. Since oxaloacetic acid arises in the course of the breakdown of citric acid by further oxidation of α-ketoglutaric and succinic acids, the role of citric acid can be described by the following cycle:

In this cycle "triose" reacts with oxaloacetic acid to form citric acid and in the further course of the cycle oxaloacetic acid is regenerated. The net effect of the cycle is the complete oxidation of "triose."

The conversion of citric into oxaloacetic acid passes through the following intermediate stages:

COOH·CH₂·C(OH)·CH₂·COOH COOH	citric acid
↓	↓
COOH·CH₂·CH·CH(OH)·COOH COOH	iso-citric acid (Wagner Jauregg and Rauen)[1]
↓	↓
COOH·CH₂·CH·CO·COOH COOH	oxalosuccinic acid (Martius and Knoop)[2]
↓	↓
COOH·CH₂·CH₂·CO·COOH	α-ketoglutaric acid (Martius and Knoop)[3]
↓	↓
COOH·CH₂·CH₂·COOH	succinic acid
↓↑	↓↑
COOH·CH:CH·COOH	fumaric acid
↓↑	↓↑
COOH·CH₂·CH(OH)·COOH	l-malic acid (Green)[4]
↓↑	↓↑
COOH·CH₂·CO·COOH	oxaloacetic acid

The intermediate stages in the synthesis of citric acid are still obscure. A probable intermediate and immediate precursor of citric acid is *cis*-aconitic acid. This acid yields rapidly citric acid in muscle, liver or testis. About 12 mg. citric acid can be formed from *cis*-aconitic acid by muscle per gramme per hour. The reaction goes practically to completion.

The cycle described suggests that citric acid, or any other intermediate in the cycle promotes catalytically the oxidation of "triose." This catalytic effect can be demonstrated directly in muscle tissues. Szent-Györgyi and Stare and Baumann have already shown the catalytic action of fumarate, oxaloacetate, and succinate which is now explained. We find a catalytic effect of the same order with citric acid. The effect is more pronounced if glycogen, or hexosephosphate or α-glycerophosphate are added to the tissues. It is on the grounds of these effects that the unknown substance which condenses with oxaloacetate is termed "triose." It must be left open whether the "triose" reacts as such, or as phosphate, or in the form of another derivative [he added afterward in longhand "such as pyruvic or acetic acid"]

The 'citric acid cycle' outlines a pathway through which carbohydrate may be oxidised in animal tissues. That this way is quantitatively significant, and probably the preferential way follows from the rapid rates of formation and decomposition of citric acid in the tissues.

H.A. Krebs; W.A. Johnson[88]
Department of Pharmacology
University of Sheffield
June 10th, 1937

Krebs organized this paper as an argument rather than as a report of an investigation progressing in time. He presented most of the evidence as generalized phenomena rather than as the results of specific experiments. He even stated "findings" in the present tense, as though these were not "found" at particular times in the recent past. Only in the phrase "We have been able to confirm Martius and Knoop's results" did the historical dimension of the investigation surface in his prose. Nevertheless, except for one inversion and some omissions, the order of reasoning in the paper generally recapitulated the order of the investigation. The first paragraph described experiments confirming Martius and Knoop's results that also had formed the first phase of the investigation. The second paragraph treated the formation of citric acid, a topic which they had also taken up next experimentally. The paper naturally omitted that phase of the investigation in which Krebs probably thought that oxaloacetic or malic acid reacts with pyruvic acid to form citric acid. The experiments showing that *cis*-aconitic acid gives rise to citric acid, which Johnson performed during this stage of the work, Krebs deferred to the next section of the paper dealing with further intermediates of the cycle. The catalytic effect of citric acid came last in the paper, just as the experiments on that effect came late in the investigation. These parallels between the sequence of the reasoning in his paper and the sequence of the research, in spite of the fact that Krebs did not write in narrative form, suggest that either he wrote the paper during the course of the investigation or, as I think more likely, he wrote it so quickly following the point at which the evidence became "convincing" that he had had little time to reflect on and rearrange the order of reasoning that had guided him through the research.

One feature of this paper that does not fit this pattern is the presentation first of a "skeleton" cycle reduced to citric and oxaloacetic acid, followed by a consideration of the intermediate steps; for, as we have seen, the starting point of the investigation was the mechanism of Martius that included most of those intermediate steps. We may ask not only why Krebs chose this order, but why he presented the drastically simplified version at all. His procedure can, I believe, be explained on two levels. On a logical level, he wished to emphasize the basic cyclical form of the pathway and the functional significance of the cycle as the means to bring about "the complete oxidation of 'triose'." From this point of view the further intermediate steps appeared to be subordinate details. At a psychological level we can conjecture that the simplified cycle represented his own principal experimental and conceptual contribution. The detailed sequence consisted of the mechanism published by Martius and Knoop connected with the C_4 dicarboxylic acid pathway that had been widely discussed since the 1920s.

From the prominence Krebs gave, both in his diagram and his language, to the basic cyclic form of the pathway he presented, we may infer that the critical event in his investigative progression was a conceptual shift from the dismutation pattern with which he had sought to solve the problem of oxidative carbohydrate breakdown to the cyclic pattern with which he succeeded. This change appears to have the character of one of those well-known "Gestalt shifts" that are frequently invoked as crucial to creative scientific discovery. Clearly Krebs did make such a shift in his perspective, but the two Gestalt forms were not mutually exclusive. Although he did not discuss the matter in this compressed text, the double arrows he placed in the reactions connecting succinic, fumaric, malic, and oxaloacetic acid signified more than that he regarded these reactions as reversible. He viewed the synthesis of citric acid from oxaloacetic acid as an anaerobic oxidation requiring a corresponding reduction. The reductive equivalent he took to be the oxaloacetic-fumaric-succinic system of Szent-Györgyi. The new cyclic pattern therefore retained, for Krebs, a subordinate place for the dismutations that he had formerly expected to constitute the dominant pattern. Another echo of his older view in the new was the division of the simplified cycle into an anaerobic and an aerobic phase. This feature can be seen as a transformation of his earlier position that anaerobic oxidations make up a large portion of the "oxidative" phase of intermediate carbohydrate metabolism.

When Krebs stated that the " 'citric acid cycle' outlines a pathway through which carbohydrate may be oxidised in animal tissues," he was announcing the solution to a problem with which he had contended without success in Freiburg and Cambridge in 1933; that he had pondered often during the following years; that he had set forth to conquer in the spring of 1936 in his "great work"; and that he had described in his lecture, "The Biological Breakdown of Carbohydrates," as "of . . . fundamental interest for those who wish to study the nature of life." Several times he had thought he was approaching a solution which afterward eluded him. How did he feel when he attained a goal at which he had aimed for so long?

* * *

When I asked Krebs in 1977 whether there had been a dramatic moment when he felt excited about having solved this problem, he replied:

> I certainly was pleased, in the same way as I was earlier with the ornithine cycle . . . that I considered a major contribution. But that's really all I can say. Yes, it has something to do with my upbringing that I neither show emotion nor much more really felt so excited about it that I had to give it an outward expression. My inner satisfaction was quite sufficient.[89]

At the time Krebs said this I was not fully convinced. It seemed to me that time might have dimmed feelings more intense than satisfaction, or that his memory was influenced by a conviction that science should not be emotional. Now that I understand him somewhat better, I believe that his description was entirely in character with his personality. He undoubtedly did not shout, or tell everyone around him that he had just made a great discovery. Still, if we had seen him then, I feel certain that we would have noticed some subtler outward expressions of an excitement consistent with his reticence. His voice would have been more animated than usual, his step quickened, and those who knew him well would have sensed that he was, indeed, very pleased.

* * *

Another sign that Krebs felt more than satisfaction about the "citric acid cycle" was his rush to publish. This was not the first time. Whenever he believed that he had found something important, he was impatient to make it public. It is also clear that he was unchastened by the fact that some of the views he had recently made public had not held up. Did he have solid reasons to be confident that this idea would be more enduring? Even if we assume that Johnson performed substantially more experiments on the subject than are recorded, it had been a very short investigation, lasting about six weeks from the initial idea to submission of the paper. The demonstration of the key features of the cycle rested on relatively few experiments. For the majority of the intermediate steps he relied on the work of others, and he had not established that any of the intermediates were metabolized at a rate equal to the overall oxidative process. Yet he asserted that he had found the "preferential" pathway, and he added no qualifications about alternative paths, as he had in the Biochemical Society lecture in which he had earlier presented a different scheme to fill the same function. Both his confidence and his urgent desire to publish quickly probably derived from the same sense he had had about the ornithine cycle five years earlier: that the pieces fit together so coherently to form an overall picture, that the picture could not be wrong.

V

Just as Krebs arrived at this peak in his research activity, he was once again confronted with the difficult choice of where his professional future would lie. On the day after he sent his article to *Nature*, he traveled to London to meet Idelson and discuss the responses to the letter he had written in March outlining requirements for a physiological chemistry laboratory in the cancer institute in Jerusalem. During the conversation Idelson made specific proposals for an appointment for Krebs as director of the new department. Krebs left promising to think them over and give his decision soon.[90]

While Krebs was in London, Johnson extended the demonstration of the formation of citric acid from oxaloacetic acid to rat testis, liver, kidney, and diaphragm tissue. On his return Krebs concentrated his own experiments for the first time almost entirely on the task of gathering further evidence for the citric acid cycle. On Saturday the 12th he pursued further the "formation of succinate and α-ketoglutarate from citrate." His immediate objective now was to improve his methods for the quantitative determination of α-ketoglutarate. Operating on a large scale, he carried out the usual experiment on minced muscle with malonate and with arsenite. Determining first the succinate already present in the medium at the end of the experiment, he next tried two different agents—$KMnO_4$ and ceric sulfate, to oxidize the ketoglutarate for subsequent estimation as succinate. The permanganate gave a good result, but the ceric sulfate method appeared "not suitable, since citrate reacts."[91]

On Monday the 14th Krebs carried out two bacterial experiments, but he returned on Tuesday to the analytical methods for the ketoglutarate-succinate determinations. This time he worked out satisfactory extractive procedures for separating ketoglutarate and succinate. On the same day he tested the oxidation of α-ketoglutarate, hydroxyglutarate, and oxaloacetate in the presence of malonate. α-Ketoglutarate was oxidized, but there was "only very slow oxidation of α-hydroxyglutarate." Krebs had tested the latter apparently on the possibility that it was formed in a "side reaction," but he was too preoccupied now with the main route to be diverted very far by such possibilities. On the 16th he performed another large-scale experiment, using pigeon breast muscle and arsenite as usual, on "citric → ketoglutaric." He determined α-ketoglutarate quantitatively as the dinitrophenylhydrazone, determined the remaining citrate as the pentabromoacetone, and calculated that 85 percent of the citric acid that disappeared was found as α-ketoglutaric acid.[92] This was the strongest evidence yet that Martius's reaction took place in muscle at a rapid rate.

By the time he carried out this last experiment, Krebs had received an unexpected setback from *Nature*, in the form of the following note[93]:

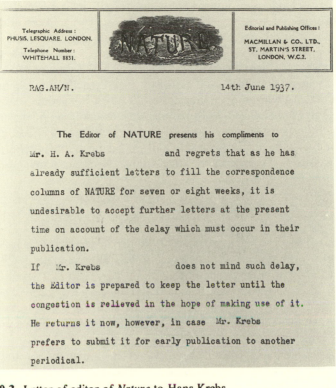

Figure 9-3 Letter of editor of *Nature* to Hans Krebs.

This was obviously a form letter that must have been sent out also to other authors submitting short papers at that time. The rapidity of the response suggests that a minimum of editorial judgment was involved. Krebs was, however, apparently shocked. As he stated still with some warmth 35 years later, "This was the first time in my career, after having published more than fifty papers, that I experienced a rejection or semi-rejection."[94] Not being prepared to wait seven or eight weeks, he treated the notice as a rejection and decided to try next a new journal published in Holland, called *Enzymologia*. This journal was less prestigious than the *Biochemical Journal* in which he normally published, but as a member of its editorial board he knew that *Enzymologia* would publish very quickly, without referees, and that the editor would not "tamper with" the article as the editor of the *Biochemical Journal*, Arthur Harden, sometimes had with his previous papers. This decision entailed writing a longer article on the citric acid cycle or completing a manuscript of that character that he may already have begun drafting during the course of the research.[95]

As Krebs worked to finish this paper, he and Johnson continued to conduct experiments intended to supply further data to incorporate into it. Johnson mainly repeated experiments on the formation of citric acid from oxaloacetic acid and from *cis*-aconitic acid and on the catalytic effect of citric acid on respiration.[96] On June 21, however, Krebs devised a new way to test his theory. He examined in minced muscle the reaction "oxaloacetic → succinic, aerobic, anaerobic + malonate." He must have reasoned out in advance the expected results. Oxaloacetic acid ought to yield succinic acid by two different routes; by oxidation through the citric acid pathway and by reduction according to the scheme of Szent-Györgyi. Malonate should block the reductive path, because it inhibited the fumaric acid ⇄ succinic acid reaction in either direction. If, therefore, the experiment were carried out aerobically, oxaloacetic acid should give succinic acid even in the presence of malonate, but anaerobically it should not. The experiment verified these predictions. Krebs commented on the result:

> Oxaloacetic + malonic → much succinic formed!! in O_2
> inhibition in N_2

He carried out two similar experiments with the same results on June 23 and 24. "This was a very crucial experiment," Krebs explained in 1977, because it provided an alternative explanation for the phenomenon that Szent-Györgyi had interpreted as an "overreduction" of oxaloacetic acid. Oxaloacetic acid *did* form succinic acid without passing through fumaric acid. It did so, however, not by leaping across that intermediate step, but by taking the long way around now provided by the citric acid cycle.[97]

By this time Krebs must have been putting the finishing touches on the article intended for *Enzymologia*. On June 26 he sent it by registered mail to the publisher in the Hague. The latter acknowledged its receipt on June 29, and added, "After having been examined by the editor your manuscript will at once be sent to the printer."[98] Bearing the same title as the elegant *Nature* article that was now destined never to appear, the *Enzymologia* version began:

> During the last decade much progress has been made in the analysis of the anaerobic fermentation of carbohydrate, but very little is so far known about the intermediate stages of the oxidative breakdown of carbohydrate. A number of reactions are known in which derivatives of carbohydrate take part and which are probably steps in the breakdown of carbohydrate; we know furthermore, from the work of Szent-Györgyi that succinic acid, fumaric acid and oxaloacetic acid play some role in the oxidation of carbohydrate, but the details of this role are obscure.
>
> In the present paper experiments are reported which throw new light on the problem of the intermediate stages of oxidation of carbohydrate; in conjunction with the work of Szent-Györgyi, Stare and Baumann and Martius and Knoop, the new experiments allow us to outline the principal steps of the oxidation of sugar in animal tissues.[99]

A comparison of these paragraphs with the first two paragraphs of his lecture to the Biochemical Society eight months earlier (see p. 345) shows that Krebs must have written his new introduction with the manuscript of his older lecture in front of him. The parallels are self-evident. Originally he had framed the introduction to his lecture that way in part because he was emulating the analysis of the anaerobic breakdown of carbohydrates in his study of the oxidative breakdown. The situation was now very different, as the aerobic pattern had turned out to be less like the anaerobic pattern than he had anticipated. The introductory contrast between knowledge of the anaerobic and aerobic breakdown was thereby reduced to a superfluous point of reference, more a residue of his earlier text than a starting point for his present argument. Despite their similarities of form and content, the two introductions also display a sharp contrast in tone. The tentative approach reflecting the unresolved problems he faced in October 1936 was transformed in the rewriting into the understated assurance of one who has now reached the goal to which he had then only aspired.

The body of the *Enzymologia* paper included expanded discussions of each of the points made in the *Nature* manuscript and tables of data for an experiment representative of each of the types that he or Johnson had performed. Krebs reorganized his discussion, however, in a manner that distanced the order of discourse from the original order of the investigation. The new order moved logically from what he now treated as the most fundamental experimental observation, the "catalytic effect of citrate on respiration," through a sequence of inferences verified by further experiments. The catalytic effect actually observed late in the investigation now became the primary step in his argument:

> Since citric acid reacts catalytically in the tissue it is probable that it is removed by a primary reaction but regenerated by a subsequent reaction. In the balance sheet no citrate disappears and no intermediate products accumulate. The first object of the study of intermediates is therefore to find conditions under which citrate disappears in the balance sheet.[100]

There followed a description of experiments employing arsenite and malonate in which citric acid added to pigeon muscle disappeared. The next logical step was to inquire what intermediate products of the oxidation accumulated under these circumstances. The next two sections, entitled "Conversion of Citric Acid into α-Ketoglutaric Acid" and "Conversion of Citric Acid into Succinic Acid," summarized the experiments which answered that question. Thus the experiments that had actually begun the investigation by confirming Martius's reaction were now transformed into the third step in the development of an argument. The fourth step began:

> The new results of the citric acid breakdown, in conjunction with previous work on the oxidation of succinic acid in tissues may be summarised by the following series:
> citric acid → α-ketoglutaric acid → succinic acid → fumaric acid →

l-malic acid → oxaloacetic acid → pyruvic acid.

If it is true that the oxidation of citric acid is a stage in the catalytic action of citric acid then it follows that citric acid must be regenerated eventually from one of the products of oxidation. We are thus led to examine whether citric acid can be resynthesised from any of the intermediates of the citric acid breakdown.[101]

The paper then described an experiment showing that citric acid can be synthesized anaerobically from oxaloacetic acid. That citric acid can be formed anaerobically in a reaction involving oxaloacetic acid was something Krebs had known long before he began this investigation. Here it appears as a question he was only "led to examine" as a consequence of all the preceding experiments.

If we were to view scientific papers as literal reports of completed investigations, then Krebs's *Enzymologia* paper would stand as a prime example of Peter Medawar's dictum that scientific papers misrepresent what scientists actually do.[102] Viewed as part of the creative process itself, however, the same paper is a classic example of the way in which the process of writing transforms an open-ended research trail into a tightly structured investigation.

The remainder of the paper dealt mainly with the conclusions Krebs drew from the preceding experimental sections concerning the "role of citric acid in the intermediate metabolism." From his *Nature* article he retained the dual presentation of the citric acid cycle first as a simplified outline scheme:

$$+ O_2 \left[\begin{array}{c} \longrightarrow \text{Oxalo-acetic acid} \\ \longleftarrow \text{citric acid} \end{array} \right] + \text{"triose"}$$

followed by a consideration of "further intermediate stages."[103]

Although the *Enzymologia* paper was in most respects a reorganized elaboration of the views and evidence sketched out in the *Nature* article, it included some significant additions, clarifications, and qualifications of the positions taken in the shorter paper. "The only hypothetical point in the [simplified] scheme," he pointed out, "is the term 'triose,' though we may consider it as certain that the substance condensing with oxaloacetic acid is related to carbohydrate." On the other hand, "Many details [of the fuller scheme] must necessarily be left open at the present time."[104] One of the details that he now left open concerned *cis*-aconitic acid. Where he had asserted that it is a "probable . . . immediate precursor of citric acid," he now acknowledged:

Martius and Knoop assume that the reaction *cis*-aconitic ⇄ citric acid is reversible and believe that it plays a role in the breakdown of citric acid. It cannot yet be said, however, whether the reaction is an intermediate step in the breakdown or in the synthesis of citric acid.[105]

Krebs now made explicit that the reversible steps shown in his diagram of the detailed scheme referred to the problem of connecting the anaerobic oxidation and reduction reactions:

> The outstanding problem in this connection is the question of the oxidative equivalent of the reduction. At least a partial answer may be given. The synthesis of citric acid as shown . . . takes place anaerobically, although it is an oxidative process. A reductive process equivalent to the oxidation must therefore occur at the same time. The reduction of oxaloacetic acid to succinic acid is the only reduction of sufficient magnitude . . . known so far to occur simultaneously with the citric acid synthesis and we assume therefore it is the equivalent for the synthesis of citric acid.[106]

This paragraph gives clear expression to the way in which Krebs's year-long preoccupation with anaerobic dismutation reactions became embedded into his initial conception of the citric acid cycle.

Krebs supported this view with one of the experiments on the effect of malonate on the aerobic and anaerobic conversion of oxaloacetic into succinic acid that he had first performed only five days before he mailed off the paper. These experiments may well therefore have been designed to fill a "gap" that he perceived only in the process of writing the paper.

At the end of his paper Krebs made a cryptic reference to "the work of Szent-Györgyi" and his conclusion that "respiration, in muscle, is oxidation of triose by oxaloacetic acid."

> In the light of our new experiments it becomes clear that Szent-Györgyi's view contained a correct conception, though the manner in which oxaloacetic acid reacts is somewhat different from what Szent-Györgyi visualised. The experimental results of Szent-Györgyi can be well explained by the citric acid cycle; we do not intend, however, to discuss this in full in our paper.[107]

As we have seen, Krebs had qualified his acceptance of Szent-Györgyi's views even before discovering the citric acid cycle. Did he still believe, as he had written Szent-Györgyi in December, that "your fundamental ideas are correct"? It is not clear whether he now had simply an explanation for the points on which he had already differed or whether his disagreement was now more basic. At any rate, he evidently did not want to go into a question that might divert attention from the main purpose of his paper.

The most important addition Krebs made to his argument in the *Enzymologia* paper was a strong case for the "quantitative significance of the citric acid cycle."

> The quantitative significance of the cycle depends on the rate of the slowest partial step, that is for our experimental conditions the synthesis of citric acid from oxaloacetic acid . . . . Since the oxidation

of triose requires 3 molecules O_2, the rate of citric acid synthesis should be one third of the rate of O_2 consumption if carbohydrate is oxidised through the citric acid cycle. We find for our conditions:

Rate of respiration $(Q_{O_2}) = -20$

Rate of citric acid synthesis $(Q_{citrate}) = +5.8$

Noting that the observed rate was "a little under the expected figure," he thought the difference could be attributed to the conditions of the experiments. Under physiological conditions the rate might be higher. "But even the observed minimum figures of the rate of the synthesis justify the assumption that the citric acid cycle is the chief pathway of the oxidation of carbohydrate in pigeon muscle."[108]

With this paragraph Krebs had essentially met the rigorous standard for demonstrating a metabolic pathway: to show that the rate of each step is at least equal to that of the overall reaction. That is probably why he was confident enough to use now the stronger phrase "chief pathway" in place of the adjective "preferential" that he had used in the *Nature* article. At the time he wrote the above statement Krebs probably had barely enough experimental data to support his claim. The measurements of the formation of citric acid from oxaloacetic acid recorded in Johnson's thesis were:[109]

Date: June	4	7	11	17	21	22
$Q_{citric\ acid}$:	4.6	6.8	3.2	4.84	2.1	5.8
			1.3		1.8	3.5

At the time he submitted the *Nature* article, therefore, Krebs may have had only a single measurement large enough to meet the requirement, and he did not make the argument in that paper. The above pattern suggests that after this time one of Krebs's priorities was to obtain more convincing data for this crucial point. Up until June 21, however, the results were not encouraging. Then, on the 22nd, Johnson attained the $Q_{citric\ acid}$ value of 5.8 that Krebs used for the *Enzymologia* paper. It may well be that he held up the completion of his paper while he waited for a second result strong enough to sustain his case. If so, that might account for the fact that this very central argument was placed so near to the end of the paper that it appears almost as an afterthought.

VI

Just three days after he sent off his paper on the role of citric acid in intermediate metabolism, Krebs wrote the following letter to Idelson:

I have carefully considered your proposals since I saw you in London, but I regret to say that I have been unable to find my way to accept them.

In planning scientific work, especially in my field, we have to think in long periods of time. I realize that it is beyond your power to give guarantees for the future, and I would not consider legal guarantees as essential if instead I could find sufficient confidence in the whole scheme. There are, however, a number of circumstances that make me doubtful and I cannot persuade myself that it is sound for me to build my future upon your proposals. It is because of this that I must ask you not to consider my name again in your future plans.

I feel I should apologise for my belated reply. You will understand that it was not easy for me to make a decision and to abandon a hope which meant much to me. I should also like to say that I do appreciate your efforts and your goodwill to make the position acceptable to me.[110]

The uncertainty about whether Krebs would cast his lot with the dream of a Jewish homeland, or remain in the England in which he felt already at home, was finally over. We may ask whether Krebs really found this decision as hard to make as his letter indicates, or whether he only found it hard to break off irrevocably a negotiation in which he had been involved for three years. Retrospectively he viewed the whole affair mainly as an alternative he had to keep open in case he would not be permitted to remain permanently in England.[111] The contemporary documents convey a different impression, suggesting that he was strongly attracted and that the Jewish identity forced on him by the actions of Nazi Germany stirred in him a desire to participate in building a Jewish society. That motivation was, however, strongly counteracted by the attachment for England that he so quickly formed. The shortcomings in the conditions offered him in Palestine perhaps eased a decision for him that would otherwise have been exceedingly difficult. I believe, therefore, that the above letter represents the resolution for Krebs of a long inner struggle.

It is probably not entirely coincidental that Krebs reached this decision soon after finishing his first publication on the citric acid cycle. Despite what his letter said, he did not customarily plan his scientific work "in long periods of time." He planned one day at a time, and he had flourished scientifically during a period in which he could not be certain where he would be for more than one year at a time. Now he not only knew that what he had just discovered would solidify his scientific position where he was, but he was aware that the consequences of the citric acid cycle were so broad that working out its implications could occupy him for years to come.[112] In Sheffield he was admirably situated for such a prospect. He had strong financial support from the Rockefeller Foundation, a laboratory very well outfitted for the kind of work he was doing, and ample space to accommodate the young investigators he could anticipate would come to work with him on this problem. These considerations must have rendered the idea of a fresh start under very different circumstances in another part of the world more removed than ever from his purposes and plans.

When Krebs discussed his Palestine offer with Margaret Fieldhouse, he never asked her opinion about it or said whether or not he was seriously considering it. Nevertheless, her presence in his life undoubtedly had become

1937   ssees gegen Hadenke

Figure 9-4 Margaret Fieldhouse in 1937. Photo courtesy of Margaret Krebs.

an important factor weighing against a departure. Their relationship had by this time reached the point at which they were beginning to contemplate marriage. They had come to realize that they shared not only common recreational interests, similar senses of humor, and the pleasure of one another's company, but deeper commitments as well. They shared a frugal style of life that was supportable on Hans's still modest salary. Both were ready to settle down and raise a family. Their differences in religious background posed no personal obstacle, for neither of them held strong religious views. Hans expected Margaret to retain her Catholic faith and was comfortable with the expectation that their children would be raised within it.[113]

After Hans asked Margaret's father for his daughter's hand, Mr. Fieldhouse said to Margaret, "I don't know what you see in him. The chap seems to have no vices." To Margaret's father, who did indulge himself in some of the ordinary vices of life, it was puzzling that anyone could be as free of them as

Hans appeared to him to be. Margaret was amused and answered, "I expect I shall find them before long." She knew already that the relentless manner in which he subordinated other considerations to the pursuit of his science would require her to make many adaptations, but she was secure in her own strength and independence and did not doubt that they would be happy together.[114]

There are few contemporary records of how Hans felt about Margaret, but we can obtain a reflected glimpse of his feelings in a letter from his sister Lise quoting from a letter he wrote a few months later announcing their engagement. Hans declared Margaret to be "the most perfectly womanly being [*das vollkommenste weibliche Wesen*]" that he had known.[115]

10

Reflections

Like the ornithine cycle, the citric acid cycle was not a revolutionary discovery. It conformed to all the accepted methodological rules for establishing intermediate metabolic reactions. It incorporated as essential substrates substances that had long been considered central to cellular respiration, and it relied upon general types of reactions—dehydrogenations, dehydrations and hydrations, decarboxylations, and dismutations—that comprised the standard repertoire invoked in metabolic reaction mechanisms. The more novel intermediates supplied by Martius were still connected by analogy to the previously known ones. Nor was the cyclic pattern novel. Morphologically and functionally it embodied features long familiar to biochemists through the widely discussed Thunberg-Knoop-Wieland scheme and its variants. Compared to Szent-Györgyi's scheme, which challenged some of these assumptions by defining the C_4 dicarboxylic acids as a carrier mechanism, the citric acid cycle appears as a smooth continuation of modes of thought and investigation that had developed over two decades in the field of intermediary metabolism. The grand achievement of Hans Krebs was not to break with these patterns, but to fulfill their promise.

Given these characteristics, it is not surprising that the citric acid cycle enjoyed a generally favorable reception from the time of its publication in 1937. Its full significance did not become obvious, however, as quickly as in the case of the ornithine cycle. There were some criticisms aimed at it from followers of Szent-Györgyi's scheme, and it took several years for Krebs's point of view to prevail over that of his friendly rival. It required many more years for the citric acid cycle—or the tricarboxylic acid cycle as it was long known because of a temporary perturbation during its subsequent history—to emerge as the centerpiece of intermediary metabolism, the final common pathway linking together manifold reaction networks. That complex story must be told elsewhere. Krebs remained central to these developments; but as soon as he published his *Enzymologia* paper, what had originated as the endeavor of an individual investigator became the public property of a scientific community. Its subsequent history must therefore belong to the story of a research specialty area, not to a scientific biography.

I

In response to questions asked him when he had acquired eminence as the discoverer of the "Krebs cycle," Hans Krebs developed a retrospective viewpoint about how it had happened. We can now examine his later statements in the light of the detailed reconstruction embodied in the foregoing narrative. In 1970 he wrote:

> I have often been asked how the work on the tricarboxylic acid cycle arose and developed. Was the concept perhaps due to a sudden inspiration and vision? It was of course nothing of the kind, but a very slow evolutionary process, extending over some five years beginning (as far as I am involved) in 1932.[1]

Krebs then outlined succinctly what he regarded as the key stages in this evolution. In some respects the much longer story told here confirms his view. He did engage himself in the general problem of the oxidative breakdown of carbohydrates as early as 1933, returned to it repeatedly for short intervals in 1934 and 1935, and began in the spring of 1936 a sustained campaign, lasting over a year, that culminated in the publication of the citric acid cycle. Even if we limit ourselves to these segments of his research trail, however, we must qualify their characterization as a "slow evolutionary process." That phrase implies continuous gradual development, whereas over the five years in question Krebs addressed himself to this problem intermittently. There are continuities linking these separated episodes, but also ruptures. During the nine months of 1933 that he included as a stage in this evolution, he worked on a "problem" that linked carbohydrate with fatty acid decomposition. The problem that he pursued through much of 1936 was concentrated on an anaerobic phase of oxidative metabolism that he viewed as the connecting link between the known reactions of glycolysis and existing theories concerning the final oxidative phase. Later that year and into the spring of 1937, he was applying Szent-Györgyi's concept of a fumarate-succinate transport system to bacterial metabolism. When he saw Martius and Knoop's paper in April 1937, he made what was in some respects a fresh start, even though he incorporated into his new conception important elements of his previous investigations. He did then acquire a new inspiration and vision. It may not have been "sudden" in the sense of a dramatic flash of insight, but it emerged within a span of days rather than of years. If we do describe the detailed trace of Krebs's investigative efforts oriented around a family of problems connected in some way to the oxidative breakdown of carbohydrates as a "slow evolution," then we must understand by that phrase something different from the common meaning of "descent with modification."

Krebs's work on these problems did not take place in isolated episodes, however, but embedded within an unbroken trail of research devoted to a wider array of problems in intermediary metabolism. In treating his investigative enterprise as a whole, we arrive at a different perspective on those elements that can be identified retrospectively as phases in the evolution of his concept of the citric acid cycle. A myriad of contingent circumstances influenced his direction at any given time, inducing him to persevere in a given line or to move to another. The "evolution" of his work on the citric acid cycle was not necessarily restricted to his work on problems of oxidative carbohydrate metabolism. Analytical techniques, conceptual analogies, accumulating experience, and a general maturation of his research style contributed in subtle ways to his capacity in 1937 to exploit the constellation of factors that he integrated into the experimental demonstration of the citric acid cycle.

Krebs also gave in 1970 a beguiling explanation for the fact that he, rather than Carl Martius, had made the discovery:

> Retrospectively, one may well ask why Martius did not arrive at the concept of the tricarboxylic acid cycle before me. Why had it not occurred to him that the reactions which he had discovered and studied may be components of the main energy-yielding process in living matter? My guess is that this was a matter of scientific outlook—of "philosophical" attitude. Influenced by his teacher, F. Arndt, Martius regarded himself at that time (so he once told me) as a "theoretical organic chemist," interested in reaction mechanisms. The oxidative degradation of citrate was for him a chemical and not a biological problem. He was therefore satisfied when he had clarified, with great ingenuity, the pathway citrate → cis-aconitate → isocitrate → α-oxoglutarate. He did not concern himself with the question of the physiological role of this pathway. Therefore he did not explore the quantitative aspects of the activity of the enzymes of the pathway or compare them with overall metabolic rates, nor did he measure the occurrence of the enzymes in tissues generally.
>
> My outlook was that of a biologist trying to elucidate chemical events in living cells. I was thus accustomed to correlating chemical reactions in living matter with the activities of the cell as a whole. By putting together pieces of information in jigsaw-puzzle manner, and by attempting to discover missing links, I tried to arrive at a coherent picture of metabolic processes. So my mind was prepared to make use of any piece of information which might have a bearing on the intermediary stages of the combustion of foodstuffs. This difference in outlook was, I believe, an important factor in determining who first stumbled on the concept of the tricarboxylic acid cycle.[2]

The strong appeal that this interpretation held for Krebs is easy to understand. It offered a rational explanation based on systematic differences rather than accidental circumstances. It enabled him to attribute his success modestly to an outlook rather than to some general superior quality of insight. It justified a "philosophical" view in which he believed, and, more subtly perhaps, it implied that he had not been handicapped by his lack of extensive training in organic chemistry. His "guess" about Martius, however, does not withstand close scrutiny. From the title of the Martius and Knoop paper alone, one can see that Martius was concerned with the *physiological* decomposition of citric acid. He studied enzymatic reactions, not merely chemical ones, and rejected a commonly discussed theory because it was based solely on chemical considerations. He was particularly interested in correlating this reaction mechanism with the physiological function of protein synthesis. He did approach the problem from the standpoint of a "theoretical organic chemist," but as a means rather than an end. In 1978 I had an opportunity to ask Carl Martius about Krebs's opinion that Martius looked at the situation as a chemist rather than a biologist. He declared that that was "very wrong." He mentioned the "Szent-Györgyi cycle" and said that he "had had enough biochemistry" to think in biochemical terms. The reason he gave for not completing the cycle himself was that he "had no proof" for the reaction synthesizing citric acid. It was just because the synthesis of citric acid that he himself had achieved in 1936 was a purely chemical one that he rejected it as evidence for the metabolic reaction, and he remained for several years skeptical that Krebs had adequately demonstrated the physiological synthesis.[3] The reason that Martius did not measure metabolic rates probably had less to do with his outlook than with the methods available to him. He did not, and perhaps in Knoop's outmoded laboratory could not, employ manometric methods to study metabolic reactions in isolated tissues.

Not surprisingly, Krebs was more perceptive about his own outlook than about the outlook of Martius. His description of the way in which he attempted to correlate chemical reactions, to put together information from various sources, in order to arrive at a coherent picture, is an accurate portrait. These objectives were, however, not unique to him during the 1930s. Others who worked on similar problems and shared that general approach would include Thunberg, Knoop, Szent-Györgyi, Quastel, Elliott, and Weil-Malherbe. We must look further for an adequate explanation of how Krebs came upon the concept. Unconsciously, I believe, he provided a more cogent clue than in his deliberate explanation. His question, why had it not occurred to Martius that the reactions "may be components of the main energy-yielding process in living matter," resonates with the outlook expressed in his lecture on the biological breakdown of carbohydrates in November 1936: that the energy-yielding process is of fundamental interest in the study of living matter (see pp. 361–362). It was in part because of his own intense interest in that problem that Martius and Knoop's paper opened up for him a new way to study the reactions that turned out to be components of the energy-yielding process.

A more general criticism of Krebs's interpretation is that the question he sought to answer is itself artificial. As he sometimes so aptly put it, scientific problems are like "pebbles on a beach." We cannot isolate single factors to explain why one scientist picks up a particular pebble and another scientist picks up a different one. The burden of the present volume is that the reasons that Krebs "stumbled on the concept" of the citric acid cycle in the late spring of

1937 involve the whole complex of events that comprised and conditioned the long investigative pathway he had followed up until that time.

In his autobiography Krebs wrote about the citric acid cycle:

> I should mention that in arriving at the concept of this cycle I was guided by my discovery five years earlier of the ornithine cycle of urea synthesis . . . . My mind was thus conditioned to watch for this type of reaction sequence in the living world.[4]

At first sight the strong resemblance between the general forms of the two cycles seems to make this recollection unassailable. Not only Krebs himself, but most commentators on his discovery of the citric acid cycle have assumed that it owed much to his earlier discovery. There is, however, no documentary support for this interpretation in anything Krebs wrote prior to the discovery of the citric acid cycle. Moreover, in the papers he published on the citric acid cycle in 1937 and the immediately following years, he nowhere drew attention to parallels between the two cycles. Even when discussing the general characteristics of the catalytic effect shown by citric acid, he did not connect this action to the catalytic effect of ornithine, but to the catalytic effects of fumaric and the other dicarboxylic acids discovered by Szent-Györgyi, by Stare and Baumann, and others.[5]

It is possible that the concept had guided Krebs privately, but that he considered it inappropriate to use such an analogy as a public argument in support of the citric acid cycle. The reconstruction of Krebs's research trail has revealed, however, no evidence that in attacking the problems of oxidative carbohydrate breakdown he followed strategies influenced in any specific way by his discovery of the ornithine cycle. It is striking that the catalytic effect of citric acid was very nearly the last phenomenon he examined in the long route that led him to the citric acid cycle.

In view of the great success of the ornithine cycle, it may seem paradoxical that he did *not* seek similar solutions to other metabolic problems. Thomas Kuhn and others have argued persuasively that to grasp an analogy between an unsolved problem and a problem previously solved is one of the most powerful sources of scientific creativity.[6] As we have seen, Krebs was strongly influenced by such an analogy, but the one that he chose was that between the solved problem of the anaerobic breakdown of carbohydrate and the unsolved problem of the oxidative breakdown. Consequently the dismutation pattern dominated his approach, to the exclusion of the cyclic pattern, until the Martius and Knoop paper transformed his perspective. To ask *why* he approached the problem in the one way rather than the other is as hazardous as to ask why Martius did not discover the citric acid cycle, but I will risk a plausible conjecture. The problem of urea synthesis may not have appeared to Krebs to be closely related to that of oxidative carbohydrate metabolism because they were functionally different processes in different subfields of intermediary metabolism. To extend the pattern of anaerobic carbohydrate breakdown into that of the oxidative phase might have appeared a far more obvious form of conceptual generalization.

An important implication of this situation is that Krebs had during the 1930s no special insights concerning the "cycle" as an underlying metabolic pattern. He was, therefore, not seeking to generalize his solution to the problem of urea synthesis. Only after he had solved a second major problem in the same manner did he begin to perceive that the features common to both solutions might hold some deeper meaning. Not until 1947, in his thoughtful article "Cyclic Processes in Living Matter" (published also in *Enzymologia*), did he discuss such meanings publicly. Then for the first time he juxtaposed the ornithine and tricarboxylic acid cycles as two paradigm examples of a larger class of metabolic cycles. "Metabolic cycles," he wrote, "thus seem to be a feature peculiar to life."[7] It was not some a priori view of the significance of cycles that led Krebs to discover the ornithine cycle and then the citric acid cycle. Rather it was his discovery of the specific cycles that led him afterward to meditate on the meaning of the common patterns they displayed.

One of the strong lessons I learned in my work with Krebs is that the historian should not discount the testimony of his or her subject without compelling reasons. It may appear that in this case I have departed from that precept. As already indicated, I believe there are good reasons for that choice. His later belief that he had been guided by the discovery of the urea cycle is readily explainable. Once he had acquired a view that the *general* pattern of the metabolic cycle was very significant, then it would seem self-evident to him that the pattern he had first seen in the ornithine cycle would have guided him toward the similar pattern he found in the reactions of oxidative carbohydrate metabolism. One of the most difficult things for anyone to reconstruct is one's own mental framework *before* one has acquired an insight that becomes basic to all of one's subsequent thought.

My doubt that Krebs was really searching for further metabolic cycles during the period between his discovery of the ornithine cycle and that of the citric acid cycle is supported by the vagueness of his retrospective statements concerning the influence of the ornithine cycle in his study of carbohydrate metabolism. In addition to the passage quoted above from his autobiography, he had written in 1970, "It was of relevance to the development of the tricarboxylic acid cycle concept that the ornithine cycle of urea formation, which I had discovered in 1932, provided a pattern of metabolic organization." Nowhere, however, did he provide specific indications of the time and circumstances under which this "pattern" directed his investigation during the next five years. When I asked him, in 1977, whether he would have been conscious *before* the discovery of the citric acid cycle that the ornithine cycle might be a type which was general enough to cause him to look for other examples, he answered, "No, I had no reason for assuming that this was a model for general metabolic processes."[8] In this case, in contrast to the hypothesis I posed concerning his reasons for testing ornithine (see Vol. I, pp. 284–287), I feel confident that, if Krebs were able to read my interpretation, he would accept my revision of his own brief historical accounts.

II

In his lecture on the biological breakdown of carbohydrates in 1936, Krebs introduced his discussion of the experimental analysis of carbohydrate breakdown with the following statement:

> The chemical analysis of these reactions has to proceed in two steps. The first is to establish the series of reactions which leads from starting material to end-product. It is obvious that, for example, the oxidation of glucose to carbon dioxide and water, is not <u>one</u> plain reaction, but the result of a number of partial steps. And to find out the intermediate stages of the reaction is our first problem.
>
> The second problem is to establish the nature of the catalysts which bring about the reactions. Obviously the cellular oxidations and fermentations are due to catalysts or enzymes. But before we can study the catalysts we have to know the reactions in detail which they catalyse. The series of reactions has therefore to be studied first.[9]

These paragraphs illuminate clearly the broad program that guided Krebs's investigative enterprise during the 1930s. Whether he took up carbohydrate, fatty acid, or amino acid metabolism, it was the same general "first problem" that he wished to solve. To him this was the logical investigative sequence. From a broader perspective it was only his subjective view of the field. Other biochemists, including his own teacher Otto Warburg, were proceeding effectively in the opposite order, elucidating general catalytic mechanisms before the full series of reactions they catalyzed was known.

If Krebs's viewpoint was a local one, it was, however, his faithful adherence to it that defined his personal investigative niche and endowed his enterprise with its powerful thrust. It was precisely in the arena in which he perceived the "first problems" to solve that he possessed the methods, the experience, and the insight to excel. Krebs's talents, skills, and his factual knowledge within these boundaries were extraordinary. His entry into biochemistry from medical school, through the specialized laboratory of Warburg, with only nominal training in more general methods of chemistry, however, were severe limits on his scope. He had neither the knowledge of mathematics and physics to emulate Warburg, of physical chemistry to emulate the approach of Roughton, nor of organic reaction mechanisms to match Martius. Even his chief, Edward Wayne, noticed that beyond the field of the simpler aliphatic compounds Krebs's knowledge of organic chemistry was sketchy.[10] Had Krebs become convinced that the catalysts of metabolic reactions should be "studied first" instead of the other way around, he would have lost much of his competitive advantage. Fortunately his sense of priorities, his abilities, and the opportunities in his field were remarkably well matched.

Within the overall boundaries of his investigative enterprise, Krebs switched easily, sometimes rapidly, from one specific "first problem" to another. Sometimes he appeared almost too flexible. Some problems he undoubtedly left unsolved when a more persistent approach might have yielded significant discoveries. His preference for continuous action—for testing one idea after another, rather than pausing to work out some ideas in depth as Martius did with the decomposition of citric acid—also limited him. On the whole, however, these aspects of his scientific style were well adapted to the conditions of his chosen problem area. Not only he himself, but the field of intermediary metabolism in general, lacked powerful heuristics able to predict metabolic pathways in advance, or even to select in advance which pathways were discoverable with the currently available methods. One was, as Krebs put it later, "feeling one's way in complete darkness."[11] His lack of deep knowledge of organic reaction mechanisms may even have been in some sense an advantage, as it began to be apparent that enzymic reactions often did not follow the courses predicted by the chemical properties of the molecules involved. He was open minded to try anything that looked plausible on simple paper chemistry. His own experiences illustrate well the capriciousness of the situations that one encountered in the field. Three steps of the ornithine cycle fell into his hands within a period of nine months. The one step he knew to be missing eluded him for four years, until he gave up on it. Hindsight shows that he could not have discovered it with the methods that had proved remarkably successful in identifying the other steps. He was able to establish the synthesis of glutamine from glutamic acid also with facility; but the application of the same methods to extend the sequence led him nowhere. When he was blocked in such a way, he had few conceptual resources available to help him foresee whether a little further effort would suffice or whether there were hidden complexities impossible to penetrate at the time. Meanwhile, there was no shortage of problems to pursue with the means at hand. Under these circumstances, Krebs's hit-or-miss approach and his preference for moving on to another problem when he felt that he was getting bogged down were probably the most effective strategies he could have devised.

To some extent these characteristics of Krebs's scientific style were permanent, but the style itself evolved as he gained experience and extended the repertoire of his methods and knowledge. In the earlier stages of the formative years covered in this volume we have seen him switching more often from one general metabolic problem to another. Later, the periods devoted to a given general problem grew longer, and the frequent shifts occurred among the subproblems within the large problem. By 1936 he had sufficient confidence to persevere in his attack on the problem of oxidative carbohydrate breakdown in the face of complexities that might well have turned him aside a year or two earlier.

"Only by trying lots of things," Krebs commented in 1979, "has one a chance of hitting the right one sometimes."[12] He had in mind mainly what one tries in the laboratory, and the foregoing narrative amply illustrates that this was his practice. As we have seen, Krebs also "tried lots of things" in print, and here too he did not always hit the "right ones." During the period we have followed he accumulated quite a list of published misses. In 1932 he asserted that amino acids are deaminated primarily in the kidneys, a claim that disappeared from his later publications on the subject. In 1936 he presented in a review article the theory that carbamino ornithine completed the urea cycle, a theory that he was forced within a little more than a year to give up. In June 1936, he published a sequence of anaerobic dismutation reactions to account for the oxidative breakdown of carbohydrates and to link these with the formation of ketone bodies. The scheme did not last through the year. During that year he gathered evidence for a broader array of dismutation reactions that he asserted in papers published in 1937 "play a role" in the oxidative breakdown of carbohydrates, fats, and the skeleton of amino acids. By 1940 he regarded these merely as "side reactions."[13] In the spring of 1937 he presented a mechanism for the conversion of pyruvic acid to ketone bodies by way of acetopyruvic acid. The experimental results were correct, but the "metabolism" of acetopyruvic acid turned out to be an artifact due to the fact that enzymes are not entirely specific to their normal substrates. Looking back on that venture, he characterized it in 1977 as a "dead-end piece of work."[14]

Alongside these misses we can assemble an impressive group of hits between 1932 and 1937: the ornithine cycle, the deamination of "nonnatural" amino acids, the identification of two intermediates in the synthesis of uric acid, the synthesis of glutamine, and, climactically, the citric acid cycle. Is the inference to be drawn that if one scatters one's shots widely enough, one is bound to score a few hits? If the hits were only minor successes, perhaps we could say that, but two of these were discoveries of sufficient magnitude to reshape the field, and a third, the synthesis of glutamine, initiated what grew into a substantial research subfield.   Underlying the major successes, the secondary contributions, and the relative failures alike was a driving investigative force, a willingness to make mistakes, and a fertility of ideas that permitted Krebs to abandon mistaken views without fear that he could not replace them with better ones. He appreciated also that his ideas were less likely to be corrected if he kept them to himself for long periods than if he subjected them to open discussion. "I think," he said in 1979, that "it is very difficult to keep to the straight and narrow unless one has help."[15] Easily annoyed though he could be with criticism he took to be unwarranted, he understood in general that scientific progress results from processes of interaction and mutual criticism among colleagues.[16] Self-reliant as he appeared to those around him, he drew heavily for his ideas on suggestions that his contemporaries published, and he himself freely put into print "suggestions" that might or might not solidify into well-confirmed theories.

A central element in Krebs's investigative success was his scientific craftsmanship. That is not to say that he was unusually skillful at performing experimental operations with his own hands. In that respect he was apparently rather ordinary, and shortly after the period covered in this book he gave up carrying out his own experiments at the bench. His craft skill consisted in a deep feeling for the way his apparatus worked, for what it could do and what it could not do, and an ingenious capacity to adapt to his own purposes a steadily expanding list of microanalytical quantitative chemical methods. The latter achievement was particularly impressive because, unlike the manometric skills

Figure 10-1 Hans Krebs in his laboratory, 1939. Photo courtesy of Philip P. Cohen.

he learned in Warburg's laboratory, he began with little training in analytical chemistry and had to learn these techniques on his own as he encountered a need for them. He also possessed a "feeling for the organism."[17] He viewed the slices and minces that he placed in his manometer flasks not as mere "surviving tissue," but as organized, living matter. Speaking about pigeon breast muscle suspensions, he remarked in 1977, "It looks like nothing in particular. But when you put it in the manometer and you see the movements, you see how alive it is." He did not mean that the material itself moved, but that the movements of the fluid in his manometers were signs of life as immediate to him as if he could see his tissues breathing. "It respires," he added, "just like you and I do . . . whether we sleep or are awake, we have to take on protein and oxygen."[18]

III

In his scientific as in his personal life, Krebs operated within strict routines. He entered and left the laboratory at exactly the same time every day. He carried out at least one set of experiments every morning and every afternoon. The format of his laboratory notebooks was uniform day after day, year after year. He neither put in overtime when he was exceptionally busy, nor took time off after a hard push. That he should prefer the same tea cakes for lunch week after week, even during his courtship, was typical. Whenever he was at home on weekends he wore brown corduroy shorts, year after year. Throughout his life he played on the piano the same Beethoven sonatas that he had learned in his youth, and continued to make mistakes in the same places.[19]

Figure 10-2 Hans Krebs in his laboratory, 1939. Photo courtesy of Philip P. Cohen.

A life lived according to routines may appear antithetical to the free spirit of creative imagination. The scientific life of Hans Krebs suggests, however, that daily outward routines can be supportive of a venturesome inner spirit. Krebs was clearly contented with his routines, but they did not appear to him as they did to those who observed him. When I mentioned to him the remarks of people who had worked in his laboratory about how strictly he organized his daily schedule, he commented, "I thought it was rather unorganized." To him it was just a matter of not wasting time.[20] Routines were to him a form of discipline that enabled him to cope with the pressures of his life, to channel a drive that might otherwise scatter itself. It was a discipline that enabled him to press on relentlessly with his work, yet reserve time to read broadly, attend the theater, play music, hike in the Black Forest or the Sheffield moors, and to reflect in solitude.

The discipline of his daily life was also a manifestation of Krebs's deep sense of responsibility and his belief in the importance of being utterly reliable. There may have been an additional underlying motivation of which he was less aware. The steady urgency of his scientific pace gave stability to a life shaped amid conditions that were anything but stable. From the time that he left for medical school following the collapse of the Germany of his childhood until the time he came to Sheffield as an established scientist, he never enjoyed the luxury of settled surrounding circumstances and an assured future. When he left the safe choice of medical practice, he prepared himself for a life in research in the face of his mentor's warning that no career opportunities in biochemistry lay before him. Hardly had he established himself in a supportive environment in

Figure 10-3 Hans Krebs in his laboratory, 1939. Photo courtesy of Philip P. Cohen.

Freiburg in which he could combine clinical medicine with part-time research than his world collapsed again from the outside. In his early years in England he bore not only the uncertainty over whether he could stay there, but anxiety over his family still in Germany. Added to all this he was still, a man in his thirties, earning only a meager income with no additional financial resources. His research was thus not only central to his life, but for the formative years we have followed, it was the greatest source of security in his life.

Perhaps Krebs also worked relentlessly because his durable optimism was not the carefree type of the individual who sees only the cheerful side of things. It was an optimism sustained more by his will than by natural temperament. His motto then and later—"Hope for the best and prepare for the worst"—is, after all, double edged.[21] One does not prepare for the worst unless one anticipates that the worst may come to pass. His generally even-tempered personality harbored also a darker underside. The most depressed point in his life came, much later than our story, when he reached retirement age and it appeared for a time that he would not be able to go on in research.[22] I think it not improbable that the energy he organized so effectively into his unflagging investigative forward movement served also, perhaps without his knowing it, to protect him from despair.

IV

It is sometimes assumed that Hans Krebs not only discovered the first two metabolic cycles that have endured, but that he introduced into biochemistry the general concept of a metabolic cycle. Influenced by such views when I began to study his early scientific career, I looked for some source of a special broad insight, or an intellectual perspective differing in some deep sense from that of his predecessors and contemporaries. As the foregoing narrative shows, nothing like that can be found. Cyclic reaction sequences were widely discussed, though not identified as a generalizable form, at the time he entered the field. Nor did he bring to the field general ideas differing from the mainstream of contemporary thought. Most of his leads for investigative problems came from the current literature.

Krebs's assessment of himself was consistent with this view. In several perceptive comments during a conversation in 1977 he said that "one of my abilities [has been] to make use of available information and to bring it together in a coherent story." That ability "implies . . . first of all, the intellectual capacity to see the connection, but it also implies being aware of what is going on in other places." "I feel . . . that is one of my stronger points—to correlate."[23]

These qualities may appear to fall short of our image of truly creative scientists as those who see beyond the mental frameworks of their contemporaries, who forge new conceptual structures to guide their investigative paths, who embark on lonely introspective intellectual journeys of discovery. Krebs's scientific style seems on the surface devoid of far-ranging forward vision or brilliant mental leaps. There are, however, many forms of creative scientific imagination. However he may have reached it, the discovery that culminated the years of his career that we have followed was one of the great creative discoveries in the history of biochemistry.

V

Although Krebs was the driving force behind his own scientific achievements, he obtained a good deal of help along the way. Warburg supplied him not only with methods but with a model for how scientific investigations should be pursued. The successive chiefs in whose laboratories he afterward worked—Lichtwitz in Altona, Thannhauser in Freiburg, Hopkins in Cambridge, and Wayne in Sheffield—were less influential on him intellectually, but they gave him crucial institutional and moral support. He was, in one way or another, fortunate in what he received from each of these men. It was not to luck alone, however, that he owed this support, for he possessed qualities of both intellect and personality that attracted others to him. He was not only able, but had an integrity that won their trust. He too was trusting, and the mutual trust that he established wherever he went contributed in no small way to his scientific success.

VI

The shaping influence of Krebs's contributions in the 1930s to intermediary metabolism extended beyond the specific reaction sequences that he established. The ornithine cycle and the citric acid cycle, together with the Embden-Meyerhof pathway of anaerobic carbohydrate metabolism, brought much closer to reality the long-held vision of unbroken reaction chains linking foodstuffs with final end products. Although Krebs did not invent the concept of the metabolic pathway, he was by the end of the 1930s emerging as its most prominent architect. His discoveries did more than those of any other single individual to give the previously fragmentary known reaction sequences the appearance of an organized structure. The arrays of structural organic formulas connected by curving arrows that became conspicuous in biochemistry textbooks after the war bear clear traces of his handiwork.

These formulas and arrows are, of course, only diagrams intended to represent a different reality. Wondering whether a form restricted by the dimensions of paper might reflect inadequately the mental conception of those processes that Krebs held, I once asked him how he visualized metabolic reactions. "My image of them," he said, "is essentially that which is on the paper."[24]

The image has proved a powerful one, and the metabolic pathway is fundamental to the modern understanding of life processes. It is also a metaphor, in which temporal processes are represented as spatial distances, a transformation so pervasive in our daily thought that we are often unaware of its character.[25]

The present volume is organized around another metaphor with certain formal similarities to the preceding one; that is, the research pathway. Although there are no arrows, we visualize a process which extends through historical time as a journey through investigative space. Other terms which we may substitute, such as research trail, line of investigation, or course of investigation, have the same property. Like all metaphors, it is suggestive but limiting. The laboratory notebooks on which much of this story is based are well suited to the construction of a research trail, because they provide a linear chronological sequence corresponding to the linear nature of a pathway. A pathway can also branch, but the branches of a research trail cannot be fitted so readily to those of a physical pathway. If we examine the mental events accompanying the linear sequence of operations recorded in the notebooks, we encounter such an intricate embedding of branching and merging problems, parallel and intermittent pursuits, that it no longer makes sense to map them conceptually onto the metaphorical pathway. Without some such metaphorical language, however, it does not seem possible to think about the investigative enterprise of science.

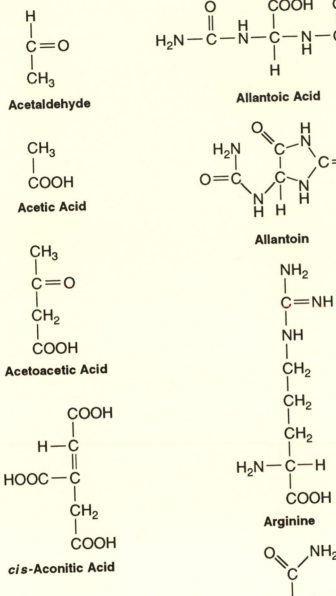

Acetaldehyde

Acetic Acid

Acetoacetic Acid

cis-Aconitic Acid

Alanine

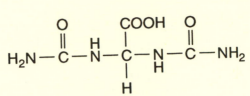

Allantoic Acid

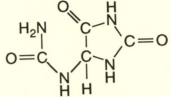

Allantoin

Arginine

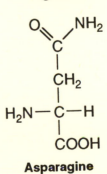

Asparagine

COOH
|
CH₂
|
H₂N—C—H
|
COOH

Aspartic Acid

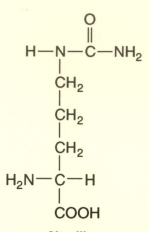

Citrulline

CH₃
|
CH₂
|
CH₂
|
COOH

Butyric Acid

CH₃
|
(CH₂)₄
|
COOH

Caproic Acid

CH₃
|
CH
‖
HC
|
COOH

Crotonic Acid

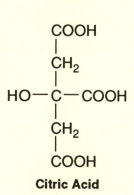

Citric Acid

COOH
|
CH
‖
HC
|
COOH

Fumaric Acid

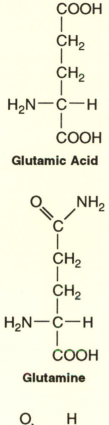

Glutamic Acid

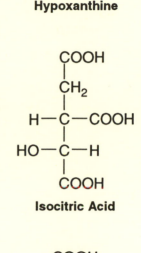

Hypoxanthine

Glutamine

Isocitric Acid

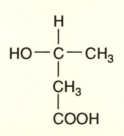

Glyoxylic Acid

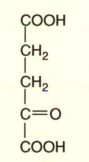

α-Ketoglutaric Acid

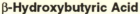

β-Hydroxybutyric Acid

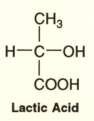

Lactic Acid

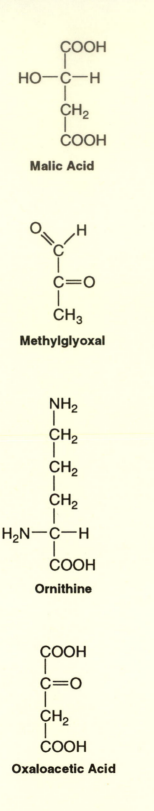

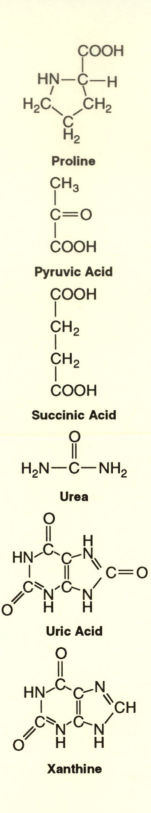

Malic Acid

Proline

Methylglyoxal

Pyruvic Acid

Succinic Acid

Ornithine

Urea

Uric Acid

Oxaloacetic Acid

Xanthine

NOTES

Introduction

1. F.L. Holmes, "The Fine Structure of Scientific Creativity," *History of Science*, 19 (1981): 60-70.
2. F.L. Holmes, *Between Biology and Medicine: The Emergence of Intermediary Metabolism* (Lectures delivered at the International Summer School of the History of Science, Uppsala, July 1990) (Berkeley: Office of History of Science and Technology, 1992).
3. Hans Krebs, with Anne Martin, *Reminiscences and Reflections* (Oxford: Clarendon Press, 1981); Sir Hans Kornberg and D.H. Williamson, "Hans Adolf Krebs, 1900-1981," *Biographical Memoirs of Fellows of the Royal Society*, 30 (1984): 351-385; Frederic L. Holmes, "Krebs, Hans Adolf," in *Dictionary of Scientific Biography*, Suppl. 2, ed. F.L. Holmes, (New York: Charles Scribner's Sons, 1990), 17: 496-506.

Chapter 1

1. Hans Krebs, *Reminiscences and Reflections* (Oxford: Clarendon Press, 1981), p. 83.
2. *Ibid.*, pp. 84-85; Hermann Blaschko-FLH, Sept. 10, 1976 (Tape 6, pp. 6-7).
3. Krebs, *Reminiscences*, p. 84.
4. HAK to F.G. Hopkins, London, July 22, 1933, KC.
5. HAK-FLH, Aug. 21, 1977 (Tape 41, p. 21).
6. Krebs, *Reminiscences*, pp. 84-85; Allan (K.A.C.) Elliott, "An Unorthodox Career," *Bulletin of the Canadian Biochemical Society*, 17 (1980): 12-15.
7. Hans Krebs, autobiographical draft, A. 1096-1098, KC, Chapter 9, p. 5. In the published version the phrase is "without any further thought." Krebs, *Reminiscences*, p. 85; HAK-FLH, Aug. 8, 1978 (Tape 79, p. 19).
8. Krebs, *Reminiscences*, p. 85; Helene Nauheim to HAK, June 29, 1933, KC.
9. HAK-FLH, July 9, 1977 (Tape 19, p. 1), Aug. 11, 1977 (Tape 40, p. 9).
10. Rudolph A. Peters to HAK, Oxford, June 27, 1933, KC, published in Krebs, *Reminiscences*, p. 85.
11. Joseph C. Aub to HAK, June 30, 1933; HAK to J.C. Aub, July 15, 1933, KC.
12. Ella Sachs Plotz Foundation to HAK, June 27, 1933, KC.
13. Robert E. Kohler, "Walter Fletcher, F.G. Hopkins, and the Dunn Institute of Biochemistry: A Case Study in the Patronage of Science," *Isis*, 69 (1978): 331-355; floor plan, "University of Cambridge School of Biochemistry," Frederick Gowland Hopkins Papers, Catalog No. ADD 7620, Cambridge University Library.
14. Ernest Baldwin, "Hopkins, Frederick Gowland," in *Dictionary of Scientific Biography, ed. C.C. Gillispie* (New York: Charles Scribner's Sons, 1970-1980), 6: 498-502; Marjory Stephenson, "Sir F.G. Hopkins' Teaching and Scientific Influence," in *Hopkins and Biochemistry 1861-1947*, ed. Joseph Needham and Ernest Baldwin, (Cambridge: W. Heffer, 1949), pp. 29-38.
15. *Ibid.*; J. Needham, "Opening Address," in "A Symposium on Biochemistry and Nutrition," *Proc. Roy. Soc. Lond.*, Ser. B, 156 (1962): 289-294.
16. Krebs, *Reminiscences*, pp. 92-93.
17. HAK-FLH, Aug. 1, 1977 (Tape 31, p. 20).
18. HAK to Siegfried Thannhauser, July 16, 1933, KC.
19. Vernon Booth-FLH, Aug. 5, 1977 (Tape 37, pp. 1, 12).

20. Krebs, *Reminiscences*, p. 85.

21. Krebs, Notebook "10," pp. 260-261.

22. *Ibid.*, p. 261.

23. *Ibid.*, pp. 262-266.

24. *Ibid.*, pp. 267-270; HAK-FLH, May 5, 1977 (Tape 13, p. 16).

25. *Ibid.*, pp. 271-279; HAK-FLH, May 5, 1977 (Tape 13, p. 17). The analytic methods Krebs utilized are described in Norman Lowther Edson, "Ketogenesis-Antiketogenesis. I. The Influence of Ammonium Chloride on Ketone-Body Formation in Liver," *Biochem. J.*, 29 (1935): 2082-2086.

26. Krebs, Notebook "10," pp. 283-285.

27. *Ibid.*, pp. 288-291.

28. *Beilsteins Handbuch der organischen Chemie*, ed. Bernhard Prager and Paul Jacobson, 4th ed., Vol. 3 (Berlin: Julius Springer, 1921), pp. 560, 562, 789-790; HAK-FLH, May 6, 1977 (Tape 13, pp. 23-24, Tape 15, p. 1).

29. Krebs, Notebook "10," pp. 292-293; *Beilsteins Handbuch*, 3: 560.

30. HAK-FLH, May 5, 1977 (Tape 13, p. 19).

31. HAK-FLH, May 6, 1977 (Tape 15, p. 2).

32. Krebs, Notebook "10," p. 293.

33. *Ibid.*, p. 294.

34. *Beilsteins Handbuch*, 3: 560.

35. Krebs, Notebook "10," p. 296; HAK-FLH, May 6, 1977 (Tape 13, pp. 23-24).

36. Krebs, Notebook "10," pp. 297-298.

37. *Ibid.*, pp. 300-301.

38. *Ibid.*, pp. 300-303.

39. *Ibid.*, p. 304.

40. *Ibid.*, p. 305; HAK-FLH, May 5, 1977 (Tape 13, p. 21).

41. Krebs, Notebook "10," pp. 306-307.

42. *Ibid.*, pp. 308-310.

43. Program, "Third International Congress for Experimental Cytology," (Cambridge: Walter Lewis, 1933); HAK-FLH, Aug. 11, 1977 (Tape 40, pp. 7-8).

44. Krebs, Notebook "10," pp. 312-315.

45. *Ibid.*, p. 317; HAK to Hermann Blaschko, Freiburg, May 31, 1933, KC; O. Meyerhof and D. McEachern, "Über anaerobe Bildung und Schwund von Brenztraubensäure in der Muskulatur," *Biochem. Z.* 260 (1933): 417-445; I. Banga and A. Szent-Györgyi, "Über das Co-Ferment der Milchsäureoxydation," *Z. physiol. Chem.*, 217 (1933): 39-43; I. Banga, K. Laki, and A Szent-Györgyi, "Über die Oxydation der Milchsäure und der β-Oxybuttersäure durch den Herzmuskel," *Ibid.*, pp. 43-51.

46. Krebs, Notebook "10," pp. 317-318.

47. *Ibid.*, pp. 319-320.

48. HAK-FLH, May 6, 1977 (Tape 15, p. 5).

49. Krebs, Notebook "10," pp. 326-333.

50. *Ibid.*, pp. 334-341; Meyerhof and McEachern, "Anaerobe Bildung und Schwund von Brenztraubensäure," p. 444.

51. *Ibid.*, p. 341; Bruno Mendel, Milly Bauch, and Frida Strelitz, "Brenztraubensäure im Stoffwechsel tierischer Zellen," *Klin. Woch.*, 10 (1931): 118-119.

52. H.A. Krebs, "Größe der Atmung und Gärung in lebenden Zellen," in *Handbuch der Biochemie des Menschen und der Tiere*, ed. Carl Oppenheimer, 2nd. ed., Suppl. Vol. 1, pt. 2 (Jena: Gustav Fischer, 1933), pp. 890-891.

53. Otto Rosenthal, "Versuche über die Aktivierung der anaeroben Gärung von Lebergewebe durch Brenztraubensäure, Acetaldehyd und Methylenblau," *Biochem. Z.*, 244 (1932): 133-156.

54. Krebs, Notebook "10," p. 344.

55. HAK-FLH, May 6, 1977 (Tape 15, pp. 4-5).

56. Krebs, Notebook "10," pp. 345-348.

57. *Ibid.*, p. 348.

58. *Ibid.*

59. *Ibid.*, pp. 349-351.

60. *Ibid.*, pp. 352–353.
61. *Ibid.*, pp. 354–356.
62. *Ibid.*, pp. 357–358.
63. *Ibid.*, pp. 359–361.
64. *Ibid.*, pp. 362–363.
65. *Ibid.*, pp. 365–367.
66. *Ibid.*
67. *Ibid.*, p. 365.
68. *Ibid.*, pp. 368–371.
69. *Ibid.*, pp. 372–391.
70. *Ibid.*, p. 388.
71. *Ibid.*, pp. 389, 392–393, 398–399, 407–411.
72. Hans Krebs, "Urea Formation in the Animal Body," *Ergebnisse der Enzymforschung*, 3 (1934): 247–264. Data of the type Krebs gathered in these experiments is presented on pp. 249, 256–259.
73. Krebs, "Urea Formation," p. 261; Krebs, Notebook "10," pp. 413–414.
74. Henry Borsook and Geoffrey Keighley, "The Energy of Urea Synthesis," *Proc. Natl. Acad. Sci.*, 19 (1933): 626–631, 720–725.
75. Krebs, Notebook "10," p. 415.
76. Krebs, "Urea Formation," p. 256.
77. *Ibid.*
78. Borsook and Keighley, "Urea Synthesis," pp. 720–724.
79. Krebs, Notebook "10," pp. 394–395.
80. *Ibid.*, pp. 396–397.
81. *Ibid.*, pp. 401–406.
82. *Ibid.*, p. 412.
83. *Ibid.*, pp. 416–418.
84. *Ibid.*, pp. 419–420.
85. *Ibid.*, pp. 421–424.
86. *Ibid.*, pp. 425–445.
87. HAK–FLH, May 2, 1977 (Tape 12, pp. 18, 21), May 3, 1977 (Tape 12, pp. 24, 27, 31), May 4, 1977 (Tape 13, pp. 4, 6, 8).
88. *Ibid.*, May 2, 1977 (Tape 12, p. 21), May 3, 1977 (Tape 13, p. 8).
89. *Ibid.*, Aug. 3, 1978 (Tape 74, p. 10).
90. HAK–FLH, May 3, 1977 (Tape 12, p. 31).
91. H.A. Krebs, "The History of the Tricarboxylic Acid Cycle," *Perspectives in Biology and Medicine*, 14 (1970): 155.
92. For further discussion of this question, see Chapter 10.
93. Juda Hirsch Quastel, "A Short Autobiography," in *Selected Topics in the History of Biochemistry: Personal Recollections*, ed. G. Semenza (Comprehensive Biochemistry Vol. 35) (Amsterdam: Elsevier, 1983), pp. 137–152; Quastel–FLH, March 29, 1984.
94. Juda Hirsch Quastel and A.H.M. Wheatley, "Biological Oxidations in the Succinic Acid Series," *Biochem. J.*, 25 (1931): 117–128.
95. Juda Hirsch Quastel and A.H.M. Wheatley, "Oxidations by the Brain," *Ibid.*, 26 (1932): 725–744; "The Effects of Amines on Oxidations of the Brain," *Ibid.*, 27 (1933): 1609–1613; "Narcosis and Oxidations of the Brain," *Proc. Roy. Soc. Lond.*, Ser. B, 112 (1932): 60–79.
96. Juda Hirsch Quastel and A.H.M. Wheatley, "Oxidation of Fatty Acids in the Liver," *Biochem. J.*, 27 (1933): 1753–1762.
97. *Ibid.*, p. 1753; Quastel–FLH, March 29, 1984. There are some problems with this account. Quastel's paper gives, as the date of this congress, August 1932. The Third International Congress of Experimental Cytology took place, however, in August 1933. Quastel is not listed on the program. Nevertheless, on balance the identification with the 1933 meeting is most plausible. It is unlikely that Quastel would have waited until September 1933 to submit for publication a paper that he had read more than a year earlier.
98. Quastel and Wheatley, "Fatty Acids," p. 1759.

Chapter 2

1. Hans Krebs, "The Physiological Role of the Ketone Bodies" (The Third Hopkins Memorial Lecture), *Biochem. J.*, 80 (1961): 231–233; *Reminiscences and Reflections* (Oxford: Clarendon Press, 1981), pp. 86–90; Norman W. Pirie–FLH, Aug. 10, 1978 (Tape 83, untranscribed).

2. HAK–FLH, May 6, 1977 (Tape 15, pp. 6–8), July 8, 1977 (Tape 19, pp. 26–27); Pirie–FLH, Aug. 10, 1978 (Tape 83); N.W. Pirie, "Glutathione," in "A Symposium on Biochemistry and Nutrition," *Proc. Roy. Soc. Lond.*, Ser. B, 156 (1962): 306–311.

3. HAK–FLH, May 6, 1977 (Tape 15, p. 6), July 8, 1977 (Tape 19, pp. 23–25); Robert E. Kohler, "Innovation in Normal Science: Bacterial Physiology," *Isis*, 76 (1985): 162–181.

4. Pirie–FLH, Aug. 10, 1978 (Tape 83); HAK–FLH, May 16, 1979 (Tape 96, p. 32), March 20, 1980 (Tape 118, p. 11); Marjory Stephenson to E. Mellanby, Oct. 10, 1935, Medical Research Council archives 2036/2.

5. *Ibid.*; W.E. Van Heyningen–FLH, May 2, 1977 (Tape 16, pp. 7, 15); Antoinette Pirie–FLH, May 3, 1977 (Tape 16, p. 1); HAK–FLH, Dec. 22, 1979 (Tape 112, p. 12); Krebs, *Reminiscences*, p. 89.

6. Krebs, *Reminiscences*, p. 89; Van Heyningen–FLH, May 2, 1977 (Tape 16, p. 4); HAK–FLH, May 15, 1979 (Tape 95, pp. 14–18), Jan. 9, 1981 (Tape 129, p. 14).

7. See Kohler, "Innovation," pp. 162–164, 166.

8. See Robert E. Kohler, *From Medical Chemistry to Biochemistry* (Cambridge: Cambridge University Press, 1982), pp. 81–84.

9. HAK–FLH, May 6, 1977 (Tape 15, p. 6). Wording slightly rearranged.

10. HAK–FLH, Aug. 10, 1977 (Tape 39, p. 14).

11. Hans Weil-Malherbe, "Roots and Broods," unpublished autobiographical memoir, pp. 37–42; Hans Weil-Malherbe–FLH, Sept. 2, 1977 (Tape 44, pp. 6–9).

12. Weil-Malherbe, "Roots," pp. 42–43; Weil-Malherbe–FLH, Dec. 8, 1986 (Tape 156, pp. 2–3).

13. Weil-Malherbe, "Roots," pp. 41, 43; Weil-Malherbe–FLH, Dec. 8, 1986 (Tape 156, p. 2).

14. Weil-Malherbe–FLH, Sept. 2, 1977 (Tape 44, pp. 10–11); Hans Weil to HAK, June 21, 1933, Aug. 17, 1933, J.815, KC.

15. Weil-Malherbe–FLH, Sept. 2, 1977 (Tape 44, p. 11–12). The quotation is Weil-Malherbe's recollection of what Krebs wrote. HAK–FLH, July 8, 1977 (Tape 18, p. 16).

16. Weil-Malherbe–FLH, Sept. 2, 1977 (Tape 44, p. 12); Weil to HAK, Sept. 28, 1933, Oct. 13, 1933, J.815, KC.

17. Weil-Malherbe, "Roots," p. 44; Weil-Malherbe–FLH, Sept. 2, 1977 (Tape 44, p. 12), Dec. 8, 1986 (Tape 156, p. 4).

18. Robert A. Lambert to H.A. Krebs, Paris, Dec. 3, 1933, Rockefeller Foundation Archives, RG1.1 Series 401a.

19. Robert A. Lambert to Joseph C. Aub, Paris, Dec. 16, 1933, *Ibid.*

20. D.P. O'Brien to A.V. Hill, Oct. 17, 1933, *Ibid.*

21. H.A. Krebs to R.A. Lambert, Cambridge, Dec. 15, 1933, *Ibid.*; HAK–FLH, May 16, 1979 (Tape 96, p. 15).

22. Werner Schuler, "Die Harnsäuresynthese im Vogelorganismus," *Klin. Woch.*, 12 (1933): 736–738; HAK–FLH, Aug. 10, 1978 (Tape 81, pp. 8–9).

23. Th. Benzinger and H.A. Krebs, "Über die Harnsäuresynthese im Vogelorganismus," *Klin. Woch.*, 12 (1933): 1206; HAK–FLH, Aug. 10, 1978 (Tape 81, pp. 8–9).

24. Werner Schuler and Wilhelm Reindel, "Die Harnsäuresynthese im Vogelorganismus," *Z. physiol. Chem.*, 221 (1933): 212.

25. S. Thannhauser to HAK, Freiburg, Dec. 21, 1933, J.716, KC.

26. HAK to S. Thannhauser, Cambridge, Dec. 30, 1933, *Ibid.*

27. Summarized in St. J. Przylecki, "La dégradation de l'acide urique chez les vertébrés," *Archives internationales de physiologie*, 24 (1924–25): 238–240, 237.

28. F. Battelli and L. Stern, "Untersuchungen über die Urikase in den Tiergeweben," *Biochem. Z.*, 19 (1909): 219–253.

29. Przylecki, "L'acide urique," pp. 238–263, 317–355.

30. Richard Fosse, *L'Urée* (Paris: Presses Universitaires, 1928), pp. 183–198.

31. R. Fosse and A. Brunel, "Un nouveau ferment," *C. r. Acad. Sci.*, 188 (1929): 426–428; "Sur le ferment producteur d'acide allantoïque par hydration de l'allantoïne. Sa présence dans le règne animal," *Ibid.*, pp. 1067–1069; R. Fosse, A. Brunel, and P. De Graeve, "Sur l'allantoïnase et l'origine de l'acide allantoïque chez les végétaux," *Ibid.*, 189 (1929): 716–717.

32. K. Felix, Fr. Scheel, and W. Schuler, "Die Urikolyse," *Z. physiol. Chem.*, 180 (1929): 90–106; W. Schuler, "Die Urikolyse," *Ibid.*, 208 (1932): 237–248; "Die Oxydation der Harnsäure in alkalischer Lösung," *Ibid.*, pp. 248–257; W. Schuler and W. Reindel, "Die Urikolyse," *Ibid.*, 215 (1933): 258–266.

33. Hans Kleinmann and Heinz Bork, "Untersuchungen über Urikolyse. III.," *Biochem. Z.*, 262 (1933): 20–40.

34. Krebs, Notebook "10," p. 449.

35. *Ibid.*, pp. 450–454, 458.

36. Franz Bielschowsky to HAK, Oct. 17, 1933, and two undated letters without return address, J.62, KC.

37. S. Thannhauser, Freiburg, undated, J.716, KC; Akten Siegfried Thannhauser, Badisches Ministerium des Kultus und Unterrichts, G.L.A. 235, No. 2584 General landesarchiv, Karlsruhe.

38. H. Blaschko to HAK, 19 letters, dated Oct.-Dec. 1983; one letter from Blaschko's mother to HAK, Berlin, Dec. 21, 1933, J.68, KC.

39. Hans Adolf Krebs, "Urea Formation in the Animal Body," *Ergebnisse der Enzymforschung*, 3 (1934): 247–264; Krebs, *Reminiscences*, p. 88.

40. HAK–FLH, Sept. 7, 1976 (Tape 4, p. 19).

41. Krebs, Notebook "10," pp. 459–491.

42. *Ibid.*, p. 470; H.A. Krebs and H. Weil, "Untersuchungen über die Urikolytischen Fermente Uricase, Allantoinase, Allantoicase," in *Problémes de Biologie et de Médecine* (Moscow, 1935), p. 506 (page proofs for a volume which was never published), KC.

43. Krebs, Notebook "10," pp. 473–474; Krebs and Weil, "Urikolytischen Fermente," p. 502.

44. Krebs, Notebook "10," pp. 483–488; Krebs and Weil, "Urikolytischen Fermente," p. 505.

45. Krebs, Notebook "10," pp. 497–519.

46. University of Cambridge, School of Biochemistry, "Biochemical Tea Club, Lent Term 1934," Frederick Gowland Hopkins Papers, Cambridge University Library.

47. Krebs, Notebook "10," pp. 521–522.

48. *Ibid.*, pp. 525–527.

49. *Ibid.*, pp. 528–529.

50. HAK to Siegfried Thannhauser, March 11, 1934, J.716, KC.

51. H.A. Krebs and H. Weil, "Untersuchungen über die Urikolytischen Fermente (Uricase, Allantoinase, Allantoicase)" typed manuscript, identical to page-proof by same title (see note 42), KC, p. 17.

52. HAK to Thannhauser, March 11, 1934, KC.

53. Theodor Benzinger to HAK, Oct. 12, 1933, Dec. 29, 1933, J.50, KC.

54. The envelope contains a letter from "Frau Baumgartner" to Krebs, describing her chronic ill health. KC.

55. Hermann Blaschko to HAK, Jan. 25, 1934, J.68, KC.

56. HAK to Thannhauser, March 11, 1934, KC.

57. Diary of R.A. Lambert, excerpt from Saturday, March 17, Rockefeller Archives, RG1.1 Series 401a.

58. Krebs, Notebook "10," pp. 530–549.

59. Ernst Frankel to HAK, March 17, 1934, J.217, KC.

60. Joseph C. Aub to HAK, March 19, 1934, KC.

61. J.L. Magnes to HAK, March 23, 1934, A.425, KC.

62. HAK to Joseph Aub, April 6, 1934, KC.

63. HAK to J.L. Magnes, April 3, 1934, A.425, KC.

64. Krebs, Notebook "11," pp. 1–11.

65. *Ibid.*, p. 13; Krebs and Weil, "Urikolytischen Fermente" (typed manuscript), p. 6.

66. Krebs, Notebook "11," p. 16; Krebs and Weil, *Ibid.*, pp. 5–6.

67. A postcard acknowledging receipt of the manuscript in Moscow is dated May 8, 1934.

68. Krebs and Weil, *Ibid.*, p. 6.

69. *Ibid.*, pp. 16–17; Hans Weil-Malherbe has pointed out to me that the figures drawn in the manuscript are in his handwriting. He does not recall, however, having participated in writing the text of the paper. Hans Weil-Malherbe to FLH, August 27, 1986; Weil-Malherbe–FLH, Dec. 8, 1986 (Tape 156, p. 4).

70. HAK–FLH, July 9, 1977 (Tape 19, pp. 11–12).

71. *Ibid.*, p. 12.

72. Krebs, Notebook "11," pp. 20–21.

73. HAK–FLH, July 4, 1977 (Tape 17, p. 16).

74. Krebs and Weil, "Urikolytischen Fermente," p. 20; Hans Weil-Malherbe to FLH, Aug. 27, 1986; Weil-Malherbe–FLH, Dec. 8, 1986 (Tape 156, p. 6). See also Sept. 2, 1977 (Tape 44, p. 13). One year later Keilin published a paper describing the coupling of the oxidation of alcohol to the uricase system: D. Keilin and E.F. Hartree, "Coupled Oxidation of Alcohol," *Proc. Roy. Soc. Lond.* Ser. B, 119 (1936): 141–159.

75. Thannhauser Akten, Hauptstaatsarchiv, Karlsruhe.

76. Thannhauser to HAK, Paris, April 21, 1934, J.716, KC.

77. F. Bielschowsky to HAK, June 12, 1934, J.62, KC. Bielschowsky was somewhat hostile to Thannhauser, claiming in several letters to Krebs that Thannhauser had appropriated some of his own ideas for a lecture Thannhauser gave in Spain. Krebs's opinion of this question has not survived, and there does not appear to be sufficient evidence available to evaluate the validity of Bielschowsky's grievance.

78. Werner F. Kümmel, "Die Ausschaltung rassisch und politisch missliebiger Ärzte," in *Ärzte im Nationalsozialismus*, ed. Fridolf Kudlien (Cologne: Kiepenheuer and Witsch, 1985), pp. 63–76; HAK–FLH, Aug. 12, 1977 (Tape 41, p.4).

79. Georg Krebs to HAK, Hildesheim, April 5, 1934, A.718, KC.

80. Lise Daniel to HAK, March 15, 1934, A.733, KC.

81. Georg Krebs to HAK, Feb. 26, 1934, A.718, KC.

82. Lise Daniel to HAK, March 19, 1934, A.733, KC.

83. Georg Krebs to HAK, April 5, 1934, A.718, KC.

84. *Ibid.*

85. *Ibid.*

86. J.L. Magnes to HAK, April 5, 1934, A.425, KC.

87. HAK to J.L. Magnes, April 14, 1934, *Ibid.*; J.L. Magnes to HAK, April 17, 1934, *Ibid.*

88. HAK to J.L. Magnes, April 29, 1934, *Ibid.*; J.L. Magnes to HAK, May 9, 1934, *Ibid.*

89. Krebs, Notebook "11," pp. 22–45; HAK–FLH, Aug. 1, 1977 (Tape 31, p. 19). On Barcroft's study see especially Joseph Barcroft, "The Mammal Before and After Birth," *The Irish Journal of Medical Science*, July 1935, pp. 289–301.

90. Redcliffe N. Salaman to HAK, May 7, 1934, May 10, 1934, A.427, KC; Barnet Litvinoff, ed., *The Essential Chaim Weizmann: The Man, the Statesman, The Scientist* (New York: Holmes and Meier, 1982), pp. 1–6, 108–110.

91. Hans Krebs, *Reminiscences and Reflections* (Oxford: Clarendon Press, 1981), pp. 90–91.

92. Litvinoff, *Weizmann*, pp. 6, 111–115.

93. David Nachmansohn to HAK, May 17 [1934], J.476, KC.

94. Walter Auerbach to HAK, May 22, 1934, KC.

95. Bruno Mendel to HAK, May 13, 1934, J.485, KC.

96. Georg Krebs to HAK, May 26, 1934, A.718, KC.

97. "Staff Meeting," University School of Biochemistry, Dec. 2, 1933, Jan. 10, 1934, Feb. 1, 1934, Feb. 14, 1934, Archives of the Biochemical Department, University Archives, Cambridge.

98. *Ibid.*, May 23, 1934, May 30, 1934.

99. F.G. Hopkins to Directors of Studies in Natural Sciences, Cambridge, May 31, 1934, *Ibid.*

100. "Part I Biochemistry, Michaelmas Term 1934," (list of lectures with initials of lecturers), *Ibid.*; same for "Lent Term, 1935."

101. Board of Biology B, Meeting of June 4, 1934, Min V 68, University Archives, Cambridge.

102. Krebs, *Reminiscences*, p. 88; Vernon Booth to HAK, June 18, 1934, J.84, KC; Vernon Booth–FLH, Aug. 5, 1977 (Tape 37, pp. 1–2).

103. Lise Daniel to HAK, June 7, 1934, A.733, KC.

104. Krebs, Notebook "11," pp. 47-59.

105. B. Gözsy and A. Szent-Györgyi, "Über den Mechanismus der Hauptatmung des Taubenbrust-muskels," Z. physiol. Chem., 224 (1934): 1-10.

106. Ibid., pp. 3-5, 7-8.

107. Ibid., p. 5; A.V. Szent-Györgyi, "Über den Mechanismus der Succin- und Paraphenylendi-aminoxydation. Ein Beitrag zur Theorie der Zellatmung," Biochem. Z., 150 (1924): 195-210.

108. Krebs, Notebook "11," p. 60.

109. Ibid., p. 61.

110. Ibid., pp. 62-63.

111. Erno Annau, "Über den chemischen Verlauf und die physiologischen Bedingungen der Acetonkörperbildung aus Brenztraubensäure," Z. physiol. Chem., 224 (1934): 141-149.

112. Krebs, Notebook "11," p. 65. Norman Edson, to whom Krebs assigned the continuation of this investigation, attributed its starting point to Annau's paper. See Norman Lowther Edson, "Ketogenesis–Antiketogenesis.I. The Influence of Ammonium Chloride on Ketone-Body Formation in Liver," Biochem. J., 29 (1935): 2082.

113. Krebs, Notebook "11," p. 65.

114. Ibid., pp. 66-68.

115. HAK to L. Halberstaedter, April 29, 1934, A.426, KC.

116. D. Nachmansohn to HAK, June 24, 1934, J.476, KC.

117. D. Nachmansohn to HAK, July 4, 1934, J.476, KC.

118. HAK to J. Magnes, July 1, 1934, A.425, KC.

119. Georg Krebs to HAK, June 18, 1934, June 28, 1934, July 4, 1934, A.718-719; Lise Daniel to HAK, June 19, 1934, June 20, 1934, July 4, 1934, A.733, KC; HAK–FLH, May 18, 1979 (Tape 97, p. 4), Jan. 9, 1981 (Tape 129, p. 11).

120. Krebs, Notebook "11," pp. 69-76.

121. HAK–FLH, July 8, 1977 (Tape 18, pp. 3-6); Hans Weil-Malherbe–FLH, Sept. 2, 1977 (Tape 44, p. 14).

122. Hans Weil-Malherbe–FLH, Sept. 2, 1977 (Tape 44, p. 17).

123. H. Blaschko–FLH, unrecorded conversation, Jan. 10, 1992; HAK to H. Blaschko, April 1980.

124. Allan (K.A.C.) Elliott, "An Unorthodox Career," Bulletin of the Canadian Biochemical Society, 17 (1980): 14-16; Malcolm Dixon and David Keilin, "An Improved Method for the Measurement of Tissue Respiration," Biochem. J., 27 (1933): 86-95.

125. Elliott, "Unorthodox Career," p. 16; Kenneth Allan Caldwell Elliott and Elmer Frederick Schroeder, "The Metabolism of Lactic and Pyruvic Acids in Normal and Tumour Tissue," Biochem. J., 28 (1934): 1920-1939.

126. K.A.C. Elliott to HAK, June 11, 1934, J.175, KC.

127. Ibid., Sept. 28, 1934; Elliott and Schroeder, "Lactic and Pyruvic Acids," p. 1938.

128. Charles Amos Ashford and Eric Gordon Holmes, "Further Observations on the Oxidation of Lactic Acid by Brain Tissue," Biochem. J., 25 (1931): 2028-2049.

129. HAK–FLH, July 9, 1977 (Tape 20, pp. 2-5).

130. Elliott, "Unorthodox Career," p. 15.

131. See above, p. 65.

132. Georg Krebs to HAK, Aug. 22, 1934, A.719; Lise Daniel to HAK, June 7, 1934, A.733, KC.

133. Katherina Holsten to HAK, Jan. 14, 1934, May 5, 1934, A.993, KC.

134. Ibid., June 1, 1934, June 20, 1934. The contents of Krebs's letters to Katrina are inferred from references to them in her letters to him.

135. Ibid., June 28, 1934, July 12, 1934, [postmark] August 6, 1934, August 7, 1934, August 16, 1934.

136. J.L. Magnes to HAK, July 16, 1934, A.425, KC.

137. HAK to J.L. Magnes, Aug. 11, 1934, A.425, KC.

Chapter 3

1. Lise Daniel to HAK, June 7, 1934, Aug. 15, 1934, A.733; E.W. Krebs to HAK, Aug. 24, 1934, A.746, KC.
2. Lise Daniel to HAK, Aug. 15, 1934, *Ibid.*
3. Georg Krebs to HAK, Aug. 22, 1934, A.719, *Ibid.*
4. E.W. Krebs to HAK, Aug. 22, 1934, *Ibid.* This change resulted from his transfer from the design department to the research and development department for electric motors and generators. Wolfgang Krebs to FLH, Nov. 7, 1991.
5. E.W. Krebs–FLH, May 4, 1988 (Tape 160, p. 3).
6. *Ibid.*, p. 1.
7. For this and the following paragraphs, *Ibid.*, May 4, 1988 (Tape 160, pp. 4–5), July 4, 1989 (Tape 164, pp. 21–25), Jan. 5, 1992 (Tape 170, pp. 1–15); E.W. Krebs, "Neue polumschaltbare Wicklung," *AEG-Mitteilungen*, 1934, Heft 12.
8. Krebs, "Wicklung," p. 4 (pagination of reprint).
9. E.W. Krebs–FLH, May 4, 1988 (Tape 160, p. 6), with stylistic modifications later suggested by him.
10. HAK–FLH, Dec. 17, 1979 (Tape 106, p. 6).
11. HAK–FLH, Aug. 12, 1977 (Tape 41, p. 5).
12. Gordon A. Craig, *Germany: 1866–1945* (New York: Oxford University Press, 1980), pp. 632–633; Ian Kershaw, *Popular Opinion and Political Dissent in the Third Reich: Bavaria, 1933–1945* (Oxford: Clarendon Press, 1983), pp. 224–257; Detlev J.K. Peukert, *Inside Nazi Germany: Conformity, Opposition and Racism in Everyday Life* trans. R. Deveson (New Haven: Yale University Press, 1987), p. 58; E.W. Krebs–FLH, Sept. 15, 1986 (Tape 154, p. 22).
13. E.W. Krebs–FLH, Sept. 15, 1986 (Tape 153, pp. 9–11, Tape 154 pp. 14–15).
14. Lise Daniel to HAK, Aug. 23, 1934, A.733, KC.
15. Hans Adolf Krebs, "Untersuchungen über den Stoffwechsel der Aminosäuren im Tierkörper," *Z. physiol. Chem.*, 217 (1933): 216–219.
16. Krebs, Notebook "11," pp. 80–98.
17. Hans Adolf Krebs, "Weitere Untersuchungen über den Abbau der Aminosäuren im Tierkörper", *Z. physiol. Chem.*, 218 (1933): 157–159; Krebs, Notebook "11," p. 101.
18. *Ibid.*, pp. 102–103.
19. *Ibid.*, p. 104.
20. W.F. Geddes and Andrew Hunter, "Observations upon the Enzyme Asparaginase," *J. Biol. Chem.*, 77 (1928): 197–229.
21. Krebs, Notebook "11," pp. 105–108.
22. *Ibid.*, p. 111.
23. *Ibid.*, p. 112.
24. Krebs, Notebook "10," pp. 226, 229–230; Notebook "11," p. 113.
25. Krebs, Notebook "11," p. 115.
26. *Ibid.*, p. 116.
27. *Ibid.*, p. 117.
28. *Ibid.*
29. Hans Adolf Krebs, "The Synthesis of Glutamine from Glutamic Acid and Ammonia, and the Enzymatic Hydrolysis of Glutamine in Animal Tissues," *Biochem. J.*, 29 (1935): 1951.
30. Krebs, Notebook "11," pp. 118–119.
31. *Ibid.*, pp. 120–121.
32. *Ibid.*, p. 123.
33. *Ibid.*, pp. 124–129.
34. *Ibid.*, p. 133.
35. Krebs, "Synthesis of Glutamine," p. 1953.
36. Krebs, Notebook "11," pp. 131–132.
37. HAK to David Nachmansohn, Aug. 24, 1934, J.476, KC.
38. *Ibid.*
39. *Ibid*; HAK–FLH, Aug. 6, 1977 (Tape 39, p. 4).

40. HAK to Nachmansohn, Aug. 24, 1934, J.476, KC.
41. *Ibid.*
42. David Nachmansohn to HAK, Aug. 28, 1934, J.476, KC.
43. *Ibid.*
44. HAK–FLH, Aug. 11, 1977 (Tape 40, p. 17), Aug. 8, 1978 (Tape 79, p. 23); Hans Krebs, *Reminiscences and Reflections* (Oxford: Clarendon Press, 1981), p. 87.
45. Krebs, *Reminiscences*, p. 92.
46. *Ibid.*, p. 88; HAK–FLH, July 9, 1977 (Tape 19, p. 1), Aug. 1, 1977 (Tape 32, p. 1); Antoinette Pirie–FLH, May 3, 1977 (Tape 16, p. 15).
47. Krebs, *Reminiscences*, p. 92.
48. HAK–FLH, May 2, 1977 (Tape 12, p. 10). In a somewhat different version of apparently the same incident, Krebs described the student in 1978 as one whom he had been supposed to supervise. HAK–FLH, March 1, 1978 (Tape 66, p. 18).
49. *Ibid.*, May 2, 1977 (Tape 12 pp. 10–11), May 6, 1977 (Tape 15, p. 7), Aug. 1, 1977 (Tape 31, pp. 17–18).
50. Bruno Mendel to HAK, Aug. 17, 1934, Sept. 28, 1934, J.458, KC.
51. HAK–FLH, Aug. 11, 1977 (Tape 40, p. 17).
52. Katherina Holsten to HAK, Aug. 22, 1934, A.994, KC.
53. *Ibid.*, Sept. 2, 1934.
54. *Ibid.*, Aug. 16, 1934, Sept. 2, 1934.
55. *Ibid.*, Sept. 2, 1934.
56. *Ibid.*, Sept. 15, 1934.
57. *Ibid.*, Aug. 16, 1934.
58. Georg Krebs to HAK, Feb. 26, 1934, April 5, 1934, April 7, 1934, May 26, 1934, A.718; Oct. 20, 1934, A.719, KC.
59. Katherina Holsten to HAK, Sept. 15, 1934, A.994, KC.
60. Krebs, Notebook "11," pp. 134–141.
61. *Ibid.*, pp. 142–143.
62. Geddes and Hunter, "Asparaginase," pp. 227–228.
63. Wolfgang Grassmann and Otto Mayr, "Zur Kenntnis der Hefeasparaginase," *Z. physiol. Chem.*, 214 (1933): 189–190.
64. Krebs, Notebook "11," pp. 143–144.
65. *Ibid.*, pp. 145–146.
66. *Ibid.*, pp. 148–151.
67. *Ibid.*, p. 151.
68. Katherina Holsten to HAK, Oct. 1, 1934, A.995, KC.
69. Albert Charles Chibnall and Roland Gordon Westall, "The Estimation of Glutamine in the Presence of Asparagine," *Biochem. J.*, 26 (1932): 122–132.
70. Krebs, Notebook "11," p. 147; Thierfelder and E. von Cramm, "Über glutaminhaltige Polypeptide und zur Frage ihres Vorkommens im Eiweiß," *Z. physiol. Chem.*, 105 (1919): 58–82.
71. *Ibid.*, pp. 154–155.
72. David Gooding, "How do Scientists Reach Agreement about Novel Observations," *Studies in the History and Philosophy of Science*, 17 (1986): 219.
73. Hans Weil-Malherbe and Hans Adolf Krebs, "The Conversion of Proline into Glutamic Acid in Kidney," *Biochem. J.*, 29 (1935): 2079.
74. Krebs, Notebook "11," pp. 156–157.
75. *Ibid.*, p. 162.
76. Hans Weil to HAK, undated [Sept. 1934], J.815, KC; Hans Weil-Malherbe–FLH, Sept. 2, 1977 (Tape 44, p. 15); Hans Weil-Malherbe, "Roots and Broods," unpublished autobiographical memoir, pp. 4–5.
77. HAK–FLH, July 8, 1977 (Tape 18, p. 19).
78. E.S. London, A.K. Alexandry and S.W. Nedswedski, "Die Frage der Beteiligung des Ornithins, des Citrullins und des Arginins am normalen Prozeß der Angiostomie," *Z. physiol. Chem.*, 227 (1934): 233. The issue of the journal containing this article was printed on Sept. 18.
79. *Ibid.*, pp. 233–241.

80. *Ibid.*, p. 241.

81. *Ibid.*, pp. 235–236.

82. Siegfried Thannhauser to HAK (undated), J.716, KC.

83. H.A. Krebs, "Bemerkungen zu einer Arbeit von E.S. London, A.K. Alexandry and S.W. Nedswedski: 'Über die Harnstoffbildung in der Leber'," *Z. physiol. Chem.*, 230 (1934): 278–279.

84. HAK to E.S. London, Oct. 1, 1934, J.876, KC.

85. *ibid.*

86. Franz Knoop to HAK, October 9, 1934, KC.

87. E.S. London, "Antwort auf obige Bemerkungen von A. Krebs," *Z. physiol. Chem.*, 230 (1934): 279.

88. HAK to E.S. London, Oct. 23, 1934, J.876, KC.

89. Frederic L. Holmes, *Claude Bernard and Animal Chemistry* (Cambridge, Mass.: Harvard University Press, 1974), pp. 1–2, 445–455.

90. Hans Adolf Krebs and Kurt Henseleit, "Untersuchungen über die Harnstoffbildung im Tierkörper," *Z. physiol. Chem.*, 210 (1932): 35.

91. Frederick Gowland Hopkins, "Atomic Physics and Vital Activities," *Nature*, 130 (1932): 871.

92. HAK–FLH, Aug. 3, 1978 (Tape 74, p. 12).

93. Krebs, Notebook "11," pp. 158–206.

94. Georg Krebs to HAK, Oct. 20, 1934, A.719, KC.

95. Krebs, *Reminiscences*, pp. 87–88; HAK–FLH, July 4, 1977 (Tape 17, p. 5), July 5, 1977 (Tape 17 [2], pp. 2–3), Aug. 1, 1977 (Tape 31, p. 15), Aug. 8, 1978 (Tape 79, p. 23).

96. Katherina Holsten to HAK, Nov. 23, 1934, A.995, KC.

97. F.G. Hopkins to E.J. Wayne, July 23, 1935; D. Keilin to E.J. Wayne, July 20, 1935, Archives of the University of Sheffield; HAK–FLH, Aug. 1, 1977 (Tape 31, p. 16).

98. HAK to Daniel O'Brien, Sept. 26, 1934; D. O'Brien to HAK, Sept. 29, 1934, A.428, KC.

99. D. Nachmansohn to HAK, Sept. 15, 1934, Sept. 22, 1934, J.476; Ch. Weizmann to HAK, Sept. 30, 1934, A.427, KC.

100. HAK to D. Nachmansohn, Nov. 4, 1934, J.476, KC.

101. HAK to Ch. Weizmann, Nov. 25, 1934, A.427, KC.

102. Georg Krebs to HAK, Nov. 4, 1934, A.719, KC.

103. *Ibid.*

104. Lise Daniel to HAK, Nov. 10, 1934, A.733, KC.

105. Craig, *Germany*, pp. 585–591, 601.

106. HAK–FLH, May 6, 1977 (Tape 15, pp. 8–10), July 30, 1977 (Tape 31, p. 11), Aug. 1, 1977 (Tape 32, p. 2), Aug. 11, 1977 (Tape 40, p. 1), Dec. 2, 1979 (Tape 112, p. 13), Jan. 9, 1981 (Tape 129, p. 11); E.W. Krebs–FLH, Aug. 3, 1977 (Tape 35, p. 19); A.J.P. Taylor, *English History*, 1914–1945 (Harmondsworth: Penguin Books, 1965), pp. 447–448. Krebs saved the newspaper clippings, and I examined them in a box in which he still kept them during the time that I visited him in Oxford.

107. HAK–FLH, May 6, 1977 (Tape 15, p. 9), Aug. 6, 1977 (Tape 39, p. 3).

108. S. Thannhauser to HAK, Freiburg (undated), J.716, KC.

109. Georg Hohmann, "Zum Andenken an Professor Dr. Siegfried Thannhauser," *Münchener Medizinische Wochenschrift*, 105 (1963): 357–359.

Chapter 4

1. Hans Weil-Malherbe, "Roots and Broods," unpublished autobiographical memoir, p. 46.

2. Hans Krebs, *Reminiscences and Reflections* (Oxford: Clarendon Press, 1981), p. 87; Vice-Chancellor, University of Cambridge, to HAK, Nov. 22, 1934; London *Times*, "Cambridge Degrees," Dec. 2, 1934, KC; HAK–FLH, March 20, 1980 (Tape 118, p. 5).

3. Hans Krebs, Notebook "11," pp. 194–195.

4. Dorothy Moyle Needham, "A Quantitative Study of Succinic Acid in Muscle: Glutamic and Aspartic Acid as Precursors," *Biochem. J.*, 24 (1930): 208–227.

5. Eric Holmes, *The Metabolism of Living Tissues* (Cambridge: Cambridge University Press, 1937), p. 130.

6. Krebs, Notebook "11," pp. 202–203.

7. *Ibid.*, pp. 204–206. The Q values recorded in the notebook are already corrected for the blank without substrate.

8. *Ibid.*, pp. 208–211; Holmes, *Metabolism*, p. 130.

9. Krebs, Notebook "11," pp. 225, 227, 234; Hans Adolf Krebs, "The Synthesis of Glutamine from Glutamic Acid and Ammonia, and the Enzymic Hydrolysis of Glutamine in Animal Tissues," *Biochem. J.*, 29 (1935): 1953–1954.

10. Krebs, Notebook "11," pp. 218–219, 244–245; Krebs, "Synthesis of Glutamine," pp. 1960–1962.

11. Krebs, Notebook "11," pp. 236, 251.

12. *Ibid.*, pp. 232, 235, 241–243, 246, 248–250, 252–256.

13. Juda Quastel to HAK, Dec. 19, 1934, A.429, KC.

14. HAK to J. Quastel, Dec. 22, 1934, A.429, KC.

15. HAK to J. Quastel, Dec. 27, 1934, A.429, KC.

16. J. Quastel to HAK, Jan. 2, 1935, A.429, KC.

17. Krebs, Notebook "11," pp. 260–264.

18. *Ibid.*, pp. 265, 279, 281; Notebook "11a," p. [6].

19. Krebs, Notebook "11," pp. 270–271; Notebook "11a," p. 1.

20. Krebs, Notebook "11," pp. 268–281; Notebook "11a," pp. 1–[7].

21. Katherina Holsten to HAK, Jan. 23, 1935, A.996, KC; "University of Cambridge School of Biochemistry: Biochemical Tea Club," Jan. 16, 1935, Biochem. 414, Archives, Cambridge University Library; HAK–FLH, Aug. 1, 1977 (Tape 31, p. 17).

22. David Nachmansohn to HAK, Jan. 1935, Feb. 1, 1935, J.477, KC.

23. Hermann Blaschko–FLH, Sept. 10, 1976 (Tape 6, p. 5).

24. [Anon.], "Protokoll ueber eine Besprechung mit Dr. Weiszmann am 10 Februar 1935," J.477, KC.

25. HAK to D. Nachmansohn, March 10, 1935. The original of this letter was given to me by Dr. Edith Nachmansohn. A copy has been deposited in the KC.

26. D. Nachmansohn to HAK, Feb. 17, 1935, J.477, KC.

27. HAK to D. Nachmansohn, March 10, 1935.

28. Katherina Holsten to HAK, Feb. 16, 1935, Feb. 17, 1935, A.996, KC.

29. H. Raistrick, "The Biochemical Society: Meeting of February 15, 1935," printed program; H.A. Krebs, "Deamination of Amino Acids in Mammalian Tissues," H.A. Krebs, "Synthesis of Glutamine in Mammalian Tissues," *Chemistry and Industry*, 54 (1935): 172–173.

30. Krebs, Notebook "11a," pp. 12–144; J.T. Saunders (Secretary General of the Faculties, University of Cambridge) to HAK, March 8, 1935, inserted at end of Notebook "11a".

31. Lise Daniel to HAK, March 13, 1935, A.734, KC.

32. Krebs, Notebook "11a," p. 32.

33. *Ibid.*, p. 33.

34. *Ibid.*, p. 34.

35. *Ibid.*, p. 35.

36. *Ibid.*, pp. 36–37. The reasoning behind the use of lowered temperature is adapted from Hans Adolf Krebs, "Deamination of Amino Acids," *Biochem. J.*, 29 (1935): 1636.

37. Krebs, Notebook "11a" p. 38. In his published account Krebs utilized the data from this particular experiment to rule out the argument. See Krebs, "Deamination," pp. 1635–1636.

38. Krebs, Notebook "11a," p. 39.

39. *Ibid.*, pp. 40–43.

40. *Ibid.*, p. 44.

41. *Ibid.*, pp. 45–85.

42. Katherina Holsten to HAK, April 1, 1935, A.996, KC.

43. *Ibid.*

44. Krebs, Notebook "11a," pp. 86–88.

45. *Ibid.*, p. 89.

46. Franz Knoop to HAK, Oct. 9, 1934, KC.

47. Krebs, Notebook "11a," p. 92.

48. *Ibid.*, p. 93.
49. *Ibid.*, pp. 97–98.
50. *Ibid.*, p. 99.
51. *Ibid.*, pp. 100–131; Krebs, "Deamination," p. 1625.
52. Krebs, Notebook "11a," p. 101.
53. *Ibid.*, pp. 86–131.
54. Lise Daniel to HAK, May 16, 1935, A.734, KC.
55. Krebs, Notebook "11a," p. 137.
56. *Ibid.*, pp. 138–139.
57. Krebs, "Deamination," pp. 1636–1637.
58. *Ibid.*, p. 1637; Krebs, Notebook, "11a," pp. 152–155.
59. Krebs, "Deamination," p. 1637.
60. *Ibid.*, pp. 1639–1642.
61. HAK–FLH, July 5, 1977 (Tape 17, p. 5).
62. *Ibid.*, July 4, 1977 (Tape 17, p. 15).
63. *Ibid.*, Tape 17, pp. 14–15.
64. *Ibid.*, Tape 17, p. 15.
65. HAK–FLH, Sept. 7, 1976 (Tape 4, p. 26).
66. Hopkins discussed these general ideas for many years. Two lectures expressing his point of view close to the time that Krebs was in Cambridge are in Frederick Gowland Hopkins, *Chemistry and Life*: The Fourth Gluckstein Lecture, 1932 (London, 1933), especially pp. 17–19, and "Some Chemical Aspects of Life," *Hopkins and Biochemistry: 1861–1947*, ed. Joseph Needham and Ernest Baldwin (Cambridge: Heffer, 1949) pp. 248–249.
67. HAK–FLH, July 4, 1977 (Tape 17, p. 6)
68. Krebs, "Deamination," p. 1620.
69. *Ibid.*, p. 1630.
70. Krebs, Notebook "11a," inserted in back. I have corrected what appear to be a few purely typographical errors.
71. Krebs, "Deamination," p. 1642.
72. HAK–FLH, July 4, 1977 (Tape 17, p. 12).
73. HAK–FLH, July 5, 1977 (Tape 17, p. 1).
74. *Ibid.*, July 4, 1977 (Tape 17, p. 13).
75. *Ibid.* The conversation is given as Krebs remembered it. Undoubtedly the original was spoken in German.
76. R. H[ill], D.M. N[eedham], and M. S[tephenson], "Memorandum on Part II," Biochem. 414, Archives of the University of Cambridge.
77. E. H[olmes], "Memorandum on Proposed Part II Courses," *Ibid.*
78. Minutes of Staff Meeting, April 30, 1935, *Ibid.*
79. *Ibid.*; HAK–FLH, conversation that I cannot locate in transcripts.
80. "Post-War Needs of the Department of Biochemistry," Biochem. 412, Archives of the University of Cambridge.
81. HAK to D. Nachmansohn, April 28, 1935, author's collection. Copy, J.477, KC.
82. D. Nachmansohn (undated), J.477, KC.
83. Krebs, *Reminiscences*, pp. 93–94; Edward Wayne–FLH, May 5, 1977 (Tape 14, pp. 4, 10). I follow here Wayne's account. Krebs gave a different account of the background: "My name had been suggested to him, as I learned much later, by Charles Harrington and George Pickering with whom Wayne had been associated at University College, London."
84. Krebs, *Reminiscences*, p. 94; Edward Wayne to FLH, Aug. 26, 1986.
85. Krebs, Notebook "11a," pp. 176–177.
86. *Ibid.*, pp. 177–178.
87. *Ibid.*, pp. 179–181.
88. *Ibid.*, p. [182].
89. *Ibid.*, p. 183.
90. *Ibid.*, p. 184.
91. *Ibid.*, p. 188.

92. *Ibid.*

93. H. Raistrick, "The Biochemical Society; Meeting of June 7, 1935," printed program; H.A. Krebs, "Synthesis of Glutamine in Animal Tissues," *Chemistry and Industry*, 13 (1935): 597; Hermann Blaschko–FLH, Sept. 10, 1976 (Tape 6, p. 6).

94. Krebs, Notebook "11a," pp. 191–206.

95. Krebs, "Synthesis of Glutamine," p. 1951.

96. *Ibid.*

97. *Ibid.*, p. 1952.

98. *Ibid.*, pp. 1952–1953

99. See Imre Lakatos, "Falsification and the Methodology of Scientific Research Programmes," in *Criticism and the Growth of Knowledge*, ed. Imre Lakatos and Alan Musgrove (Cambridge: Cambridge University Press, 1970), p. 138.

100. Krebs, "Synthesis of Glutamine," pp. 1953–1966.

101. *Ibid.*, pp. 1966–1967.

102. *Ibid.*, p. 1968.

103. *Ibid.*

104. *Ibid.*

105. *Ibid.*

106. Hans Weil-Malherbe to FLH, Nov. 4, 1986.

107. HAK–FLH, Dec. 22, 1979 (Tape 112, p. 3).

108. HAK–FLH, July 4, 1977 (Tape 17, p. 5).

Chapter 5

1. Hans Weil-Malherbe, "Roots and Broods," unpublished autobiographical memoir, pp. 46–47; Hans Weil-Malherbe–FLH, Dec. 8, 1986 (Tape 156, pp. 10–11); *Nature*, 135 (1935): 468; HAK–FLH, July 8, 1977 (Tape 18, p. 20).

2. Hans Weil-Malherbe and Hans Adolf Krebs, "The Conversion of Proline to Glutamic Acid in Kidney," *Biochem. J.*, 29 (1935): 2077–2081. After his move to Newcastle, Weil adopted the hyphenated name Weil-Malherbe, adding his wife's maiden name, in accord with a common Swiss custom. Weil-Malherbe, "Roots and Broods," p. 47.

3. Weil-Malherbe and Krebs, "Proline," p. 2081; HAK–FLH, July 8, 1977 (Tape 18, p. 20); Weil-Malherbe–FLH, telephone conversation, July 22, 1987.

4. Weil-Malherbe, "Roots," p. 47; Weil-Malherbe–FLH, Dec. 8, 1986 (Tape 156, p. 11); Weil to HAK, July 27, 1935, July 31, 1935, J.815, KC.

5. Weil to HAK, July 31, 1935.

6. *Ibid.*; Weil-Malherbe, "Roots," p. 47; Weil-Malherbe–FLH, Dec. 8, 1986 (Tape 156, p. 11); Hans Weil-Malherbe, "The Metabolism of Glutamic Acid in Brain," *Biochem. J.*, 30 (1936): 665–676.

7. Norman Lowther Edson, "The Influence of Ammonium Chloride on Ketone-Body Formation in Liver," *Ibid.*, 29 (1935): 2082–2094.

8. Norman Lowther Edson, "Ketogenesis from Amino-Acids," *Ibid.*, pp. 2498–2505.

9. Werner Schuler and Wilhelm Reindel, "Die Harnsäuresynthese im Taubenorganismus - eine Purinsynthese," *Z. physiol. Chem.*, 234 (1935): 63–82.

10. Norman Lowther Edson, Hans Adolf Krebs, and Alfred Model, "The Synthesis of Uric Acid in the Avian Organism: Hypoxanthine as an Intermediary Metabolite," *Biochem. J.*, 30 (1936): 1381–1382.

11. Hans Krebs, Notebook "11a," pp. 218–219.

12. *Ibid.*, p. 220.

13. *Ibid.*, p. 221.

14. M. Neber. "Über die Aminosäuresynthese aus Ketosäure und die Harnstoffsynthese in der Leber," *Z. physiol. Chem.*, 234 (1935): 83–96.

15. Krebs, Notebook "11a," pp. 222–228.

16. Alfred Model to HAK, Feb. 21, 1934, June 19, 1934, Dec. 16, 1934, May 2, 1935, June 10, 1935, J.469, KC.

17. Reactions as represented in Eric Holmes, *The Metabolism of Living Tissues* (Cambridge: University Press, 1937), p. 60.

18. Norman Edson to HAK, July 23, 1935, J.168, KC.

19. W.M. Gibbons, "The University of Sheffield: Appointment of Lecturer," June 25, 1935; H.A. Krebs, "Application for the Lectureship in Pharmacology," Archives of the University of Sheffield. The application is undated, but the acknowledgment of receipt of the application from the registrar of the University of Sheffield is dated July 10, 1935.

20. Katherina Holsten to HAK, June 6, 1935, June 15, 1935, June 24, 1935, July 5, 1935. A.996, KC.

21. *Ibid.*, Aug. 7, 1935, Aug. 21, 1935, A.997, KC.

22. *Ibid.*, Aug. 1, 1935, Aug. 7, 1935, Aug. 21, 1935.

23. Edward Wayne to HAK, July 16, 1935, KC.

24. *Ibid.*, Aug. 1, 1935.

25. F.G. Hopkins to Edward Wayne, July 23, 1935; David Keilin to Edward Wayne, July 20, 1935, Archives of the University of Sheffield. Copies of these letters were kindly presented to me by the registrar of the university when I visited there in August 1977. Upon my return to Oxford I showed them to Hans Krebs, who had not previously seen them.

26. Norman Lowther Edson to HAK, July 23, 1935; HAK–FLH, Aug. 10, 1978 (Tape 82, p. 15).

27. Edward James Morgan, "The Distribution of Xanthine Oxidase," *Biochem. J.*, 20 (1926): 1289.

28. Edson to HAK, July 23, 1935; Schuler and Reindel, "Harnsäuresynthese," pp. 64–65.

29. Gerald Holton, *The Scientific Imagination: Case Studies* (Cambridge: Cambridge University Press, 1978), p. 71.

30. HAK–FLH, Aug. 10, 1978 (Tape 82, p. 27).

31. HAK–FLH, Aug. 10, 1978 (Tape 82, p. 11).

32. On the first three days, Aug. 2–4, Krebs carried out three experiments involving guanase and casein hydrolysate. Their relation to the following experiments are not entirely clear to me, and I have not discussed them. See Krebs, Notebook "11a," pp. 232–235.

33. Neber, "Aminosäuresynthese," pp. 83–96.

34. *Ibid.*, p. 93.

35. Krebs, Notebook "11a," pp. 236–242.

36. *Ibid.*, pp. 245–249.

37. *Ibid.*, pp. 250–251.

38. H.A. Krebs, "Metabolism of Amino Acids and Related Substances,"*Annual Review of Biochemistry*, 5 (1936): 254.

39. Marjory Stephenson, "Formic Hydrogenlyase," *Ergebnisse der Emzymforschung*, 6 (1937): 139–156; Robert E. Kohler, "Innovation in Normal Science: Bacterial Physiology," *Isis*, 76 (1985): 175–179.

40. HAK–FLH, July 8, 1977 (Tape 19, pp. 23–24).

41. Stephenson, "Formic Hydrogenlyase,"p. 145; Krebs, Notebook "11a," p. 236. My assumption that this experiment initiated the investigation of the reversibility of hydrogenlyase must remain tentative, since there may well have been prior experiments on the question by Stephenson or by D.D. Woods.

42. *Ibid.*, p. 236.

43. *Ibid.*, p. 243.

44. Ernest F. Gale and Paul Fildes, "Donald Devereux Woods: 1912-1964,"*BiographicalMemoirs of Fellows of the Royal Society*, 11 (1965): 203–205; HAK–FLH, July 8, 1977 (Tape 19, pp. 23–24), May 16, 1979 (Tape 96, p. 33).

45. Krebs, Notebook "11a," pp. 260-261; Stephenson, "Formic Hydrogenlyase," pp. 148–155; Kohler, "Innovation," pp. 177–179; Donald Devereux Woods, "The Synthesis of Formic Acid by Bacteria," *Biochem. J.*, 30 (1936): 516.

46. Woods, "Synthesis of Formic Acid," p. 515.

47. Krebs (untitled undated carbon copy), KC.

48. Q.H. Gibson, "Francis John Worsley Roughton: 1899-1972,"*BiographicalMemoirs of Fellows of the Royal Society* 19 (1973): 563–564, 573–574; F.J.W. Roughton, "Recent Work on Carbon Dioxide Transport by the Blood," *Physiol. Rev.*, 15 (1935): 241–296.

49. Roughton, "Carbon Dioxide Transport," pp. 252–277.

50. Gibson, "Roughton," p. 564; HAK–FLH, July 8, 1977 (Tape 18, pp. 1–2), May 15, 1979 (Tape 95, p. 32).

51. Roughton, "Carbon Dioxide Transport," pp. 263–264; HAK to F.J.W. Roughton, Feb. 28, 1948, Roughton papers, Archives of the American Philosophical Society.

52. Krebs, "Metabolism of Amino Acids," pp. 261–262.

53. HAK–FLH, July 5, 1977 (Tape 17, p. 11).

54. *Ibid.*, Aug. 1, 1978 (Tape 72, p. 17).

55. *Ibid.*, May 16, 1979 (Tape 95, p. 6).

56. *Ibid.*, July 5, 1977 (Tape 17, p. 12).

57. See H.H. Mitchell and T.S. Hamilton, *The Biochemistry of the Amino Acids* (New York: Chemical Catalogue Company, 1929), p. 72; Roughton, "Carbon Dioxide Transport," pp. 269–270.

58. These considerations are inferred from subsequent discussions by Krebs and Roughton. See Krebs, "Metabolism of Amino Acids," pp. 261–262; Hans Adolf Krebs and Francis John Worsley Roughton, "Carbornithine as an Intermediate in the Synthesis of Urea by the Liver," unpublished handwritten manuscript, 1937, Roughton Papers, Archives of the American Philosophical Society, Philadelphia. See also Roughton, "Carbon Dioxide Transport," p. 269.

59. HAK–FLH, July 5, 1977 (Tape 17, p. 9).

60. Krebs, Notebook "11a," p. 255.

61. *Ibid.*, pp. 256–257.

62. HAK–FLH, May 3, 1977 (Tape 12, p. 30), July 5, 1977 (Tape 17, pp. 9–10).

63. Krebs, Notebook "11a," pp. 252–254, 264–266.

64. *Ibid.*, p. 269; Jakob Parnas, "Ueber fermentative Beschleunigung der Cannizaroschen [sic] Aldehydumlagerung durch Gewebssäfte," *Biochem. Z.*, 28 (1910): 274–294; F. Battelli and L. Stern, "Die Aldehydase in den Tiergeweben," *Ibid.*, 29 (1910): 130–151; Heinrich Wieland and Karl Frage, Über den Mechanismus der Oxydationsvorgänge, *Z. physiol. Chem.*, 186 (1929): 195–204.

65. HAK–FLH, July 9, 1977 (Tape 19, pp. 13–15).

66. Krebs, Notebook "11a," p.[269].

67. Parnas, "Cannizaroschen Aldehydumlagerung," pp. 274–287.

68. Krebs, Notebook "11a," pp. [269–270].

69. HAK–FLH, April 26, 1977 (Tape 9, p. 12). Krebs made this remark in connection with such an experience that had happened to him the night before our conversation.

70. Krebs, Notebook "11a," pp. 260 ff.

71. HAK–FLH, Aug. 10, 1978 (Tape 82, p. 18).

72. David Nachmansohn to HAK, Aug. 19, 1935, J.477, KC.

73. Bruno Mendel to HAK, Aug. 25, 1935, J.458, KC.

74. David Nachmansohn to HAK, Sept. 2, 1935, J.477, KC.

75. HAK to Nachmansohn, Sept. 3, 1935. Original given to the author by Mrs. Nachmansohn.

76. Edward Wayne to HAK, Sept. 3, 1935, Sept. 6, 1935, KC.

77. Lise Daniel to HAK, Sept. 5, 1935, A.734, KC; HAK–FLH, May 14, 1979 (Tape 94, p. 30).

78. *Ibid.*, March 13, 1935.

79. *Ibid.*, Sept. 25, 1935; HAK–FLH, Aug. 12, 1977 (Tape 41, p. 5).

80. Gordon A. Craig, *Germany: 1866–1945* (New York: Oxford University Press, 1980), p. 634.

81. Lise Daniel to HAK, Sept. 25, 1935.

82. See *Ibid.*, March 13, 1935.

83. *Ibid.*, Sept. 25, 1935, Adolf Daniel, note added to Lise's letter to HAK.

84. Wayne to HAK, Sept. 6, 1935, KC; personal visit to Sheffield by author, August 1977.

85. Wayne to HAK, Sept. 6, 1935; Hans Krebs, *Reminiscences and Reflections* (Oxford: Clarendon Press, 1981), p. 94.

86. Krebs, *Reminiscences*, pp. 94–95; HAK–FLH, Aug. 9, 1977 (Tape 36 (2), pp. 3, 6–7).

87. *Ibid.* My characterization of Edward and Nan Wayne is drawn in part from my own experience, when I came to their home 40 years later to talk about Hans Krebs.

88. Sir Edward Wayne–FLH, May 5, 1977 (Tape 14, pp. 4, 19).

89. Krebs, *Reminiscences*, p. 94; HAK–FLH, Aug. 9, 1977 (Tape 36 (2), p. 3).

90. Edward Wayne to HAK, Oct. 1, 1935, KC.

91. Craig, *Germany*, p. 634.

92. Katherina Holsten to HAK, Aug. 21, 1935, Sept. 6, 1935, A.997, KC.

93. *Ibid.*, Sept. 19, 1935.

94. *Ibid.*, Sept. 26, 1935.

95. Krebs, Notebook "lla," pp. 272–278.

96. Krebs, "Metabolism of Amino Acids," p. 261.

97. Nachmansohn to HAK, Sept. 19, 1935, Oct. 6, 1935, J.477, KC; HAK to Nachmansohn, Sept. 30, 1935, author's collection.

98. HAK to Nachmansohn, Sept. 30, 1935; Nachmansohn to HAK, Oct. 6, 1935.

99. Krebs, Notebook "11a," pp. 279–288.

100. Edward Wayne to HAK, Oct. 1, 1935, KC.

101. Krebs, *Reminiscences*, p. 94.

102. E. Wayne to HAK, Oct. 4, 1935, Oct. 11, 1935; Mrs. F.M. Hall to HAK, Oct. 5, 1935, KC.

103. Krebs, *Reminiscences*, p. 94.

104. Edward Wayne–FLH, May 5, 1977 (Tape 14, p. 9).

105. Katherina Holsten to HAK, Oct. 14, 1935, A.997, KC.

106. *Ibid.* Despite the reference to his "new chief," Edward Wayne does not recall having known about Katrina (unrecorded conversation, June 1986).

107. HAK–FLH, Aug. 11, 1977 (Tape 40, p. 3); E. Wayne–FLH, May 5, 1977 (Tape 14, p. 12); E. Wayne to HAK, Oct. 11, 1935, KC.

108. Krebs, Notebook "11b," pp. 1–4.

109. Norman Lowther Edson and Hans Adolf Krebs, "Micro-Determination of Uric Acid," *Biochem. J.*, 30 (1936): 732.

110. Krebs, Notebook "11b," pp. 5–6.

111. *Ibid.*, pp. 10–12.

112. *Ibid.*, p. 13.

113. HAK to Hermann Blaschko, Oct. 26, 1935, letter given to author by Mrs. Nachmansohn.

114. *Ibid.*; HAK to David Nachmansohn, Nov. 17, 1935, letter given to author by Mrs. Nachmansohn; "The Part II Class Practical Work," Biochem 415, Archives of the Cambridge University Biochemistry Department; Minutes of meeting of May 27, 1935, *Ibid.*; HAK–FLH, Aug. 10, 1978 (Tape 82, p. 26).

115. Alfred Model to HAK, Sept. 16, 1935, KC.

116. N.L. Edson, H.A. Krebs, and A. Model, "The Synthesis of Uric Acid in the Avian Organism," *Chemistry and Industry*, 13 (1935): 1026–1027.

117. *Ibid.*

118. Norman Lowther Edson, Hans Adolf Krebs, and Alfred Model, "The Synthesis of Uric Acid in the Avian Organism: Hypoxanthine as an Intermediary Metabolite," *Biochem. J.*, 30 (1936): 1381.

119. HAK–FLH, Aug. 10, 1978 (Tape 82, pp. 21–22).

120. Krebs and Roughton, "Carbornithine."

121. *Ibid.*; J.K.W. Ferguson and F.J.W. Roughton, "The Direct Chemical Estimation of Carbamino Compounds of CO_2 with Haemoglobin," *J. Physiol.*, 83 (1935): 68–82.

122. HAK–FLH, May 3, 1977 (Tape 12, p. 30), July 4, 1977 (Tape 17, pp. 9–10), July 8, 1977 (Tape 18, pp. 1–2), Aug. 3, 1978 (Tape 74, pp. 15–16), May 15, 1979 (Tape 95, pp. 27–31).

123. Krebs, "Metabolism of Amino Acids," pp. 261–262.

124. HAK–FLH, July 5, 1977 (Tape 17, p. 10).

125. See the very interesting discussion of the "notion of credibility" in Bruno Latour and Steve Woolgar, *Laboratory Life: The Social Construction of Scientific Facts*, 2nd ed. (Princeton; Princeton University Press, 1986) pp. 187–208. For a cogent discussion of the presentation of speculation in the scientific literature, see Greg Myers, "Scientific Speculation and Literary Style in a Molecular Genetics Article," *Science in Context*, 4 (1991): 321–346.

126. Krebs, *Reminiscences*, p. 96.

127. Lise Daniel to HAK, Nov. 11, 1935, A.734, KC.

128. *Ibid.*, Dec. 5, 1935.

129. HAK to D. Nachmansohn, Nov. 17, 1935, letter given to author by Mrs. Nachmansohn.

130. Krebs, *Reminiscences*, p. 96.

131. Lise Daniel to HAK, Aug. 5, 1935, A.734, KC.

132. E.W. Krebs–FLH, Sept. 15, 1986 (Tape [153], p. 9).

133. *Ibid.*, (Tape [153], pp. 11–12).

134. Lise Daniel to HAK, Dec. 5, 1935, A.734, KC.

135. Nachmansohn to HAK, Oct. 6, 1935, Nov. 21, 1935, J.477, KC.; HAK to Nachmansohn, Nov. 17, 1935, HAK to Blaschko, Oct. 26, 1935, letters given to author by Mrs. Nachmansohn. Since the letter from Krebs to Blaschko is found among the papers of Nachmansohn, it is possible that Krebs sent it first to Nachmansohn and that it was never forwarded to Blaschko.

136. HAK to Nachmansohn, Nov. 17, 1935, letter given to author by Mrs. Nachmansohn.

137. Nachmansohn to HAK, Nov. 21, 1935, J.477, KC.

138. HAK to Blaschko, Dec. 3, 1935, letter given to author by H. Blaschko. Original now in KC.

139. Nachmansohn to HAK, Dec. 15, 1935, J.477, KC.

140. Chaim Weizmann to HAK, Dec. 16, 1935, KC.

141. Home Office, Whitehall, to HAK, Nov. 14, 1935, KC.

142. W. Schuler and W. Reindel, "Die Urikolyse," *Z. physiol. Chem.*, 215 (1933): 258-266.

143. *Ibid.*, p. 262; Krebs, Notebook "11b," p. 20.

144. Krebs, Notebook "11b," pp. 21–24.

145. *Ibid.*, pp. 25–32; Edson and Krebs, "Uric Acid," p. 734.

146. Edson and Krebs, "Uric Acid," p. 735. The evidence that Krebs had completed his work on the development of the method at this point is that both the "example" of the procedure and much of the data eventually published in the uric acid paper are taken from these pages of his notebook. *Ibid.*, p. 734.

147. HAK–FLH, Aug. 10, 1978 (Tape 82, p. 28).

148. Krebs, Notebook "11b," pp. 33–37; Claude Fromageot and Pierre Desnuelle, "Eine neue Methode zur Bestimmung der Brenztraubensäure," *Biochem. Z.*, 279 (1935): 174–183.

149. Krebs, Notebook "11b," p. 38.

150. *Ibid.*, pp. 38–39.

151. *Ibid.*, p. 40.

152. *Ibid.*, p. 41.

153. *Ibid.*, p. 42.

154. *Ibid.*, p. 43.

155. E. Holmes, *Metabolism*, p. 60.

156. *Ibid.*; Krebs, Notebook "11b," p. 44.

157. Krebs, Notebook "11b," pp. 45–47.

158. *Ibid.*, pp. 48–49.

159. *Ibid.*, pp. 49, 52–53.

160. *Ibid.*, p. 54.

161. *Ibid.*, p. 55.

162. *Ibid.*, pp. 56–57.

163. *Ibid.*, p. 58.

164. HAK–FLH, Aug. 10, 1978 (Tape 82, p. 28).

165. *Ibid.*, Aug. 11, 1978 (Tape 84, pp. 1–2).

166. Krebs, Notebook "11b," pp. 50–51.

167. *Ibid.*

168. *Ibid.*, pp. 60–61; HAK–FLH, Aug. 11, 1978 (Tape 84, p. 2).

169. *Ibid.* The final comment is retrospective. The remainder are laboratory notes.

170. Krebs, Notebook "11b," pp. 63–66.

171. Edson, Krebs, and Model, "Synthesis of Uric Acid," p. 1382.

172. Kathrina Holsten to HAK, Dec. 4, 1935, A.997, KC.

173. *Ibid.*, Dec. 12, 1935.

174. *Ibid.*, Dec. 15, 1935, Dec. 25, 1935.

Chapter 6

1. Hans Krebs, *Reminiscences and Reflections* (Oxford: Clarendon Press, 1981), p. 97; E. Wayne to HAK, Oct. 11, 1935; Leonard V. Eggleston, typed summary of his career, B.29, KC; HAK–FLH, Aug. 9, 1977 (Tape 36 (2) p. 7), March 16, 1980 (Tape 117, p. 4); W.A. Johnson–FLH, Jan. 17, 1978 (Tape 51, p. 4).

2. HAK–FLH, Aug. 13, 1977 (Tape 42, p. 3). Krebs noted in this conversation, as well as in his *Reminiscences*, pp. 95–96, that he offered two series of lectures, one for chemists and one for students in honors physiology. It is not clear, however, whether or not he introduced both series during his first year at Sheffield.

3. Quoted in Krebs, *Reminiscences*, p. 96.

4. *Ibid.*

5. HAK–FLH, July 9, 1977 (Tape 20, p. 8).

6. Krebs, *Reminiscences*, p. 97; HAK–FLH, May 16, 1979 (Tape 96, p. 25).

7. H.A. Krebs to F.G. Hopkins (draft), Jan. 9, 1936, KC. I have lightly edited the text, mainly by omitting phrases that Krebs wrote down and crossed out.

8. Krebs, *Reminiscences*, p. 97.

9. *Ibid.*, p. 96.

10. Georg Krebs to HAK, Nov. 13, 1936, A.721, KC.

11. HAK–FLH, July 9, 1977 (Tape 20, p. 9).

12. Edward Wayne–FLH, May 5, 1977 (Tape 14, p. 10).

13. Krebs, *Reminiscences*, p. 95.

14. *Ibid.*; Edward and Nan Wayne–FLH, May 5, 1977 (Tape 14, pp. 4, 5, 8, 12–13, 19).

15. Krebs, *Reminiscences*, p. 95; Edward and Nan Wayne–FLH, May 5, 1977 (Tape 14, pp. 17–19)

16. H.A. Krebs, "Metabolism of Amino Acids and Related Substances," *Annual Review of Biochemistry*, 5 (1936): 254.

17. F. Knoop, "Über den physiologischen Abbau der Säuren und die Synthese einer Aminosäure im Tierkörper," *Z. physiol. Chem.*, 67 (1910): 495–498; F. Knoop and Ernst Kertess, "Das Verhalten von α-Aminosäuren und α-Ketonsäuren im Tierkörper," *Ibid.*, 71 (1911): 252–265.

18. Knoop, "Abbau der Säuren," p. 498.

19. *Ibid.*, pp. 499–502; Knoop and Kertess, "α-Aminosäuren und α-Ketonsäuren," pp. 261–264.

20. F. Knoop and Jose Garcia Blanco, "Über die Acetylierung von Aminosäuren im Tierkörper," *Ibid.*, 146 (1925): 267–275.

21. Vincent du Vigneaud and Oliver J. Irish, "The Role of the Acetyl Derivative as an Intermediary Stage in the Biological Synthesis of Amino Acids from Keto Acids," *J. Biol. Chem.*, 109 (1935): xciv.

22. Krebs, "Metabolism of Amino Acids," pp. 254–255.

23. Krebs, Notebook "11b," p. 67.

24. *Ibid.*, pp. 68–69.

25. *Ibid.*, pp. 70–71.

26. *Ibid.*, pp. 72–73.

27. *Ibid.*, pp. 75–76.

28. *Ibid.*, pp. 77–79.

29. *Ibid.*, pp. 80–83.

30. Edward J. Wayne to E. Mellanby, Jan. 30, 1936, 1951, Medical Research Council Archives.

31. HAK–FLH, May 16, 1979 (Tape 97, p. 35). Notes on unrecorded conversation with Edward Wayne, May 18, 1979.

32. Krebs, Notebook "11b," pp. 84–85.

33. *Ibid.*, pp. 86–87.

34. *Ibid.*, pp. 88–89.

35. *Ibid.*, pp. 90–107.

36. HAK to Norman Edson, Feb. 12, 1936, J.168, KC.

37. Norman Lowther Edson and Hans Adolf Krebs, "Micro-Determination of Uric Acid," *Biochem. J.*, 30 (1936): 732–735.

38. Juda Hirsch Quastel and Arnold Herbert Maurice Wheatley, "Acetoacetic Acid Breakdown in the Kidney," *Biochem. J.*, 29 (1935): 2773-2786.

39. Krebs, Notebook "11b," pp. 109-110. See also Hans Adolf Krebs and William Arthur Johnson, "Metabolism of Ketonic Acids in Animal Tissues," *Biochem. J.*, 31 (1937): 651-653, Deodata Krüger and Erich Tschirch, "Die Blaufärbung des basischen Lanthanacetats mit Jod. Eine hochempfindliche Reaction auf Acetat-Ion," *Ber. Deutsch. Chem. Gesell.*, 62 (1929): 2776-2783.

40. Krebs, Notebook "11b," pp. 112-121.

41. *Ibid.*, p. 120; Krebs and Johnson, "Metabolism of Ketonic Acids," p. 647.

42. Krebs, Notebook "11b," pp. 122-124.

43. *Ibid.*, pp. 125-126.

44. *Ibid.*, pp. 127-133.

45. HAK-FLH, July 23, 1977 (Tape 26, p. 12).

46. Krebs, Notebook "11b," pp. 135-139.

47. E.J. Wayne to Secretary of the Medical Research Council, March 3, 1936, Archives of the Medical Research Council.

48. HAK-FLH, May 16, 1979 (Tape 97, p. 38).

49. Wayne to MRC, March 3, 1936.

50. Krebs, Notebook "11b," pp. 140-141.

51. O. Meyerhof and W. Kiessling, "Über den Hauptweg der Milchsäurebildung in der Muskulatur," *Biochem. Z.*, 283 (1936): 83-113.

52. Krebs, Notebook "11b," p. 119.

53. *Ibid.*, pp. 144-145.

54. *Ibid.*, pp. 146-147.

55. *Ibid.*, pp. 148-151. In the first of these two ratios Krebs pyruvate wrote "pyruvate/lactate:1/1.86," an obvious slip. See also B.F. Avery and A. Baird Hastings, "A Gasometric Method for the Determination of Lactic Acid in the Blood," *J. Biol. Chem.*, 94 (1931): 273-280.

56. Krebs, Notebook "11b," p. 149.

57. Krebs and Johnson, "Metabolism of Ketonic Acids," p. 646.

58. E. Annau, I. Banga, B. Gözsy, St. Husak, K. Laki, B. Straub, and A. Szent-Györgyi, "Über die Bedeutung der Fumarsäure für die tierische Gewebsatmung," *Z. physiol. Chem.*, 236 (1935): 1-5.

59. *Ibid.*, pp. 6-11.

60. *Ibid.*, pp. 4, 6. When I examined Krebs's copy it was contained in a large reprint collection in his laboratory.

61. HAK-FLH, July 9, 1977 (Tape 20, pp. 1-2).

62. HAK-FLH, July 15, 1977 (Tape 23, p. 5), July 21, 1977 (Tape 24, pp. 1, 3, 16), July 29, 1977 (Tape 30, p. 21).

63. See David Nachmansohn to HAK, Jan. 17, 1936, J.478, KC. Nachmansohn answered Krebs's inquiry about where a Latapie apparatus could be obtained in Paris.

64. Annau, et al., "Bedeutung der Fumarsäure," pp. 54-56.

65. Krebs, Notebook "11b," p. 152.

66. *Ibid.*, pp. 153-154.

67. See Krebs and Johnson, "Metabolism of Ketonic Acids," p. 653.

68. Krebs, Notebook "11b," pp. 155-161.

69. *Ibid.*, p. 162.

70. H.A. Krebs, "The History of the Tricarboxylic Acid Cycle," *Perspectives in Biology and Medicine*, 14 (1970): 156.

71. Nachmansohn to HAK, Jan. 17, 1936, J.478, KC.

72. Georg Krebs to HAK, Feb. 26, 1936, A.721, KC.

73. Gerhard L. Weinberg, *The Foreign Policy of Hitler's Germany: Diplomatic Revolution in Germany, 1933-36* (Chicago: University of Chicago Press, 1970), pp. 239-263.

74. Nachmansohn to HAK, March 9, 1936, J.478, KC.

75. Weinberg, *Foreign Policy*, pp. 257-259.

76. Krebs, Notebook record of journey to Palestine, entry of March 15, 1936, A.435, KC.

77. *Ibid.*, March 21, 1936.

78. Weinberg, *Foreign Policy*, p. 261.

79. Krebs, *Reminiscences*, p. 99; HAK–FLH, May 6, 1977 (Tape 15, p. 9). A self-professed optimist, Krebs did not readily acknowledge a pessimistic side in his personality, traces of which can nevertheless be found frequently in passing remarks. There is little direct expression of a more pessimistic outlook growing specifically out of the militarization of the Rhineland. In his autobiography he stated only in general terms that "during the middle 1930s the expansion of Hitler's powers and his threats to world peace became an increasing source of great anxiety." In conversations about that situation, however, he characteristically mentioned the Rhineland occupation as an instance of this expansion and of the failure of Britain and France to intervene. The most extensive reflection of an underlying pessimism concerning the course of world affairs is found in a diary that Krebs kept during World War II (A.80, KC).

80. Krebs, Notebook, journey to Palestine, March 21–23, 1936.

81. Georg Krebs to HAK, Feb. 26, 1936, A.721, KC; E.W. Krebs–FLH, Sept. 15, 1986 (Tape 153, p. 9).

82. HAK–FLH, Jan. 9, 1981 (Tape 129, p. 11). I have edited this passage slightly. Krebs's eloquent description began as a response to my question about the "later" times, including this meeting when he saw his father, but shaded into a summary of the whole experience of the advent of the Nazis on his father.

83. Krebs, Notebook, journey to Palestine, March 26–31; D. Nachmansohn to HAK, "Thursday" (otherwise dated only "1936?"), J.478, KC.; Georg Krebs to HAK, April 6, 1936, A.721, KC.

84. Krebs, Notebook, journey to Palestine, April 1, 1936.

85. *Ibid.*, April 2–4, 1936.

86. *Ibid.*, April 5–9, 1936; Nachmansohn to HAK, Feb. 21, 1937, J.479, KC; HAK–FLH, May 16, 1979 (Tape 96, p. 17).

87. Krebs, *Reminiscences*, p. 91; Nachmansohn to HAK, March 9, 1936; HAK–FLH, Aug. 6, 1977 (Tape 39, p. 2), Aug. 10, 1977 (Tape 39, p. 6).

88. Georg Krebs to HAK, April 6, 1936, A.721, KC.

89. Krebs, *Reminiscences*, p. 91; HAK–FLH, Aug. 6, 1977 (Tape 39, p. 3); Nachmansohn to HAK "Wednesday" (otherwise undated), J.478, KC.

90. HAK to N.L. Edson, April 24, 1936, J.168, KC.

91. F.H.K. Green to E. Wayne, March 11, 1936, March 13, 1936, March 24, 1936; E.J. Wayne to F.H.K. Green, March 12, 1936, March 25, 1936, 1951 Archives of the Medical Research Council.

92. Krebs, *Reminiscences*, p. 97; HAK–FLH, Aug. 9, 1977 (Tape 36 (2), pp. 3–4), March 20, 1980 (Tape 118, p. 1); W.A. Johnson–FLH, Jan 17, 1978 (Tape 52, p. 30), Jan. 19, 1978 (Tape 53, p. 18). In 1978 Johnson remembered Krebs's absence as a "three months" tour in the United States, probably mixing the Palestine journey up with a trip Krebs took to the United States in 1938.

93. W.A. Johnson–FLH, Jan. 17, 1978 (Tape 51, pp. 2–8); HAK–FLH, Feb. 16, 1978 (handwritten notes).

94. HAK–FLH, *Ibid.*

95. W.A. Johnson–FLH, Jan. 17, 1978 (Tape 51, p. 2), Jan. 19, 1978 (Tape 53, pp. 14–15).

96. W.A. Johnson–FLH, Jan. 17, 1978 (Tape 51, p. 6, Tape 52, pp. 22, 33).

97. Edward Wayne–FLH, May 5, 1977 (Tape 14, p. 11).

98. HAK–FLH, Feb. 16, 1978 (handwritten notes).

99. W.A. Johnson–FLH, Jan. 19, 1978 (Tape 52, pp. 5–6, Tape 53, p. 15).

100. HAK–FLH, Feb. 16, 1978 (handwritten notes). The personality traits of Krebs described here as Johnson experienced them remained throughout his professional career. Similar incidents were frequently mentioned to me by workers in Krebs's laboratory when I visited it between 1976 and 1981. Forewarned about his insistence on punctuality, I made certain to appear precisely on time for all appointments with him, and it was clear that the somewhat misleading impression he gained of me as a very punctual person was important to the confidence he acquired in my project. I also found out how tolerant he was of mistakes when I once lost an early document belonging to him on the way back from having it copied. When I explained the situation he simply said, "If there is a copy, then it doesn't matter."

101. Krebs, Notebook "11b," pp. 164–165; S.W. Clausen, "A Method for the Determination of Small Amounts of Lactic Acid," *J. Biol. Chem.*, 52 (1922): 263–280.

102. Krebs, Notebook "11b," pp. 166–167.

103. This explanation is included in a detailed interpretation of the experiment in Krebs and Johnson, "Metabolism of Ketonic Acids," pp. 650–651. There it is connected to further arguments that Krebs may not yet have formulated at this point; what I have given, however, seems implied in the statement he made in his notebook.

104. *Ibid.*

105. Krebs, Notebook "11b," p. 169.

106. *Ibid.*, pp. 170–173.

107. *Ibid.*, pp. 173–175; William Arthur Johnson, "Studies in the Intermediate Metabolism of Carbohydrates," Ph.D. thesis, University of Sheffield, 1938 (personal annotated copy of W.A. Johnson), p. 15; see also Krebs and Johnson, "Metabolism of Ketonic Acids," p. 651.

108. Krebs, Notebook "11b," pp. 176–177.

109. *Ibid.*, pp. 178, 180–183.

110. *Ibid.*, pp. 185–187. We have seen that in the formative experiment of April 28, the quantities of "free CO_2" and "total CO_2" produced by pyruvate were the same, but that in the repeat experiment of April 30 the latter was larger. In Krebs and Johnson, "The Metabolism of Ketonic Acids," p. 655, such differences are explained in terms of a fraction of the CO_2 formed remaining as bicarbonate. I am suggesting here that Krebs was already employing the same type of reasoning in this experiment on the CO_2:pyruvate ratio that he later summarized in the published paper.

111. Krebs, Notebook "11b," p. 192.

112. *Ibid.*, pp. 189–192.

113. *Ibid.*, pp. 149, 190–191, 193.

114. *Ibid.*, pp. 195–197.

115. *Ibid.*, pp. 198–199.

116. Rudolph A. Peters, "The Biochemical Lesion in Vitamin B_1 Deficiency: Application of Modern Biochemical Analysis in Its Diagnosis," *The Lancet*, 230 (May 23, 1936), pp. 1161–1162.

117. *Ibid.*, pp. 1162–1163.

118. Krebs, Notebook "11b," pp. 200–201.

119. *Ibid.*, pp. 203–204.

120. *Ibid.*, p. 205.

121. *Ibid.*, p. 206. This experiment is dated "27.V.36," whereas the previous one is dated "28.5.36." Krebs sometimes put down wrong dates in his notebook, and normally entered his experiments in chronological order, and I have here followed the latter order rather than the dates given.

122. For a review of the current state of the question see Hans Weil-Malherbe, "Formation of Succinic Acid," *Biochem. J.*, 31 (1937): 299–300.

123. Krebs, Notebook "11b," pp. 208–210.

124. E. Wayne–FLH, May 5, 1977 (Tape 14, p. 7).

125. HAK–FLH, Aug. 13, 1977 (Tape 42, p. 7).

126. HAK to N. Edson, Feb. 12, 1936, March 20, 1936, J.168, KC.

127. *Ibid.*, March 17, 1936.

128. N. Edson to HAK, April 28, 1936, J.168, KC.

129. HAK to N. Edson, May 4, 1936.

130. HAK to N. Edson, July 23, 1936; N. Edson to HAK, Aug. 24, 1936. In these two letters, written later in the summer, Krebs and Edson mentioned these terms as though they were by then accustomed to referring that way to Krebs's investigation.

Chapter 7

1. Lise Daniel to HAK, Jan. 6, 1936, Feb. 14, 1936, April 23, 1986, A.735, KC; E.W. Krebs–FLH, Sept. 15, 1986 (Tape 153, pp. 11–12, Tape 154, pp. 15, 38); E.W. Krebs to F.L. Holmes, Feb. 8, 1988; E.W. Krebs, "Summary of My Comments on Prof. Holmes's Manuscript," n.d. handwritten memorandum; allgemeine Elektricitaets-Gesellschaft, "Vorläufiges Abgangszeugnis," Nov. 15, 1935, personal file, E.W. Krebs.

2. Lise Daniel to HAK, May 7, 1936 A.735, KC; A. Leppett to E.W. Krebs, May 7, 1936; personal file, E.W. Krebs.

3. Lise Daniel to HAK, Nov. 11, 1935, Dec. 5, 1935, Jan. 6, 1936, Feb. 14, 1936, April 23, 1936, May 7, 1936, A.735, KC.

4. Krebs, Notebook "11b," pp. 211–215.

5. Ibid., p. 211.

6. Ibid., p. 215.

7. Ibid., pp. 216–217.

8. Ibid., pp. 220–221.

9. Ibid., pp. 222–223.

10. Ibid., p. 224. See also Hans Adolf Krebs and William Arthur Johnson, "Metabolism of Ketonic Acids in Animal Tissues," Biochem. J., 31 (1937): 653.

11. Krebs, Notebook "11b," p. 225; Notebook "12," pp. 1–2.

12. HAK–FLH, July 9, 1977 (Tape 20, p. 16).

13. Ibid., (Tape 20, pp. 14–15).

14. Krebs, Notebook "12," pp. 3–4. His choice of gonococcus probably was influenced by the fact that Guzman Barron, whose work he followed regularly, had recently published a paper entitled "The Oxidation of Pyruvic Acid by Gonococcus" (J. Biol. Chem., 113 (1936): 695–715). See HAK–FLH, July 9, 1977 (Tape 20, p. 15).

15. HAK–FLH, July 9, 1977 (Tape 20, p. 17).

16. Krebs, Notebook, "12," pp. 5–7.

17. HAK to N. Edson, June 10, 1936, J.168, KC.

18. Krebs, Notebook "12," p. 8.

19. Ibid., pp. 10–11.

20. Ibid., pp. 16–19; Krebs and Johnson, "Metabolism of Ketonic Acids," pp. 645–646; HAK–FLH, July 9, 1977 (Tape 20, p. 22).

21. Krebs, Notebook "12," p. 15; HAK–FLH, July 9, 1977 (Tape 20, p. 22).

22. Krebs, Notebook "12," pp. 20–21.

23. Ibid., pp. 23–30.

24. N. Edson to HAK, June 6, 1936; HAK to N. Edson, June 17, 1936, J.168, KC.

25. Krebs, Notebook "12," pp. 31–43.

26. HAK–FLH, July 13, 1977 (Tape 21, p. 9).

27. Krebs, Notebook "12," pp. 44–49.

28. Ibid., pp. 52–53; William Arthur Johnson, "Studies in the Intermediate Metabolism of Carbohydrates," unpublished Ph.D. dissertation, Sheffield, 1938, pp. 108–109. The protocol for this experiment is hand-dated in Johnson's copy "8.7.36." The data indicate that it is the same experiment recorded in Krebs's notebook for "3.7.36."

29. Johnson, "Intermediate Metabolism," pp. 15, 17.

30. H.A. Krebs, "Intermediate Metabolism of Carbohydrates," Nature, 138 (1936): 288.

31. Ibid.

32. Steven Benner, "The Tricarboxylic Acid Cycle," Yale Scientific, 50 (1976): 6–7.

33. HAK–FLH, July 13, 1977 (Tape 21, pp. 17–18).

34. Krebs, "Intermediate Metabolism of Carbohydrates," pp. 288–289.

35. See Volume 1, especially pp. 4–5.

36. HAK to N. Edson, June 30, 1936, J.168, KC.

37. Ibid., July 8, 1936.

38. D. Nachmansohn to HAK, June 28, 1936, J.478, KC.

39. Esco Foundation for Palestine, Inc., Palestine: A Study of Jewish, Arab, and British Policies, Vol. 1 (New Haven: Yale University Press, 1947), pp. 792–798.

40. Lise Daniel to HAK, May 7, 1936, Adolf Daniel to HAK, June 12, 1936, A.735, KC.

41. Norman Rose, Chaim Weizmann: A Biography (New York: Viking, 1986), pp. 310–311.

42. Nachmansohn to HAK, May 7, 1936, June 28, 1936, J.478, KC.

43. E.W. Krebs–FLH, Sept. 15, 1986 (Tape 154, pp. 15–17), May 4, 1988 (Tape 160, p. 16). When he read my account of Hans Krebs's departure from Germany in 1933 (see Vol. I, pp. 418–434), Wolf Krebs mentioned that he found "many parallels to my own" departure that he had not hitherto recognized.

44. *Ibid.*; E.W. Krebs to FLH, Feb. 8, 1988.

45. HAK to N. Edson, July 8, 1936, J.168, KC; Krebs, Notebook "12," pp. 54–57.

46. *Ibid.*, pp. 58–77, 79–84, 87.

47. Krebs, "Intermediate Metabolism of Carbohydrates," p. 289.

48. Johnson, "Intermediate Metabolism," pp. 21–22, 108. Johnson presented this experiment in his thesis with the equation shown. The thesis was written later, but it is a reasonable inference, in view of the generic equation in the *Nature* article, that Krebs had already formulated it this way when Johnson performed the experiment.

49. Krebs, Notebook "12," pp. 64–67, 80–81.

50. *Ibid.*, pp. 78, 82.

51. HAK to N. Edson, July 23, 1936, J.168, KC.

52. HAK to S. Elsden, July 23, 1936, J.177, KC.

53. [L. Eggleston], Notebook "13," (unpaginated) July 25, 1936–July 31, 1936. As he gained experience Eggleston gradually came to work in the manner of a research assistant rather than a technician. He remained with Krebs until his death in 1974.

54. W.A. Johnson–FLH, Jan. 17, 1978 (Tape 51, pp. 1–5).

55. Johnson, "Intermediate Metabolism," pp. 17, 39.

56. Krebs, Notebook "12," p. 89.

57. HAK–FLH, July 13, 1977 (Tape 21, pp. 21–22), July 14, 1977 (Tape 22, pp. 1–3).

58. Krebs, Notebook "12," pp. 90–91.

59. *Ibid.*, pp. 92–94.

60. HAK–FLH, July 14, 1977 (Tape 22, pp. 6–7).

61. *Ibid.* See also Hans Adolf Krebs, "The Role of Fumarate in the Respiration of *Bacterium coli commune*," *Biochem. J.*, 31 (1937): 2096–2097.

62. See J. Tikka, "Über den Mechanismus der Glucosevergärung durch *B. coli*," *Biochem. Z.*, 279 (1935): 264–288.

63. HAK–FLH, July 14, 1977 (Tape 22, p. 7).

64. M.G. Sevag and N. Neuenschwander-Lemmer, "Über die Dehydrierung von Milchsäure durch Staphylococcen," *Biochem. Z.*, 286 (1936): 7–12; Krebs, Notebook "12," p. 226.

65. *Ibid.*, pp. 96–103.

66. *Ibid.*, p. 104. See W.H. Schopfer, "Les vitamines cristallisés B comme hormones de croissance chez un microorganisme (Phycomyces)," *Archiv für Mikrobiologie*, 5 (1934): 511–549, 6 (1935): 139–140.

67. Krebs, Notebook "12," p. 229; Akiji Fujita and Takeshi Kodama, "Untersuchungen über Atmung und Gärung pathogener Bakterien," *Biochem. Z.*, 269 (1934): 367–374.

68. Thorsten Thunberg, "Zur Kenntnis des intermediären Stoffwechsels und der dabei wirksamen Enzyme," *Skandinavisches Archiv für Physiologie*, 40 (1920): 41–42, 53–54, 67; T. Thunberg, "Zur Kenntnis der Spezifität der Dehydrogenasen," *Biochem. Z.*, 258 (1933): 48–60.

69. Krebs, Notebook "12," p. 105. See HAK–FLH, July 14, 1977 (Tape 22, pp. 10–11).

70. Krebs, Notebook "12," pp. 106–107.

71. *Ibid.*, pp. 108–109.

72. *Ibid.*, pp. 110–114. See H.A. Krebs, "Dismutation of Pyruvic Acid in *Gonococcus* and *Staphylococcus*," *Biochem. J.*, 31 (1937): 667–668.

73. Krebs, Notebook "12," p. 115; HAK–FLH, July 14, 1977 (Tape 22, p. 11).

74. Krebs, Notebook "12," p. 117–120; HAK–FLH, July 15, 1977 (Tape 22, pp. 12–13).

75. Krebs, Notebook "12," pp. 121–125.

76. *Ibid.*, pp. 126–148; HAK–FLH, July 15, 1977 (Tape 22, pp. 15–20).

77. Johnson, "Intermediate Metabolism," pp. 9, 36–39.

78. D. Nachmansohn to HAK, August 22 [1936], J.477, KC. The italicized sentence is written in English, the rest in German.

79. [L. Eggleston], Notebook "13," (unpaginated) Aug. 28, 29, 1936.

80. D. Nachmansohn to HAK, July 19, 1936, J.478, KC.

81. *Ibid.*, Aug. 7, 1936, J.478, KC.

82. Joseph S. Fruton, *Molecules and Life: Historical Essays on the Interplay of Chemistry and Biology* (New York: Wiley-Interscience, 1972), pp. 334–337; E. Negelein to HAK, Aug. 24, 1936, KC; Krebs, Notebook "12," pp. 133, 149.

83. *Ibid.*, p. 150.

84. *Ibid.*, p. 151; HAK–FLH, July 15, 1977 (Tape 23, p. 1).

85. See Tikka, "Glucosevergärung durch B. Coli," p. 279, and Sevag and Neuenschwander-Lemmer, "Dehydrierung von Milchsäure," pp. 7–12.

86. Krebs, Notebook "12," pp. 152–153.

87. *Ibid.*, pp. 154–157.

88. Johnson, "Intermediate Metabolism," p. 43.

89. Krebs, Notebook "12," pp. 158–159; HAK to K. Myrbäck, Aug. 20, 1936; Myrbäck to HAK, Sept. 2, 1936, KC.

90. A. Szent-Györgyi, "Über die Bedeutung der Fumarsäure für die tierische Gewebsatmung," *Z. physiol. Chem.*, 236 (1935): 10–11; I. Banga, "Einfluss der C_4-Dicarbonsäuren auf die Gewebsatmung," *Ibid.*, pp. 20–21; F.B. Straub, "Mikrofumarsäurebestimmung und ihre Anwendung," *Ibid.*, pp. 42–48.

91. HAK–FLH, July 16, 1977 (Tape 23, p. 5). I have constructed this statement by combining two similar statements that Krebs made in succession.

92. Krebs, Notebook "12," p. 160.

93. *Ibid.*, p. 161.

94. *Ibid.*, pp. 164–165. Retrospectively Krebs readily explained this experiment when he reexamined it in 1977. The inhibition of succinoxidase is due to a competition between succinate and malonate. Increasing the concentration of succinate relative to malonate will tend to "overcome the inhibition." When fumarate is added, the concentration of "succinate will rise only gradually, and therefore the effect of malonate will be greater." HAK–FLH, July 16, 1977 (Tape 23, p. 6).

95. Krebs, Notebook "12," pp. 166–167.

96. Johnson, "Intermediate Metabolism," pp. 39–41, 121.

97. Krebs, Notebook "12," pp. 168–169.

98. *Ibid.*, pp. 170–171.

99. *Ibid.*, pp. 172–174; S.R. Elsden to HAK, July 21, 1936, Aug. 23, 1936; HAK to S.R. Elsden, July 23, 1936, Aug. 25, 1936, J.177, KC.

100. Krebs, Notebook "12," p. 173. See HAK–FLH, July 16, 1977 (Tape 23, p. 14).

101. Krebs, Notebook "12," pp. 175–180.

102. Flora Jane Ogston and David Ezra Green, "The Mechanism of the Reaction of Substrates with Molecular Oxygen," *Biochem. J.*, 29 (1935): 1983–2012; D.E. Green, "α-Glycerophosphate Dehydrogenase," *Ibid.*; 30 (1936): 629–644.

103. Krebs, Notebook "12," pp. 181–183.

104. *Ibid.*, p. 188.

105. *Ibid.*, pp. 184–187.

106. *Ibid.*, pp. 189–190.

107. *Ibid.*, p. 191; Marjory Stephenson and Leonard Hubert Stickland, "Hydrogenase: A Bacterial Enzyme Activating Molecular Hydrogen," *Biochem. J.*, 25 (1931): 205–214.

108. HAK to M. Stephenson, Oct. 6, 1936, J.687, KC.

109. Krebs, Notebook "12," pp. 192–196.

110. *Ibid.*, pp. 197–199.

111. *Ibid.*, p. 200; Harlan Goff Wood and Chester Hamlin Werkman, "The Utilisation of CO_2 in the Dissimilation of Glycerol by the Propionic Acid Bacteria," *Biochem. J.*, 30 (1936): 48–53; HAK–FLH, July 16, 1977 (Tape 23, pp. 17–18).

112. Krebs, Notebook "12," pp. 202–205.

113. *Ibid.*, pp. 206–207.

114. *Ibid.*, pp. 211–212.

115. *Ibid.*, p. 212.

116. *Ibid.*, pp. 213–215.

117. *Ibid.*, p. 216; HAK to Stephenson, Oct. 6, 1936, J.687, KC.

118. Krebs, Notebook, "12," p. 218.

119. Krebs, Notebook "12," pp. 219–221.

120. Johnson, "Intermediate Metabolism," pp. 37, 39–40, 48–49.

121. HAK to M. Stephenson, Sept. 30, 1936, J.687, KC.

122. N. Edson to HAK, Sept. 17, 27, 1936; HAK to N. Edson, Sept. 25, 1936, J.168, KC.

123. Lise Daniel to HAK, Sept. 5, 1936, A.735, KC.

Chapter 8

1. Guy Drummond Greville, "Fumarate and Tissue Respiration," *Biochem. J.*, 30 (1936): 877–887; F.J. Stare and C.A. Baumann, "The Effect of Fumarate on Respiration," *Proc. Roy. Soc. Lond.* Ser. B, 121 (1936): 338–357; Jean March Innes, "The Role of the 4-Carbon Dicarboxylic Acids in Muscle Respiration," *Chemistry and Industry*, 14 (1936): 840; J.M. Innes, *Biochem. J.*, 30 (1936): 2040–2048.

2. Maurice Jowett and Juda Hirsch Quastel, "Effects of Hydroxymalonate on the Metabolism of Brain," *Biochem. J.*, 31 (1937): 275–281; Juda Quastel-FLH, March 29, 1964 (Tape 168, p. 20); HAK–FLH, July 25, 1977 (Tape 27, p. 3).

3. Hans Weil to HAK, Feb. 28, 1936, J.816, KC.; Hans Weil-Malherbe, "The Metabolism of Glutamic Acid in Brain," *Biochem. J.*, 30 (1936): 665–676.

4. Hans Weil to HAK, Feb. 28, 1936; H. Weil-Malherbe, "Carbohydrate Metabolism," *Nature*, 138 (1936): 551.

5. H. Weil-Malherbe, *Ibid.*, pp. 551–552.

6. *Ibid.*, p. 552.

7. F. Knoop and C. Martius, "Über die Bildung von Citronensäure," *Z. physiol. Chem.*, 242 (1936): I; Carl Martius-FLH, Aug. 16, 1978 (Tape 85, pp. 1–8). See also C. Martius, "How I Became a Biochemist," in *Of Oxygen, Fuels, and Living Matter*, ed. G. Semenza (New York: John Wiley, 1982), pp. 1–3.

8. Knoop and Martius, "Bildung von Citronensäure."

9. Krebs, Notebook "14," pp. 0–2; HAK–FLH, July 21, 1977 (Tape 24, pp. 11–12).

10. Krebs, Notebook "12," p. 172; Notebook "14," pp. 4–5; William Arthur Johnson, "Studies in the Intermediate Metabolism of Carbohydrates," unpublished Ph.D. thesis, University of Sheffield, 1938, p. 43.

11. Krebs, Notebook "14," pp. 6–7.

12. *Ibid.*, pp. 10–11.

13. M. Stephenson to HAK, Oct. 4, 1936, J.687, KC.

14. HAK to M. Stephenson, Oct. 6, 1936, J.687, KC.

15. Krebs, Notebook "14," pp. 8–9.

16. *Ibid.*, p. 12.

17. *Ibid.*, p. 13.

18. HAK–FLH, July 21, 1977 (Tape 24, p. 16).

19. Krebs, Notebook "14," pp. 15–18.

20. Johnson, "Intermediate Metabolism of Carbohydrates," pp. 48–49, 75.

21. HAK–FLH, July 13, 1977 (Tape 21, p. 21).

22. H.A. Krebs and W.A. Johnson, "The Oxidative Breakdown of Carbohydrates," *Chemistry and Industry*, 14 (1936): 840.

23. Krebs, Notebook "14," p. 19.

24. *Ibid.*, pp. 20–21.

25. *Ibid.*, p. 22; HAK–FLH, July 22, 1977 (Tape 24, p. 6).

26. Krebs, Notebook "14," p. 23.

27. *Ibid.*, pp. 24–26.

28. *Ibid.*, p. 27.

29. *Ibid.*, pp. 28–31.

30. [Eggleston], Notebook "13," unpaginated, October 13–15, 1936.

31. [Krebs], "The Oxidative Breakdown of Carbohydrates," H.100, KC. There is at the top of this manuscript a penciled date "1937," obviously added later. The manuscript can, however, be identified confidently as the paper Krebs delivered at the Biochemical Society on October 17, 1936, on the following grounds: (1) the title; (2) the identity of an abstract attached to the manuscript with

an abstract of the delivered paper in *Chemistry and Industry*, 14 (1936): 840; and (3) the general internal contents being consistent with the point at which Krebs's experimentation had reached in October 1936.

32. [Krebs], "Oxidative Breakdown of Carbohydrates," p. 1.

33. *Ibid.*

34. *Ibid.*, pp. 1–5.

35. *Ibid.*, pp. 6–7.

36. *Ibid.*, p. 7.

37. See Albert Szent-Györgyi, "Über den Mechanismus der Hauptatmung des Taubenmuskels," *Z. physiol. Chem.*, 224 (1934): 1–10.

38. [Krebs], "Oxidative Breakdown of Carbohydrates," pp. 7–9.

39. *Ibid.*, p. 9.

40. *Ibid.*, pp. 9–10.

41. *Ibid.*, p. 11.

42. *Ibid.*

43. *Ibid.*

44. *Ibid.*, p. 12.

45. *Ibid.*, pp. 13–14.

46. The Biochemical Society, program for 180th meeting, Oct. 17, 1936, Cambridge. Krebs's copy is inserted in Krebs, Notebook "14," p. 32; H. Weil-Malherbe, "Pyruvic Acid Oxidation in Brain," *Chemistry and Industry*, 14 (1936): 838.

47. Weil-Malherbe, "Pyruvic Acid Oxidation," p. 838.

48. Weil-Malherbe made this connection explicit two months later in his paper, "Formation of Succinic Acid," *Biochem. J.*, 31 (1937): 299, 309–310.

49. J.M. Innes, "The Role of the 4-Carbon Dicarboxylic Acids," p. 840.

50. [Krebs], "Oxidative Breakdown of Carbohydrates," p. 14.

51. Biochemical Society, program; HAK–FLH, July 22, 1977 (Tape 24, p. 13).

52. Norman Edson to HAK, Oct. 11, 1936, J.168, KC; HAK to F.J.W. Roughton, Oct. 9, 1936, Roughton Papers, American Philosophical Society Library.

53. Krebs, Notebook "14," pp. 33–38; HAK–FLH, July 23, 1977 (Tape 25, pp. 1–2).

54. Krebs, Notebook "14," pp. 39–77; Johnson, "Intermediate Metabolism of Carbohydrates," pp. 37–43.

55. Krebs, Notebook "14," p. 78.

56. *Ibid.*, p. 79.

57. *Ibid.*, pp. 80–88.

58. *Ibid.*, p. 89.

59. [Krebs], untitled typewritten manuscript, penciled notation "Medico Chir Soc 19/11 36," H.98, KC, pp. 1–2.

60. *Ibid.*, pp. 2–7.

61. *Ibid.*, pp. 7–9.

62. *Ibid.*, pp. 9–10.

63. *Ibid.*, pp. 10–12.

64. Weil-Malherbe, "Formation of Succinic Acid," pp. 299–300, 309–310.

65. Kenneth Allan Caldwell Elliott, Marjorie Pickard Benoy, and Zelma Baker, "The Metabolism of Lactic and Pyruvic Acids in Normal and Tumour Tissues: II," *Biochem. J.*, 29 (1935): 1937.

66. F.G. Hopkins to HAK, Nov. 5, 1936; HAK to F.G. Hopkins, Nov. 2, 1936, Nov. 6, 1936, J.289, KC; E.W. Krebs to FLH, Nov. 15, 1987; E.W. Krebs–FLH, May 4, 1988 (Tape 160, p. 7), Jan. 5, 1992 (Tape 170, pp. 7–10).

67. Krebs, Notebook "14," pp. 96–97.

68. F. Challenger to HAK, Aug. 12, 1936, Aug. 14, 1936, Nov. 22, 1936; [Krebs], "Biological Breakdown of Carbohydrates," typewritten manuscript, H.99, KC. The correspondence is attached to the manuscript.

69. *Ibid.*, pp. 10–16.

70. *Ibid.*, pp. 1–2.

71. *Ibid.*, pp. 4–6.

72. F. Challenger to HAK, Nov. 22, 1936; The Chemical Society, announcement of meeting of Nov. 26, 1936 inserted in Krebs, Notebook "14," p. 96; HAK–FLH, July 22, 1977 (Tape 24, pp. 15–16); Krebs, Notebook "14," pp. 98–99. It is an example of the whimsical side of Krebs's memory that he recalled in 1977 Challenger's remark, but did not remember what he himself had talked about.

73. HAK to F.G. Hopkins, Nov. 2, 1936; Institute of Chemistry: Birmingham and Midlands Section, announcement of meeting of Nov. 25, 1936 inserted in Krebs, Notebook "4," p. 104; University of Cambridge School of Biochemistry, Biochemical Tea Club, Michaelmas Term, 1936, Biochem. 414, University of Cambridge Archives.

74. Johnson, "Intermediate Metabolism of Carbohydrates," pp. 40–45.

75. Krebs, Notebook "14," pp. 111–113, 115–116.

76. HAK–FLH, July 25, 1977 (Tape 26, p. 13).

77. E. Annau, I. Banga, A. Blazsó, V. Bruckner, K. Laki, F.B. Straub, and A. Szent-Györgyi, "Über die Bedeutung der Fumarsäure für die tierische Gewebsatmung," Z. physiol. Chem., 244 (1936): 105–152; HAK to A. Szent-Györgyi, Dec. 15, 1936, J.702, KC. When I examined Krebs's copy of the reprint it was in his reprint collection in his laboratory.

78. A. Szent-Györgyi, "Einleitung, Übersicht, Methoden," in E. Annau, et al., "Bedeutung der Fumarsäure," pp. 105–110.

79. Ibid., pp. 111–113.

80. Ibid., p. 115.

81. F.B. Straub, "Decarboxylierung der OES durch Muskelgewebe," in Ibid., pp. 140–141.

82. Krebs, Notebook "14," p. 127.

83. I. Banga, "Über Oxydation der FS. und Reduktion der OES. durch zerkleinertes Muskelgewebe," in E. Annau, et al., "Bedeutung der Fumarsäure," p. 133.

84. Krebs, Notebook "14," pp. 125, 128.

85. HAK to A. Szent-Györgyi, Dec. 15, 1936, J.702, KC.

86. Ibid.

87. Ibid.

88. Hans Adolf Krebs, "Intermediary Hydrogen-Transport in Biological Oxidations," in Perspectives in Biochemistry: Thirty-one Essays presented to Sir Frederick Gowland Hopkins, ed. Joseph Needham and David E. Green (Cambridge: University Press, 1938), p. 151.

89. Joseph Needham and David Green to HAK, May 1936, J.490, KC.

90. A. Szent-Györgyi to HAK, Dec. 20, 1936, J.702, KC.

91. A. Szent-Györgyi to H.H. Dale, Nov. 26, 1931; H.H. Dale papers, Archives of the Royal Society of London; Albert Szent-Györgyi, "Lost in the Twentieth Century," Annual Review of Biochemistry, 32 (1963): 7.

92. Szent-Györgyi to H.H. Dale, Sept. 12, 1931, Archives of the Royal Society. This is typical of numerous comments on his own research style by Szent-Györgyi in his letters to Dale.

93. A.J.P. Taylor, English History, 1914–1945 (Harmondsworth: Penguin Books, 1965), pp. 489–496; HAK–FLH, Aug. 11, 1977 (Tape 40, p. 5).

94. Lady Margaret Krebs–FLH, June 30, 1982 (Tape 152, pp. 11–12), June 29, 1982, notes on untaped conversation.

95. Ibid.; HAK–FLH, July 29, 1977 (Tape 30, p. 24), Jan. 9, 1981 (Tape 129, p. 19); Margaret Krebs–FLH, May 7, 1987 (Tape 158, p. 1).

96. Margaret Krebs–FLH, May 7, 1987 (Tape 158, p. 17).

97. Ibid., pp. 1–2.

98. Ibid., pp. 2–3.

99. Ibid., pp. 12, 20, Jan. 3, 1992 (Tape 169, pp. 1–2); HAK–FLH, Jan. 9, 1981 (Tape 129, pp. 19–21).

Chapter 9

1. HAK to F.J.W. Roughton, Dec. 9, 1936, Roughton papers, American Philosophical Society Library. My account assumes that Krebs carried out the intention stated in this letter.

2. H.A. Krebs and F.J.W. Roughton, "Carbornithine as an Intermediate in the Synthesis of Urea by the Liver," handwritten manuscript, *Ibid.*

3. *Ibid.*

4. F.J.W. Roughton to HAK, Jan. 9, [1937], J.624, KC.

5. F.J.W. Roughton to HAK (undated), J.624, KC.

6. HAK to F.J.W. Roughton, Jan. 12, 1937, Roughton papers.

7. J.C. Poggendorff, *Biographisch-LiterarischesHandwörterbuch der Exacten Naturwissenschaften,* Vol. VIIb, pt. 3 (Berlin: Akademie-Verlag, 1970), p. 1652.

8. Krebs, Notebook "14," pp. 150–151.

9. *Ibid.*, pp. 152–177; William Arthur Johnson, "Studies in the Intermediate Metabolism of Carbohydrates," unpublished Ph.D. dissertation, University of Sheffield, 1938, pp. 9, 12, 15, 18.

10. Hans Adolf Krebs and William Arthur Johnson, "Acetopyruvic Acid ($\alpha\gamma$-Diketovaleric Acid) as an Intermediate Metabolite in Animal Tissues," *Biochem. J.*, 31 (1937): 772.

11. Johnson, "Intermediate Metabolism of Carbohydrates," pp. 27, 30, 78.

12. *Ibid.*, pp. 26–27, 31; Krebs, Notebook "14," p. 174.

13. Krebs, Notebook "14," pp. 175–191; Hans Adolf Krebs, "The Role of Fumarate in the Respiration of *Bacterium coli commune*," *Biochem. J.*, 31 (1937): 2102–2103.

14. HAK to J.H. Quastel, Feb. 1, 1937, J.585, KC.

15. That Quastel felt generally unappreciated by Krebs was clearly evident during a conversation I held with Quastel in Vancouver, B.C., on   March 29, 1984.

16. R. Schoenheimer to HAK, Feb. 4, 1937, J.642, KC.

17. Joseph S. Fruton, *Molecules and Life: Historical Essays on the Interplay of Chemistry and Biology* (New York: Wiley-Interscience, 1972), pp. 456–464.

18. W.S. Cohen to HAK, Feb. 5, 1937, Feb. 8, 1937; D. Nachmansohn to HAK, Feb. 21, 1937, J.479, KC.

19. HAK to H. Blaschko, Jan. 24, 1937. See also HAK to Blaschko, April 25, 1937, April 30, 1937, May 26, 1937, and June 14, 1937, J.71, KC.

20. HAK to Blaschko, Jan. 24, 1937. See also Nachmansohn to HAK, Jan. 6, 1937, Jan. 12, 1937, J.479, KC.   Nachmansohn's letters indicate that there were some extenuating circumstances prompting his desire to publish quickly.

21. Johnson, "Intermediate Metabolism of Carbohydrates," pp. 9, 18, 29–32; Krebs, Notebook "14," pp. 197–203.

22. A. Neuberger to F.J.W. Roughton, Jan. 18, 1937, Roughton papers, American Philosophical Society Library.

23. HAK to F.J.W. Roughton, Feb. 12, 1937, *Ibid.*

24. HAK–FLH, July 5, 1977 (Tape 17, p. 11), July 8, 1977 (Tape 18, p. 1).

25. *Ibid.*, May 15, 1979 (Tape 95, pp. 27–34), May 16, 1979 (Tape 95, pp. 3–7).

26. HAK to R.M. Archibald, March 22, 1946, KC.

27. HAK–FLH, May 19, 1979 (Tape 97, pp. 10–11).

28. Margaret Krebs–FLH, June 30, 1982 (Tape 152, p. 10), May 7, 1988 (Tape 158 p. 8).

29. Margaret Krebs–FLH, Jan. 8, 1992 (Tape 169, pp. 3–4).

30. *Ibid.*; Margaret Krebs–FLH, notes of unrecorded conversation, Jan. 6, 1992.

31. *Ibid.*, June 30, 1982 (Tape 152, p. 11), May 7, 1988 (Tape 158, p. 4).

32. *Ibid.*, May 7, 1988 (Tape 158, p. 4). An undated, annotated typewritten copy of the script for *The Happy Journey* is among Krebs's personal papers. Krebs took part in at least one other play. *Arrows*, the newspaper of the Union of Students of the University of Sheffield, reported in the issue of December 1937 on a one-act play "E. and O.E." by Crawshay Williams performed by members of the staff. His reviewer included the judgment that "as the solicitor, Dr. Krebs was charming. He added quite an old-world touch to a macabre situation." In 1977 Krebs was still able to recount the plot of *The Happy Journey* in detail. HAK–FLH, Aug. 12, 1977 (Tape 41, p. 7).

33. Hans Adolf Krebs and William Arthur Johnson, "Metabolism of Ketonic Acids in Animal Tissues," *Biochem. J.*, 31 (1937): 645.

34. *Ibid.*, pp. 646–652.

35. *Ibid.*, p. 653.

36. *Ibid.*, p. 654.

37. *Ibid.*, pp. 656–658.

38. *Ibid.*, pp. 658–659.

39. Krebs, Notebook "14," pp. 96–102, 109, 114, 118, 122–124, 132–142.

40. Hans Adolf Krebs, "Dismutation of Pyruvic Acid in *Gonococcus* and *Staphylococcus*," *Biochem. J.*, 31 (1937): 661–671.

41. Krebs, Notebook "14," pp. 133–137.

42. Krebs, "Dismutation of Pyruvic Acid," p. 667.

43. HAK–FLH, July 16, 1977 (Tape 23, p. 8).

44. Krebs, Notebook "15," pp. 11–29.

45. *Ibid.*, p. 30; HAK–FLH, July 28, 1977 (Tape 28, p. 1).

46. Krebs, Notebook "15," p. 31.

47. *Ibid.*, pp. 32–33.

48. *Ibid.*, pp. 34–35, 39–40.

49. Krebs and Johnson, "Acetopyruvic Acid," pp. 772–773.

50. *Ibid.*, pp. 774–777.

51. D. Nachmansohn to HAK, Feb. 21, 1937, J.479, KC.

52. Ch. Weizmann to HAK, March 2, 1937, KC.

53. HAK to V. Idelson, March 24, 1937, KC.

54. Margaret Krebs–FLH, May 7, 1988 (Tape 158, pp. 5–6).

55. *Ibid.*

56. *Ibid.*, p. 13.

57. Margaret Krebs–FLH, Jan. 8, 1992 (Tape 169, pp. 1–3).

58. Krebs, Notebook "15," pp. 43–46; HAK–FLH, July 28, 1977 (Tape 28, p. 5).

59. Krebs, Notebook "15," pp. 47–88; Johnson, "Intermediate Metabolism of Carbohydrates," pp. 50, 125.

60. Carl Martius, "Über den Abbau der Citronensäure," *Z. physiol. Chem.*, 247 (1937): 104–106; C. Martius–FLH, Aug. 16, 1978 (Tape 85, pp. 9–13).

61. Martius–FLH, *Ibid.*; Martius, "Abbau der Citronensäure," pp. 106–109.

62. Martius–FLH, *Ibid.* (Tape 85, p. 14).

63. C. Martius and F. Knoop, "Der physiologische Abbau der Citronensäure: Vorläufige Mitteilung," *Z. physiol. Chem.*, 246 (1937): I–II.

64. Martius, "Abbau der Citronensäure," p. 108; Martius–FLH, Aug. 16, 1978 (Tape 85, p. 19).

65. HAK–FLH, July 28, 1977 (Tape 29, p. 12).

66. *Ibid.* (Tape 29, p. 15). In our discussion in 1978 Krebs suggested that there were three initial questions, the other one being whether citric acid is resynthesized. He did not then recall, however, that in April 1937 he believed he already had evidence for such a synthesis.

67. W.A. Johnson–FLH, Jan. 17, 1978 (Tape 51, pp. 15–16).

68. W.A. Johnson, "Intermediate Metabolism of Carbohydrates," pp. 51, 56–57, 131.

69. *Ibid.*, p. 57.

70. Krebs, Notebook "15," p. 89; Hans Weil-Malherbe, "Formation of Succinic Acid," *Biochem. J.*, 31 (1937): 304, 306–307; HAK–FLH, July 28, 1977 (Tape 29, p. 14).

71. Krebs, Notebook "15," pp. 90–91.

72. HAK to H. Blaschko, April 25, 1937; HAK to D.D. Woods, April 14, 1937, J.853, KC.

73. HAK–FLH, July 28, 1977 (Tape 29, p. 15).

74. Krebs, Notebook "15," pp. 100–101.

75. Krebs, Notebook "15," pp. 102–112; Johnson, "Intermediate Metabolism of Carbohydrates," pp. 40, 58–59, 93, 133.

76. H.A. Krebs and W.A. Johnson, "The Role of Citric Acid in Intermediate Metabolism in Animal Tissues," typewritten manuscript, dated June 10, 1937, KC, p. 2.

77. W.A. Johnson–FLH, Jan. 17, 1978 (Tape 51, pp. 15–16). Johnson identified the visitor as "a Swede that came to work for us." The person who best fit this description, Ake Örstrom, did not begin work in the laboratory until 1938. In further discussion, when he found difficulty in identifying the visitor, Johnson expressed some doubt about his recollection of the story in general. The incident has, however, a feeling of authenticity, which makes me believe that either the visitor was someone else or Örstrom came for a preliminary visit at that time.

78. Johnson, "Intermediate Metabolism of Carbohydrates," pp. 48, 50.

79. Krebs, Notebook "15," p. 116; HAK–FLH, July 28, 1977 (Tape 29, p. 19).
80. Krebs, Notebook "15," p. 117; HAK–FLH, July 28, 1977 (Tape 29, pp. 20–21).
81. Johnson, "Intermediate Metabolism of Carbohydrates," p. 48.
82. *Ibid.*, pp. 54–55.
83. *Ibid.*, pp. 48, 51, 53.
84. *Ibid.*, p. 53.
85. HAK–FLH, July 29, 1977 (Tape 29, p. 7).
86. *Ibid.*, Aug. 13, 1977 (Tape 42, p. 7).
87. HAK to H. Blaschko, June 14, 1937, J.71, KC; HAK–FLH, Aug. 13, 1977 (Tape 42, p. 6).
88. H.A. Krebs and W.A. Johnson, "The Role of Citric Acid in Intermediate Metabolism in Animal Tissues," typed manuscript, KC. Hans Weil-Malherbe has drawn my attention to the incongruity of the figure of 100 mg of citric acid per gram of citric acid formed per hour from oxaloacetate, "while the presumed intermediary, *cis*-aconitic acid, forms only 12 mg per gr per hr.," and asks "should it be 10 mg of citric acid?" (Hans Weil-Malherbe to FLH, n.d. The figures are, however, correct as stated in the manuscript.)
89. HAK–FLH, July 29, 1977 (Tape 30, p. 9).
90. V. Idelson to HAK, June 2, 1937; HAK to V. Idelson, June 29, 1937, KC.
91. Johnson, "Intermediate Metabolism of Carbohydrates," p. 48; Krebs, Notebook "15," pp. 135–137.
92. *Ibid.*, pp. 138–145; HAK–FLH, July 29, 1977 (Tape 30, p. 19).
93. Editor of *Nature* to H.A. Krebs, June 14, 1937, KC.
94. Hans Krebs, *Reminiscences and Reflections* (Oxford: Clarendon Press, 1981), pp. 98–99.
95. HAK–FLH, July 27, 1977 (Tape 28, p. 4), July 29, 1977 (Tape 30, p. 23).
96. Johnson, "Intermediate Metabolism of Carbohydrates," pp. 48, 52, 54, 64, 135.
97. Krebs, Notebook "15," pp. 148, 150, 152; HAK–FLH, July 29, 1977 (Tape 30, p. 21).
98. W. Junk to HAK, June 29, 1937 (inserted in Notebook "15" at p. 154), KC.
99. H.A. Krebs and W.A. Johnson, "The Role of Citric Acid in Intermediate Metabolism in Animal Tissues," *Enzymologia*, 4 (1937): 148.
100. *Ibid.*, p. 150.
101. *Ibid.*, p. 151.
102. P.B. Medawar, *The Art of the Soluble: Creativity and Originality in Science* (Harmondsworth: Penguin Books, 1967), p. 169.
103. Krebs and Johnson, "The Role of Citric Acid," p. 153.
104. *Ibid.*, p. 153.
105. *Ibid.*
106. *Ibid.*, p. 154.
107. *Ibid.*, p. 155.
108. *Ibid.*
109. Johnson, "Intermediate Metabolism of Carbohydrates," p. 48.
110. HAK to V. Idelson, June 29, 1937, KC.
111. HAK–FLH, Aug. 6, 1977 (Tape 39, p. 3).
112. *Ibid.*, Aug. 11, 1977 (Tape 40, p. 3), Aug. 8, 1978 (Tape 79, p. 18).
113. M. Krebs–FLH, June 30, 1982 (Tape 152, pp. 9–10), May 7, 1988 (Tape 158, p. 19), Jan. 8, 1992 (Tape 169, p. 6); HAK–FLH, Jan. 9, 1981 (Tape 129, p. 19).
114. M. Krebs–FLH, May 7, 1988 (Tape 158, pp. 9–12) and notes of unrecorded conversation, June 29, 1982.
115. Lise Daniel to HAK, Feb. 26, 1938, A.737, KC.

Chapter 10

1. H.A. Krebs, "The History of the Tricarboxylic Acid Cycle," *Perspectives in Biology and Medicine*, 14 (1970): 154.
2. *Ibid.*, pp. 166–167.
3. C. Martius–FLH, Aug. 16, 1978 (Tape 85, pp. 18–19). When I mentioned Martius's comments to Krebs in 1979, Krebs replied, "I think when he says he saw no possibility [to prove the

physiological synthesis of citric acid], that is only another way of saying that he hadn't learned enough biochemistry." Krebs also pointed out that it was Martius himself who had told him that he "regarded himself as a theoretical organic chemist." Martius must since then "have changed his views." HAK–FLH, Dec. 18, 1979 (Tape 106, p. 8).

4. Hans Krebs, *Reminiscences and Reflections* (Oxford: Clarendon Press, 1981), p. 116.

5. See, for example, H.A. Krebs, "The Intermediate Metabolism of Carbohydrates," *The Lancet*, Sept. 25, 1937, pp. 736–738.

6. Thomas S. Kuhn, *The Structure of Scientific Revolutions*, 2nd. ed. (Chicago: University of Chicago Press, 1970), p. 189.

7. H.A. Krebs, "Cyclic Processes in Living Matter," *Enzymologia*, 12 (1947): 88–100.

8. Krebs, "Tricarboxylic Acid Cycle," p. 161; HAK–FLH, April 28, 1977 (Tape 10–11, p. 3). Krebs did add the proviso, "But it was certainly at the back of my mind when it came to the formulation of the TCA cycle," an assertion that can neither be verified nor disproved.

9. [Krebs] "The Biological Breakdown of Carbohydrates," typewritten manuscript, H.99, KC, pp. 6–7.

10. E. Wayne to FLH, July 19, 1986.

11. HAK–FLH, May 16, 1979 (Tape 97, p. 38).

12. *Ibid.*

13. Hans Adolf Krebs and Leonard Victor Eggleston, "The Oxidation of Pyruvate in Pigeon Breast Muscle," *Biochem. J.*, 34 (1940): 442.

14. HAK–FLH, July 25, 1977 (Tape 27, p. 6).

15. *Ibid.*, May 19, 1979 (Tape 97, p. 10).

16. See HAK–FLH, May 19, 1979 (Tape 97, p. 9, Tape 98, p. 13), March 20, 1980 (Tape 118, p. 10).

17. See Evelyn Fox Keller, *A Feeling for the Organism: The Life and Work of Barbara McClintock* (New York: Freeman, 1983). Keller uses this phrase to characterize the scientific style of McClintock.

18. HAK–FLH, July 22, 1977 (Tape 24, p. 4).

19. M. Krebs–FLH, June 30, 1982 (Tape 151, p. 7). Notes of unrecorded conversation, same date.

20. HAK–FLH, Aug. 12, 1977 (Tape 41, p. 15), Aug. 4, 1981 (Tape 143, p. 11).

21. HAK to H. Blaschko, June 14, 1937, KC; Krebs, *Reminiscences*, p. 83.

22. M. Krebs–FLH, June 25, 1982, notes of unrecorded conversation.

23. HAK–FLH, July 9, 1977 (Tape 20, pp. 7–8).

24. *Ibid.*, July 25, 1977 (Tape 26, p. 1).

25. See George Lakoff and Mark Johnson, *Metaphors We Live By* (Chicago: University of Chicago Press, 1980).

INDEX

Acetic acid: exp. of Krebs on formation of acetoacetic from, 11, 13, 19–26, 30, 33-34; on metabolic fate of, 322, 347, 357-358; on H_2 from, 391–392

Acetoacetic acid metabolism: exp. of Krebs on, 15–16, 20–36; passim, 245–250, 376; exp. of Quastel on, 40–42, 247; exp. of Annau on, 80; exp. of Weil-Malherbe on, 352-353

Acetopyruvic acid: proposed as intermediate in pyruvic acid condensation, 376–378, 386-387, 391, 393–394, 432; method for determination of, 377–378

Aconitic acid: as intermediate in citric acid metabolism 13, 396–397, 405–406, 410-411, 416, 418

Amino acids: synthesis of, 149–151, 238–245, 250–251; metabolism of, 237–240. *See also*, Deamination

d-Amino acid deaminase: discovery by Krebs, 99, 101, 157–163, 223

Anaerobic oxidations, 330–331, 335–337. *See also*, Dismutation reactions

Analogy: in discovery of citric acid cycle, 428-429

Annau, Erno, co-worker of Szent-Györgyi: and formation of ketone bodies, 80–81, 83, 240, 285

Anti-Semitism in Nazi Germany, 54–55, 68-69, 96–97, 129–130

Arndt, Fritz: teacher of Martius, 426

Ashford, Charles Ames: and respiratory oxidations, 86–87

Aub, Joseph: offer of position for Krebs, 4, 49–50, 63–64

Auerbach, Walter: friend of Krebs, 73, 89

Avery, B.F.: method for lactic acid determination, 253, 273

Bacterial metabolism, 44, 190–193, 322–328, 351, 355–356, 376–379; anaerobic reactions of pyruvic acid in, 287, 296, 305-306, 342–344, 388–391; dismutation reactions in, 309–311, 314–316, 319, 324–327, 356, 391–393; reactions of fumaric acid in, 314–315, 324–327, 335–340; formation of propionic acid in, 324; reactions of H_2 in, 190–193, 322–323, 326–328, 338–339, 344, 392–393

Baldwin, Ernest: and dynamic biochemistry, 43; preparation of d-ornithine, 354

Banga, Ilona, co-worker of Szent-Györgyi: on conserving action of fumaric acid, 254; on disappearance of oxaloacetic acid, 366

Barcroft, Joseph: 8, 84, 109; studies of fetal respiration, 71

Battelli, Federico: and respiration in isolated tissues, 38, 39; and uric acid metabolism, 51, 57, 64; and Cannizzaro reaction, 201

Baumann, Carl A: and catalytic role of fumarate, 332, 411, 416, 428

Bawden, Frederick: on crystallization of viruses, 353

Benner, Steven: on metabolic cycles, 295

Benzinger, Theodor: and uric acid synthesis, 50, 59–60, 179–181, 212, 214, 225

Bergmann, Max: preparation of d-acetylglutamic acid 240

Bernal, John Desmond: on crystallization of viruses, 353

Bernard, Claude: on primacy of physiology, 124

Bielschowsky, Franz: corr. with Krebs, 54, 68

Biochemistry: teaching of at Cambridge, 73-75, 127, 163–164, 213; discipline of, 164

Blaschko, Hermann: 3; illness, 55; arrival in Cambridge, 84; and Palestine, 140, 219-220, 258–259; support of by Krebs, 380, 402, 409

Booth, Vernon: 9, 75, 157

Bork, Heinz: and uric acid metabolism, 53

Borsook, Henry: on ornithine effect, 32–33

Brinkmann, E: 278 *See also* Toenniesson-Brinkmann scheme

Buchner, Eduard: discovery of cell-free fermentation, 124, 362

Bucky, Ursula, 112

Butyric acid: scheme for metabolic decomposition of, 17–18

Cambridge Biochemical Laboratory: ethos of, 5–7, 43–46, 108–109

Cancer Institute in Jerusalem: efforts to recruit Krebs for, 63–64, 69–71, 81–83, 89–91, 127–128, 135, 380, 394, 414; visit of Krebs to, 263

Cannizzaro reaction: 200–203, 314, 379